ADVANCES IN ROCK-SUPPORT AND GEOTECHNICAL ENGINEERING

ADVANCES IN ROCK-SUPPORT AND GEOTECHNICAL ENGINEERING

SHUREN WANG

PAUL C. HAGAN

CHEN CAO

AMSTERDAM • BOSTON • HEIDELBERG • LONDON
NEW YORK • OXFORD • PARIS • SAN DIEGO
SAN FRANCISCO • SINGAPORE • SYDNEY • TOKYO

Butterworth-Heinemann is an imprint of Elsevier

Butterworth-Heinemann is an imprint of Elsevier
The Boulevard, Langford Lane, Kidlington, Oxford OX5 1GB, United Kingdom
50 Hampshire Street, 5th Floor, Cambridge, MA 02139, United States

Copyright © 2016 Tsinghua University Press Ltd. Published by Elsevier Inc. All rights reserved.

No part of this publication may be reproduced or transmitted in any form or by any means, electronic or mechanical, including photocopying, recording, or any information storage and retrieval system, without permission in writing from the publisher. Details on how to seek permission, further information about the Publisher's permissions policies and our arrangements with organizations such as the Copyright Clearance Center and the Copyright Licensing Agency, can be found at our website: www.elsevier.com/permissions.

This book and the individual contributions contained in it are protected under copyright by the Publisher (other than as may be noted herein).

Notices
Knowledge and best practice in this field are constantly changing. As new research and experience broaden our understanding, changes in research methods, professional practices, or medical treatment may become necessary.

Practitioners and researchers must always rely on their own experience and knowledge in evaluating and using any information, methods, compounds, or experiments described herein. In using such information or methods they should be mindful of their own safety and the safety of others, including parties for whom they have a professional responsibility.

To the fullest extent of the law, neither the Publisher nor the authors, contributors, or editors, assume any liability for any injury and/or damage to persons or property as a matter of products liability, negligence or otherwise, or from any use or operation of any methods, products, instructions, or ideas contained in the material herein.

Library of Congress Cataloging-in-Publication Data
A catalog record for this book is available from the Library of Congress

British Library Cataloguing in Publication Data
A catalogue record for this book is available from the British Library

ISBN: 978-0-12-810552-8

For information on all Butterworth-Heinemann publications
visit our website at https://www.elsevier.com/

Publisher: Jonathan Simpson
Acquisition Editor: Simon Tian
Editorial Project Manager: Vivi Li
Production Project Manager: Susan Li
Designer: Maria Inês Cruz

Typeset by TNQ Books and Journals

Contents

Biography vii
Preface ix
Acknowledgments xi

1. Rock Testing

1. Instability Characteristics of a Single Sandstone Plate 1
2. Instability Characteristics of Double-Layer Rock Plates 11
3. Rupture and Energy Analysis of Double-Layer Rock Plates 18
4. Double-Layer Rock Plates With Both Ends Fixed Condition 27
5. Viscoelastic Attenuation Properties for Different Rocks 32
6. Cutting Fracture Characteristics of Sandstone 39
7. Energy Dissipation Characteristics of Sandstone Cutting 46
8. Fracture Properties on the Compressive Failure of Rock 52

References 58
Further Reading 60

2. Rockbolting

1. Mathematical Derivation of Slip Face Angle 61
2. A Mechanical Model for Cone Bolts 73
3. Effect of Introducing Aggregate Into Grouting Material 86
4. Optimizing Selection of Rebar Bolts 90
5. Poisson's Ratio Effect in Push and Pull Testing 102
6. Study on Rockbolting Failure Modes 112
7. Steel Bolt Profile Influence on Bolt Load Transfer 128
8. Tensile Stress Mobilization Along a Rockbolt 141

References 146
Further Reading 149

3. Grouted Cable

1. Load Transfer Mechanism of Fully Grouted Cable 151
2. Theoretical Analysis of Load Transfer Mechanics 159
3. Impacting Factors on the Design for Cables 168
4. Mechanical Properties of Cementitious Grout 177
5. Anchorage Performance Test of Cables 187
6. Axial Performance of a Fully Grouted Modified Cable 196
7. Sample Dimensions on Assessing Cable Loading Capacity 203

References 212
Further Reading 215

4. Tunnel Engineering

1. Construction Optimization for a Soft Rock Tunnel 217
2. Water Inrush Characteristics of Roadway Excavation 225
3. Lining Reliability Analysis for Hydraulic Tunnel 233
4. Disturbance Deformation of an Existing Tunnel 239
5. Energy Dissipation Characteristics of a Circular Tunnel 250
6. Pressure-Arch Evolution and Control Technique 256
7. Skewed Effect of the Pressure-Arch in a Double-Arch Tunnel 266

References 276
Further Reading 279

5. Slope Engineering

1. Three-Dimensional Deformation Effect and Optimal Excavated Design 281
2. Stability Analysis of Three-Dimensional Slope Engineering 289

3. Fracture Process Analysis of Key Strata in the Slope   295
4. Parameters Optimization of the Slope Engineering   303
5. Key Technologies in Cut-and-Cover Tunnels in Slope Engineering   310
6. Potential Risk Analysis of a Tailings Dam   319
7. A New Landslide Forecast Method   326
References   331
Further Reading   333

6. Mining Geomechanics

1. Analytical Analysis of Roof-Bending Deflection   335
2. Analytical Solution of the Roof Safe Thickness   345
3. Catastrophe Characteristics of the Stratified Rock Roof   350
4. Pressure-Arch Analysis in Coal Mining Field   357
5. Analysis of Accumulated Damage Effects on the Roof   363
6. Tunnel and Bridge Crossing the Mined-Out Regions   372
7. Pressure-Arch Analysis in Horizontal Stratified Rocks   381
8. Pressure-Arch in a Fully Mechanized Mining Field   392
References   400
Further Reading   402

Index   403

Biography

Shuren Wang, PhD, Professor at School of Civil Engineering, Henan Polytechnic University, Jiaozuo 454003, China. Contact details: +86 15738529570; w_sr88@163.com.

His research interests primarily includes the challenging areas of mining engineering, geotechnical engineering, rock mechanics, and numerical simulation analysis. His research projects have been supported by National Natural Science Foundation of China (51474188; 51074140; 51310105020), National Natural Science Foundation of Hebei Province of China (E2014203012), Science and Technology Department of Hebei Province of China (072756183), and so forth. He is the recipient of six state-level and province-level awards and the 2015 Endeavor Research Fellowship provided by the Australian Government. He has authored more than 90 academic papers and books. These include 85 articles in peer-reviewed journals, six monographs, and numerous textbooks. He has been authorized four patents and one software of intellectual property right in China.

Paul C. Hagan, PhD, Associate Professor, Head of School of Mining Engineering, University of New South Wales (UNSW), Sydney, NSW 2052, Australia. Contact details: +61 2 9385 5998; p.hagan@unsw.edu.au.

His principal research interests include mine geotechnical engineering and other areas in mining engineering. He has more than 30 years of experience within the mining industry and university sectors. Prior to his appointment to UNSW in 1998, he worked locally and

internationally in the coal, gold, and iron ore sectors in a range of operational, management, technical, and research roles. He has been the principal research investigator of leading projects in the rock-cutting research facility. The research has made significant advances in the application of acoustic emissions in monitoring and control of rock-cutting machines and in the determination of controlling factors associated with abrasivity testing of rock. He has authored 80 peer-reviewed papers, 110 project reports, and three patents.

Chen Cao, PhD, Research Fellow at School of Civil, Mining, and Environmental Engineering, University of Wollongong, NSW 2530, Australia. Contact details: +61 2 4221 7945; ccao@uow.edu.au.

After he graduated from Xi'an Jiaotong University, China with a bachelor's degree in science, he worked in computer engineering and had nine years of experience in the field of construction and management. He obtained a Bachelor of Engineering degree in 2009 and received a PhD degree with distinguished award for his PhD thesis from the University of Wollongong in 2013. He has been mainly involved in rock mechanics, mining engineering, and numerical simulation, and has made many breakthroughs at dealing with the problems with complex geological and engineering conditions, especially in large shear displacement by mining pressure and other factors on rock bolting and anchor support. He has authored more than 20 academic papers in the research field.

Preface

There have been significant advances in rock mechanics and understanding of the behavior of rock with developments in science and engineering. This has occurred at the same time as there has been greater demand for the utilization of underground space that has in many cases pushed the limits in the engineering design of underground excavations while there has been the continual need to improve safety and reduce the cost of excavation. It is imperative, then, that research continues which will provide the knowledge necessary to underpin the design and development of new excavation techniques.

The book summarizes and enriches the latest research results on the theory of rock mechanics, analytical methods, innovative technologies, and its applications in practical engineering. The book is divided into six chapters including such features as Chapter 1: Rock Testing (Shuren Wang Sections 1–7; Paul Hagan Section 8); Chapter 2: Rock Bolting (Chen Cao Sections 1–7; Paul Hagan Section 8); Chapter 3: Grouted Anchor (Paul Hagan); Chapter 4: Tunneling Engineering (Shuren Wang); Chapter 5: Slope Engineering (Shuren Wang); and Chapter 6: Mining Geomechanics (Shuren Wang). This book is innovative, practical, and rich in content, which can be of great use and interest to the researchers undertaking various geotechnical engineering and rock mechanics, teachers and students in the related universities, as well as on-site technicians.

The material presented in this book contributes to the expansion of knowledge related to rock mechanics. The authors, through their extensive fundamental and applied research over the past decade, cover a diverse range of topics from the microbehavior of rock and rock properties through the interaction of large-scale rock masses and its effect on surface subsidence, mechanics of rock cutting, techniques to improve the strength and integrity of rock structures in surface and underground excavations, and improvement in approaches to modeling techniques used in engineering design.

Shuren Wang, PhD
Professor at School of Civil Engineering, Henan Polytechnic University, China
Paul C. Hagan, PhD
Associate Professor and Head of School of Mining Engineering, University of New South Wales, Australia
Chen Cao, PhD
Research Fellow at School of Civil, Mining and Environmental Engineering, University of Wollongong, Australia

Acknowledgments

The authors are pleased to acknowledge the support received from various organizations, including the National Natural Science Foundation of China (51474188; 51074140; 51310105020); the Natural Science Foundation of Hebei Province of China (E2014203012); the China Scholarship Council (CSC); the Hebei Provincial Office of Education (2010813124); 2015 Endeavor Research Fellowship and Program for Taihang Scholars; International Cooperation Project of Henan Science and Technology Department (162102410027); Doctoral Fund of Henan Polytechnic University (B2015-67); Opening Project of Key Laboratory of Deep Mine Construction; Provincial Key Disciplines of Civil Engineering of Henan Polytechnic University; and the School of Mining Engineering, University of New South Wales.

While it is not possible to name them all, the authors are particularly thankful to Prof. Manchao He, Prof. Meifeng Cai, Prof. Ji'an Wang, Prof. Youfeng Zou, Prof. Xiaolin Yang, Prof. Xiliang Liu, Prof. Zhaowei Liu, Prof. Zhengsheng Zou, Prof. Jianhui Yang, Prof. Yanbo Zhang, and other related people. A number of postgraduate students, namely Jianhang Chen, Li Li, Chunliu Li, Haiqing Zhang, Yan Cheng, Baowen Hu, Chengguo Zhang, Peipei Liu, Huihui Jia, Hu Wang, Mengshi Chang, Zhongqiu Wang, Ning Li, Dianfu Xu, Yongguang Wang, Junqing Su, Huaiguang Xiao, Yanhai Zhao, Chunyang Li, and others assisted in the design, construction, and commissioning of the test facility and with the experimentation; their contributions to the book are acknowledged and appreciated. The untiring efforts of Mr. Kanchana Gamage, Dr. Juninchi Kodama, Dr. Mojtaba Bahaaddini, and Dr. Hossein Masoumi during equipment design and laboratory testing programs are gratefully appreciated. Thanks and apologies to others whose contributions we have overlooked.

Rock Testing

1. INSTABILITY CHARACTERISTICS OF A SINGLE SANDSTONE PLATE

1.1 Introduction

In China, numerous shallow mined-out areas have been left due to the disordered mining by the private coal mines. It is of important theoretical and practical value for the roof stability evaluation and disaster forecasting to research the deformation rupture, instability mechanism, and failure mode of the rock roof in the mined-out areas.

The studies on the instability of the rock roof in the mining field have been a main topic both for scholars in China and abroad. For example, according to elastic thin plate theory, Wang et al. (2006) analyzed the fracture instability characteristics of the roof under different mining distances in the mining work face. Wang et al. (2008a) analyzed the rheological failure characteristics of the roof in the mined-out areas through combining the thin plate and rheology theories. Pan et al. (2013) had conducted the analytical analysis of the variation trend of the bending moment, the deflection, and the shear force of the hard roof in the mining field. This research is inclined to adopt traditional analytic methods to probe into the roof stability. New theories and methods have been used in recent years. Zhao et al. (2010) utilized the catastrophe theory to set up vertical deformation model of the overlapping roof in the mined-out areas, and put forward the criteria for evaluating the roof stability. Wang et al. (2013c) analyzed the chaos and stochastic resonance phenomenon produced in the roof during the evolutionary process of the rock beam deformation. Meanwhile, some numerical computation methods were applied in discussing the mechanical response of rock plate or beam. Wang et al. (2008a) analyzed the blast-induced stress wave propagation and the spalling damage in a rock plate by using the finite-difference code. Nomiko et al. (2002) researched the mechanical response of the multijointed roof beams using two dimensional distinct element code. Mazor et al. (2009) examined the arching mechanism of the blocky rock mass deformation after the underground tunnel being excavated using the discrete element method. Cravero and Iabichino (2004) discussed the flexural failure of a gneiss slab from a quarry face by virtue of linear elastic fracture mechanics (LEFM) and finite element method (FEM).

In summary, though many research achievements have been made, the most results still lack laboratory testing and need to be verified. In addition, some numerical calculations were conducted based on the continuum mechanics, which could not reflect the spatial heterogeneity and the anisotropic effect of the roof in the mining field. Only a few researchers utilized the discrete element methods to study the macromechanical response of the rock plate, and did not further explore the microscopic damage of the rock plate. Therefore, a new loading device was developed to study the rock-arch instability characteristics of the plate, and particle flow code (PFC) was used to further probe into the microscopic damage of the rock plate under the concentrated and the uniform loading, respectively.

1.2 Loading Experiment of Rock Plate

1.2.1 Samples of Rock Plate

The rock samples used in the test were Hawkesbury sandstone, which obtained from Gosford Quarry in Sydney, Australia. The quartz sandstones which contained a small quantity of feldspars, siderite, and clay minerals were formed in marine sedimentary basin of the mid-Triassic, and located on the top of coal-bearing strata. The surface of specimen exhibited local red rather than white because of the content and distribution of iron oxide.

For the single-layer roof of the mined-out areas, it could be classified into two categories according to the thickness: the thin plate and the thick plate. And the roof was always made up of various combinations of the thin plates and the thick plates. Thus, according to the definition of the thin plate and the thick plate in elastic mechanics, the specimen size of the thick plate was deigned to 190 mm × 75 mm × 24 mm (length, width, and thickness) and that of the thin plate was deigned to 190 mm × 75 mm × 14 mm (length, width, and thickness). The specimens were obtained by cutting the same sandstone in the laboratory of School of Mining Engineering, University of New South Wales. The physical—mechanical parameters of rock plates were shown in Table 1.1.

1.2.2 Loading Equipment

The MTS-851 rock mechanics testing machine was selected as loading equipment, and the load was controlled by vertical displacement and loading rate was set 1×10^{-2} mm/s (Potyondy and Cundall, 2004). The vertical force and displacement occurred in the process of the test and were automatically recorded in real time by a data acquisition system.

As shown in Fig. 1.1, the concentrated and the uniform-loading test sets mainly consisted of three parts. The top was a point-loading for the concentrated loading or an assembly of the steel balls for the uniform loading. The middle was a loading framework which included four bolts with nuts connecting the steel plates on both sides, and the lateral pressure cell was placed between the deformable steel plate and the thick steel plate so as to monitor the horizontal force. The capacity of the lateral pressure cell Low Pressure X Type (LPX) was 1000 kg. The bottom was a rectangle steel foundation, the rotatable hinge supports were set on both sides of the loading framework to maintain connecting with the steel plates.

1.2.3 Acoustic Equipment and Data Acquisition System

To monitor the cracks initiated and identify the failure location of the rock plate, the USB Acoustic Emission (AE) Nodes were used in the test. The USB AE Node is a single channel AE digital signal processor with full AE hit and time based features. In the test there were four USB AE nodes being connected to a USB hub for multichannel operation (Fig. 1.2). All these AE nodes were made in MISTRAS Group, Inc., in the United States.

TABLE 1.1 Physical and Mechanical Parameters of Rock Plates

Name	Density (kg/m³)	Elastic Modulus (GPa)	Poisson Ratio	Cohesion (MPa)	Friction Angle (Degrees)	Tensile Strength (MPa)	Compression Strength (MPa)
Sandstone	2650	2.7	0.20	2.8	45	0.95	13.5

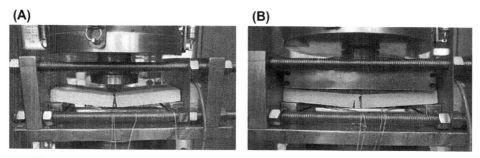

FIGURE 1.1 Loading experiment for the rock plate. (A) Concentrated loading. (B) Uniform loading.

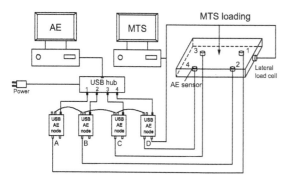

FIGURE 1.2 Mechanics Testing System (MTS) connection with acoustic emission monitoring system diagram.

1.3 Experiment Results and Analysis

1.3.1 Characteristic of Force–Displacement Curve

As shown in Fig. 1.3, the vertical force-displacement curves appeared two peaks under both the concentrate loading and the uniform loading, and the second peak value is higher than the first one. The thin rock plate showed the similarity cases in the test with the thick plate; only the peak values of the vertical and the horizontal force were lower than that of the thick one. In general, the curves of

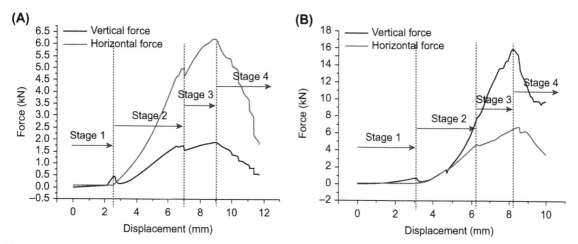

FIGURE 1.3 Force–displacement curves under different loading conditions. (A) Concentrated loading. (B) Uniform loading.

the force—displacement could be classified as four mechanical response stages as follows (Fig. 1.1A):

Stage 1: The rock plate was in the small deformation elastic stage. With the vertical force slowly increasing, the vertical displacement grew gradually. On the contrary, the horizontal force showed a slight decrease, which was mainly caused by the slight horizontal shrink of the rock plate during the loading process.

Stage 2: The rock plate produced a brittle rupture and formed the rock-arch structure. As the vertical displacement went to about 2.5 mm, the vertical force appeared to first increase abruptly and then drop sharply in a small interval, which indicated the rock plate producing a brittle rupture. Subsequently, the rock-arch structure was formed under the vertical and the horizontal reaction forces, and the horizontal force started to increase.

Stage 3: The rock-arch structure began to bear loads and produced deformation. With the vertical force increasing, the middle hinge point of the rock-arch structure moved down, and the two flanks of the rock-arch rotated around the hinge point, respectively. Such kinds of motion would stretch the rock-arch structure in the horizontal direction and squeezed the plate in two sides, and the horizontal force showed a significant growth.

Stage 4: The hinged rock-arch structure became unstable. With the vertical force continuously increasing, the middle hinged point of the rock-arch structure moved down constantly, and when the hinged point exceeded the horizontal line formed by the hinged point and two ends of the plate, the rock-arch structure became thoroughly unstable.

Under the uniform loading, the damage and fracture extent of the rock plate was more serious than that under the concentrated loading, especially at the two ends of the rock plate (Fig. 1.1B). As shown in Fig. 1.3, the load—displacement curve showed similarity with the concentrated loading, and the peak value of the vertical force was greater than that under the concentrated loading.

1.3.2 Acoustic Characteristic of the Rock-Plate Failure

As shown in Fig. 1.4, in the beginning of the stage two, the AE hits under the uniform loading were greater than that under the concentrated loading, which was about 5000 and 4500, respectively. In Stage 3 and Stage 4, the AE hits were also greater and more evenly distributed under the uniform loading compared with the concentrated loading, which was about 5000 and 3000, respectively.

As shown in the AE location map (Figs. 1.5 and 1.6), the results showed obvious differences in the initial crack position and the cracks distribution of the rock plate under different loading conditions. When the rock-arch structure went into instability, there also showed the differences in the damage extent and scope between the two loading methods. All in all, the results of AE hits and location showed the over-damage extent and scope of the rock plate caused by the uniform loading were more serious than that under the concentrated loading condition.

1.4 Numerical Simulations of the Loading Test

1.4.1 Parameters Calibration of the Rock Plate

The rock plate was treated as the porous and solid material that consisted of particles and cement bodies. The force—displacement curve was simulated under the concentrated loading using the three-dimensional particle flow code (PFC3D).

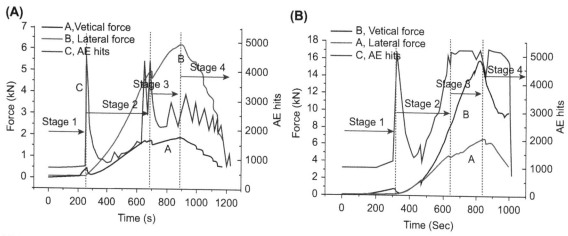

FIGURE 1.4 Acoustic emission hits and force–displacement curves under different loading conditions. (A) Concentrated loading. (B) Uniform loading.

Before the numerical simulation model could be built, the microparameters needed to be adjusted repeatedly and finalized until the macromechanical parameters calculated were consistent with the physical macromechanical parameters.

The microparameters required to be adjusted were as follows: ρ is ball density, R_{min} is minimum ball radius, R_{ratio} is ball size ratio, $\bar{\lambda}$ is parallel-bond radius multiplier, E_c is ball–ball contact modulus, $\bar{E}_c$ is parallel-bond modulus, k_n/k_s is ball stiffness ratio, $\bar{k}_n/\bar{k}_s$ is parallel-bond stiffness ratio, μ is ball friction coefficient, $\bar{\sigma}_c$ is parallel-bond normal strength, and $\bar{\tau}_c$ is parallel-bond shear strength. The microparameters required to be adjusted are listed in Table 1.2.

1.4.2 The Computational Model

Take the thick plate 190 mm × 75 mm × 24 mm (length, width, and thickness) as an example to show how to build the numerical calculation model.

First, a parallelepiped specimen consisting of arbitrary particles confined by six frictionless walls was generated by the radius expansion method. Second, the radii of all particles were changed uniformly to achieve a specified isotropic stress so as to reduce the magnitude of locked-in stresses that would develop after the subsequent bond installation. In this paper the isotropic stress was set to 0.1 MPa. Third, the floating particles that had less than three contacts were eliminated. Fourth, the parallel bonds were installed throughout the assembly between all particles that were in near proximity to finalize the specimen. Finally, the loading devices were installed on the rock plate as shown in Fig. 1.7.

A square wall with sides 10 mm was made on the top of the rock plate as the concentrated loading, and the loading rate was set to 0.01 m/s (The loading rate could be regarded as the quasistatic loading). The two cylinder walls were placed on the right and left at the bottom respectively as supporting base. The two walls located on both sides could install the initial horizontal force at the specified value. During the loading, the cracks generated in the rock plate were monitored in real time. The red cracks represented the tensile fracture, and the black ones represented the shear fracture.

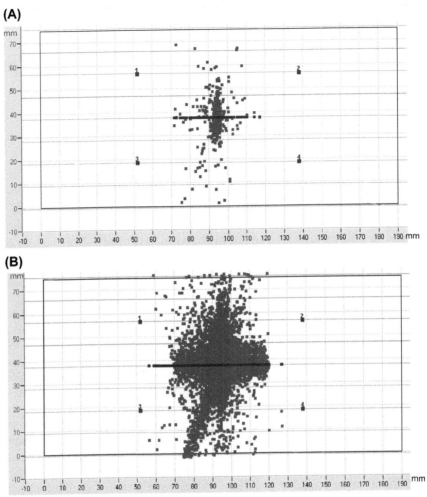

FIGURE 1.5   Acoustic emission location of rock plate under concentrated loading. (A) Initial cracks. (B) Ultimate cracks.

1.4.3 Analysis of Numerical Simulation Results

As shown in Fig. 1.8, since the interaction forces among the particles were simplified in PFC3D, there were some differences in the vertical force–horizontal force–displacement simulated curves compared with the physical experimental results, but the variation trend of the curves was basically the same for two cases, so the physical experimental results confirmed the numerical credibility.

In the elastic deformation stage (Fig. 1.9A), the displacement vector field described that a slight elastic deformation produced in the rock plate, and at the same time there was no crack generated in this stage. In the brittle rupture

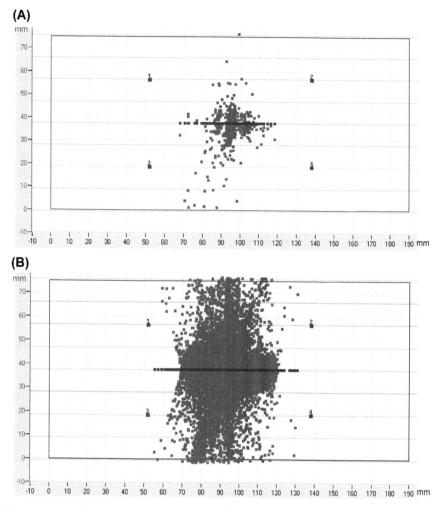

FIGURE 1.6 Acoustic emission location of rock plate under uniform loading. (A) Initial cracks. (B) Ultimate cracks.

TABLE 1.2 Microparameters of the Model in PFC3D

ρ (kg/m³)	R_{min} (mm)	R_{ratio}	μ	$\bar{\lambda}$	E_c (GPa)	$\bar{E}_c$ (GPa)	k_n/k_s	$\bar{k}_n/\bar{k}_s$	$\bar{\sigma}_c$ (MPa)	$\bar{\tau}_c$ (MPa)
2650	1.2	1.66	0.5	1.0	2.7	2.8	1.8	1.8	16	16

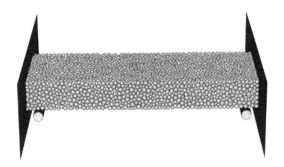

FIGURE 1.7 Computational model and its boundaries.

stage (Fig. 1.9B), there was many tensile cracks produced in the rock plate, and these tensile cracks formed a tensile failure plane in the rock plate. In the rock-arch bearing load stage (Fig. 1.9C), the shearing and tension cracks emerged in the hinged plane and both ends of the rock plate. In the rock-arch instability stage (Fig. 1.9D), the rock-arch structure had a large deformation, and parts of the particles in the hinged plane of both sides had escaped from the rock plate mainly due to the squeezing fracture.

As shown in Fig. 1.10, the number of shear cracks obeyed the S-figure curve during the whole mechanical response process, which was also applicable to the tensile cracks only after the brittle rupture. When the vertical displacement reached around 1.0 mm, the number of the tensile cracks surged to 300. As the displacement varied in the interval 1.0–2.5 mm, the crack development kept almost unchanged. However, with the displacement continuously increasing, the number of both shearing and tension cracks kept increasing, the hinged planes and both ends of the rock plate showed the mixture of shearing and tensile cracks. As rock-arch structure went into instability, the number of cracks still kept significant increasing until the displacement reached to 6 mm.

1.5 Factors Sensitive Analysis of Rock-Arch Instability

1.5.1 Material Parameter Effect

As shown in Fig. 1.11, with the friction coefficient of the particles increasing, the peak values of the vertical force and the horizontal force of the rock-arch structure also increased. This was mainly because the friction growth enhanced the peak strength of the rock material, namely after breakage of the parallel bond, the strength of the rock material often contributed to the contact friction of the particles.

1.5.2 Geometry Size Effect

As shown in Fig. 1.12, the length, width, and thickness of the rock plate changed, respectively, to reveal the size effect on the instability of the rock-arch structure. With the length of the rock plate increasing, the peak values of the vertical and the horizontal force were gradually decreased, and the whole variation interval was small. With the width and the thickness of the rock plate increasing, the peak values of the vertical and the horizontal force showed obvious growth. In short, the response of the rock-arch structure instability was more sensitive to the width and thickness compared with the length.

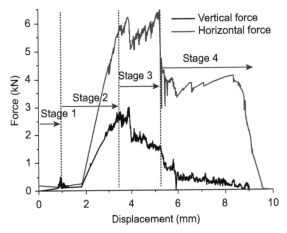

FIGURE 1.8 Force–displacement relationship curves.

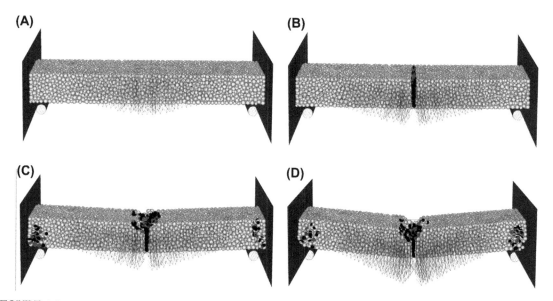

FIGURE 1.9 Rock-arch instability process under the concentrated loading. (A) Elastic stage. (B) Brittle rupture stage. (C) Bearing loading stage. (D) Rock-arch instability stage.

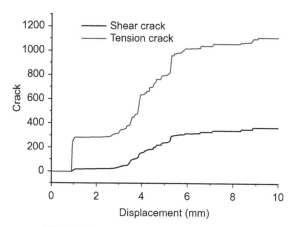

FIGURE 1.10 Crack-displacement curves.

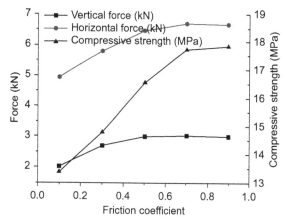

FIGURE 1.11 Force-friction coefficient curves.

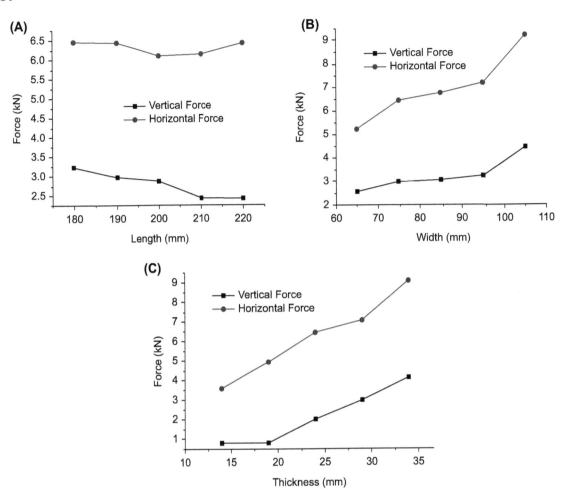

FIGURE 1.12 The forces variation with the rock plate geometry parameters. (A) Length effect. (B) Width effect. (C) Thickness effect.

1.5.3 Loading Rate and Initial Horizontal Force Effect

As shown in Fig. 1.13A, when the loading rate exceeded 10 mm/s, with the loading rate increasing, the peak values of the vertical and the horizontal force showed the linear growth trend, and the amplitude of that variation was small. When the loading rate was in the interval of 1.0–10 mm/s, the peak values were almost unchanged, therefore such loading rate could be regarded as the quasistatic loading.

As shown in Fig. 1.13B, when the initial horizontal force was less than 2.0 kN, with the initial horizontal force increasing, the vertical and the horizontal force of the rock plate showed the nonlinear fluctuating growth trend. When the initial horizontal force was larger than 2.0 kN, with the initial horizontal force

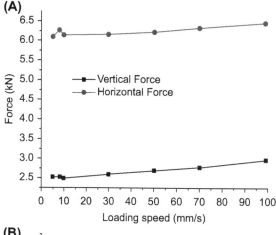

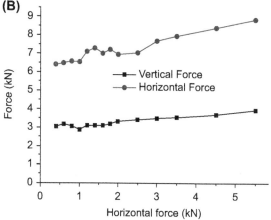

FIGURE 1.13 Forces versus the loading speed and the initial horizontal force. (A) Loading speed effect. (B) Initial horizontal force.

increasing, the vertical and the horizontal force would show the linear growth trend.

1.6 Conclusions

Under the both concentrated and uniform loading, there were elastic deformations, brittle ruptures, rock-arch bearing loads, and rock-arch instability four stages in the force-displacement curves. The peak value of the vertical force under uniform loading was greater than that under the concentrated loading. The number of AE hits and AE location showed the damage extent and scope of the rock plate under the uniform loading were greater than that under the concentrated loading.

The simulation results showed that the tensile cracks of the rock plate were dominating during the concentrated loading. The numerical test results showed the instability of the rock-arch structure was more sensitive to the width and thickness compared with the length. The loading rate could be regarded as the quasistatic loading when that was less than a critical value. The vertical and the horizontal force would show the growth trend with the initial horizontal force increasing.

To obtain the more precise simulation results in calculating by using PFC3D compared with the physical experimental results, the more precise description for the numerical model and the interaction forces among the particles should be improved.

2. INSTABILITY CHARACTERISTICS OF DOUBLE-LAYER ROCK PLATES

2.1 Introduction

In the past three decades, numerous shallow mined-out areas had been left due to disordered mining by private coal mines in China. With the decrease of available land resources, a number of industrial and civil buildings, and many structures, such as expressways, bridges, and tunnels had to cross the mined-out areas. The stress of overlying strata in shallow mined-out areas was redistributed after the coal being mined, and then ground subsidence and destruction of these buildings appeared, which seriously affected the building plan, construction, and operation of the structures built above the mined-out areas (Tong et al., 2004).

The layered sedimentary strata structure was the common structure type of the rock roof in the coal mined-out areas. Considering the effect of the bedding structure, the characteristics of transverse isotropy and the interaction between layers, the deformation feature, failure mode, and fracture mechanism of the layered roof were more complicated. Thus, this is a difficult problem that needs to be urgently solved, specifically to study the roof deformation, failure process, and catastrophe mechanism in engineering practice.

At present, most scholars usually regard the roof of a mined-out area as the rock beam to do mechanical analysis and the relevant tests (Zhang, 2009; Zhang et al., 2010b; Swift and Reddish, 2002; Nomikos et al., 2002; Diederichs and Kaiser, 1999). The processing method simplified the analytic process but has apparent limitations, which cannot reflect the spatial effect and the anisotropic of the roof; thus, the reliability of the results cause unavoidable doubt. On the other hand, most analysis and design are based on the thin plate theory while the practical ratio of length to thickness of the roof could not meet the requirement of the thin plate theory (He et al., 2007; Wang et al., 2011a; Cravero and Iabichino, 2004). In practical mining engineering, the major deformation and failure of the roof of the mined-out areas are mostly performed on rock-plate types; furthermore, they are usually expressed in the layered rock plates. Therefore, it is necessary to develop and design a new loading device which could simulate the process of the deformation of thin plates, thick plates, or thin—thick combined plates, which is the key to study the deformation and failure process and catastrophe mechanism of the roof of a mined-out area (Wang et al., 2010; Wang and Jia, 2012).

To overcome the disadvantages of these hypothesis and relevant tests about elastic rock beam, a new loading device was developed to conduct the loading and AE test of four classes of double-layer rock plates, which is of great theoretical meaning and practical value for revealing the deformation process and the catastrophe mechanism of the combined rock plates (Zhao et al., 2007; Wu et al., 2008; Miao et al., 2009; Chen et al., 2011).

2.2 Experimental Design

2.2.1 Sandstone Samples and Test Programs

The rock plate samples in the test were Hawkesbury sandstone, which were obtained from Gosford Quarry in Sydney, Australia. According to the definition of the thin plate and thick plate in elastic mechanics, the specimen size of the thick plate was deigned to 190 mm × 75 mm × 24 mm (length, width, and thickness) and that of the thin plate was deigned to 190 mm × 75 mm × 14 mm (length, width, and thickness). Each kind of rock plate was prepared for at least three tests.

Under the concentrated load, the double-layer rock plates contacted closely and performed the bending deformation together; the frictional resistance caused by slide between these two plates could reflect the mechanical effects of the interaction to some extent.

In the test, these double-layer rock plates were classified into four categories: upper thin plate and lower thick plate; upper thick plate and lower thin plate; double-layer thin plates and double-layer thick plates. Each rock plate was produced for three sets, and the comparative tests were done under the same conditions.

2.2.2 Loading Devices and Loading Modes

As shown in Fig. 1.14, the new loading device for rock plates bending test had already been authorized the utility model patent by State Intellectual Property Office of the PRC (ZL201120284625.7).

The device consisted of three parts: the top was the circular board used for the concentrated load; the middle was a loading framework which included four bolts with nuts connecting the steel plates on both sides; and the bottom was a rectangle steel foundation, and the

FIGURE 1.14 Loading device schematic for rock-plate bending test.

rotatable hinge supports were set to connect the steel plates on both sides of the loading framework (Fig. 1.14).

The overlying rock above the layered roof was usually treated as the uniform load, which created a key problem for how to apply the uniform load continuously on the layered roof in the process of the test. To simplify the stress condition reasonably, the uniform load was identified with the concentrated load through analyzing. The loading circular plate where the steel ball was embedded (Fig. 1.15) was selected to match the contactor of Mechanics Testing System (MTS) test machine.

2.2.3 Ends Design and Connection Method

Whether the design of double-layer rock plate's ends was reasonable or not was a key point of the test.

In the test, the boundary restraint state of these two ends was reasonably simulated to avoid issues of whether stiffness was too high or too small, which revealed the deformation process and fracture mechanism.

Generally, the constraint forms of rock plate's ends was classified into three categories: hinged-support boundary, fixed supported boundary, and unconstrained boundary. To simulate the real boundary restraint of the roof of the mined-out areas, this test selected the first boundary constraint form that the displacement along the width of the rock plate was restricted,

FIGURE 1.15 Loading circular plate.

and the ends along the length of the rock plate were simply supported. Some special designs were used; for example, the curved groove along the length of the hinge bearing was designed to place a cylindrical ball. When the special designs were used, the movable bearing should be put on the fixed base in which the cylindrical steel was built to cause the free bending of the rock plate's end along one direction to achieve the restraint of the hinge bearing (Fig. 1.16).

2.2.4 Test Loading and Data Acquisition System

The loading device selected was MTS-851 rock mechanics testing machine, and the load was controlled by axial displacement and the loading rate was 1×10^{-2} mm/s (Fig. 1.17). The measure of axial load value and axial displacement value could be performed in real time by data acquisition system automatically.

The bridge modules of AE data acquisition were from American National Instruments and these models were Ni 9237, Ni 9205, and Ni 9201, which were based on Labview software platform and AE signals could be analyzed (Fig. 1.18). In the test, the sampling frequency

FIGURE 1.16 Hinge bearing design.

FIGURE 1.18 Module for acoustic emission data acquisition.

FIGURE 1.17 MTS-851 rock mechanics testing machine.

of AE monitoring data was set to 50 kS/s, and the cumulative value of AE event could be read at regular intervals.

2.3 Test Procedure

Under the concentrated load, the loading process of fracture instability of the double-layer rock plates were as follows:

First, assemble the spontaneously developed loading device in order.

Second, according to combining form, place the sandstone plates on the rotatable bearing of both ends of the loading framework.

Third, regulate four nuts to keep the steel plates of both sides of the loading framework vertical and make the steel plates clamp the double-layer rock plates.

Fourth, place the loading device equipped with specimens on the loading cushion of the MTS-851 rock mechanics testing machine horizontally.

Fifth, place the loading circular plate on the rock plates and adjust the position of the steel ball of the loading plate to the centroid of sandstone plates.

Finally, place the AE probe on one side of rigid contact with tape and apply the load on the rock plates with MTS-851 rock mechanics testing machine to conduct the test.

2.4 Test Results and Analysis

2.4.1 Characteristics of Load–Displacement Curves

Through 12 sets of experiments, it was found that the load–displacement curves had common features. As shown in Fig. 1.19, there were

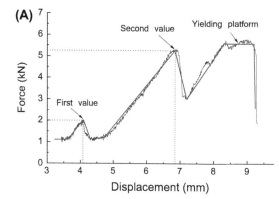

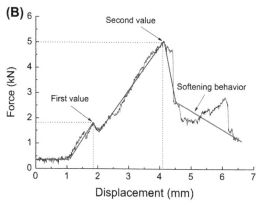

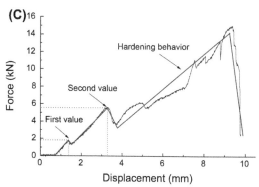

FIGURE 1.19 The load–displacement relationship curves. (A) Case 1. (B) Case 2. (C) Case 3.

generally four distinct mechanical response stages in the load–displacement curves of double-layer rock plates, taking the combination of upper thin plate and lower thick plate as an example.

The first stage: The horizontal initial stage of load–displacement curve. This stage mainly included contact adjustment between the loading contactor and sandstone plates, which was initial-load (or initial-displacement) adjustment.

The second stage: When the applied load reached to the first peak (approximately 2.0 kN), there were smaller vertical displacements in upper thin plate: the fracture developed and then the peak dropped. There was a linear relationship between the load increment and displacement increment, and this stage showed the brittle failure feature.

The third stage: with the vertical displacement increasing, the applied load showed a significant growth trend. After reaching the second peak, there was a through-and-through fracture in the upper thin plate and lag fracture in the lower thick plate. Then the load dropped again. This stage showed the brittle failure feature too.

The fourth stage: as the load continued to increase after the second peak (approximately 5.2 kN), there were integral fractures, instability occurred in three sets of specimens, and the load dropped at last. The loads varied greatly depending on variation of displacement and the double-layer rock plates of the integral fracture and instability showed plastic deformation feature.

Although three sets of double-layer rock plates had the same mechanical behavior before the second load peak, after that the combined specimens showed three different types of mechanical behavior of the sandstone plates: the first set showed postpeak yielding behavior, the second set showed postpeak hardening behavior, and the third set showed postpeak softening behavior. Overall, bearing capacity of the double-layer rock plates was greater than that of single-layer rock plates.

The other sandstone plates, such as the upper thick plate and the lower thin plate, double thick plates and double thin plates, had the same features of load–displacement curve as the above.

2.4.2 Load-Time Acoustic Emission Event Rate Curves

AE is a phenomenon of transient elastic wave generation due to a rapid release of strain energy caused by a structural alteration in a solid material. The AE event is closely related to the deformation and fracture mechanism.

As shown in Fig. 1.20, the force−time AE event rate curves of double-layer rock plates showed the features as follows: At horizontal initial stage of the load−displacement curves, when the load was small, there were few or no AE events (Fig. 1.20B and C). When the load was larger, there were a few AE events because of machining accuracy and the interaction between the sandstone plates, but the force−time AE event rate was low.

When the applied load reached to the second peak and before the integral fracture and instability occurrence, there were more AE events and the higher AE event rate occurred due to the macro crack propagation and coalescence of rock plates (Fig. 1.20A and C). By contrast, the load varied little with the displacement, micro cracks in the rock plates increased gradually, and the crack propagation and coalescence did not occur in prepeak phase, so AE events were kept at a constant level and the AE event rate was relatively stable. But in the phase between the second peak and the load peak dropping, there were a number of the crack propagations and coalescences occurred because the load varied significantly with the displacement, so there were more AE events and a higher AE event rate (Fig. 1.20C), which suggested that the crack propagations and macro coalescences were remarkable.

As shown in Fig. 1.20B, in the phase between the second peak and the load dropping there were more AE events and higher AE event rates because of the crack propagation and coalescence. It was clear that for many reasons, such as uneven mineral components of the rock, the error of machining accuracy and bad

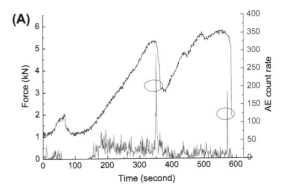

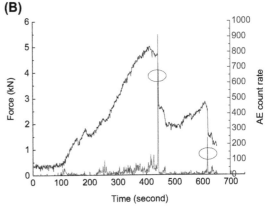

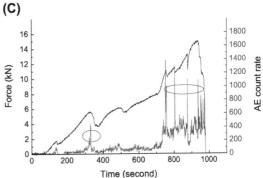

FIGURE 1.20 The load-time acoustic emission event rate curves. (A) Case 1. (B) Case 2. (C) Case 3.

connection between the double-layer rock plates all caused the AE complications. The test results indicated that the AE events were not only in the prepeak phase, but also in the postpeak phase. Although the AE could not explain accurately, it could be generally an assistant method to monitor the initiation, development, fusion, and coalescence of inner micro crack of the specimens, and it revealed the failure mechanism and the complicated mechanism of the rock plates. The other sandstone plates, such as the upper thick plate and the lower thin plate, double thick plates and double thin plates, had the same features of force-time AE event rate curves as that mentioned earlier.

2.4.3 Fracture Instability Models of Double-Layer Rock Plates

As shown in Fig. 1.21, the cracks propagated from top to bottom along the cross-section in the middle of the upper thin plate under the concentrated load. As the load increased, the coalescence phenomenon occurred finally. The lower plate was cracked after the upper thin plate, the cracks propagated from top to bottom along the cross section in the middle as the upper thin plate, too, and both of them had the same position and direction of crack initiation.

As shown in Fig. 1.22, the surface crack of the upper thin plate of sandstone plates distributed radially under the concentrated load, and which showed the plate-fracture feature; the lower thick plate showed the beam-fracture feature along the main crack in middle of the thick plate. Due to the end effects of rock plates, there showed three sets of fractures in Fig. 1.22C.

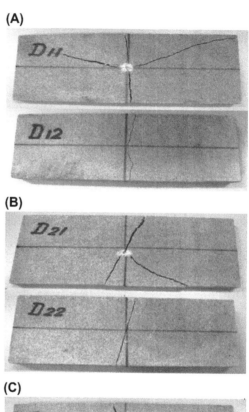

FIGURE 1.22 The surface crack distribution map of the rock plates. (A) Case 1. (B) Case 2. (C) Case 3.

FIGURE 1.21 The cross-sectional crack propagation map.

FIGURE 1.23 The cross-sectional arched crack distribution map. (A) The upper thick and the lower thin plates. (B) Double layer thick plates.

The experimental study showed that the double-layer rock plates, nearly all the upper plates had the radial plate-fracture feature and nearly all the lower plate had the beam-fracture feature in different plate combinations. Particularly noteworthy was that shown in Fig. 1.23; two upper thick plates showed the arch-fracture feature in the two sets of rock plates, which showed that as the thickness increased, the chance of having arch-fracture in the upper plate would increase, too.

2.5 Conclusions

The fracture instability process of double-layer rock plates test was conducted by the new loading device, and the device is simple, economic, and operable.

There were four distinct mechanical response stages in the force—displacement curves of double-layer rock plates. The double-layer rock plates had the same mechanical behaviors before the second load peak, and three types of mechanical behavior of the double-layer rock plates arose after the second load peak, which were postpeak yielding, postpeak hardening, and postpeak softening.

There were many different AEs among the adjustment phase, the prepeak phase, and the postpeak phase of force—time AE event rate curves; it could reveal the mechanical mechanism for the crack generating, propagation, and coalescence of double-layer rock plates in the process of fracture and instability.

The experimental results showed that there were three failure modes of beam fracture, arch destruction, and plate-fracture damage. And the results mentioned earlier were instructive to study on the stability and risk prediction of the roof of the mined-out areas.

The combined plates of more than three layers and that with weak intercalated layer will be further studied in the follow-up tests.

3. RUPTURE AND ENERGY ANALYSIS OF DOUBLE-LAYER ROCK PLATES

3.1 Introduction

There are the obvious characteristics of the layered rock structure in the coal-bearing strata, so it is of important theoretical significance and practical value to study the deformation and fracture characteristics of the layered roof in mining field considering the layers interaction.

It is the main topic for scholars in China and abroad to research the instability of the rock roof in the mining field. For example, Zhang et al. (2000) established the mechanical model of a combination system for rock mass with the interlayer, and analyzed the failure and instability mechanism of rock mass with the interlayer. Wang et al. (2006) analyzed the fracture instability characteristics of the roof under different mining distances in the mining work face. Li et al. (2014) researched the deformation and failure mechanism of deep roadway with intercalated coal seam in a roof. Yang et al. (2009) put forward the mechanical modal of the thick-hard rock layers and analyzed the modal's features. Wang et al. (2008a,b) analyzed the rheological failure characteristics of the roof in the mined-out areas through combining the thin plate and rheology theories, and the related studies (Milošević et al., 2013; Mi et al., 2014).

In thermodynamics, entropy is a physical quantity to be used to describe the degree of disorder or the uniformity of energy distribution of the system. Now the concept of entropy is widely used in the fields of classical statistical mechanics, ecology, and economy, and some mature theories have been formed (Teng et al., 2012; Li et al., 2012). So the strain energy entropy can be defined to describe the stress state change of the layered roof system in the mining field (Yin et al., 2013).

In summary, though many research achievements have been made, only a few researchers researched on the macro-mechanical response of the rock plates considering the layer effect, and did not further explore the microscopic damage of the rock plates. Therefore, a new loading device was developed to study the rock-arch instability characteristics of the rock plates, and PFC was used to further probe into the microscopic damage of the double-layer rock plates under the concentrated load.

3.2 Instability Experiment on Double-Layer Rock Plates

3.2.1 Sandstone Plates

The rock plate samples in the test were Hawkesbury sandstone obtained from Gosford Quarry in Sydney, Australia (Wang et al., 2013a,b,c). According to the definition of the thin plate and thick plate in elastic mechanics, the specimen size of the thick plate was deigned to 190 mm × 75 mm × 24 mm (length, width, and thickness) and that of the thin plate was deigned to 190 mm × 75 mm × 14 mm (length, width, and thickness). Double-layer sandstone plates of four groups were combined as upper thin and lower thick plates; upper thick and lower thin plates; two thick plates; and two thin plates. Each group of sandstone samples was prepared for at least six plates.

3.2.2 Loading Equipment

The MTS-851 rock mechanics testing machine was selected as loading equipment, and the load was controlled by vertical displacement and loading rate was set 1×10^{-2} mm/s, and all the samples were tested according to International Society for Rock Mechanics (ISRM) standards. The vertical force and displacement occurred in the process of the test were automatically recorded in real time by data acquisition system.

The double-layer sandstone plates and the test device were shown in Fig. 1.24. The capacity of the lateral load cell LPX is 1000 kg.

3.2.3 Test Results and Analysis

As shown in Fig. 1.25, for the upper thick and lower thin plates, double-layer sandstone plates all displayed beam-style rupture failure under the concentrated load, and the lower thin plate showed another rupture in the left end due to the end effect.

As shown in Fig. 1.26A, for the upper thick and lower thin plates under the concentrated load, the vertical force—displacement curve appeared two peaks under the concentrate load and the second peak value was higher than the first one. The maximum peak value of the vertical force was lower than that of the horizontal force. In general, the curves of the force—displacement

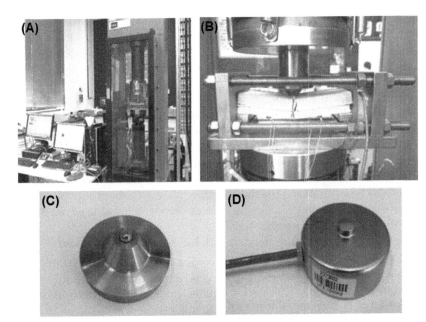

FIGURE 1.24 Loading experiment for double-layer sandstone plates. (A) Loading test system. (B) Double-layer plates rupture failure. (C) Concentrated load component. (D) Lateral load cell.

could be classified as four mechanical response stages as follows:

Stage 1: The double-layer rock plates were in the small deformation elastic stage. With the vertical force slowly increasing, the vertical displacement grew gradually. On the contrary, the horizontal force showed the slight decrease, which was mainly caused by the slight horizontal shrink of the double-layer rock plates during the loading process.

Stage 2: The double-layer rock plates produced brittle rupture and formed the rock-arch structure. As the vertical

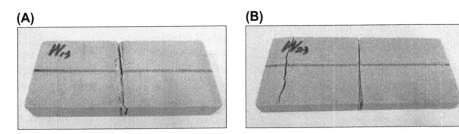

FIGURE 1.25 Double-layer sandstone plates failure under the concentrated load. (A) Upper thick plate. (B) Lower thin plate.

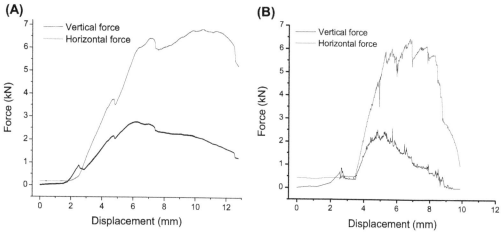

FIGURE 1.26 Force–displacement relationship curves under the concentrated load. (A) Experimental curves. (B) Simulated curves.

displacement went to about 2.5 mm, the vertical force appeared in the first peak and the lower thin plate, producing a brittle rupture, and the lower rock-arch structure was formed. As the vertical load and horizontal force continued to increase, the horizontal force displayed the first peak, and the upper rock plate showed brittle fracture. At the same time, the vertical force appeared the second peak, the fractured double-layer rock plates started to form the overlapped hinge structures.

Stage 3: The rock-arch structure began to bear loads and produced deformation. With the vertical force increasing, the hinge points of the double rock-arch structures moved down, the two wings of each rock-arch structure rotated around the hinge point, respectively. Such kinds of motion would stretch the rock-arch structure in the horizontal direction and squeezed each plate in two sides, thus the horizontal force showed a significant growth.

Stage 4: The instability of the hinged rock-arch structures happened. With the vertical force continuously increasing, the hinged points of the rock-arch structure moved down constantly. When the hinged points exceeded the horizontal lines formed by each hinged point and two wings of each plate, the rock-arch structures went into instability thoroughly.

3.3 Simulation Rupture of the Double-Layer Rock Plates

3.3.1 Parameters Calibration of the Rock Plates

The rock plate was treated as the porous and solid material that consisted of particles and cement bodies. The force–displacement curve was to be simulated under the concentrated load using the PFC3D. Before the numerical simulation, the computational model and the microparameters needed to be adjusted repeatedly, and until the calculated micromechanic parameters of the model were consistent with the physical macromechanic parameters.

The microparameters that required adjustment were shown in Table 1.2.

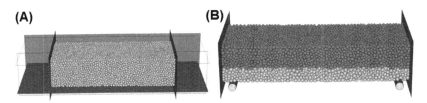

FIGURE 1.27 The process of generation of rock plates. (A) Initial generation rock. (B) Simulated rock plates.

3.3.2 The Computational Model

Taking the upper thick plate and the lower thin plate as an example, the following steps were showed how to build the computational model in PFC3D.

First, a parallelepiped specimen consisting of arbitrarily particles confined by six frictionless walls was generated by the radius expansion method.

Second, the radii of all particles were changed uniformly to achieve a specified isotropic stress so as to reduce the magnitude of locked-in stresses that would develop after the subsequent bond installation. In this paper the isotropic stress was set to 0.1 MPa.

Third, the floating particles that had less than three contacts were eliminated. The parallel bonds were installed throughout the assembly between all particles that were in near proximity to finalize the specimen.

Fourth, a single-layer rock plate specimen was generated as shown in Fig. 1.27A. Then the double-layer rock plates were generated where the upper was the thick plate and the lower was the thin one after the level joint plane being set according to the actual size.

Finally, a square wall with sides was made on the top of the rock plates as the concentrated loading. The two cylinder walls were paced on the right and left at the bottom of the model, respectively, as a supporting base, shown in Fig. 1.27B.

The value of initial horizontal force was set 0 N, and the loading rate was set to 0.01 m/s. During the loading, the cracks generated in the rock plates were monitored in real time. The red cracks represented the tensile fracture, and the black ones represented the shear fracture.

3.3.3 Analysis of Numerical Simulation Results

As shown in Fig. 1.26B, under the concentrated load, the simulated variation trend of the vertical force—horizontal force—displacement curves was basically the same as the physical experimental results, which confirmed the numerical credibility.

As shown in Fig. 1.28, under the concentrated load, the lower thin plate of the double-layer plates produced brittle fracture from bottom to top and formed the hinged rock-arch structure. At the same time, the tensile cracks also emerged in the upper thick plate from bottom to top (Fig. 1.28A). As the vertical load continued to increase, the double-layer rock plates formed the overlapped hinged rock-arch structure. The compression-shear damage effect appeared on both ends of the upper rock plate at the bottom (Fig. 1.28B). With the further increase of the vertical load, the double-layer rock-arch structures began to bear load, the hinge point of the rock-arch structure moved down, the two wings of the rock-arch rotated around the middle transfixion crack, and which made the transfixion crack damage sharply (Fig. 1.28C). When the increased vertical load exceeded the maximum load of the rock-arch structure, the rock-arch structure began to show the unload damage

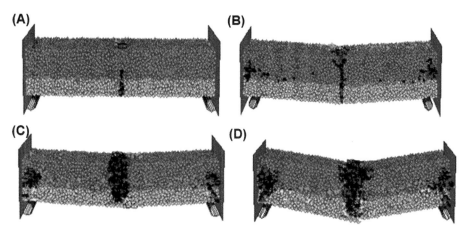

FIGURE 1.28 The deformation and failure process of double-layer rock plates. (A) Brittle rupture stage. (B) Hinged arch stage. (C) Bearing loading stage. (D) Rock-arch instability stage.

phenomenon, and both ends at the bottom of the rock plates displayed the compression-shear damage.

Fig. 1.29 showed that two points were installed in the double-layer rock plates to monitor the AE, and the average values of two points within each scope of the sphere were regarded as the variation of AE in the process of loading.

As shown in Figs. 1.30 and 1.31, when the displacement of the double-layer rock plates reached about 2.7 mm, the number of AE increased sharply around the first peak in the vertical force–displacement curves, the tensile, and shear fracture produced in the lower rock plate. As the displacement varied in the interval 2.7–4.5 mm, the number of AE and the cracks kept almost unchanged, the curves appeared in a relatively stable platform. When the displacement reached the interval 4.5–9.0 mm, the number of AE and the cracks kept increasing.

The vertical force curve went up first and then down, and the horizontal force curve increased to the nearby maximum peak platform and fluctuated, thus the rock-arch structure showed to

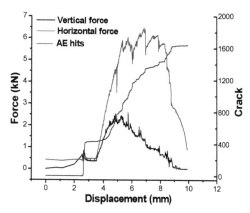

FIGURE 1.29 The sketch of monitoring locations of acoustic emission.

FIGURE 1.30 Force–acoustic emission displacement curves.

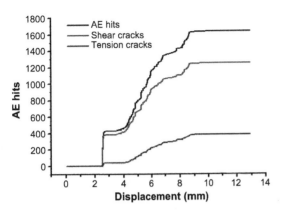

FIGURE 1.31 Crack number–acoustic emission displacement curves.

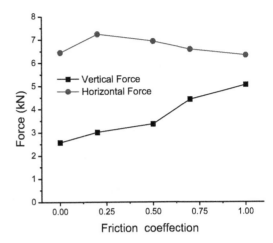

FIGURE 1.32 Force-friction coefficient curves.

bear loading. When the vertical displacement exceeded about 9.0 mm, the rock-arch structure came into the instability stage, the curves of the vertical force and horizontal force decreased rapidly, and the number of AE and the cracks became stable. As shown in Fig. 1.31, the double-layer rock plates displayed tensile and shear composite failure mode during the process of its deformation, fracture, and instability stage, and the shear failure was the main failure model.

3.4 Numerical Test on Rupture Characteristics of Rock Plates

3.4.1 Friction Coefficient Effect of the Layer

As shown in Fig. 1.32, with the layer friction coefficient between two rock plates increased from 0, 0.25, 0.5, 0.75 to 1.0, the maximum vertical force was gradually increasing from 2.5 to 5.0 kN, and the vertical force peak curve presented an upward trend. The horizontal force peak curve appeared to increase initially and decrease afterward, and the maximum horizontal force did not change and stabilized at 6.5 kN. The maximum horizontal force was greater than that of the vertical force. When the friction coefficient reached 1.0, the maximum horizontal force was closed to the vertical force. The result showed that the integrity and stability of the double-layer rock plates increased with the layer friction coefficient increase between two rock plates.

3.4.2 Cohesive Strength Effect of the Layer

As shown in Fig. 1.33, with the cohesive strength between two rock plates increased from 0.5, 1.0, 1.5 to 2.0 MPa, the maximum vertical force was gradually increasing from 2.0 to

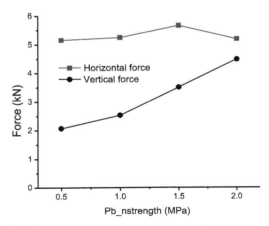

FIGURE 1.33 Force-pb_nstrength coefficient curves.

4.5 kN, and the vertical force peak curve presented an upward trend. The horizontal force peak curve appeared to increase initially and decrease afterwards, and the maximum horizontal force did not change obviously and stabilized at 5.2 kN. The maximum horizontal force was greater than that of the vertical force. When the cohesive strength reached 2.0 MPa, the maximum horizontal force was closed to the vertical force. The result showed that the integrity and stability of the double-layer rock plates increased with the layer cohesive strength increase between two rock plates.

3.4.3 Size Effect of the Rock Plates Thickness

As shown in Fig. 1.34, with the combination changing of the double-layer rock plates from the two thin plates, the upper thin and the lower thick plates, the upper thick and the lower thin plates and two thick plates, the maximum vertical force and the maximum horizontal force curves presented an upward trend. The combination of the upper thin and the lower thick plates and the upper thick and the lower thin plates had little influence on the maximum vertical force, but had obvious influence on the maximum horizontal force. The stability and resistance of deformation ability of the rock-arch structure of the upper thin and the lower thick plates was better than the upper thick and the lower thin plates. In short, with the thickness of the double-layer rock plates increasing, the maximum vertical force and the maximum horizontal force showed obvious growth trend, the size effects of the rock plates thickness were remarkable.

3.5 Energy Dissipation Characteristic of Double-Layer Rock Plates

3.5.1 Definition of Strain Energy Entropy

If the particle strains energy of the rock plates can be expressed:

$$u_i = \int_{V_i} \sigma_{ij}\varepsilon_{ij}dV_i \quad (1.1)$$

where V_i was the volume of the unit i, ε_{ij} was the element strain of the unit i, σ_{ij} was the element stress of the unit i.

The unit strain energy of the double-layer rock plates were added together, then the total strain energy of the double-layer rock plates system was

$$U = \sum_{i=1}^{n} u_i \quad (1.2)$$

Supposed that $P_i = u_i/U$, ($i = 1, 2, ..., n$), and p_i met the requirement as follows:

$$\sum_{i=1}^{n} P_i = 1 \quad (1.3)$$

where $P_i \geq 0$ ($i = 1, 2, ..., n$) is the percentage of the strain energy u_i of the ith element in the total strain energy U, which showed the strain energy distribution of the double-layer rock plates system. So the strain energy entropy could be defined as follows:

$$H(x) = -\sum_{i=1}^{n} P_i \ln P_i \quad (1.4)$$

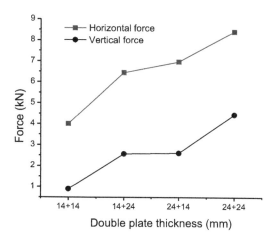

FIGURE 1.34 Force-plate thickness curves.

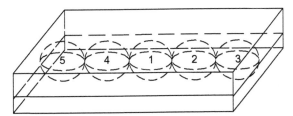

FIGURE 1.35   Monitoring locations of strain energies.

Eq. (1.4) could transform the strain energy of the double-layer rock plates system into the entropy value; that is, the strain energy entropy could be used to describe the strain energy distribution of the double-layer rock plates system to reflect the change process of the system state.

3.5.2 Energy Dissipation Characteristics of the Plates System

As shown in Fig. 1.35, five monitoring points were installed in the double-layer rock plates, and each monitoring site could detect the strain energy change within the scope of the sphere.

As shown in Fig. 1.36, when the displacement of the double-layer rock plates was less than 2.7 mm, the strain energy curve of each monitoring point did not change significantly. When the displacement reached the interval 2.7–8.5 mm, the strain energy curves of all monitoring points showed the trend of significant growth, especially monitoring site 1, the strain energy changed significantly, which showed that the local position of the double-layer rock plates deformed and destructed violently. When the vertical displacement exceeded 9.0 mm, the double-layer rock-arch structures came into instability stage, and the strain energy curves became stable.

As shown in Fig. 1.37, when the vertical displacement was less than 2.7 mm, the double-layer rock plates were the elastic deformation stage, the strain energy entropy curve of the rock plates system kept almost unchanged. When the vertical displacement reached about 2.7 mm, the strain energy entropy curve suddenly dropped, at this time the lower rock plate fractured. As the displacement varied in the interval 2.7–4.5 mm, the strain energy entropy curve went up first and then down, the system of the double-layer rock plates experienced an instability and chaos state.

3.6 Conclusions

The double-layer rock plates produced brittle fracture and formed the double hinged rock-arch structures under the concentrated load. With the increase of cohesive strength and friction coefficient of the layer between two rock plates, the

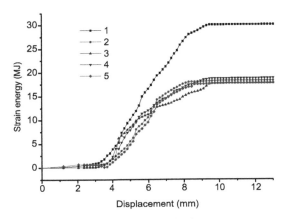

FIGURE 1.36   Strain energy–displacement curves.

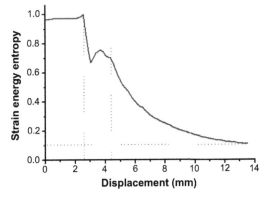

FIGURE 1.37   Strain energy entropy–displacement curves.

integrity and stability of the double-layer rock plates increased. With the thickness of the rock plate increasing, the peak values of the vertical force and the horizontal force showed obvious growth, the size effect of the rock plate thickness was remarkable. The monitoring results showed that the strain energy and strain energy entropy were consistent with the fracture instability process of the double-layer rock plates, and the strain energy entropy could characterize the steady state of the whole system of the double-layer rock plates.

4. DOUBLE-LAYER ROCK PLATES WITH BOTH ENDS FIXED CONDITION

4.1 Introduction

The roof of mined-out areas is composed of layered rock plates in general. The instability of the layered rock plates is influenced by many factors, such as the rock particle size, the rock-plate boundary conditions, the rock-plate temperature, and so forth.

Some scholars have conducted a lot of research on the stability of the layered rock plates. Gou (2008) considered the double-bedded rock slope as the folded beams structure and analyzed the deformation and failure mechanism of the double layered rock slope, discussed the buckling behavior and the bifurcation characteristic of the laminated beams by using the slab buckling theory and proposed the disaster criterion of the rock structure. Liu et al. (2013) revealed the anisotropy of mechanical properties for the layered rock, and found that there were internal compression shear failure and shear surface failure based on the uniaxial tests under different bedding angle being measured. Bao et al. (2013) studied the whole failure process of the layered rock under the uniaxial tension, including surface crack initiation, nucleation, extending, and crack saturation.

Pan et al. (2013) conducted the analytical analysis of the hard roof deflection of the mined-out areas. Wang et al. (2011a,b) researched the fracture evolution of the brittle rock plate under the impact load. Wang et al. conducted the loading test of the single and double-layer rock plates using the patented loading device, and proved that there were four mechanical response phases in the load—displacement curve (Wang and Jia, 2012, 2013a,b,c). Wang et al. (2011a,b) researched the rheological failure characteristics of the pillar-roof in the mined-out areas. Cravero and Iabichino (2004) discussed the flexure instability characteristics of the gneiss plate by using LEFM and FEM, and the related studies (Kang et al., 2013), etc.

In summary, the research as mentioned in this section usually simplified rock plate as elastic rock beam or elastic thin plate based on the continuum mechanics, therefore the theory, method, and content of the research needed to be further developed and deepened. In the paper, using self-prepared PFC3D, the numerical loading tests of double-layer rock plates were performed under the concentrated load, meanwhile, the sensitive factors of fracture instability of double-layer rock plates were analyzed. Thus the research results would provide reference and guiding significance for the stability evaluation on the roof of the mined-out areas.

4.2 Numerical Experiment of Double-Layer Rock Plates

4.2.1 Microparameters of the Rock Plates

Before building the computational model, the microparameters of the rock plate should be determined. By repeatedly adjusting the microparameters of the model until the numerical experiment results were consistent with the physical test results (Wang et al., 2013a,b,c). Fig. 1.38 was the result of double-layer rock plates with both ends hinged under the concentrated load which was conducted at School of

FIGURE 1.38 The laboratory test of double-layer rock plates under the concentrated load.

expansion method (Fig. 1.39A). Then, the level joint plane was set in the double-layer rock plates, and the upper thickness was 24 mm and the lower thickness was 14 mm (Fig. 1.39B). After removing the wall around the rock plate, on the top of the upper plate a square wall was built as the concentrated load whose side length was 1.0 mm, and meanwhile the two cylinder walls were respectively placed under the right and left of the bottom of the lower plate as the supporting base (Fig. 1.39C).

The loading rate was set to 0.01 m/s. During the loading process of the rock plates, to study the crack growth, the appearing cracks were monitored in real time. The red cracks represented the tensile fracture, and the black ones represented the shear fracture.

4.2.3 Fracture Characteristics of Double-Layer Rock Plates

As shown in Fig. 1.40, under the concentrated load, the double-layer rock plates with the fixed boundary condition had four mechanical response phases: the elastic deformation stage, the brittle fracture of upper thick plate arching stage, two rock-arch bearing stage, and two rock-arch failure stage. In the elastic stage in

Mining Engineering, University of New South Wales, Australia. The microparameters that required to be adjustment are listed in Table 1.2.

4.2.2 Building the Computational Model

Taking the upper thick plate and the lower thin plate as an example, to illustrate how to build the numerical calculation model, the special idea of the model was as follows:

First of all, the model size of the rock plate was 190-mm long, 75-mm wide, and 38-mm high, and the model was generated by the radius

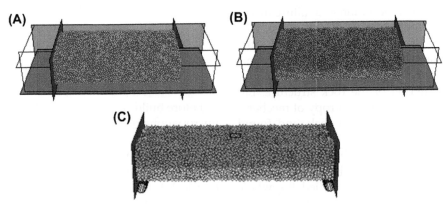

FIGURE 1.39 Building the computational model. (A) The initial model. (B) The generated double-layer plates. (C) The computational model.

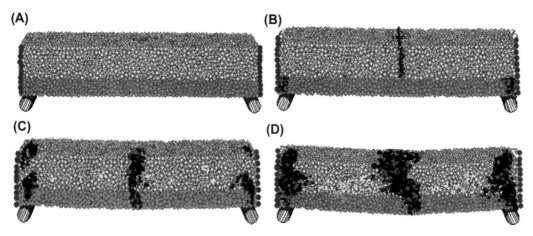

FIGURE 1.40 The deformation and failure process of double-layer rock plates. (A) The elastic deformation stage. (B) The upper thick plate arching stage. (C) Two rock-arch bearing stage. (D) Two rock-arch failure stage.

Fig. 1.40A, the double-layer rock plates produced the small deformation and there was no crack appearing at the moment. At the brittle fracture stage in Fig. 1.40B, the upper thick plate showed a transfixion crack and formed a hinged rock-arch structure, while there was shear and tension cracks appearing on both ends of the lower thin plate. With the vertical force increase, the crack number of upper plate increased, and the fracture zone produced in the central position of the lower plate, then there formed two hinged rock-arch structures, and the double-layer rock plates stepped into two rock-arch bearing stage in Fig. 1.40C. At this moment, the shear and tension cracks formed on the junction plane of the upper and the lower plate, but the tension cracks were major. At two rock-arch failure stage in Fig. 1.40D, the larger deformation appeared and the cracks significantly increased.

As shown in Fig. 1.41, under the concentrated load, the relation curves of double-layer rock plates which were about the vertical force, the crack number and the displacement could be divided into four stages: the elastic deformation stage A was corresponding to the stage with the displacement from 0 to 1.8 mm, and in stage A with the increase of vertical displacement the vertical force curve presented the linear growth trend, but the crack number was zero. The brittle fracture stage B was corresponding to the stage with the displacement from 1.8 to 4.3 mm, and in stage B the vertical displacement continued to increase and the vertical force curve and the crack number curve showed a rising trend, at the moment the tension crack was major, the

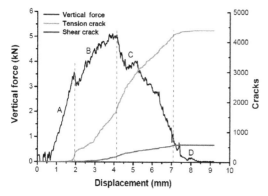

FIGURE 1.41 The vertical force-crack number–displacement relation curves.

growth of shear crack was not obvious. The rock-arch structure bearing stage C was corresponding to the stage with the displacement from 4.3 to 7.2 mm. In stage C, a through fractured zone appeared on the double-layer rock plates, formed two hinged rock-arch structures, the vertical force presented a step downtrend, and both the tension cracks and shear cracks had an obvious growth trend. The instability stage D was the stage with the displacement exceeding 7.2 mm, and in stage D the vertical force of the double-layer rock plates significantly declined and the cracks in the number of tension and shear maintained stability.

4.3 Sensitive Factors Analysis of Double-Layer Rock Plates

4.3.1 Analysis of Rock Particle Radius Changing

As shown in Fig. 1.42, under the boundary condition of both ends fixed and both sides free, the vertical force curves went up first and then down. The influence factors of the result were mainly the number of particles and the size of particles in particles aggregate. When the particle radius of rock plates increased from 1.0 to 1.4 mm, the increase of particles bond energy held a leading post, therefore, the vertical force of rock plates showed a trend of gradual increase; When the particle radius increased from 1.4 to 1.8 mm, the increase of particles bond energy was not obvious, and the number of particles significantly decreased that resulted in the decrease of vertical force.

4.3.2 Geometry Size Effect Analysis of Rock Plates

As shown in Fig. 1.43, when the total thickness of rock plates was a certain fixed value of 38 mm, with the thickness of the lower and the upper rock plate being changed, the maximum vertical force of double-layer rock plates produced a significant change. According to the order of the maximum vertical force from small to large, the combination forms were in turn the upper thin 14 mm and the lower thick 24 mm, the same thickness 19 mm of the upper plate and the lower plate, the upper thick 24 mm and the lower thin 14 mm, the whole piece of thick plate 38 mm. With the combination forms of rock plates being changed, the maximum vertical force of double-layer rock plates increased as shown in Fig. 1.44.

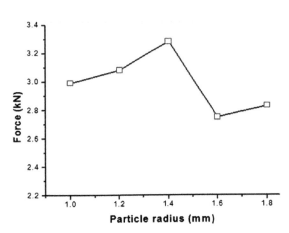

FIGURE 1.42 The maximum vertical force–particle radius relation curve.

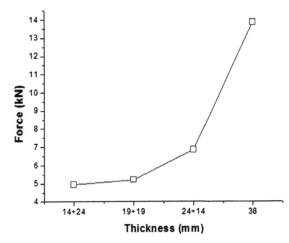

FIGURE 1.43 Maximum vertical force–rock plates thickness curves.

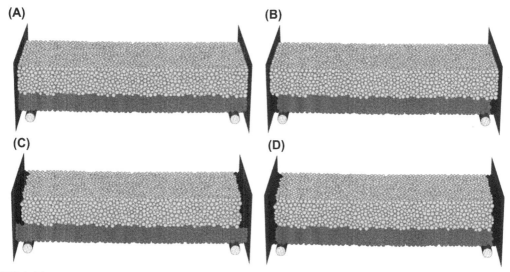

FIGURE 1.44 The boundary conditions change on both long sides of the rock plates. (A) Two hinged rock plates. (B) The lower fixed and the upper hinged plates. (C) The upper fixed and the lower hinged plates. (D) Two rock plates fixed.

4.3.3 Analysis of Boundary Conditions Change Effect

As shown in Fig. 1.44, taking the combination form of the upper thick and the lower thin plates as an example, under the condition of double-layer rock plates on both short sides free, the rest both sides were in turn two hinged rock plates in Fig. 1.44A, the lower fixed and the upper hinged in Fig. 1.44B, the upper fixed and the lower hinged in Fig. 1.44C, and the two plates fixed in Fig. 1.44D. As shown in Fig. 1.45, with the boundary conditions being changed, the maximum vertical force of the rock plates did not change significantly, whereas the maximum horizontal force markedly decreased with the articulated degree increased.

4.3.4 Cohesive Strength Effect Between Two Rock Plates

As shown in Fig. 1.46, with the cohesive strength of the layer between double-layer rock plates increasing from 0, 0.5, 1.0, 1.5, 2.0, 5.0, 7.5 to 10 MPa, the maximum vertical force curve

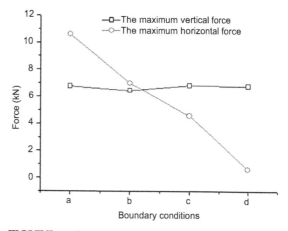

FIGURE 1.45 The vertical–horizontal force: the boundary conditions effect.

showed a decrease first and then increase, and finally tended to be stable. As the cohesive strength varied from 0 to 1.0 MPa, the maximum vertical force showed first an increase and then a fast decrease, indicating that the peak value existed in the curve; as the cohesive strength varied from 1.0 to 10 MPa, the interaction between

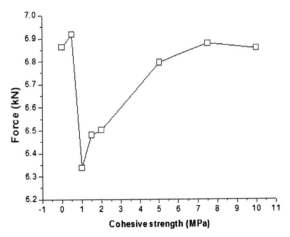

FIGURE 1.46 Maximum vertical force-layer cohesive strength between rock plates.

the upper and the lower rock plates gradually increased, the maximum vertical force gradually increased and then tended to a stable value, and the amplitude of fluctuation was limited.

4.4 Conclusions

The fracture nonlinear mechanical behavior of double-layer hinged rock-arch structures could be revealed by using PFC3D technique. The number of tension and shear cracks, crack initiation sequence, and crack propagation process could be visualized through the PFC. The sensitivity tests of influent factors on the limiting load on the double-layer rock plates were carried out, and the factors which affected the maximum vertical force from large to small were rock plate thickness, rock plate temperature, rock plate particle size, layer cohesive strength between two plates and the boundary conditions change, meanwhile, the maximum horizontal force varied depending on the boundary conditions of two rock plates change. The research results were of great significance in evaluating the roof stability of the mined-out areas.

5. VISCOELASTIC ATTENUATION PROPERTIES FOR DIFFERENT ROCKS

5.1 Introduction

Under the dynamic alternating loading, the studies on the properties of the storage modulus, the loss modulus and damping with temperature variation are of important theoretical significance and practical value. For example, the earthquake wave attenuation is closely related to the characteristics of the energy storage and dissipation for different rocks, and the damping and dynamic mechanical parameters of rock are essential for the earthquake response analysis and site safety evaluation. In addition, the nonlinear viscoelastic characteristics of the different rocks are key scientific issues for the risk estimate and the siting of repositories for disposal of high-level radioactive wastes.

Although as early as 1972, Metravib, a French company, developed the dynamic mechanical analyzer, and conducted the test and analysis of the dynamic mechanical parameters for a variety of materials. It was found that the dynamic mechanical parameters of rocks being analyzed at different temperatures were rarely reported through the vast literature. Bagde and Petros (2005) found that the loading frequency as well as the amplitude was of great significance and influenced the sandstone behavior in dynamic cyclic loading conditions under dynamic uniaxial cyclic loading, and the dynamic axial stiffness of the rock reduced with the loading frequency and amplitude. Vlastislav and Ivan (2008) proved that the attenuation quality factor of rock is related to the attenuation coefficient measured along a profile in the direction of the energy-velocity vector in an isotropic dissipative medium by test. Yun et al. (2010) verified the time-temperature superposition principle for hot-mix asphalt with growing damage and permanent strain at different confining pressures in both the tension and compression stress states.

Jackson et al. (2011) developed a laboratory equipment which was allowed both torsional and flexural oscillation measurements at submicrostrain amplitudes, and conducted a series of studies of viscoelastic and poroelastic behavior of rocks, and so on.

Chinese scholars have made advances on dynamic mechanical parameters of rocks. Yin et al. (2007) studied the influence of high temperature to the dynamic characteristics of sandstone in laboratory. Liu et al. (2008b) carried out the uniaxial cyclic loading and unloading tests to study the damping characteristics of the red argillaceous siltstones. Zhang et al. (2010a,b) studied the thermal damage properties and analyzed the stress—strain curve, peak stress, modulus of elasticity of marble at high temperature. Xi et al. (2011) researched the variations of traveling wave energy attenuation, imaginary modulus, elastic modulus, and elastic wave velocity with temperature or frequency of the saturated sandstones under uniaxial cyclic loading. Zhang and Gao (2012) carried out tests on the red sandstones under cyclic loading—unloading uniaxial compression with different loading rates, and obtained the evolution and distribution laws of energy with stress in the red sandstones. Zhu et al. (2012) studied the time-temperature equivalence effect for granite based on uniaxial creep test.

5.2 Test Equipment and Experiment Principle

5.2.1 Test Equipment and Materials

Experiments were performed by DMA/SDTA861e dynamic thermomechanical analyzer which was imported from Switzerland under sine wave loading. The dynamic load of the sine wave was 0.3 N, the temperature was controlled from −60°C to 300°C, and the heating speed was 3°C/min. Three rock samples of granite, sandstone, and mudstone were tested under three-point bending mode loading with temperature changing at frequency of 0.1, 1, and 10 Hz, respectively. Test equipment and samples are shown in Figs. 1.47 and 1.48.

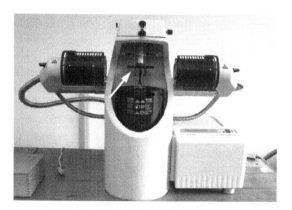

FIGURE 1.47 Photo of test equipment.

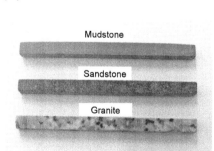
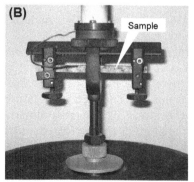

FIGURE 1.48 Sketch of rock samples and its installation position. (A) Rock samples. (B) The position of samples.

TABLE 1.3 The Basic Parameters of the Rock Samples

Rocks	Number	Density (kN/m³)	Length × Width × Height (mm³)
Granite (red meat)	G1	26.02	80.11 × 6.21 × 5.10
	G2	25.69	80.17 × 6.23 × 5.10
	G3	25.59	80.20 × 6.19 × 5.11
	G4	26.13	80.25 × 6.21 × 5.07
Sandstone (dark red)	S1	23.56	80.14 × 6.23 × 5.23
	S2	23.72	80.10 × 6.15 × 5.18
	S3	23.83	80.10 × 6.19 × 5.23
Mudstone (brown gray)	M1	22.74	80.03 × 6.05 × 5.02
	M2	22.68	80.05 × 6.03 × 5.04

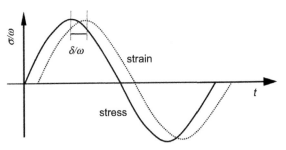

FIGURE 1.49 Relationship curves of stress-time and strain-time.

The dry granite, sandstone, and mudstone were respectively selected at room temperature and were processed into rectangular columnar samples (80-mm length, 6-mm width, and 5-mm height). For each rock, at least two samples were processed. The basic parameters of the rock samples are shown in Table 1.3.

5.2.2 Experiment Principle

Dynamic thermomechanical analysis method is a testing technique, that is, under the controlled temperature conditions, used to measure the relationship between dynamic modulus, energy loss, and temperature of the test materials under the dynamic loading.

The experiment principle is that when elastic material is applied to a sine alternating luffing stress, there will be a lagging phase angle δ ($0 < \delta < \pi/2$) between the strain response and the stress of the test material (Fig. 1.49, ω is the angular frequency), then the elastic material is called viscoelastic material. If the viscoelastic material is expressed with two moduli, the complex modulus M^* of the material consists of a storage and loss component, and it can be written as $M^* = M' + iM''$. The storage modulus M' can reflect the elastic properties of the viscoelastic materials, symbolizing the material stiffness, which means the storage capacity of deformation energy of the sample by force. And the loss modulus M'' can reflect the viscous properties of the viscoelastic material, representing the damping, which means the consumption capacity of energy of the samples by deformation. Therefore, the loss factor $\tan\delta$ of the material is introduced, which is the ratio of loss and storage component, $\tan\delta = M''/M'$, representing the dynamic mechanical behavior of the viscoelastic materials.

5.3 Results and Discussion

5.3.1 Analysis of the Storage Modulus

Fig. 1.50 is the load–displacement curves for three rock samples. As shown in Fig. 1.49, the displacement curves of three rock samples showed a linear increase with the load increasing. Since the slope of the load–displacement curve represents stiffness of the rock sample, it can be seen there are obvious differences on stiffness of three rocks in the trend of load–displacement curves change, and according to the stiffness from large to small, the order is granite, sandstone, and mudstone, which meets conventional understanding.

5. VISCOELASTIC ATTENUATION PROPERTIES FOR DIFFERENT ROCKS   35

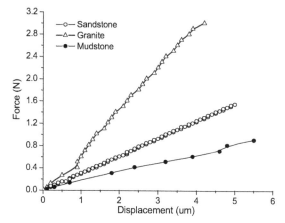

FIGURE 1.50 The load–displacement curves of three rocks.

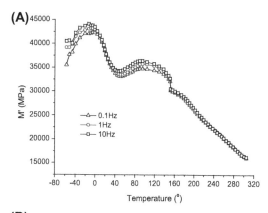

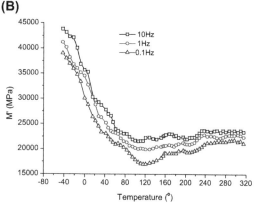

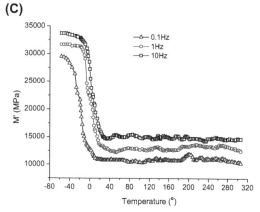

FIGURE 1.51 The storage moduli curves of different rocks with temperature change. (A) Granite. (B) Sandstone. (C) Mudstone.

As shown in Fig. 1.51, it can be seen that the storage moduli curves of three rock samples decrease rapidly with temperature increasing, and increase with the frequency increasing. But there are two significant differences in curves of the storage moduli of different rocks with temperature change.

The reduced rate of the storage moduli of different rocks with temperature change was significantly different. The curves of the storage moduli for sandstone and mudstone was from steep to flat, and the transition temperatures was nearly 120°C and 30°C, and for granite, the curve was not flat and without the transition temperature until 300°C. The reduced rate of the storage moduli from slow to fast was granite, sandstone, and mudstone, which are consistent with the results in Fig. 1.51.

The increased amplitude of the storage moduli of different rocks with frequency increase was significantly different, and the order of the increased amplitude from small to large was granite, sandstone, and mudstone. The results showed that, under the alternating stress, granite has better impact strength and rigidity relative to the sandstone and mudstone with temperature change.

5.3.2 Analysis of the Loss Modulus

Fig. 1.52 is the curves of the loss moduli with temperature change for three rocks. As shown in Fig. 1.52, the loss moduli of different rocks decrease with temperature increasing and frequency increasing, and the curves of loss moduli show an obvious fluctuation.

The order from weak to strong for the curves of the loss moduli fluctuation is granite, sandstone, and mudstone. Overall, under the alternating stress, with temperature increasing the curves change amplitude of the loss moduli for granite is small relative to sandstone and mudstone, and it means granite has a lower energy dissipation performance.

5.3.3 Analysis of the Loss Factor

Fig. 1.53 is the curves of the loss factors with temperature change for three rocks, reflecting the characteristics of the internal loss of rock samples caused by the molecular friction with temperature change. As shown in Fig. 1.53, the loss factors of granite, sandstone, and mudstone first increase and then decrease with temperature increasing, and decrease with frequency increasing.

According to the order of magnitude of the loss factors from small to large, the order is granite, sandstone, and mudstone. Thus, under the alternating stress, with temperature increasing, the curves change amplitude of the loss factor for granite is smaller, and it means granite with lower energy dissipation performance.

5.3.4 Analysis of Microstructure Characteristics

With the test temperature and the loading stress changing, the storage modulus, the loss modulus, and the loss factor for granite, sandstone, and mudstone are closely related with the microstructure characteristics of different rocks. Figs. 1.54, 1.55, and 1.56, are micrographs of scanning electron microscope for three rock samples respectively (magnification of 100 times, 1000 times, and 5000 times).

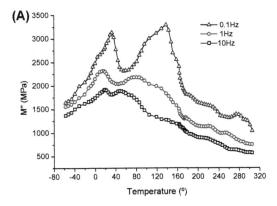

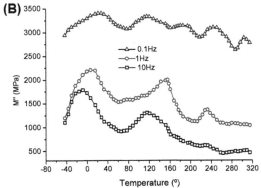

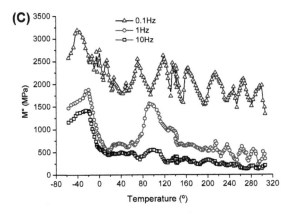

FIGURE 1.52 The loss moduli curves of different rocks with temperature change. (A) Granite. (B) Sandstone. (C) Mudstone.

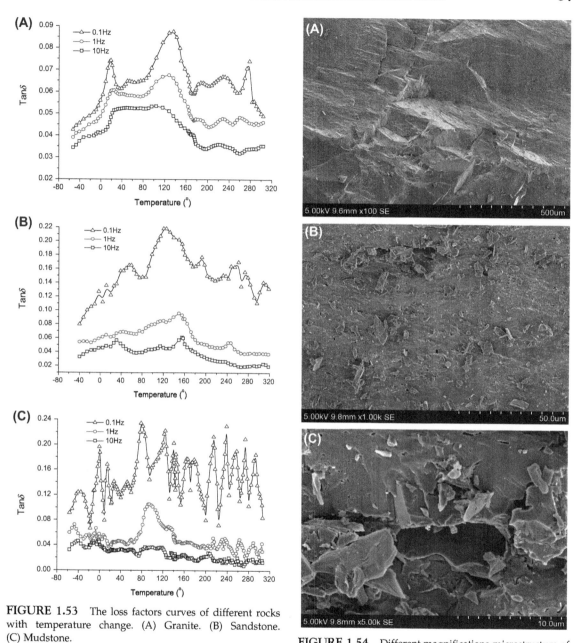

FIGURE 1.53 The loss factors curves of different rocks with temperature change. (A) Granite. (B) Sandstone. (C) Mudstone.

FIGURE 1.54 Different magnifications microstructure of the granite sample. (A) 100 times. (B) 1000 times. (C) 5000 times.

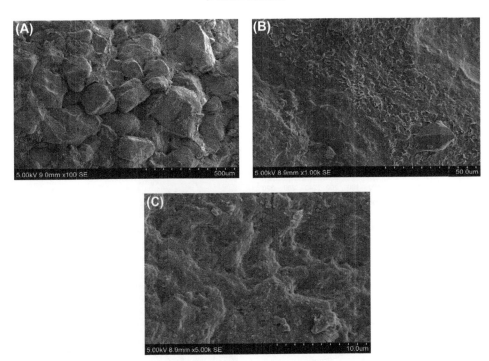

FIGURE 1.55 Different magnifications microstructure of the sandstone sample. (A) 100 times. (B) 1000 times. (C) 5000 times.

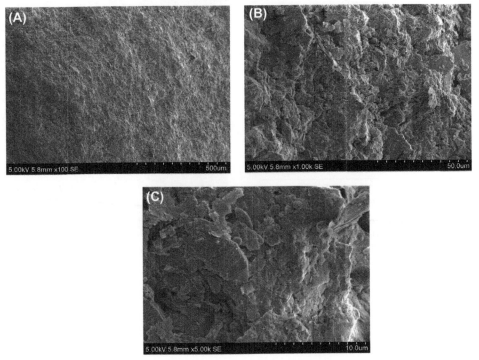

FIGURE 1.56 Different magnifications microstructure of the mudstone sample. (A) 100 times. (B) 1000 times. (C) 5000 times.

As shown in Fig. 1.54, the granite sample showed dense block structures, with a few joints, tiny holes, and flaky aggregates at the microscopic level. The sandstone sample was showed granular, with cracks and pores, which was cemented as floc distribution patterns (Fig. 1.55). The integrity of the mudstone sample was good, with microscopic porosity and scaly aggregates (Fig. 1.56). Due to the microstructural differences of the three rocks, the thermodynamic mechanical properties of these three rocks differed significantly.

5.4 Conclusions

The storage moduli of the three rocks decrease rapidly with increasing temperature, and increase with the frequency increasing for the same rock. But there are two significant differences in the curves of the storage moduli of different rocks with temperature change. Under the alternating stress, granite has better impact strength and rigidity relative to the sandstone and mudstone with temperature change.

The loss moduli of different rocks decrease with temperature increasing and frequency increasing, and the curves of the loss moduli show obvious fluctuation. Under the alternating stress, with temperature increasing, the curves fluctuation of the loss moduli for granite, sandstone, and mudstone are from weak to strong. The curve change amplitude of the loss modulus for granite is small relative to sandstone and mudstone, and it means granite with lower energy dissipation performance.

The loss factors of granite, sandstone, and mudstone first increase and then decrease with temperature increasing, and decrease with frequency increasing. Under the alternating stress, with temperature increasing, the curve change amplitude of the loss factor for granite is smaller, and it means granite with lower energy dissipation performance.

With the test temperature and the loading stress changing, the thermodynamic mechanical properties for granite, sandstone, and mudstone are closely related with the microstructure characteristics of different rocks. It is because of the microstructural differences of three rocks, which make the thermodynamic mechanical properties of three rocks different significantly.

6. CUTTING FRACTURE CHARACTERISTICS OF SANDSTONE

6.1 Introduction

Rock-breaking is usually caused by the process of initiation crack, development, expansion, aggregation and transfixion of micro fissures in the rock under dynamic loading. In today's mining, tunneling, quarrying, and construction industries, both rock cutting and drilling processes are widely used. Due to the degree of difficulty, energy consumption and broken effects during the rock-breaking are important parameters and measurement indices, and they are usually used to evaluate the drilling, blasting, mining, and mineral processing. The key to guiding rock-breaking design is to obtain the reasonable parameters and indices through establishing a simple experimental system, which can be used to evaluate and analyze the degree of rock-breaking.

Many studies show that broken rocks have a good space—time effect and energy fractal structure (Wang and Gao, 2007; Carpinteri and Puzzi, 2009; Zhu et al., 2011). The fractal theory is widely used in rock mechanics at present. Particularly, some theories have been established from previous research on rock damage, broken block distribution, and energy dissipation, and these theories can provide the basis for the study of rock-breaking (Liu et al., 2008a,b; Feng et al., 2009; Li and Roland, 2009). However, we found that many scholars focused on theoretical analyses and interpretation of experimental phenomena and rock-breaking effect in the rock crushing, and they paid less attention to the test content, test standard, and test methods of the rock-breaking index (Alberto and Nicola, 2002; Tivadar and Vass, 2011).

A test method of mechanical shock rock-breaking was used in the School of Mining Engineering, University of New South Wales (UNSW), which is easy to operate and has a good visual effect. However, the test method has some defects: these indicators of specific energy and rock-breaking production per unit length have characteristics of data dispersion and poor statistical regularity. Furthermore, these indicators do not reflect well the size and distribution of rock fragmentation, and these indicators don't specifically express the integrity and complexity of the rock structure. Thus, some new evaluation indicators should be put forward to improve the test method.

6.2 Fractal Analysis of Rock Fragmentation

Fractal analysis has become a new subject branch since the conception of fractal geometry was proposed by B. B. Mandelbrot, a French scientist, in 1975. At present, fractal geometry has become a powerful tool to study the irregularity of nature sciences, engineering technology, and social sciences.

In 1992, Tyler et al. put forward and developed a fractal model of the three dimensional space particle as follows (Tyler and WheatCraft, 1992):

$$V(r > R) = C_v \left(1 - \left(\frac{R}{R_L}\right)\right)^{3-D} \quad (1.5)$$

where $V(r > R)$ is the volume whose particle size is bigger than R. C_v, and λ_v are the constants representing particle shape and size, respectively. R is the particle size, and D is the fractal dimension.

Afterwards, based on the hypothesis that the particle density is the same, Tyler et al. transformed Eq. (1.5) and got Eq. (1.6) as follows:

$$\frac{M(r < R)}{M_T} = \left(\frac{R}{R_L}\right)^{3-D} \quad (1.6)$$

The Eq. (1.6) is the fractal model of particle mass. Where, $M(r < R)$ is the particle volume whose size is smaller than R. M_T is the gross mass of particle. R is the particle size, and R_L is the maximum size of the particle.

Taking the logarithm of both sides of the Eq. (1.6), respectively:

$$\lg \frac{M(r < R)}{M_T} = (3 - D)\lg \frac{R}{R_L} \quad (1.7)$$

It can be seen from Eq. (1.7), $\lg \frac{M(r < R)}{M_T} \propto (3 - D)\lg R$.

Therefore, the rock-breaking tests should show if the factual dimension D need to be calculated. First, the particle gradation sizing and the average diameter of particles in each grade interval of the broken debris should be obtained. Second, the cumulative mass percentage of the rest debris, which is less or equal to the size of each interval, should be calculated. Finally, in the double logarithmic coordinate axis, it is easy to calculate the slope b of the fitting curves and the factual dimension D of the rock debris can be obtained according to $b = 3 - D$.

6.3 Sandstone Cutting Test

6.3.1 Test Equipment and Sandstone Samples

The cutting test is conducted by using the linear rock-cutting machine in the laboratory of School of Mining Engineering (UNSW). As shown in Fig. 1.57, the dimensions of the cutting machine are 1829 mm long, 1118 mm wide, and 1321 mm high, respectively. For past research work at UNSW, the original cutter head was

FIGURE 1.57 Testing machine and test system.

modified to enable a dynamometer to be attached, with the cutting tool holder fixed to the dynamometer. The cutting tests were performed using a tungsten carbide cutting bit, mounted in the tool holder on the dynamometer. A triaxial solid plate dynamometer was used to obtain the normal, lateral, and cutting forces that the cutting bit was subjected to during each of the test cuts. During cutting the cutting head moves in a straight line, at a constant height and velocity.

All the sandstone samples are obtained from the field adjacent to hynds seam in Hunter Valley, New South Wales. For each cutting test, the dynamic monitor fixed on the cutter monitored the voltage change data along the cutting direction and vertical cutting direction (Fig. 1.58A); the mechanical impact of cutting force and normal force curves can be shown on the computer screen in real time through the conversion software program. For each sandstone sample, the cutting test should be operated for four times in order, as shown in Fig. 1.58B.

6.3.2 Test Methods and Test Procedures

First, each cylindrical core sample with diameter Ø 76 mm should be fixed horizontally in the planer with a flat clamp.

Second, install the new standard cutting tool, which is a 12.5 mm wide chisel with front rake angle 0°, back clearance angle 5°, and cemented tungsten carbide.

Third, adjust the cutter depth of 5 mm through raising or lowering the lathe height. A flat tray is prepared for collecting cutting debris, and the testing machine must be covered with transparent plastic sheeting to prevent debris scattered everywhere.

Fourth, start the testing machine to cut the sandstone sample at a constant speed of 150 mm/s (Fig. 1.58A).

Fifth, as shown in Fig. 1.58B, the sandstone sample should be rotated counterclockwise by 180° after the first cutting is finished, then the second cutting can be operated along the same direction as the first cutting. After the second

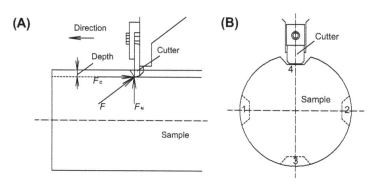

FIGURE 1.58 Schematic diagram of the cutting machine. (A) Side view of the test sample. (B) Front view of the test sample.

cutting it should be rotated counterclockwise by 90° according to the position of the first cutting and the third cutting should be conducted. Finally, the sandstone sample should be rotated counterclockwise by 180° according to the third cutting, and then the fourth cutting should be completed.

6.4 Test Results and Fractal Analysis

6.4.1 *Cutting Debris and Screening*

Fig. 1.59 shows that four sandstone samples are tested, respectively. For each rock sample and for each impact cutting, the cutting debris must be collected, bagged, numbered, and weighted. From the test results, we can find that sample (a) and (d) are short, the recovery ratio of sample (a) and (d) are low, the cleavage and cranny developed, and then the integrity is poor. On the contrary, sample (b) and (c) are long, the recovery ratio are high, the cleavage and cranny are not developed, and the integrity is good.

According to the size characteristics of the generated debris in the test, the selected aperture sizes of the meshes from large to small are as follows: 11.20, 6.70, 3.33, 1.18, and 0.43 mm

FIGURE 1.60   Test sieves.

(Fig. 1.60). After each screening, the quality of the stranded fragments of each sieve aperture must be measured by the weighing balance and recorded (Fig. 1.61). Then the mass distribution of the broken rock can be calculated and statistically analyzed.

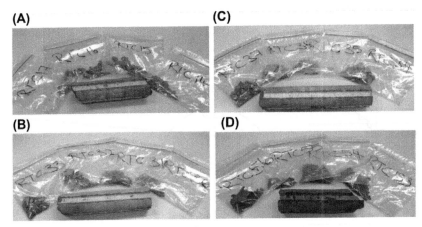

FIGURE 1.59   Schematic diagram of sandstone samples and debris. (A) No. 377542. (B) No. 377554. (C) No. 370526. (D) No. 377590.

6.4.2 Fractal Analysis of Cutting Debris

After four cutting test samples are completed, the rock-breaking debris should be sieved according to the different aperture sieves. The sieve cumulative quality and the percentage content of the statistical results are listed in Tables 1.4 and 1.5.

As shown in Table 1.5 and Fig. 1.62, the sandstone samples are divided into the two groups according to the quality distribution characteristics of the four sample debris: Large pieces in the (a) and (d) sample debris are higher, small pieces are lower, and the fragmentation distribution is uneven. On the contrary, large pieces in the (b) and (c) sample debris are lower, small pieces are higher, and the fragmentation distribution is even.

FIGURE 1.61 Weighting apparatus.

TABLE 1.4 Statistics Table for the Mass of Rock Debris

Sample Number	Measuring Sequence	Sieve Cumulative Quality of the Different Mesh (g)				
		<11.20	<6.70	<3.33	<1.18	<0.43
No. 377542	RTC14	10.0467	6.433	3.7188	1.8925	0.8224
	RTC15	17.2644	9.6957	6.3077	3.555	1.7557
	RTC16	18.7572	11.2675	7.8113	4.6267	2.5954
	RTC17	17.2802	10.7951	6.6582	3.5791	1.4408
No. 377554	RTC30	15.9262	11.2805	7.9609	4.8692	2.7126
	RTC31	18.5819	12.6303	8.2047	4.839	2.6993
	RTC32	25.5429	13.2302	9.3069	5.3627	3.1669
	RTC33	15.018	11.226	7.38	4.731	2.733
No. 370526	RTC34	20.7606	14.3551	9.8903	6.3439	3.6254
	RTC35	20.6098	13.9659	10.1951	6.5624	3.7902
	RTC36	23.1703	16.1592	10.9214	6.8578	4.0479
	RTC37	29.3711	18.3478	11.8751	7.7839	4.5557
No. 377590	RTC53	17.7805	12.3753	7.1072	3.2909	1.3714
	RTC54	37.0674	20.1786	10.3365	4.4435	1.6507
	RTC55	12.0535	9.574	5.6733	2.9413	1.3641
	RTC56	13.2883	8.5421	5.4678	2.6397	1.2472

TABLE 1.5 Statistics Table for Percentage Content of Rock Debris

Sample Number	Measuring Sequence	Sieve Cumulative Quality Percent of the Different Mesh (%)				
		<11.20	<6.70	<3.33	<1.18	<0.43
No. 377542	RTC14	43.85	28.08	16.23	8.26	3.59
	RTC15	44.75	25.13	16.35	9.21	4.55
	RTC16	41.63	25.01	17.34	10.27	5.76
	RTC17	43.47	27.16	16.75	9.00	3.62
	Mean value	**43.42**	**26.34**	**16.67**	**9.19**	**4.38**
No. 377554	RTC30	37.25	26.39	18.62	11.39	6.35
	RTC31	39.57	26.90	17.47	10.31	5.75
	RTC32	45.12	23.37	16.44	9.47	5.59
	RTC33	36.55	27.32	17.96	11.51	6.65
	Mean value	**39.63**	**25.99**	**17.62**	**10.67**	**6.08**
No. 370526	RTC34	37.76	26.11	17.99	11.54	6.59
	RTC35	37.39	25.34	18.50	11.90	6.88
	RTC36	37.89	26.42	17.86	11.21	6.62
	RTC37	40.83	25.51	16.51	10.82	6.33
	Mean value	**38.47**	**25.84**	**17.71**	**11.37**	**6.61**
No. 377590	RTC53	42.41	29.52	16.95	7.85	3.27
	RTC54	50.31	27.39	14.03	6.03	2.24
	RTC55	38.14	30.29	17.95	9.31	4.32
	RTC56	42.61	27.39	17.53	8.46	4.00
	Mean value	**43.37**	**28.65**	**16.62**	**7.91**	**3.46**

So, through the core cutting tests, the statistical analysis of the sieve cumulative quality not only directly reflect rock fragmentation distribution, but also reflect rock joints development, as well as the integrity of the rock mass structure.

As shown in Fig. 1.63, Under mechanical shock loading conditions, between the sieve cumulative quality percent and the screening aperture of the four sample debris, there are significant fractal relationships in double logarithmic axes. Moreover, the experimental data and the fitting lines have high linear correlation degrees. Similarly, according to the fractal dimension values of the four sample debris, these sandstone samples are divided into two groups: The fractal dimensions of the (a) and (d) sample debris are smaller, and the mean values are 2.319 and 2.226, respectively. The fractal fitting curves have wide distribution areas and have discrete characteristics. On the contrary, the fractal dimensions of the samples (b) and (c) debris are larger, the mean values

6. CUTTING FRACTURE CHARACTERISTICS OF SANDSTONE

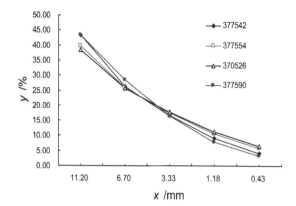

FIGURE 1.62 Curves of debris mass and its size.

are 2.446 and 2.481, respectively. The fractal fitting curves have narrow distribution areas and have intensive characteristics. So, the core cutting tests show that the values of the fractal dimensions of the broken debris not only quantitatively reflect the rock-breaking extent and rock hardness, but also can reflect the dissipation energy and the integrity of the core structure.

6.5 Conclusions

The rock-breaking fragmentation and fractal features are significant under the mechanical shock-loading conditions for engineering practice.

The broken debris fractal dimension of the structural integrity specimens is larger, the range of the fractal dimension is smaller, and the fractal curves have the intensive feature. In contrast,

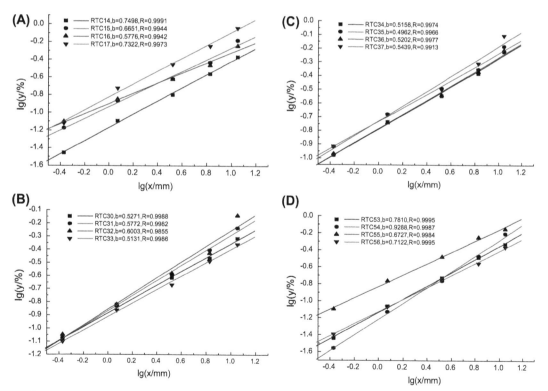

FIGURE 1.63 Curves of fractal dimensions of different samples. (A) No. 377542. (B) No. 377554. (C) No. 370526. (D) No. 377590.

that of the poor structural integrity specimens is smaller, the range of the fractal dimension is larger, and the fractal curves have the discrete feature.

The broken debris fractal dimension is larger, and the broken debris size distribution is more even; the broken debris fractal dimension is smaller, and the broken debris size distribution is less even.

The results show that the fractal dimensions not only reflect the quality of the broken debris and the distribution of the particle size quantitatively, but also can reflect the structural integrity and the complexity of specimens indirectly. Thus, the fractal dimension is the ideal test indicator to assess and analyze the rock-breaking degree.

7. ENERGY DISSIPATION CHARACTERISTICS OF SANDSTONE CUTTING

7.1 Introduction

Rock-cutting usually caused the process of initiation crack, development, expansion, aggregation, and transfixion of micro fissures in the rock under dynamic loading (Lin and Chen, 2005). For the difficulty degree and energy dissipation of rock-cutting, they are not only the important indicators to evaluate the drilling, blasting, mining, beneficiation process, but also the key parameters to optimize the rock fragment and improve the production capacity.

At present, many domestic scholars have done lots of work in evaluating the difficulty of rock-cutting and energy dissipation, and achieved fruitful research results (Hu et al., 2002; Gao et al., 2013). For example, Li et al. (1992) had conducted the impact crushing tests on four kinds of typical rocks under different loading waveforms, and found that the amount of energy absorbed by the rock was the most under the bell-wave loading. You and Hua (2002) conducted the conventional triaxial loading test, and proved that the amount of energy absorbed by the broken rock was increased in a linear manner with increasing confining pressure. Xia et al. (2006) had carried out the impact experiments of man-made rock with different porosities with the Split Hopkinson Pressure Bar, the characteristic of energy dissipation and the influence of rock porosity on energy dissipation were investigated during the impact process, and the relationship between the energy dissipation of critical failure and rock porosity was also analyzed. Kong et al. (2012) researched the mechanism of sand energy dissipation under the cyclic loading using PFC, and proved that the smaller the sand porosity, the more system energy dissipation, and the energy dissipation improved with the confined pressure increasing. For research on the specific energy of rock breaking, the main representative theories were put forward by foreign scholars, such as the new-surface theory, the similarity theory, and the crack theory (Guo et al., 2008). Goktan and Yilmaz (2005) researched the relationship between the rock brittleness index and rock-cutting efficiency, and found that the rock-cutting efficiency improved with the rock brittleness increasing under other conditions remained unchanging. Copur et al. (2003) developed a set of new empirical indices and a new interpretation method of macroscale indentation tests, which were useful for predicting the cutting efficiency and mechanical properties of rocks. Adebayo (2008) made an assessment of the cuttability of granite and limestone, and found that the cutting rate increased in a linear manner with the growth of rebound hardness value, uniaxial compressive strength, and silica content. Dey and Ghose (2011) proposed a new index which could reasonably predict the production efficiency of the surface coal mining. Tumac et al. (2007) tried to evaluate the cuttability of rocks by using the rock hardness and the uniaxial compressive strength index.

In summary, though many research achievements have been obtained for evaluating the difficulty of rock breaking and energy dissipation, there is still a lot of work to be researched in depth due to the nonuniformity of rock material, the difference of geological conditions, and the complexity of engineering conditions. Therefore, the numerical calculation model of rock-cutting was built based on the test by using the linear rock-cutting equipment at the UNSW in Australia. The numerical simulation of rock-cutting energy dissipation was conducted under different conditions, which was of important theoretical significance and practical value in guiding similar engineering to reduce the cost of drilling and blasting and improve the production efficiency.

7.2 Specific Energy Analysis of Cutting Debris

In the cutting experiment, the specific energy (SE) is the energy or work required to cut unit volume of the rock, then SE is calculated as follows:

$$\text{SE} = \frac{F_c}{V} \quad (1.8)$$

where F_c is the mean cutting force (kN), and V is the volume of excavated material per unit length of cutting (m^3/km). The unit of SE is usually MJ/m^3.

$$V = LS \quad (1.9)$$

where L is the distance the pick travels (m), and S is the unit cross-sectional area along the pick travels (m^2).

In a given rock the SE is used as a measure of the efficiency of a cutting system with lower values indicating higher efficiency, and what's more, in the context of cuttability assessment, it can be used to both compare the cuttability of different rocks and indicate approximately the potential excavation rate for a particular machine type in a given rock (Roxborough, 1987).

Table 1.6 showed the mean cutting force, the mean normal force and the SE of rock-cutting on the basis of the experimental data statistics by using the linear rock-cutting equipment.

As seen from Figs. 1.64 and 1.65, with the curves of the mean values changed from high to low, it indicated that the strength of the four types of sandstones changed from high to low. That is, the SE of rock-cutting increased with the growth of rock strength, and vice versa.

7.3 Computational Model and Model Process

7.3.1 The Computational Model

As shown in Figs. 1.66 and 1.67, the numerical calculation model was built by using PFC according to the engineering mechanical model. The horizontal velocity of the calculation model was zero, the top surface was free or the confining pressure boundary. The cutter simulated by using two mutually perpendicular wall sections was moved to the right at a certain velocity, the back-rake angle, and the cutting depth to simulate the cutting process.

The rock specimen was treated as the porous material that consisted of particles and the cemented bodies. Before the numerical test, the micromechanical parameters should be adjusted repeatedly until these parameters were consistent with the physical experiment results.

The micromechanical parameters required to be adjusted were listed in Table 1.2.

7.3.2 The Modeling Process

Step 1: The specimen was rectangular (width × height = 100 mm × 50 mm) and confined by three frictionless walls on the bottom, the left, and the right sides, which was generated by the radius expansion method.

Step 2: The radii of all particles were changed uniformly to achieve a specified isotropic stress so as to reduce the magnitude of

TABLE 1.6 Parameter Statistics of the Rock-Cutting Experiment

Sample	Cutting Sequence	Cutting Force (kN)	Normal Force (kN)	Distance (mm)	Specific Quantity (m³/km)	Specific Energy (MJ/m³)
377542	RTC14	1.56	1.62	180	0.0534	29.28
	RTC15	1.56	1.54	326	0.0738	21.17
	RTC16	1.78	1.87	486	0.0793	22.47
	RTC17	2.51	2.68	611	0.0615	40.78
	Mean	**1.85**	**1.93**	**401**	**0.067**	**28.43**
377554	RTC30	1.77	1.62	180	0.0711	24.96
	RTC31	1.82	1.63	198	0.0666	27.29
	RTC32	1.80	1.63	208	0.0671	26.79
	RTC33	1.28	1.25	190	0.0665	26.79
	Mean	**1.67**	**1.53**	**194**	**0.068**	**26.46**
370526	RTC34	1.31	1.09	218	0.0640	20.47
	RTC35	1.21	1.01	437	0.0566	21.32
	RTC36	1.37	1.06	647	0.0720	19.00
	RTC37	1.35	1.01	863	0.0682	19.77
	Mean	**1.31**	**1.04**	**541**	**0.065**	**20.14**
377590	RTC53	0.75	0.58	162	0.0940	7.95
	RTC54	1.15	0.78	322	0.1489	7.75
	RTC55	0.79	0.62	458	0.0517	15.32
	RTC56	0.81	0.63	593	0.0897	8.98
	Mean	**0.88**	**0.65**	**384**	**0.096**	**10.00**

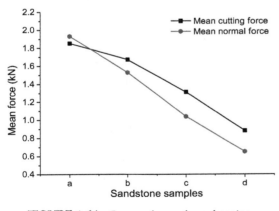

FIGURE 1.64 Curves of mean force changing.

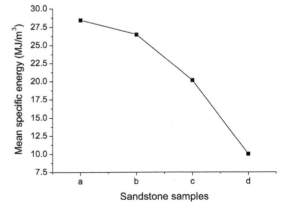

FIGURE 1.65 Curves of mean specific energy changing.

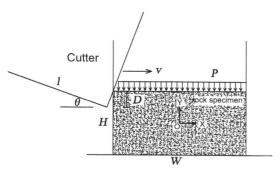

FIGURE 1.66 Engineering mechanic model of rock cutting.

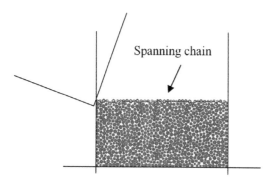

FIGURE 1.67 Computational model of rock cutting.

locked-in stresses that would develop after the subsequent bond-installation. In this paper the isotropic stress was set to 0.1 MPa.

Step 3: The floating particles that have less than three contacts were eliminated. Then the parallel bonds where installed throughout the assembly among all particles. The cutter was represented by two wall segments (both of length $l = 80$ mm) that were mutually perpendicular with a back-rake angle of 20°. The cutter was moved horizontally across the rock at a velocity of 0.5 m/s and at a cutting depth of 5 mm.

Step 4: To simulate the compressive stress on the rock surface, a pressure-application algorithm is developed for PFC by continually identifying a connected chain of particles on the rock surface and applying the pressure to those particles as the cutting process proceeds.

Step 5: In the process of rock-cutting simulation, the calculation model could trace energy data through the built-in energy function in PFC, and record the crack data through the built-in crack function so as to monitor the variation of SE and AE.

Step 6: The variation law of SE and AE would be analyzed statistically with the change of the cutting speed, the cutting depth, and the confined pressure.

7.4 Rock-Cutting Energy Characteristics

7.4.1 Specific Energy Variation with the Cutting Velocity

As shown in Fig. 1.68, the SE of rock-cutting was proportional to the cutting velocity. The SE under the confined pressure was greater than that without the confined pressure at the same cutting velocity. What was more, as a result of the effect of the confined pressure, the curve slope was greater; that is, the SE of rock-cutting increased faster with the cutting velocity increasing under the confined pressure condition.

Fig. 1.68A showed that the SE of the hard rock increased the fastest, followed by the medium hard rock, and the last was the soft rock with the cutting velocity increasing without the confined pressure. Fig. 1.68B showed that the slope of curve, the value of SE, and the amount of energy consumption increased under the confined pressure being set to 10 MPa.

7.4.2 Specific Energy Variation with the Cutting Depth

As shown in Fig. 1.69A, with the cutting depth increasing, the SE curves of three kinds of sandstones presented the slow-growth trend without the confined pressure, while the medium hard rock began to decline when the cutting depth

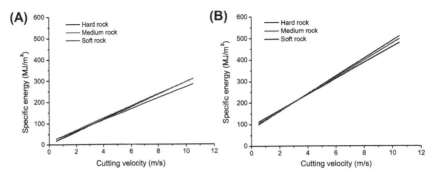

FIGURE 1.68 Specific energy-cutting velocity variation curves. (A) Without the confined pressure. (B) With the confined pressure $P = 10$ MPa.

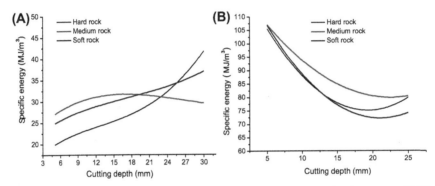

FIGURE 1.69 Specific energy-cutting depth variation curves. (A) Without the confined pressure. (B) With the confined pressure $P = 10$ MPa.

was at 20 mm. As shown in Fig. 1.69B, under the confined pressure, with the cutting depth increasing, the SE curves decreased at first and then increased. There was a minimum value of SE which was called the optimal rock-cutting depth. The optimal rock-cutting depth of hard rock was about 18 mm, that of the medium hard rock was about 24 mm, and that of the soft rock was about 21 mm.

Compared Fig. 1.69A and B, we could find that the cutting depth should not be too deep for hard rock and soft rock, while the cutting depth of the medium hard rock could increase appropriately.

7.4.3 Acoustic Emission Variation with Cutting Velocity

The AE is commonly defined as transient elastic waves within a material caused by the release of localized stress energy. The rock AE is directly related to the generation of internal micro cracks (damage). A micro crack occurs along with a release of strain energy, that is, there is an AE. Therefore, the rock AE can be simulated by recording the number of micro cracks. The crack function defines a link damage between particles as a crack generation in PFC. As a result, the rock AE can be simulated by the crack function in PFC.

As shown in Fig. 1.70, the AE was directly proportional to the cutting velocity. Without the confined pressure, there was an obvious difference among the growth rate of the AE three kinds of sandstones, while under the confined pressure, the growth rate of the AE were almost the same.

As shown in Fig. 1.70A, with the cutting velocity increasing, the AE growth rate of the hard rock increased the fastest, followed by the medium hard rock, and the last was the soft rock. When the cutting velocity was less than 2.0 m/s, the growth rate of AE was soft rock > medium hard rock > hard rock; when the cutting velocity was more than 2.0 m/s, the growth rate of AE was hard rock > medium hard rock > soft rock. As shown in Fig. 1.70B, after being set the confined pressure at 10 MPa, the growth rate of AE of three kinds of sandstones were almost the same.

7.4.4 Acoustic Emission Variation with the Cutting Depth

As shown in Fig. 1.71, with the cutting depth increasing, the confined pressure only had a slight influence on the AE growth rate of the

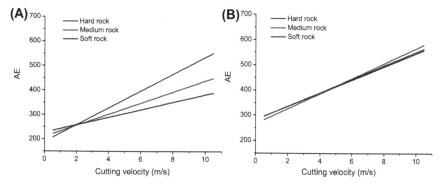

FIGURE 1.70 Specific energy–acoustic emission variation curves. (A) Without the confined pressure. (B) With the confined pressure $P = 10$ MPa.

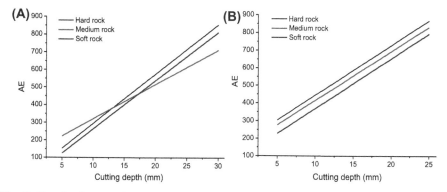

FIGURE 1.71 Cutting depth–acoustic emission variation curves. (A) Without the confined pressure. (B) With the confined pressure $P = 10$ MPa.

medium hard rock, and it had no effect on that of the hard rock and the soft rock. In a word, no matter whether the confined pressure existed or not, the hard rock had the same AE growth rate as that of the soft rock.

As shown in Fig. 1.71A, without the confined pressure, the number of AE was proportional to the cutting depth. The AE growth rate of the hard rock was equal to the soft rock and greater than the medium hard rock. The curves in Fig. 1.71B showed that three kinds of sandstones AE were proportional to the cutting depth, and had the same slope—namely, the growth rates were equal under the confined pressure. The AE of three kinds of sandstones had the following relationship: the hard rock was the greatest, followed by the medium hard rock and the soft rock.

7.5 Conclusions

The SE of rock-cutting was proportional to the cutting velocity. With the cutting depth increasing, the SE curves of three kinds of sandstones presented the slow-growth trend without the confined pressure, while the medium hard rock began to decline with the cutting depth at a certain value. Under the confined pressure, with the cutting depth increasing, the SE curves decreased at first and then increased, and there was an optimal depth of rock cutting.

After applying the confined pressure, the SE curves of hard rock and soft rock demonstrated S-shaped nonlinear change at the high value, while the medium hard rock changed greatly than the others. AE was directly proportional to the cutting velocity and the cutting depth.

8. FRACTURE PROPERTIES ON THE COMPRESSIVE FAILURE OF ROCK

8.1 Introduction

Failure of rock subjected to compressive loading has been extensively studied and the results tend to be reproducible to some extent depending on the degree of homogeneity of the rock and sample preparation. The ISRM standard for the UCS test was developed in part to ensure a consistent test method is used in the determination of the compressive strength of rock. The standard defines recommended dimensions and a slenderness ratio of test specimens as well as condition of the core. As a result, failed test specimens often exhibit a characteristic shear failure plane that intersects the axis of loading.

The uniaxial compressive strength of rock is an important parameter that is being used in many engineering designs. However, the uniaxial compression test is usually a tedious and expensive exercise to be conducted. In addition, the test requires high standard quality samples with regular geometry, which cannot always be extracted from weak rocks (Mohamed et al., 2008). For these reasons, an alternative method of predicting the uniaxial compressive strength of intact rock is very attractive to engineers. So far the strength of intact rock specimen of homogenous rocks, such as granite, sandstone, etc. had been well published, as well as the findings that were based on the study of a single rock type. Nevertheless, little research has been undertaken on nonhomogenous rock to investigate the uniaxial compressive strength behavior of sedimentary rock mass with discontinuities (Mohamed et al., 2008). A study was carried out to investigate the influences of fractures on rock strength by conducting uniaxial compressive tests on samples with segments.

The main purpose of the study presented herein is to determine the relationship between the uniaxial compressive strength of rock samples and fracturing frequency. These fractures are horizontally parallel with the specimen end face and equally distributed along the specimens, which are to be used to create the segments. "Intermediate layer" are used to simulate different conditions of discontinuities. The ultimate objective of the study is attempting to establish a meaningful correlation between the rock strength and the discontinuities. To accomplish this, 57 standard uniaxial compression tests have been conducted using these samples, and results are analyzed accordingly to each type of failure.

8.2 Experiment Design

Based on the standard test procedure, a new experiment procedure was adopted to fit the purpose of this study. The size of each specimen will be prepared in right circular cylinder at a diameter of 54 mm. ISRM (1979) recommended the height-to-diameter ratio should be between 2.5 and 3.0, therefore the height of the specimen is 135 mm. There are a total of five sets of sandstone specimens, and each set has more than nine replications. The first set of samples are in standard size (54 mm × 135 mm) and will be labeled as Set A. The second set of specimens (Set B) have the same geometry of Set A, however, they are sliced into half to create equal segments and fractures. The end surface will be flattened by using a surface grinder. As suggested by the ISRM Commission (1979), the end surface should be flat to 0.02 mm. The third set of specimens is similar to Set B, but this set is cut into three equal segments, which is named as Set C. Set D is cut into quarters with a consistent diameter and height. Set E is sliced into five equivalent segments. Fig. 1.72 displays how the samples were extracted.

The experiment program consists of three stages; all five sets of specimens will be undertaken in an uniaxial compressive test. The first phase of the experiment program is to conduct tests on each set without putting on any intermediate layers. The sets with segments will be tested by stacking them up. The second phase is to carry out tests on each set with intermediate layer in between each contact faces (excluding both ends). Paper is used as the intermediate layer, as paper itself has such little compressive strength. The use of an intermediate layer is to

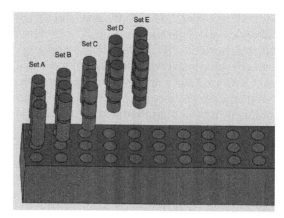

FIGURE 1.72 Illustration of how the samples were extracted.

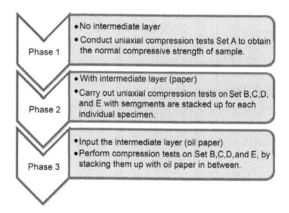

FIGURE 1.73 Even diagram of the testing stages.

reduce the friction between contact faces and to simulate the different conditions of discontinuities. The final stage of this program is to further reduce the friction by putting down oil paper. Fig. 1.73 illustrates the procedure of this experiment program.

8.3 Test Sample Preparation

8.3.1 Sample Collection

The sample was sourced from Gosford Quarrying, which is located at 300 Johnston St,

Annandale, Sydney. Due to the size and weight limitations, the most suitable sample was chosen and transported to Rock Mechanics Laboratory. A specification sheet was obtained from the Gosford Quarrying store, which gives a general idea of the characteristics of the sample. The sandstone is in a brown and banded color, and primarily names as Mount White Brown. Its geological name is Argillaceous Quartz Sandstone, which is formed in the Triassic age. Based on the specification sheet, the sample is described as medium-grained quartz sandstone with a predominantly argillaceous matrix. The concentration and distribution of iron oxides influence the nature of the color banding and density of color. The bulk density of this sandstone is approximately 2.27 t/m^3 with 4.4% of absorption. The modulus of rupture is 8.9 MPa in dry condition and 2.5 MPa when is wet. The compressive strength is around 37 MPa (dry) and 22 MPa (wet).

8.3.2 Specimen Preparation

Specimens were drilled by using the RD900 Core Drill, which is available in the laboratory. Once the samples were drilled, they were cut by the Small-Diameter Diamond Saw. To preserve the nature of the materials, the specimens were put into a LABEC Oven for 24 h at 170°. This process is to dry out the water from the specimens that absolved during drilling and cutting.

Due to the availability of the core bits in the Rock Mechanical Laboratory, only a 44 mm core drill is available. Therefore, the original specimen size was adopted to fit this circumstance. The diameter of the specimen was drilled at 44 mm and the height is still 2.5 times the diameter. Table 1.7 summarizes the geometry size of the actual specimen prepared.

8.3.3 Procedure

Once all the specimens have been prepared, the test is ready to commence. A platen was placed in the center of the set and the specimen is placed right on top of the platen. Both upper and bottom platens were used to reduce the end effect and is able to adjust the specimen to parallel to the ram. The specimens, the platens, and spherical seat were accurately adjusted to be centered with respect to one another and to the loading machine. As only a qualified person is allowed to operate the machine, the author was trained by a supervisor.

TABLE 1.7 Actual Number and Geometry of Samples Prepared

Set Number	Height/Diameter (mm/mm)	No. of Replications
Set A	110/44	5
Set B	2 × 55/44	15
Set C	3 × 33.5/44	13
Set D	4 × 27.5/44	15
Set E	5 × 22/44	14

Personnel protection equipment is used during the test. Prior to the experiment, the release and load valves are shut as well as the fine adjustment valve. The ram is fully lowered while starting up the machine and while setting the load range. As the normal compressive strength of samples is above 50 kN, the load range was set at 100 kN. The maximum load needle was set at zero and rested on the load needle. Fast action handle was used to bring sample close to the top platen, while the release valve was set at zero.

Once the top end of the specimen was parallel to the top platen, the load valve was opened to start compression. Manual adjustment was made by turning the load valve to ensure the load card speed is followed with the load needle. After failure, the load valve will shut and then gradually shut the center valve. The results were recorded and the procedure was repeated for all experiments.

8.4 Results and Analysis

8.4.1 Clean Fracture Surface

Readings obtained from the Avery machine are in kN, which were converted into stress using the loading divided by the area of contact face. The average results were calculated with the standard deviation. Table 1.8 shows the uniaxial compressive strength of Set A samples. The average strength is approximately 33.1 MPa. The standard deviation is 4.32, which is relatively high. This variation was caused by the different composition in each sample. Because the samples are not perfectly homogenous, the uniaxial compressive strength varies slightly.

Table 1.9 summarizes the results obtained from Phase 1. Considering the magnitude of the standard deviation, the results of Phase 1 are generally consistent. Compared with the given compressive strength from the specification sheet (37 MPa), Set A has a slightly lower average uniaxial compressive strength. This may be caused by the moisture content increased during drilling and cutting. Based on Shakoor and Barefield (2009), the moisture content has significant influence on the compressive strength. Another reason is the geometry of the specimen was not preferable to a standard compression test. The required diameter is 54 mm, however, the actual diameter of the specimen is 44 mm. Smaller-sized specimens are less resistant to the compression, causing the sample to fail before the original peak strength.

Fig. 1.74 was plotted to determine the uniaxial compressive strength of specimens with respect to change in number of segments. As seen from the graph, the strength decreases as the number of segments increases. It is because the fractures have weakened the structure of the specimen, resulting in the decrease of strength. In addition, fractures increased the number of contact face which increased the end effects. According to studies of Brady and Brown (2004), the end

TABLE 1.9 Result Summary of Phase 1 Clean Fracture Surface

Set	Test Number	Segments	UCS (MPa)	Standard Deviation
B	1	1/2	29.07	1.48
	2	1/2	28.93	
	3	1/2	31.56	
C	1	1/3	25.65	2.37
	2	1/3	22.36	
	3	1/3	21.04	
D	1	1/4	22.49	2.24
	2	1/4	22.36	
	3	1/4	18.54	
E	1	1/5	15.85	2.07
	2	1/5	17.82	
	3	1/5	19.99	

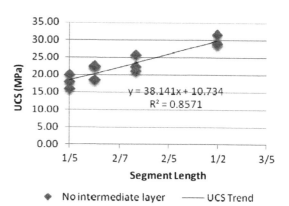

FIGURE 1.74 Uniaxial Compressive Strength (UCS) results of Phase 1 against the segment length.

TABLE 1.8 Uniaxial Compressive Strength (UCS) Results of Standard Sample Size

Test No.	1	2	3	4	5
UCS (MPa)	33.00	34.52	25.97	36.96	35.77
Standard deviation	4.32				

effects are quite significant to the compressive strength of rocks. Because of the friction, the specimen will be restrained near its ends and prevented from deforming uniformly. As a result, the axial stress may not be a principal stress and the stress may not be axial. Overall, the uniaxial compressive strength has a liner relationship with respect to the number of segments.

8.4.2 Intermediate Layer (Paper)

Results derived from Phase 2 have also been summarized and analyzed as shown in Table 1.10. Set B and Set D in Phase 2 have relative consistency, as the standard deviations of results are very small. The results of Set C and Set E have a greater degree of variation.

A similar graph was plotted based on the summarized Phase 2 results. As can be seen from Fig. 1.75, the uniaxial compressive strength decreases dramatically from Set B to Set C. This was caused by the addition fracture,

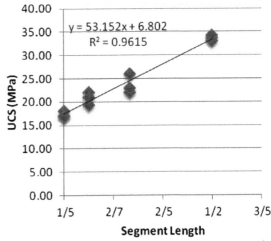

FIGURE 1.75 UCS versus Segment Number (intermediate layer: paper).

which further weakened the structure of the specimen. However, between Set C and Set E the compressive strength remains relatively constant. This is because the additional intermediate layers have increased the friction between the contact faces. And due to the reduction in friction, the effects of the fracture have been reduced. The intermediate layers have also reduced the end effects, and therefore the change of compressive strength is less significant.

8.4.3 Failure Mechanism

The failure mechanism of sandstone under compression test is studied by investigating the failure patterns. Photos were taken on the specimen after the failure and the most representative photos were selected for analysis. These failures are the most common ones appeared during the compression test. As seen from Fig. 1.76, Set A has a typical shear failure as it was under a standard uniaxial compression test.

The most common failure that occurred in Phase 2 for Set B, C, D, and E is the longitudinal

TABLE 1.10 Result Summary of Phase 2 Intermediate Layer (Paper)

Set	Test Number	Segments	UCS (MPa)	Standard Deviation
B	4	1/2	33.00	0.63
	5	1/2	33.67	
	6	1/2	34.26	
C	4	1/3	22.01	2.1
	5	1/3	23.05	
	6	1/3	26.14	
D	4	1/4	22.24	1.35
	5	1/4	21.01	
	6	1/4	19.33	
E	4	1/5	17.50	0.58
	5	1/5	18.71	
	6	1/5	17.01	

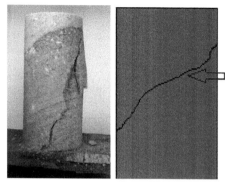

FIGURE 1.76 Typical failure of Set A occurred.

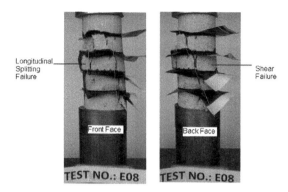

FIGURE 1.78 Typical combined longitudinal splitting and shear failure occurred in Phase 3.

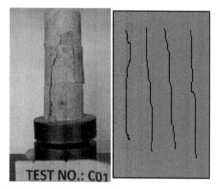

FIGURE 1.77 Typical failure occurred in Phase 2 (longitudinal splitting failure).

splitting failure. Fig. 1.77 illustrates an example of splitting failure. The fractures created increasing end effects and caused the cracking initiated from the contact surfaces. As the crack growth from both ends and the contact faces, these cracking jointed together formed a longitudinal splitting failure.

In the final stage, longitudinal splitting failure and shear failure both appeared. As seen from the Fig. 1.78, the cracking started from the contact surface, which shows an evidence of the lateral tensile stress caused by the fluid pressure. Due to the stress condition change and further reduction in friction, the fracturing deformed in new failure pattern.

8.5 Conclusions

The average uniaxial compressive strength of Set A is approximately 33.3 MPa, which is close to the original compressive strength. A linear relationship was discovered between the strength and the number of segments. The uniaxial compressive strength of the sandstone specimens decreased with an increase in the number of fractures. While the intermediate layers were inputted during the test, the liner relation was changed to curvilinear due to the effects of fractures and reduction in friction. The oil paper complicated the stress conditions; it is difficult to determine the different natures of the relationships between the uniaxial compressive strength and the friction at this stage. Generally, a negative correlation between the strength of sandstone and fractures is expected. Considering the complexity of stress conditions, it can be concluded that the number of fractures and the friction should not be used alone to predict the uniaxial compressive strength of sandstones. In conclusion, more research should be conducted on the other aspects of sandstone.

It is recommended that despite obtaining no meaningful correlation through this study, the author would like to recommend that this study be considered as a starting point of the project in a large scope. A simple prediction model of the uniaxial compressive strength of sandstone would benefit the mining industry, and the author would like to suggest that the research should be continued.

References

Adebayo, B., 2008. Evaluation of cuttability of selected rocks in South-Western Nigeria. Assumption University Journal of Technology 12 (2), 126—129.

Alberto, C., Nicola, P., 2002. A fractal comminution approach to evaluate the drilling energy dissipation. International Journal of Numerical and Analytical Methods in Geomechanics 26, 499—513.

Bagde, M.N., Petros, V., 2005. Fatigue properties of intact sandstone samples subjected to dynamic uniaxial cyclical loading. International Journal of Rock Mechanics and Mining Sciences 42 (2), 237—250.

Bakun-Mazor, D., Hatzor, Y.H., Dershowitz, W.S., 2009. Modeling mechanical layering effects on stability of underground openings in jointed sedimentary rocks. International Journal of Rock Mechanics and Mining Sciences 46 (2), 262—271.

Bao, C.Y., Tang, C.A., Tang, S.B., Cai, M., Yu, Q., 2013. Research on formation mode and mechanism of layered rock surface fractures under uniaxial tension load. Chinese Journal of Rock Mechanics and Engineering 32 (3), 474—482.

Brady, B.H.G., Brown, E.T., 2004. Rock mechanics: for underground mining. Kluwer Academic Publishers, Dordrecht.

Carpinteri, A., Puzzi, S., 2009. The fractal-statistical approach to the size-scale effects on material strength and toughness. Probabilistic Engineering Mechanics 24 (1), 75—83.

Chen, Y.L., Wei, Z.A., Xu, J., Tang, X.J., Yang, H.W., Li, S.C., 2011. Experimental research on the acoustic emission characteristics of rock under uniaxial compression. Journal of China Coal Society 36 (Suppl. 2), 237—240.

Copur, H., Bilgin, N., Tuncdemir, H., Balci, C., 2003. A set of indices based on indentation tests for assessment of rock cutting performance and rock properties. Journal of the South African Institute of Mining and Metallurgy 103 (9), 589—599.

Cravero, M., Iabichino, G., 2004. Analysis of the flexural failure of an overhanging rock slab. International Journal of Rock Mechanics and Mining Sciences 41 (Suppl. 1), 241—246.

Dey, K., Ghose, A.K., 2011. Review of cuttability indices and a new rockmass classification approach for selection of surface miners. Rock Mechanics and Rock Engineering 44 (5), 601—611.

Diederichs, M.S., Kaiser, P.K., 1999. Tensile strength and abutment relaxation as failure control mechanisms in underground excavations. International Journal of Rock Mechanics and Mining Sciences 36 (1), 69—96.

Feng, Z.C., Zhao, Y.S., Zhao, D., 2009. Investigating the scale effects in strength of fractured rock mass. Chaos, Solitons and Fractals 41 (5), 2377—2386.

Goktan, R.M., Yilmaz, N.G., 2005. A new methodology for the analysis of the relationship between rock brittleness index and drag pick cutting efficiency. Journal of the South African Institute of Mining and Metallurgy 105 (10), 727—733.

Gou, Z., 2008. Research on buckling behavior of bedding slope with bilayer rocks. Journal of Shandong University of Science and Technology (Natural Science) 27 (6), 21—23.

Guo, C.Y., Xie, X.F., Wu, X.H., Yao, W.J., Zhang, C.Y., 2008. The relationship among rock crushing energy, the protodyakonov coefficient and rock strength. Journal of Chongqing Jianzhu University 30 (6), 28—31.

Guo, L.J., Yang, Y.H., Hua, Y.H., 2013. Test and analysis on distortion and damage of rock under impact loading. Journal of Water Resources and Architectural Engineering 11 (6), 31—34.

He, G.L., Li, D.C., Zhai, Z.W., Tang, G.Y., 2007. Analysis of instability of coal pillar and stiff roof system. Journal of China Coal Society 32 (9), 897—901.

Hu, L.Q., Li, X.B., Zhao, F.J., 2002. Study on energy consumption in fracture and damage of rock induced by impact loadings. Chinese Journal of Rock Mechanics and Engineering 21 (Suppl. 2), 2304—2308.

International Society for Rock Mechanics Commission (ISRM Commission), 1979. Suggested methods for determining the uniaxial compressive strength and the deformability of rock materials. International Journal of Rock mechanics and Mining Sciences 16, 135—140.

Jackson, I., Schijns, H., Schmitt, D.R., Mu, J.J., Delmenico, A., 2011. A versatile facility for laboratory studies of viscoelastic and poroelastic behaviour of rocks. Review of Scientific Instruments 8 (6), 1—8.

Kang, H.Z., Jia, K.W., Ma, W.H., 2013. Experimental study on compressive strength and elastic modulus of ferrous mill tailing concrete. Journal of Engineering Science and Technology Review 6 (5), 123—128.

Kong, L., Su, Q., Wang, Y.C., Peng, R., 2012. Mesoscopic mechanics and energy dissipation analysis of sand deformation under slow cyclic loading. Chinese Journal of Underground Space and Engineering 8 (2), 268—273.

Li, K.W., Roland, N., 2009. Experimental study and fractal analysis of heterogeneity in naturally fractured rocks. Transport in Porous Media 78 (2), 217—231.

Li, X.B., Lai, H.H., Gu, D.S., 1992. Energy consumption of ore rocks under different loading waveforms. The Chinese Journal of Nonferrous Metals 2 (4), 10—14.

REFERENCES

Li, Z.W., Yang, D.W., Lei, H.M., 2012. Evapotranspiration model based on the maximum entropy production principle. Journal of Tsinghua University (Science and Technology) 52 (6), 785–790.

Li, W.T., Wang, Q., Li, S.C., Wang, D.C., Huang, F.C., Zuo, J.Z., Zhang, S.G., Wang, H.T., 2014. Deformation and failure mechanism analysis and control of deep roadway with intercalated coal seam in roof. Journal of China Coal Society 39 (1), 47–56.

Lin, D.N., Chen, S.R., 2005. Experimental study on damage evolution law of rock under cyclical impact loadings. Chinese Journal of Rock Mechanics and Engineering 24 (22), 4094–4098.

Liu, C.X., Jiang, J.Q., Liu, F.S., Wang, S.H., 2008a. Fractal study of scale effect in microscopic, mesoscopic and macroscopic states for fracture mechanism of rock materials. Rock and Soil Mechanics 29 (10), 2619–2622.

Liu, J.F., Xie, H.P., Xu, J., Yang, C.H., 2008b. Experimental study on damping characteristics of rock under cycling loading. Chinese Journal of Rock Mechanics and Engineering 27 (4), 712–717.

Liu, Y.S., Fu, H.L., Wu, Y.M., Rao, J.Y., Yin, Q., Yuan, W., 2013. Experimental study of elastic parameters and compressive strength for transversely isotropic rocks. Journal of Central South University (Science and Technology) 44 (8), 3398–3404.

Mi, G.F., Li, C.Y., Gao, Z., 2014. Application of numerical simulation on cast-steel toothed plate. Engineering Review 34 (1), 1–6.

Miao, S.J., Lai, X.P., Zhao, X.G., Ren, F.H., 2009. Simulation experiment of AE-based localization damage and deformation characteristic on covering rock in mined-out area. International Journal of Minerals, Metallurgy and Materials 16 (3), 255–260.

Milošević, B., Mijalković, M., Petrović, Ž., Hadžimujović, M., Mladenović, B., 2013. Comparative analysis of limit bearing capacity of frames depending on the character of the load. Technical Gazette 20 (6), 1001–1009.

Mohamed, Z., Mohamed, K., Chun, C.G., 2008. Uniaxial compressive strength of composite rock material with respect shale thickness ratio and moisture content. The Electronic Journal of Geotechnical Engineering 13 (Bund. A), 1–10.

Nomikos, P.P., Sofianos, A.I., Tsoutrelis, C.E., 2002. Structural response of vertically multi-jointed roof rock beams. International Journal of Rock Mechanics and Mining Sciences 39 (1), 79–94.

Pan, Y., Gu, S.T., Qi, Y.S., 2013. Analytic solution of tight roof's bending moment, deflection and shear force under advanced super charger load and supporting resistance before first weighting. Chinese Journal of Rock Mechanics and Engineering 32 (8), 1545–1553.

Potyondy, D.O., Cundall, P.A., 2004. A bonded-particle model for rock. International Journal of Rock Mechanics and Mining Sciences 41, 1329–1364.

Roxborough, F.F., 1987. The role of some basic rock properties in assessing cuttability. In: Proceedings of Seminar on Tunnels. Wholly Engineered Structures, Sydney, I. E. Aust./AFCC, April, pp. 1–22.

Shakoor, A., Barefield, E.H., 2009. Relationship between uncofined compressive strength and degree of saturation for selected sandstones. Environmental & Engineering Geoscience 1, 29–40.

Swift, G.M., Reddish, D.J., 2002. Stability problems associated with an abandoned ironstone mine. Bulletin of Engineering Geology and the Environment 61 (3), 227–239.

Teng, S.H., Lu, M., Zhang, J., Tan, Z.G., Zhuang, Z.W., 2012. Entropy theory and information granularity in information systems. Computer Engineering & Science 34 (4), 94–101.

Tivadar, T.M., Vass, I., 2011. Relationship between the geometric parameters of rock fractures, the size of percolation clusters and REV. Mathematical Geosciences 43 (1), 75–97.

Tong, L.Y., Liu, S.Y., Qiu, Y., Fang, L., 2004. Current research state of problems associated with mined-out regions under expressway and future development. Chinese Journal of Rock Mechanics and Engineering 23 (7), 1198–1202.

Tumac, D., Bilgin, N., Feridunoglu, C., Ergin, H., 2007. Estimation of rock cuttability from shore hardness and compressive strength properties. Rock Mechanics and Rock Engineering 40 (5), 477–490.

Tyler, S.W., WheatCraft, S.W., 1992. Fractal scaling of soil particle size distributions: analysis and limitations. Soil Science Society of America Journal 56 (2), 362–369.

Vlastislav, C., Ivan, P., 2008. Quality factor Q in dissipative anisotropic media. Geophysics 73 (4), 63–75.

Wang, L., Gao, Q., 2007. Fragmentation prediction of rock based on damage energy dissipation. Journal of China Coal Society 32 (11), 1170–1174.

Wang, S.R., Jia, H.H., 2012. Analysis of creep characteristics of shallow mined-out areas roof under low stress conditions. Applied Mechanics and Materials 105–107, 832–836.

Wang, H.W., Chen, Z.H., Du, Z.C., Li, J.W., 2006. Application of elastic thin plate theory to change rule of roof in underground stope. Chinese Journal of Rock Mechanics and Engineering 25 (Suppl. 2), 3769–3774.

Wang, J.A., Shang, X.C., Liu, H., Hou, Z.Y., 2008a. Study on fracture mechanism and catastrophic collapse of strong roof strata above the mined area. Journal of China Coal Society 33 (8), 850–855.

Wang, Z.L., Li, Y.C., Wang, J.G., 2008b. Numerical analysis of blast-induced wave propagation and spalling damage in a rock plate. International Journal of Rock Mechanics and Mining Sciences 45 (4), 600–608.

Wang, S.R., Jia, H.H., Wu, C.F., 2010. Determination method of roof safety thickness in the mined-out regions under dynamic loading and its application. Journal of China Coal Society 35 (8), 1263–1268.

Wang, J.A., Li, D.Z., Shang, X.C., 2011a. Mechanics analysis on creep fracture of strong roof strata above mined-out area. Journal of University of Science and Technology Beijing 33 (2), 142–147.

Wang, Y.X., Cao, P., Yin, T.B., 2011b. Simulation research for impact damage fracture evolution of brittle rock plate under impact loading. Journal of Sichuan University (Engineering Science Edition) 43 (6), 85–90.

Wang, S.R., Hagan, P., Cheng, Y., 2013a. Experimental research on the instability characteristics of double-layer rock plates based on MTS-AE system. Applied Mathematics & Information Sciences 7 (1L), 339–345.

Wang, S.R., Liu, X.M., Li, D.J., Li, C.L., Cao, C., 2013b. Dynamic thermo-mechanical analysis of viscoelastic attenuation properties for different rocks. Disaster Advances 6 (13), 331–340.

Wang, S.R., Wang, H., Hu, B.W., 2013c. Analysis of catastrophe evolution characteristics of the stratified rock roof in shallow mined-out areas. Disaster Advances 6 (S1), 59–64.

Wu, S.X., Zhang, S.X., Shen, D.J., 2008. An experimental study on Kaiser effect of acoustic emission in concrete under uniaxial tension loading. China Civil Engineering Journal 41 (4), 31–39.

Xi, D.Y., Du, Y., Xi, J., Yi, L.K., Xu, S.L., 2011. Viscoelastic properties of saturated sandstones under fatigue loading. Chinese Journal of Rock Mechanics and Engineering 30 (5), 865–870.

Xia, C.J., Xie, H.P., Ju, Y., Zhou, H.W., 2006. Experimental study of energy dissipation of porous rock under impact loading. Engineering Mechanics 23 (9), 1–5.

Yang, Y., Feng, G.C., Liang, B., 2009. Study on the mechanical deformation of overlying thick-hard rock layers after mining. Science Technology and Engineering 9 (6), 1402–1405.

Yin, T.B., Li, X.B., Zhou, Z.L., Hong, L., Ye, Z.Y., 2007. Study on mechanical properties of post-high-temperature sandstone. Chinese Journal of Underground Space and Engineering 3 (6), 1060–1063.

Yin, J.F., Wang, S.R., Wang, Z.Q., Yu, J.M., Cao, C., 2013. Disturbance effects analysis of the coal pillar-roof system caused by a sudden instability coal pillar. Disaster Advances 6 (13), 29–37.

You, Q.M., Hua, A.Z., 2002. Energy analysis on failure process of rock specimens. Chinese Journal of Rock Mechanics and Engineering 21 (6), 778–781.

Yun, T.Y., Underwood, B.S., Kim, Y.R., 2010. Time-temperature superposition for HMA with growing damage and permanent strain in confined tension and compression. Journal of Materials in Civil Engineering 22 (5), 415–422.

Zhang, Z.Z., Gao, F., 2012. Experimental research on energy evolution of red sandstone samples under uniaxial compression. Chinese Journal of Rock Mechanics and Engineering 31 (5), 953–961.

Zhang, D.L., Wang, Y.H., Qu, T.Z., 2000. Influence analysis of interband on stability of stratified rock mass. Chinese Journal of Rock Mechanics and Engineering 19 (2), 140–144.

Zhang, L.Y., Mao, X.B., Li, T.Z., 2010a. Experimental research on thermal damage properties of marble at high temperature. Journal of Mining and Safety Engineering 27 (4), 505–511.

Zhang, X.Z., Chen, H.Q., Wang, Z.Y., 2010b. On the acoustic emission characteristics of concrete fracture subjected to three-point-bending. Journal of Experimental Mechanics 25 (4), 457–462.

Zhang, X.Y., 2009. Analysis of creep damage fracture of upper roof. Journal of Liaoning Technical University (Natural Science Edition) 28 (5), 777–780.

Zhao, X.D., Li, Y.H., Yuan, R.B., Yang, T.H., Zhang, J.Y., 2007. Study on crack dynamic propagation process of rock samples based on acoustic emission location. Chinese Journal of Rock Mechanics and Engineering 26 (5), 944–950.

Zhao, Y.L., Wu, Q.H., Wang, W.J., Wang, W., Zhao, F.J., 2010. Strength reduction method to study stability of goaf overlapping roof based on catastrophe theory. Chinese Journal of Rock Mechanics and Engineering 29 (7), 1424–1434.

Zhu, W.H., Ming, F., Song, C.Z., 2011. Fractal study of rock damage under blasting loading. Rock and Soil Mechanics 32 (10), 3131–3135.

Zhu, Y.G., Liu, Q.S., Zhang, C.Y., Shi, K., 2012. Nonlinear viscoelastic creep property of rock with time-temperature equivalence effect. Rock and Soil Mechanics 33 (8), 2303–2309.

Further Reading

Chen, R., Shiau, Y.C., Tsai, C.I., 2012. Research on the performances of seismic isolation and damping in high-rise buildings. Disaster Advances 5 (4), 219–224.

Wang, S.R., Hagan, P., Cheng, Y., 2012a. Fractal characteristics of sandstone cutting fracture under mechanical shock loading conditions. Applied Mechanics and Materials 226–228, 1789–1794.

Wang, S.R., Hagan, P., Cheng, Y., Wang, H., 2012b. Experimental research on fracture hinged arching process and instability characteristics for rock plates. Chinese Journal of Rock Mechanics and Engineering 31 (8), 1674–1679.

Wang, S.R., Hagan, P., Hu, B.W., Gamage, K., Cheng, Y., Xu, D.F., 2014a. Rock-arch instability characteristics of the sandstone plate under different loading conditions. Advances in Materials Science and Engineering 2014, 1–9, 950870.

Wang, S.R., Su, J.Q., Hagan, P., 2014b. Energy dissipation characteristics of sandstone cutting under mechanical impact load. Computer Modelling & New Technologies 18 (3), 13–20.

Wang, S.R., Xu, D.F., Hagan, P., Li, C.L., 2014c. Fracture characteristics analysis of double-layer rock plates with both ends fixed condition. Journal of Engineering Science and Technology Review 7 (2), 60–65.

CHAPTER 2

Rockbolting

1. MATHEMATICAL DERIVATION OF SLIP FACE ANGLE

1.1 Introduction

There are usually two major types of rockbolts, namely the end anchored and full-length anchored rockbolts according to the anchored length. The mechanics involved in the reinforcing element subject to axial loading, also known as bond mechanism, have been extensively studied. Bond can be thought of as the shearing stress or force between a reinforcing bar and the surrounding concrete.

A recurring way of studying reinforcement performance is through the analysis of the structural failure. For deformed concrete reinforcing bars subject to axial loading, two major failure modes have been identified. One involves the radial splitting of the concrete cover by the wedge action of the bar ribs, while the other concerns the shearing of the bar against the surrounding medium. The two mechanisms are supposedly influenced by the brittle characteristics of the system, and the predominant mechanism in action depends primarily on the rib geometry, the mechanical properties of the materials, and the level of confinement.

It is generally agreed that bond strength is provided by the combined effect of chemical adhesion, friction, and mechanical interlock. The adhesion strength of the bond is normally weak and compliant. The frictional component is pressure sensitive and can only take effect after relative slip between the bar and the concrete has occurred. The mechanical interlock, which is generated by the ribs of the bar and the concrete, provides bearing force against the face of the bar ribs. A schema of each strength component in the stress—slip model is shown in Fig. 2.1 (Hong and Park, 2012).

Splitting of concrete cover can never occur for plain bars; the shear resistance of plain bars is significantly lower than that of deformed bars. Thus, it can be concluded that the rib geometry, which is determined by the rib height, spacing, and the rib face angle, as shown in Fig. 2.2, dominates the bond behavior whenever the reinforcing element is subject to axial loading (Tepfers, 1979).

The knowledge of bond mechanisms and quantification of bond strength are the first step toward engineering analysis and design of reinforced concrete structures. Understanding the failure mechanisms of the bond is essential to establishing a realistic bond stress—slip relationship for evaluating the contribution of the reinforcing element. Rib geometry studies had first been carried out over a century ago. Most of the rib face angle and bearing face angle—related studies are based on statistical methodologies. Conclusions drawn are highly dependent on the materials used, which may limit their

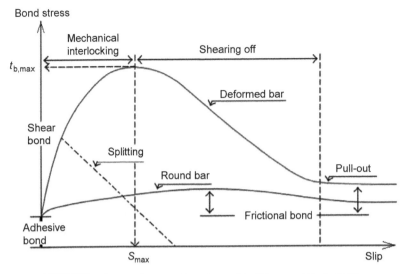

FIGURE 2.1 Bond stress–slip relationship (Hong and Park, 2012).

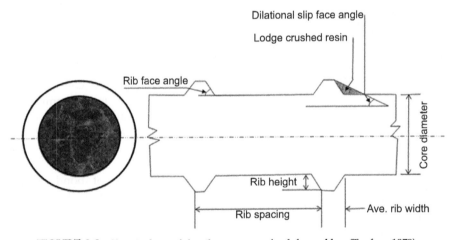

FIGURE 2.2 Terminology of the rib geometry of a deformed bar (Tepfers, 1979).

validity in other situations. This chapter presents an analytical investigation on the bond-slip face angle of deformed concrete reinforcing bars that are subject to axial loading. The authors believe that this detailed analytical study dealing with the slip face angle of the bond is the first of its kind and will be very useful to concrete technology.

1.2 Experimental Investigations in Literature

Modern reinforcing bar-rib geometry research dates back to 1913. Abrams (1913) reported tests of bond between steel and concrete and suggested that the rib face should be as close to 90 degree as possible. Menzel (1939) investigated factors influencing results of pull-out tests, and found that rib face angles ranging from 45 to 57 degree imposed a similar influence on bond-slip behavior.

Via pull-out testing of the resistance to slip in concrete of 17 different deformed reinforcing bars, Clark (1946, 1949) concluded that the rib height and the rib face angle appeared as important factors in determining bond resistance. These studies were used to establish the modern ASTM standards for deformation requirements for reinforcing bars (ASTM, 1996).

Through pull-out testing of bars with only one rib (Rehm, 1957; Lutz et al., 1966), it was found that the mechanical behavior is about the same for all bars with a rib face angle between 40 degree and 105 degree, whereas bars with rib face angles less than 30 degree showed poor bond-slip performance. This phenomenon was explained by Lutz and Gergely (1967) as that slip of a deformed bar with a large rib face angle causes crushing of the concrete in front of the ribs to form a small rib bearing angle, which in effect produces ribs with slipping face angles of 30—40 degrees. In this chapter, the bearing angle of the bar, or effective rib face angle, is termed *slip face angle*, to be distinguished from the term bar-rib *face angle*, which is the geometric angle of the rib, as shown in Fig. 2.2.

Kokubu and Okamura, 1997 conducted pull-out tests on cast-iron bars with diameter 51 mm. Results showed that the bond stress was similar up to 0.5-mm slip when the rib face angle was 45 degree or more. Skorbogatov and Edwards, 1979 further demonstrated that a change of rib face angle from 48.5 to 57.8 degree did not affect bond strength. Soretz and Holzenbein (1979) also performed pull-out tests, and results showed that changing the rib face angle from 90 to 45 degree had no significance, whereas changing it to a very small angle (<15 degree) only slightly influenced bond behavior. The splitting effect was not affected at all. These studies suggest that steep rib face—angled bars can't successfully reduce splitting stress in concrete because the formation of the concrete wedges neutralizes the effect of the different rib face angles.

Goto (1971) conducted experimental studies on the deformational behavior of concrete around deformed tension bars. Results showed that the elastic stiffness of the bond and the ring strain of the concrete would be different for bars with rib face angles 15, 30, 45, 60, and 90 degree. Murata and Kawai (1984) performed splitting pull-out tests using deformed bars to evaluate the splitting bond strength and the initial bond strength, which gave information about adhesion between the steel bar and concrete. The rib face angle in the tests was increased with 5 degree increments from 15 to 90 degree, and the results are shown in Fig. 2.3.

To study the effect of rib height and spacing on the bond strength of reinforcing bars, Darwin and Graham (1993) also observed that concrete powder formed against the loaded face of the ribs to form a slip face angle from 17 to 40 degree. The lower angles were observed on the 1.27 mm rib height, while the higher angles were observed for 1.91- and 2.54-mm ribs.

Hamad (1995) conducted a series of seven experiments to evaluate the effect of rib geometry on bond-slip characteristics of deformed bars in reinforced concrete structures. Test results indicated that bond strength varied with the bar-rib face angle, spacing, and height, and results concluded that when the rib spacing and rib height of the test bar were kept constant, the bar with a rib face angle of 60 degree developed greater bond strength and load-slip stiffness than bars with angles of 30, 45, 75, and 90 degree.

Cairns and Jones (1995) conducted an analysis of the bond strength of ribbed reinforcing bars, in which bond failure was considered as a bearing failure of the ribs on the concrete. They suggested that a ribbed bar could have bond failure by one of three mechanisms, as shown in Fig. 2.4. For failure mode IIA and mode IIB, failure takes place by splitting of the concrete cover

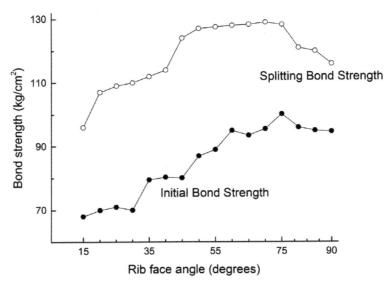

FIGURE 2.3 Effect of rib face angle on bond strength (Murata and Kawai, 1984).

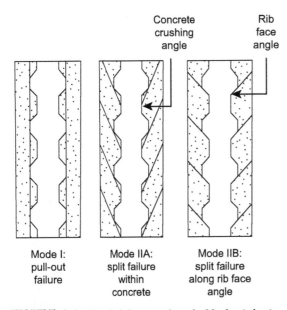

FIGURE 2.4 Bond failure modes of ribbed reinforcing bars: mode I, pullout failure; mode IIA, failure along an angle within the concrete; mode IIB, failure along rib face angle (Cairns and Jones, 1995).

along the slip face angle and the rib face angle respectively. Failure mode IIB was said to more likely occur when the face of the rib was smooth; for example, when the bar was coated with a fusion-bonded epoxy. If concrete cover of the bar was thick or if there were other restraints to splitting, mode I failure occurred with the concrete shearing on a surface running across the tops of the ribs. Results of the experimental studies showed that bond strength was dependent on the relative rib area and the cross-sectional shape of the bar ribs, but the rib face angle was not found to be significant within the range of 28–51 degree.

Idun and Darwin (1999) studied the effects of rib face angle on bond strength by measuring the friction coefficients between epoxy-coated and uncoated reinforcing steel and mortar. Results from 68 test specimens indicated that the friction coefficient varied from 0.503 to 0.627 with a mean of 0.561 between uncoated reinforcing steel and mortar. The effect of the rib face angle on the bond strength for uncoated bars was also evaluated and results are shown in Fig. 2.5.

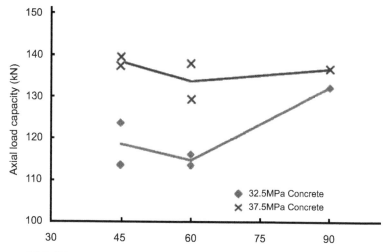

FIGURE 2.5 The influence of rib face angle (Idun and Darwin, 1999).

Choi and Lee. (2002) derived analytical expressions to predict bond strength using interfacial mechanical properties. The coefficient of friction and slipping face angle were recognized as key variables in the proposed equation. In the study, if the coefficient of friction of the slip surface was estimated to be 0.45, then the slip face angle of the bond was evaluated to be within the range of 26−33 degree using back analysis technology, as shown in Fig. 2.6.

Wu et al. (2012) investigated the bond strength between vitreous enamel-coated rebar and concrete through testing of 96 pullouts of cylinder specimens. Results showed that enamel coating increased the bond strength by approximately 15% because enamel-coated rebar was pulled out of concrete cylinders with smaller slip face angles. In addition, it was also reported that the slip face angle increased along the rebar from the loading end as illustrated in Fig. 2.7. It was proposed that this was likely due to the decreasing radial stress along the rebar.

To sum up, results from previous experimental studies suggest that rib face angle does not significantly affect bond strength within certain limits, say >45 degree, because a wedge of crushed concrete is formed in front of the rib as a rib begins to move. This wedge acts to change the rib face angle to the effective face angle of the rib. The slip face angle has been identified as one of the most important factors in bond behavior and failure mechanisms of rebar-reinforced concrete. The magnitude of the slip face angle is supposed to be related to material properties, confinement level and rib geometry; and had been estimated to fall in the ranges of 30−40 degree by Lutz and Gergely (1967), 17−40 degree by Darwin and Graham (1993), and 26−33 degree by Choi and Lee (2002). However, there is no related analytical study to explain these experimental observations and to quantify the slip face angle so as to identify the role of rib geometry in bond mechanics and to establish bond stress−slip relationships with physical meanings.

1.3 Analytical Investigation

1.3.1 Problem Description and Assumptions

One rebar profile that's subject to an axial resultant force, F, is the subject of discussion;

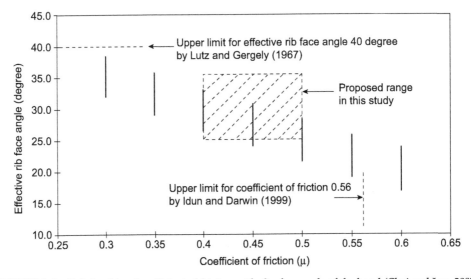

FIGURE 2.6 Relationship of coefficient of friction with slip face angle of the bond (Choi and Lee, 2002).

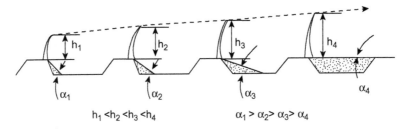

FIGURE 2.7 Varying of slip face angle along the bar (Wu et al., 2012).

the rib geometry is shown in Fig. 2.8. The problem is simplified to be axisymmetric. The initial confining pressure, p, is assumed to be compressive and equal in magnitude toward the entire rebar surface. This initial confinement would not be affected by increasing axial loading as they are perpendicular until failure occurs.

The following mathematical derivation is based on the Mohr–Coulomb failure criterion; that is, the surface inclined at an angle, i, to the axis, is assumed to be the weakness. Once the shear stress induced by axial loading along this surface reaches a critical value, relative slip will take place at this surface. The shear stress, τ, at the supposed weakness surface when failure occurs can be expressed by the following equation:

$$\tau = c + \sigma_n \cdot \tan \phi \qquad (2.1)$$

Some mechanical property parameters are also assumed according to commonly used values. In calculation examples, concrete with compressive strength of 30 MPa and internal friction angle of 37 degree is used as the default material. The coefficient of friction between steel and concrete is assumed to be 0.45, equivalent to an internal friction angle of 24 degree. The cohesion at the steel–concrete interface is assumed to be 1.5 MPa. It should be noted that calculation results drawn using these assumptions are only applicable for such conditions. Authors

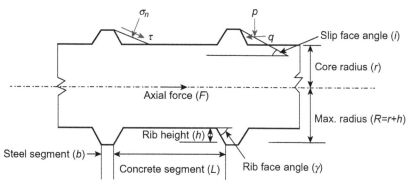

FIGURE 2.8 Notation of the studied problem.

are aware that the mechanical property parameters used in the calculation examples may be quite different to those of some previous studies; however, a detailed evaluation of these parameters and also the Mohr–Coulomb failure criterion itself are beyond the scope of this chapter. Calculation examples in this chapter are mainly for the purpose of demonstrating the practical applications of this work.

1.3.2 Governing Equations

Via stress transformation, the normal stress and shear stress of the supposed surface can be expressed as

$$\sigma_n = p\cos i + q\sin i \quad (2.2)$$

$$\tau = q\cos i - p\sin i \quad (2.3)$$

In which, q is the longitudinal stress induced by the axial load F and calculated by

$$q = \frac{F}{\frac{\pi(R^2-r^2)}{\sin i}} \quad (2.4)$$

where R and r are rib geometric parameters as shown in Fig. 2.8.

Substituting Eqs. (2.2) and (2.3) into Eq. (2.1), the axial load that causes bond slip can be expressed as

$$q = \frac{\cos\phi}{\cos(i+\phi)}\cdot c + p\cdot \tan(i+\phi) \quad (2.5)$$

This is the governing equation of the problem studied here. It should be noted that this equation is an extension of Mohr's law and Patton's equation (1966). In case of cohesion $c = 0$, it reduces to Patton's equation: $q = p\cdot\tan(i+\phi)$. Hence, this equation can be used to describe the dilational slip behaviors for Mohr–Coulomb materials. On the other hand, if the dilational angle in Eq. (2.5) reduces to zero, it returns to the form of Mohr–Coulomb's failure criterion, ie, Eq. (2.1). Thus, it can also be used to evaluate the critical state at orientations other than the loading direction.

1.3.3 The Upper Limits of Rib Face Angle and Slip Face Angle

Combining Eqs. (2.1)–(2.3) to eliminate q and τ, and rearranging $p = \sigma_n\cos i - (c + \sigma_n\cdot\tan\phi)\sin i$. This is the expression of the confining pressure just before bond slip occurs. The confining pressure is always positive (compression), so $\sigma_n\cos i - (c + \sigma_n\cdot\tan\phi)\sin i \geq 0$. As $c \geq 0$ and $i \in [0, \pi/2]$, the equation can be simplified as $\sigma_n\cos i - \sigma_n\cdot\tan\phi\cdot\sin i \geq 0$. It can be solved as

$$i \leq \frac{\pi}{2} - \phi \quad (2.6)$$

This expression gives an upper limit of the possible slipping face angles. And it demonstrates that the upper limit of the slipping angle is solely dependent upon the internal friction angle of the failure surface.

According to Eq. (2.6), for reinforced concrete members, if bond failure occurs within the concrete (failure mode IIA in Cairns and Jones, 1995) for which the internal frictional angle is 37 degree, then the rib slip face angle must be less than 53 degree. It means that shear failure can never occur for concrete beyond this angle.

The derivation procedure of Eq. (2.6) is also applicable for the case of steel–concrete interface failure (failure mode IIB IIA in Cairns and Jones, 1995). That is, bond slipping along the rib face angle, γ, must satisfy:

$$i \leq \frac{\pi}{2} - \phi_{sc} \quad (2.7)$$

where ϕ_{sc} = frictional angle of the steel–concrete interface.

For example, assume the coefficient of friction between concrete and plant steel is 0.45 (equivalent to $\phi_{sc} \approx 24$ degree), a common value for such materials. Then, the maximum rib face angle for which slipping may occur is $\gamma_{upp} = \frac{\pi}{2} - \phi_{sc} = 90$ degree $- 24$ degree $= 66$ degree.

For such materials, if the rib face angle is greater than 66 degree, then the concrete and steel interface has fixed bonding with respect to axial loading. Consequently, bond-slip behavior will be no different for reinforcing bars with rib face angles greater than 66 degree.

It should be noted that the slip angle, either within the concrete or at the steel–concrete interface, reaches its maximum value $\pi/2 - \phi$ when $c = 0$ and $p = 0$. This condition can never be satisfied in a real reinforcing system. Hence, the slipping angle will always be under this limit whenever bond slip occurs.

The theory of the upper limit of the bond-slip face angle developed in this section agrees well with experimental data conducted by Murata and Kawai (1984), shown in Fig. 2.3, for the initial bond strength; and is acceptable for the splitting bond strength. The variation should be thought as a difference between theory and practice.

1.3.4 Slip Face Angle Prediction for Known Rib Geometry

According to the analysis in the last section, the theoretical upper limit of the slip face angle is independent of rib geometry. In other words, it is universal whenever bond slip occurs. For a known bar, the range of the slip face angle can be further narrowed down.

Substituting Eq. (2.4) into Eq. (2.5), the axial force that causes bond slip can be expressed as:

$$F = \frac{\pi(R^2 - r^2)}{\sin i} \left[\frac{\cos \phi}{\cos(i + \phi)} \cdot c + p \cdot \tan(i + \phi) \right] \quad (2.8)$$

This axial load should be always less than the shear resistance of the whole cylindrical surface of the profile to ensure the occurrence of the proposed bond slip; in other words failure mode II should occur prior to failure mode I, ie, $F \leq 2\pi RL(c + p \cdot \tan \phi)$.

Therefore, the following equation can be used to estimate the domain of the proposed slip face angle for bars with known rib geometry:

$$\frac{\pi(R^2 - r^2)}{\sin i} \left[\frac{\cos \phi}{\cos(i + \phi)} c + p \cdot \tan(i + \phi) \right]$$
$$\leq 2\pi RL(c + p \cdot \tan \phi) \quad (2.9)$$

To avoid cumbersome mathematics, let $p = 0$ in this equation, so it simplifies to:

$$\sin(2i + \phi) \geq \frac{R^2 - r^2}{RL} \cos \phi + \sin \phi \quad (2.10)$$

Solve for $i \leq \pi/4$:

$$\frac{\pi}{2} - \frac{1}{2}\sin^{-1}\left(\frac{R^2 - r^2}{RL}\cos \phi + \sin \phi\right) - \frac{\phi}{2} \geq i$$
$$\geq \frac{1}{2}\sin^{-1}\left(\frac{R^2 - r^2}{RL}\cos \phi + \sin \phi\right) - \frac{\phi}{2}$$

$$(2.11)$$

The developed formula is examined by experimental data reported by Darwin and Graham (1993). The deformation patterns of machined bars are shown in Fig. 2.9. The rib face angle is 60 degree for all bars. The rib geometric parameters are listed in Table 2.1. The internal friction angle of the concrete used in the experiment is assumed to be 37 degree. The calculation results of the theoretical maximum and minimum slip face angles, denoted as i_{max} and i_{min}, respectively, are also listed in Table 2.1.

Calculation results in Table 2.1 show that the theoretical minimum slip face angle for some deformation geometric patterns is very small. From a practical viewpoint, it would be unlikely to occur. Hence, the theoretical range of slip face angles for studied rib geometries should be 15–50 degree, which has covered the measured range [17 and 40 degree].

It should be noted that the derivation of the domain of the slip face angle is based on the assumption $p = 0$. Before bond slip occurs, the confining pressure of the bar is supposed to be small, say less than 5 MPa. To investigate the effect of the confining pressure on the domain of the slip face angle, a sensitivity study has been conducted (not shown here) on p in Eq. (2.9) from 0 to 5 MPa. Results show that increasing the confining pressure result in a slight decrease of the size of the domain for the slip face angle. Therefore, the greatest domain for the slip face angle occurs when $p = 0$.

1.3.5 The Most Vulnerable Slip Face Angle

For failure mode IIA, Eq. (2.8) provides the axial load with respect to different rib face angles. Mathematical investigation of the equation has revealed that there is one stationary point at which mode IIA failure commences with the minimum axial loading regardless of the confinement level. It means that bond failure most likely occurs at a specific slip face angle.

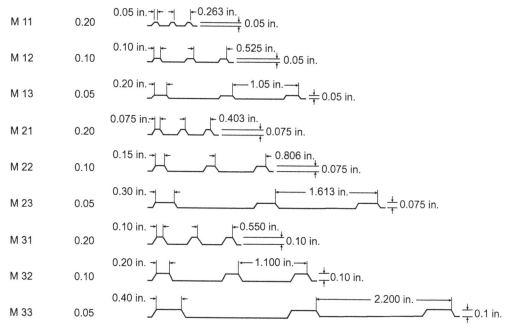

FIGURE 2.9 Rib geometric parameters of bars studied by Darwin and Graham (1993).

2. ROCKBOLTING

TABLE 2.1 Rib Geometric Parameters and Calculated Range the Most Vulnerable Slip Face Angle

Bars	R, mm (in.)	r, mm (in.)	L, mm (in.)	i_{max}, Degree	i_{min}, Degree
M11	12.7 (0.5)	11.43 (0.45)	5.41 (0.231)	35	18
M12	12.7 (0.5)	11.43 (0.45)	10.8 (0.425)	46	7
M13	12.7 (0.5)	11.43 (0.45)	21.6 (0.85)	50	3
M21	12.7 (0.5)	10.79 (0.425)	8.33 (0.328)	36	17
M22	12.7 (0.5)	10.79 (0.425)	16.7 (0.657)	46	7
M23	12.7 (0.5)	10.79 (0.425)	33.4 (1.31)	50	3
M31	12.7 (0.5)	10.16 (0.4)	11.4 (0.449)	38	15
M32	12.7 (0.5)	10.16 (0.4)	22.9 (0.902)	47	6
M33	12.7 (0.5)	10.16 (0.4)	45.7 (1.8)	50	3

This bond-slip direction, denoted as i_{vul} herein, is defined as the most vulnerable slip face angle for the reinforced concrete subject to axial load.

The stationary points can be calculated by $dF/di = 0$ for Eq. (2.8), ie,

$$\frac{d}{di}\left\{\frac{\pi(R^2 - r^2)}{\sin i}\left[\frac{\cos\phi}{\cos(i+\phi)} \cdot c + p \cdot \tan(i+\phi)\right]\right\} = 0 \quad (2.12)$$

It leads to $\frac{\cos\phi \cdot \cos(\phi+2i)}{\cos^2(i+\phi)\sin^2 i} \cdot c + \left[\frac{\cos i \sin(i+\phi)\cos(i+\phi) - \sin i}{\cos^2(i+\phi) \cdot \sin^2 i}\right] \cdot p = 0$.

Rearranging gives:

$$\frac{c}{p} = \frac{\sin i - \cos i \sin(i+\phi)\cos(i+\phi)}{\cos\phi \cdot \cos(2i+\phi)} \quad (2.13)$$

It is interesting to find that the most vulnerable slip face angle depends on the internal friction angle of the concrete and the cohesion to confinement ratio, but is irrespective of the rib geometry.

The closed-form solution of Eq. (2.13) is hard to obtain, but it can be shown that the most vulnerable slip face angle has a narrow domain. As $c/p \to 0$, Eq. (2.13) becomes:

$$2\tan i = \sin(2\phi + 2i) \quad (2.14)$$

The solution of Eq. (2.14) can be transformed to the following equation by regression with $R^2 = 0.999999$:

$$i_{vul-min} = 0.000109\phi^3 - 0.0166\phi^2 + 0.482\phi + 22.7 \quad (2.15)$$

This expression is the lower boundary of the most vulnerable slip face angle.

On the other hand, when $p/c \to 0$, Eq. (2.13) can be solved to give:

$$i_{vul-max} = \frac{\pi}{4} - \frac{\phi}{2} \quad (2.16)$$

This is the upper boundary of the most vulnerable slip face angle. As a result, the most vulnerable dilational slip face angle, i_{vul}, at which the axial force reaches its minimum value, is in the domain $[i_{vul-min}, i_{vul-max}]$. Fig. 2.10 shows

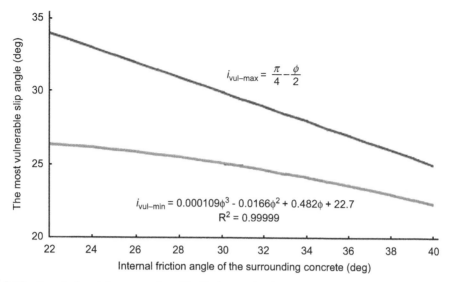

FIGURE 2.10 Boundaries of the most vulnerable slip face angles for different internal friction angle of the concrete.

the i_{vul} range for different internal friction angles of the concrete.

According to Fig. 2.10, for failure mode IIA, the most vulnerable slip face angle of reinforced concrete with internal friction angle of 37 degree is approximately 23.5 degree at low confinements and approximately 26.5 degree if the initial confinement pressure is 20 MPa, regardless of the rib geometry. As the most vulnerable slip face angle lies in a small range, from a practical viewpoint, it can be thought of as a single value of the average of its boundaries, ie, $i_{vul} \approx (i_{vul\text{-}min} + i_{vul\text{-}max})/2$. For normal conditions, $i_{vul} = (23.5 + 26.5)/2 = 25$ degree. This value is in agreement with the back analysis result of 26–33 degree (Choi and Lee, 2002).

One significance of this finding is that, once bond slip occurs, the rib face angle will no longer be involved in subsequent mechanical interactions between the rib and the concrete. Instead, the slip face angle, which satisfies Eq. (2.13), becomes an element in subsequent failure mechanisms.

1.3.6 Occurrence of Failure Modes IIA and IIB

With the aid of the concept of the most vulnerable slip face angle developed in the last section, the minimum axial load to cause failure mode IIA can be evaluated using Eq. (2.8). That is,

$$F_{vul} = \frac{\pi(R^2 - r^2)}{\sin(i_{vul})} \left[\frac{\cos\phi}{\cos(i_{vul} + \phi)} \cdot c + p \cdot \tan(i_{vul} + \phi) \right] \quad (2.17)$$

Eq. (2.8) can also be used to evaluate the axial load for failure mode IIB by using the Mohr–Coulomb parameters of the steel–concrete interface, ie,

$$F_{sc} = \frac{\pi(R^2 - r^2)}{\sin\gamma} \left[\frac{\cos\phi_{sc}}{\cos(\gamma + \phi_{sc})} \cdot c_{sc} + p \cdot \tan(\gamma + \phi_{sc}) \right] \quad (2.18)$$

Note that subscript sc stands for the steel–concrete interface.

It is reasonable that the preferred failure mode between IIA and IIB will be the one with a smaller axial loading capacity. Here is an example. For C30 concrete we have assumed $\phi = 37$ degree, and we have calculated $i_{vul} = 25$ degree. The cohesion of the concrete can be estimated as 7.5 MPa by using the formula:

$$\text{UCS} = \frac{2c \cdot \cos \phi}{1 - \sin \phi} \quad (2.19)$$

Substituting these data into Eq. (2.17) gives:

$$F_{uvl} = \frac{\pi(R^2 - r^2)}{\sin 25°} \left[\frac{\cos 37°}{\cos(25° + 37°)} \times 7.5 + p \cdot \tan(25° + 37°) \right] \quad (2.20)$$

For failure mode IIB, we have assumed the interface property parameters $\phi_{sc} = 24$ degree and $c_{sc} = 1.5$ MPa. Substituting these into Eq. (2.18) gives:

$$F_{sc} = \frac{\pi(R^2 - r^2)}{\sin \gamma} \left[\frac{\cos 24°}{\cos(\gamma° + 24°)} \times 1.5 + p \cdot \tan(\gamma + 24°) \right] \quad (2.21)$$

Let $F_{vul} = F_{sc}$, and solve γ for different confining pressures (p). Results are plotted in Fig. 2.11.

It is shown that for the illustrated condition, if the rib face angle is greater than 60 degree, failure mode IIA will occur under normal conditions. In case of rib face angles smaller than 60 degree, failure at the steel–concrete interface will occur prior to concrete failure.

Calculation also suggests that decreasing the mechanical strength of the interface, such as in cases of the proxy-coated bars or the vitreous enamel-coated bars, results in interface failure occurring at higher rib face angles. For example, in case of $\phi_{sc} = 16.5$ degree (equivalent to 2/3 of the original coefficient of friction), failure mode IIB may occur at $\gamma = 68$ degree.

In rib geometry studies, it has been well demonstrated that bond behavior will not change once the rib face angle exceeds a limit. Analytical derivations conducted in this chapter suggest that this limit is around 60 degree under normal conditions. In reality, the rib face angle is often great than 65 degree, so failure mode IIA is often the dominant failure mode of bond slip.

Some previous experiments have observed that concrete powder could lodge in front of

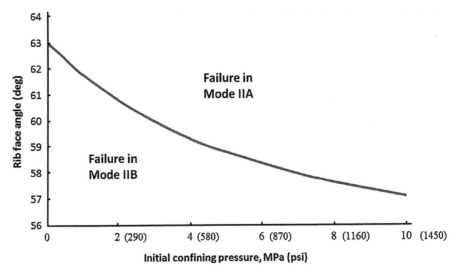

FIGURE 2.11 Boundary of failure in mode IIA and IIB.

the rib when the rib face angle is less than 60 degree. Other than variations of the mechanical properties between those tests and in the calculation, another much more likely reason is that concrete failure in such tests may be subsequent failure to the interface failure, which means that the lodged concrete powder is probably left in the postfailure reactions of the bond.

1.4 Conclusions

This study provides insight into the initial stage of the bond-slip failure of reinforced concrete members that are subject to axial loading. These conclusions are drawn based on Mohr—Coulomb's material properties.

The upper limits of the rib face angle and the slip face angle are formulated to be complementary to the internal friction angle of the failure surface (Eq. (2.6)), regardless of the confinement level and the rib geometry. This concept is in consensus with data in the literature.

Once the geometric parameters of a rib have been given, the bond-slip face angle can be determined more accurately using Eq. (2.11). This finding can be used to explain the experimental observations conducted by Darwin and Graham (1993).

Regardless of the rib geometry, there exists the most vulnerable slip face angle. In most cases, it is independent of the rib face angle and is in the domain [$i_{vul-min}$, $i_{vul-max}$], expressed by Eqs. (2.15) and (2.16). Under normal conditions, the most probable slip face angle that involves bond failure is around 25 degree.

With the aid of the concept of the most vulnerable slip face angle, the occurrences of failure modes IIA and IIB defined by Cairns and Jones (1995) have been quantified. Analytical solutions suggest that if the rib face angle is great than 60 degree, failure mode IIA will occur under normal conditions. In case of rib face angles smaller than 60 degree, failure at the steel—concrete interface will occur prior to concrete failure.

2. A MECHANICAL MODEL FOR CONE BOLTS

2.1 Introduction

Rockbolts have been widely used in civil and mining engineering applications as support and reinforcement for surrounding rock. In underground construction, the major functions of rock support can be classified as reinforcing, retaining, and holding under static loading (Kaiser et al., 1996). With better understanding of load transfer mechanisms and advances in bolting technology, various rock support products have been developed and used in practice. Among them, D-bolts, H-bolts, and cone bolts have been identified as good energy absorbing supporting elements in a support system (Cai and Champaigne, 2012; Ansell, 2006; Li, 2010; He et al., 2014). The design principle of such bolts is that these devices are able to yield to tolerate large axial displacement while still providing necessary support for the surrounding rock.

Cone bolt is an anchorage that internal fixture is similar to a wedge (Winsdor, 1997). Compared to the frictional anchors or fully grouted rockbolts, the load transfer mechanism of this type of anchor is quite different because the anchorage mainly relays on a wedge mechanism, ie, the resistance caused by the cone.

The supporting effect of cone bolts has been studied by many researchers. The concept of cone bolts was developed in South Africa (Roberts and Brummer, 1988; Malan and Basson, 1998) and they have reportedly been used in a few Australia mines (Turner, 2002). Tannant and Buss (1994) conducted pull-out testing of cone bolts for different grouting materials and found that the maximum axial displacements were 240 and 100 mm for cementitious grout and resin grout, respectively. Kilic et al. (2003) conducted an experimental study on cone geometry by pull-out tests and the results showed that anchorage strength depended on a combination of the shear and compressive strengths of the grouting material. Gaudreau et al. (2004) conducted pull tests of resin grouted cone bolts at Brunswick mine and characterized certain portions of the load–displacement curve as being the elastic stage, the yield point, and the plastic stage, as shown in Fig. 2.12. For a cone bolt subjected to dynamic load, several mechanical models have been developed (St-Pierre et al., 2009; Thompson et al., 2004; Tannant et al., 1995). For example, St-Pierre et al. (2009) developed a lumped mass dynamic model for cone bolts based on drop weight tests. Recently, Cai and Champaigne (2012) experimentally studied the influence of surface roughness of the bolt shank on load transfer mechanics and developed a new debonding agent for the bolt.

Previous studies show that the load transfer mechanism of cone bolts is different to that of conventional rockbolts. However, the cone's functions as a wedge-style anchor have not yet been clearly delineated. This chapter provides an analytical approach for the load transfer mechanisms of cone bolts subjected to tensile load. The deformational behavior of the surrounding material around the cone is formulated and the supporting effect of a cone bolt with a small conical angle is specifically considered. The validity of the proposed theory is also scrutinized using experimental data in the literature.

2.2 Problem Description and Approach

A cone bolt subjected to tensile load, as shown in Fig. 2.13, is the subject of this study. In this study, only pull-out failure of the bolt

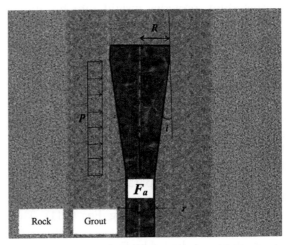

FIGURE 2.13 Schematic of a cone bolt subjected to tensile load.

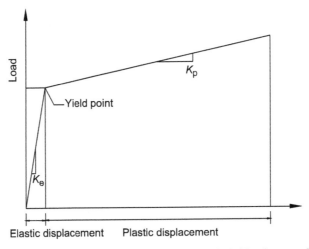

FIGURE 2.12 An idealized pullout curve of a cone bolt (Gaudreau et al., 2004).

is considered. The grout is assumed to be a Mohr–Coulomb material, and the friction between the bolt shank and the grout is neglected; ie, only resistance generated by the cone section is considered.

In this chapter, the axial displacement of the cone bolt is considered to generate confinement for the cone. In addition, for simplification, the initial average confining pressure around the cone surface, p, is assumed to be zero. Experimental results suggest that this assumption is fairly acceptable for a specimen test conducted in a laboratory (Kaiser et al., 1992).

To evaluate the interaction between the cone and the surrounding medium, the radial deformational behavior caused by the pull-out load has to be quantified. Based on the elasticity theory and plane strain analysis, for the axisymmetric problem shown in Fig. 2.14A, the radial displacement of an infinite medium can be calculated by (Yazici and Kaiser, 1992; Hyett et al., 1995):

$$u_r = \frac{1+v}{E} rp \quad (2.22)$$

where u_r = radial displacement; E = Young's modulus of the grout; v = Poisson's ratio of the grout; r = hole radius; p = internal pressure.

For a cone bolt, since only the conical section expands the borehole while slippage occurs, the confining pressure developed by the axial load should concentrate around the conical section of the bolt, as shown in Fig. 2.14B. Except for the parameters presented in Eq. (2.22), the radial stiffness in this process will also be related to the cone length and the cone face angle (i in Fig. 2.13). The analytical solution of the radial stiffness is expected to be nonlinear.

For engineering purpose, the radial stiffness of the surrounding medium of a cone bolt with known geometry can easily be found by the numerical method. Fig. 2.15 shows a Fast Lagrangian Analysis of Continua (FLAC) model that measures the radial stiffness for a specific case. Result shows that the measured average confining pressure around the cone is 11% higher than the calculated value using Eq. (2.22). For simplification, in this chapter, the radial displacement of the surrounding medium is estimated using Eq. (2.2), which is a simple modification of Eq. (2.1). However, it is worthwhile to mention that accurate estimation of radial stiffness for different situations can always be obtained via similar numerical study.

$$u_r = \frac{1+v}{E} \cdot \frac{R+r}{2} \cdot p \quad (2.23)$$

where R and r are cone radii as shown in Fig. 2.14B.

It should be noted that only the elasticity properties of the grout material are considered in Eq. (2.23), which is precise provided that the grout and the surrounding rock have similar elasticity properties to each other. However, if the difference in the elasticity properties between the grout and the rock is considerable, the equivalent elasticity properties of the confinement should be used in the equation instead. Calculation examples can be found in Cao (2012) based on the thick-walled cylinder theory for

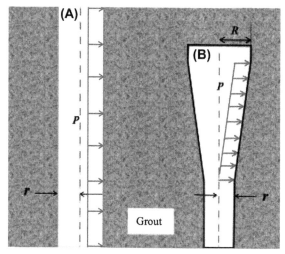

FIGURE 2.14 (A) Uniformly distributed internal pressure of an infinite medium and (B) axial load induced pressure on the cone bolt.

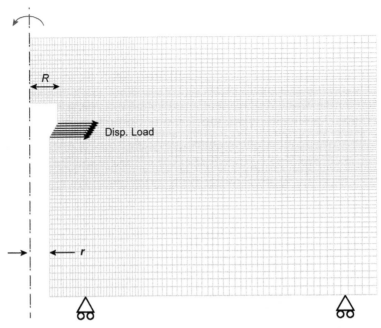

FIGURE 2.15 FLAC axisymmetric model for measurement of radial stiffness of known condition.

compound cylinders. The elasticity properties of the confinement will affect the theoretical axial displacement in subsequent calculations, and hence also the accuracy of the stiffness of the pull-out curve of the bolting system.

To identify one common failure mode, a cone bolt with 90 degree face angle is studied firstly using the stress analysis method. The pull-out stages with respect to the sequence of the partial shearing off of the grout and the corresponding load—displacement curve of the bolt are studied analytically. The validity of the developed theory is examined by comparing the calculated results to experimental data in the literature. Subsequently, the developed mechanical model is applied to normal cone bolts (ie, the cone face angle is less than 90 degree), and the results are also compared to experimental measurements.

2.3 Model Development

The mechanical model is developed according to the deformational procedure of a cone bolt subject to tensile load. The initial failure of the grout around the cone is analyzed in stage 1. The subsequent failure of grout vertex being sheared off is formulated in stage 2 until the face angle of the cone section reaches a small angle, ie, stage 3. In the last stage, the bond strength of the newly formed conical anchorage is analyzed.

2.3.1 Stage 1

Whenever an axial load, F_a, is applied to the bolt, it induces axial stress to the surrounding medium, as shown in Fig. 2.16. The average magnitude of the induced axial stress, q, can be calculated by:

$$F_a = \pi(R^2 - r^2)q \qquad (2.24)$$

where F_a = axial load on the bolt; q = axial stress induced by F_a.

For pull-out failure of cone bolts with a large face angle, experimental studies (Lutz and Gergely, 1967; Tepfers, 1979; Cairns and Jones, 1995; Jeng and Huang, 1999; Wu et al., 2012;

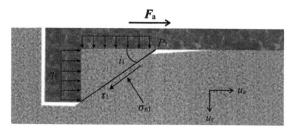

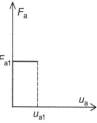

FIGURE 2.16 Terminology, concept, and the pull-out curve of the cone bolt subjected to axial load at stage 1.

Cao et al., 2013) have confirmed that the grout in front of the cone face will be sheared off and move with the cone to form a compound conical entity. The new cone angle indicates the direction of slippage and is termed as the "bearing angle" or the "effective face angle" in the literature to distinguish it from the geometric angle of the cone bolt. For the situation studied, as it has been assumed that the initial confining pressure $p_1^{ini} = 0$, the theoretical failure angle of the grout, according to Mohr–Coulomb's principle, can be expressed as:

$$i_1 = \frac{\pi}{4} - \frac{\phi_p}{2} \qquad (2.25)$$

where i_1 = failure angle in stage 1; ϕ_p = peak internal friction angle of the grout.

The axial stress to cause the failure can be expressed as:

$$q_1 = \sigma_c = \frac{2 \cdot c_p \cdot \cos \phi_p}{1 - \sin \phi_p} \qquad (2.26)$$

where q_1 is axial stress in stage 1; σ_c is uniaxial compressive strength (UCS) of grout; c_p is peak cohesion of grout.

Substituting Eq. (2.26) into Eq. (2.24), the axial load when failure occurs can be calculated by Eq. (2.27), and the load–displacement curve for this process is plotted by the red line in Fig. 2.16.

$$F_{a1} = \pi (R^2 - r^2) \cdot \frac{2 \cdot c_p \cdot \cos \phi_p}{1 - \sin \phi_p} \qquad (2.27)$$

After the initial failure, the cone slips along the bearing angle i_1. It induces radial displacement of the grout, which develops confining pressure, p, around the slipping surface. In this process, the normal stress and the shear stress along the slipping surface can be calculated using stress transformation, ie,

$$\sigma_{n1} = p_1 \cdot \cos^2 i_1 + q_1 \cdot \sin^2 i_1 \qquad (2.28)$$

$$\tau_1 = (q_1 - p_1)\sin i_1 \cos i_1 \qquad (2.29)$$

As the cone slips, it is also known that:

$$\tau_1 = c_r + \sigma_{n1} \cdot \tan \phi_r \qquad (2.30)$$

Where p_1 and q_1 are radial stress and axial stress of the grout block respectively; σ_{n1} and τ_1 are normal stress and shear stress of the failure surface respectively; c_r and ϕ_r are residual strengths of grout.

Combining Eqs. (2.28)–(2.30), the confining pressure required for the system to reach an equilibrium state can be calculated by:

$$q_1 = \frac{c_r \cdot \cos \phi_r}{\cos(\phi_r + i_1) \cdot \sin i_1} + \frac{\tan(\phi_r + i_1)}{\tan i_1} \cdot p_1^{eq} \qquad (2.31)$$

where p_1^{eq} is confining pressure at which the system reaches equilibrium.

Substituting Eq. (2.26) into Eq. (2.31), the average value of the developed confining pressure around the cone at the equilibrium point can be calculated by:

$$p_1^{eq} = \frac{\tan i_1}{\tan(\phi_r + i_1)} \cdot \left(\frac{2 \cdot c_p \cdot \cos \phi_p}{1 - \sin \phi_p} - \frac{c_r \cdot \cos \phi_r}{\cos(\phi_r + i_1) \cdot \sin i_1} \right) \qquad (2.32)$$

In this procedure, the radial displacement of the grouting material around the cone can be estimated using Eq. (2.23), ie,

$$u_{r1} = \frac{1+v}{E} \cdot \frac{R+r}{2} \cdot p_1^{eq} \qquad (2.33)$$

where u_{r1} is radial displacement of the grout around the cone.

Since slippage along i_1, the relationship between the radial displacement and the axial displacement is:

$$\tan i_1 = \frac{u_{r1}}{u_{a1}} \qquad (2.34)$$

where u_{a1} is axial displacement of the bolt.

Hence, the axial displacement can be represented by Eq. (2.35) whereas the load–displacement curve of this process is represented by the blue line in Fig. 2.16.

$$u_{a1} = \frac{1+v}{E} \cdot \frac{R+r}{2} \cdot \frac{1}{\tan(\phi_r + i_1)} \cdot \left(\frac{2 \cdot c_p \cdot \cos \phi_p}{1 - \sin \phi_p} \right.$$
$$\left. - \frac{c_r \cdot \cos \phi_r}{\cos(\phi_r + i_1) \cdot \sin i_1} \right)$$
$$(2.35)$$

2.3.2 Stage 2

As axial load increases beyond the equilibrium state formulated in stage 1, the cone section slips along the direction of the bearing angle i_1 until more surrounding grout is sheared off. Fig. 2.17 shows a diagram illustrating this process and also the stress state of a grout slice that has a bearing angle of i_2. By performing a stress analysis of the selected grout slice and introducing Mohr–Coulomb's failure criterion to the i_2 surface, it is found that the axial stress

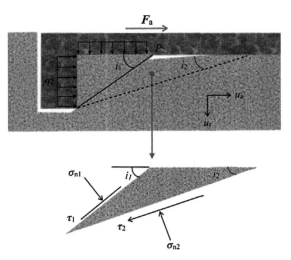

FIGURE 2.17 Stress state when failure occurs along the bearing angle i_2.

that causes a second grout failure can be expressed as:

$$q_2 = \frac{c_p \cdot \cos \phi_p}{\cos(\phi_p + i_2) \cdot \sin i_2} + \frac{\tan(\phi_p + i_2)}{\tan i_1} \cdot p_2 \qquad (2.36)$$

Where p_2 and q_2 are axial stress and radial stress acting on the i_1 block while grout failure at i_2 occurs respectively, as shown in Fig. 2.17.

As the slippage is still along the bearing angle i_1 in this process, from Eq. (2.31), we know that:

$$q_2 = \frac{c_r \cdot \cos \phi_r}{\cos(\phi_r + i_1) \cdot \sin i_1} + \frac{\tan(\phi_r + i_1)}{\tan i_1} \cdot p_2 \qquad (2.37)$$

Combine Eqs. (2.36) and (2.37) to eliminate p_2, simplify and rearrange to:

$$q_2 = \frac{2c_p \cdot \cos \phi_p \cdot \sin(\phi_r + i_1) \cdot \sin i_1 + c_r \cdot \cos \phi_r [\cos(\phi_p + 2i_2) - \cos \phi_p]}{\sin i_1 [\cos(i_1 - 2i_2) - \cos i_1]} \qquad (2.38)$$

Eq. (2.17) expresses the axial load needed to shear off a grout block at bearing angle i_2 while slipping along the bearing angle i_1. The theoretical failure angle i_2 can be calculated using $dq_2/di_2 = 0$, at which the axial load is at a minimum. For the case of $c_r = 0$, it is easy to identify that the next theoretical failure direction will occur at a bearing angle of:

$$i_2 = \frac{i_1}{2} \qquad (2.39)$$

That is, the grouting material will be sheared off at an angle approximately half of the previous one. This estimation is acceptable for engineering purposes. Accordingly, q_2, F_{a2}, and p_2^{fail} can be calculated using Eqs. (2.38), (2.24), and (2.36) respectively; and the radial displacement developed in this process and the corresponding axial displacement can be calculated using Eq. (2.33) replacing p_1^{eq} with $p_2^{\text{fail}} - p_1^{\text{eq}}$ and Eq. (2.34), respectively. The pull-out profile of this process is represented by the red line in Fig. 2.18.

After the grout is sheared off along the second bearing angle i_2, the material will lodge in front of original cone to form a new cone having half of the previous conical angle. The confining pressure of the system will be redistributed to the newly formed cone as the cone slips along that angle, as shown in Figs. 2.17 and 2.18. Therefore, the average confining pressure while slippage occurs along the new failure surface can be calculated by:

$$\frac{p_2^{\text{fail}}}{\tan i_1} = \frac{p_2^{\text{ini}}}{\tan i_2} \qquad (2.40)$$

where p_2^{ini} is the average initial confining pressure when slippage occurs along bearing angle i_2.

The axial displacement while the cone slips along bearing angle i_2 will induce radial displacement of the grout that in turn increases the confining pressure of the cone until the cone reaches another equilibrium state at this stage. The analysis of this process is similar to that for stage 1, and the axial displacement developed in this process can be calculated using Eq. (2.35) replacing i_1 with i_2. The load–displacement curve of this process is shown as the blue line in Fig. 2.18.

2.3.3 Stage 3

This stage is a repetition of the failure process of stage 2 but at a smaller failure angle, i_3, which equals to approximately half of i_2. In this stage, the confining pressure will also be redistributed to the new slipping surface and the pull-out curve can be obtained with a similar manner as in stage 2. A schematic and the axial performance of the bolt in this stage are shown in Fig. 2.19.

2.3.4 Stage 4

For most geomaterials, grout shearing off failure could only occur twice or three times as the next theoretical failure angle would be very close

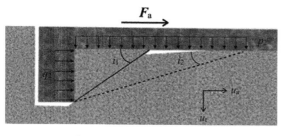

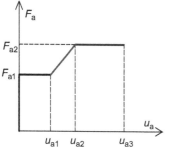

FIGURE 2.18 Concept and pull-out curve of the cone bolt subjected to axial load at stage 2.

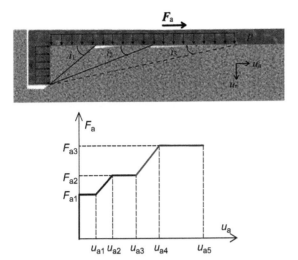

FIGURE 2.19 Concept and pull-out curve of the cone bolt at stage 3.

to the last one. For example, for a grout with a peak internal frictional angle of 22 degree, which is common for the 0.4 w:c cementitious grout (Hyett et al., 1995), the first failure angle will be around 34 degree and subsequent failure occurs at 17 degree and then at 8.5 degree. However, it is not practical that the grout will then be sheared off at 4.25 degree. Hence, the subsequent pull-out behavior of the bolt is supposed to be a small bearing angle cone (<10 degree) slipping while expanding the surrounding material.

The resistance of a small angle cone slipping while expanding the surrounding medium is supposed to have two components: friction- and cone geometry—caused resistance. For the frictional component, the effect of the cone shape can be ignored; ie, the slipping surface can be simplified to be cylindrical, as shown in Fig. 2.20. The second component is the axial resistance for a perfectly smooth cone passing through a smaller hole within a deformable medium. These two components can be calculated separately and the bond strength should be a combination of them.

There is no simple analytical model to estimate the axial resistance of a frictionless cone passing through a borehole. The formula based on the elasticity theory may overestimate the actual resistance but the calculation cost can be expected to be high if plastic deformation of the process is considered. In this chapter, numerical method is employed to deal with this problem. As there is no subsequent material failure and the contact is perfectly smooth, a computer model can be easily constructed with good accuracy by a range of commercial software. Fig. 2.21 shows an example of the result of the axial resistance of the cone using a FLAC axisymmetric model similar to Fig. 2.13 when the interface strength is set to zero.

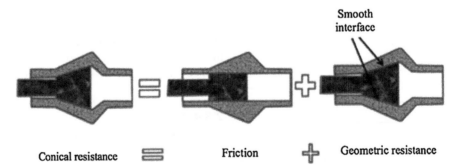

FIGURE 2.20 Mechanical model of a small angle cone slipping while expanding the surrounding material.

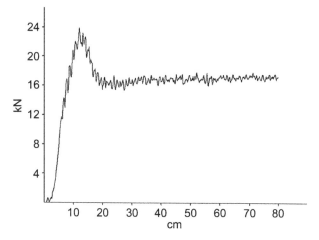

FIGURE 2.21 The numerical solution of axial resistance of a smooth cone with $R = 10$ mm, $r = 7$ mm, $i = 8.5$ degree, grout $E = 3.33$ GPa, $v = 0.3$, $c = 6.8$ MPa, $\Phi = 23$ degree and rock $E = 13.8$ GPa, $v = 0.3$, $c = 11.6$ MPa, and $\Phi = 30$ degree.

2.4 Comparison Between Theoretical and Experimental Results

Kilic et al. (2003) conducted pull-out tests on cone bolts having cone angles of 10, 15, 20, 25, 30, 60, and 90 degree. Their experimental results are used to examine the proposed mechanical model presented in this chapter. The related parameters of the experiments are shown in Table 2.2. The peak internal frictional angle of 0.4 w/c cementitious grout is estimated as 22 degree (Hyett et al., 1995), and the peak cohesion of the material can be calculated as 6.8 MPa using Eq. (2.26). The residual cohesion and residual frictional angle of the grout are estimated as 1.0 MPa and 20 degree respectively. A sensitivity study has been conducted for the assumed residual strength and will be shown later in this chapter.

According to the developed model, it is found that, $i_1 = \frac{\pi}{4} - \frac{\phi_p}{2} = 34°$, $q_1 = \sigma_c = 20.4$ MPa, $F_{a1} = \pi(R^2 - r^2)q_1 = 3.2$ kN.

Using Eq. (2.31), the confining pressure at the equilibrium point while axial load is q_1 is $p_1^{eq} = 8.6$ MPa.

The radial stiffness is estimated using Eq. (2.23); thus the radial displacement can be calculated as $u_{r1} = \frac{1+v}{E} \cdot \frac{R+r}{2} \cdot \Delta p_1 = 0.03$ mm.

TABLE 2.2 Parameters of Experiments Conducted (Kilic et al., 2003)

Grout (after 7 days)	Young's modulus (GPa)	E	3.33 ± 0.21
	Poisson's ratio[a]	v	0.2
	Uniaxial compressive strength (MPa)	σ_c	20.4 ± 2.3
	Peak cohesion (MPa)	c_p	6.9
	Peak frictional angle (degree)[a]	ϕ_p	22
	Residual cohesion (MPa)[b]	c_r	1.0
	Residual frictional angle (degree)[b]	ϕ_r	20
Bolt	Bolt radius (mm)	r	7
	Cone radius (mm)	R	10

[a] Estimated by Kaiser et al. (1992).
[b] Estimated.

Hence, by Eq. (2.34) $u_{a1} = \frac{u_{r1}}{\tan i_1} = 0.04$ mm.

These are the calculated results for the 90 degree cone bolt in stage 1, as shown in Fig. 2.22.

In stage 2, as it is known that the next failure angle (i_2) is approximately 17 degree, the axial load can be found using Eqs. (2.38) and (2.24) as $q_2 = 56$ MPa; $F_{a2} = 9$ kN.

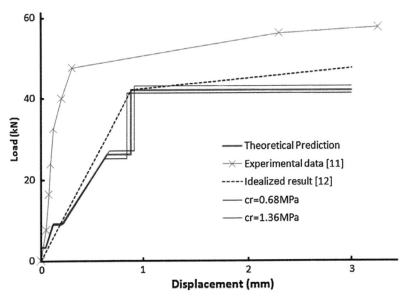

FIGURE 2.22 Pull-out curves: x = experimental measurements by Kilic et al. (2003); *solid lines* = sensitivity study; *dashed line* = idealized pull-out curve using Gaudreau's model (Gaudreau et al., 2004).

The confining pressure just before failure occurs at angle i_2 can be found via Eq. (2.36) or Eq. (2.37) as p_2^{fail} 26.1 MPa.

In this process, using Eqs. (2.33) and (2.35) $\Delta u_r = 0.05$ mm; $\Delta u_a = 0.08$ mm.

After failure occurs, the confining pressure will be redistributed around the new slip surface determined by Eq. (2.40) $p_2^{ini} = \frac{\tan i_2}{\tan i_1} \cdot p_2^{fail} = 11.8$ MPa.

Due to the development of the radial displacement along i_2, the confining pressure of that conical section increases to reach a new equilibrium state, which can be calculated using Eq. (2.32) and replacing i_1 with i_2, ie, $p_2^{eq} = 21.1$ MPa.

Accordingly, the radial and axial displacements developed in this process can be calculated as $\Delta u_r = 0.03$ mm; $\Delta u_a = 0.09$ mm.

The corresponding pull-out profile is shown in Fig. 2.22.

Using the same calculation procedure, it can be found that when $i_3 = 8.5$ degree, $p_3^{fail} = $ 64.4 MPa, and the associated $\Delta u_r = 0.13$ mm and $\Delta u_a = 0.43$ mm. It can also be found that:

$p_3^{inil} = 31.5$ MPa; $q_3 = 163$ MPa; $F_{a3} = 26.1$ kN; $p_3^{eq} = 42.8$ MPa, To reach the final equilibrium state $\Delta u_r = 0.03$ mm and $\Delta u_a = 0.23$ mm. The corresponding pull-out profile of this stage is shown in Fig. 2.22.

As the last failure angle is 8.5 degree, it is presumed that there is no grout being sheared off from then on. Consequently, the subsequent anchorage strength can be treated as a frictional component of $F_{a3} = 26.1$ kN and a cone geometric resistance of around 17 kN, which is numerically determined as shown in Fig. 2.21. Therefore, the final strength of the cone bolt will be approximately 43 kN, as shown in Fig. 2.22.

The calculation procedure presented has been repeated using $c_r = 0.68$ MPa and $c_r = 1.36$ MPa, ie, from 1/10 to 1/5 of the peak cohesion. The corresponding theoretical pull-out profiles are also illustrated in Fig. 2.22. It suggests that the

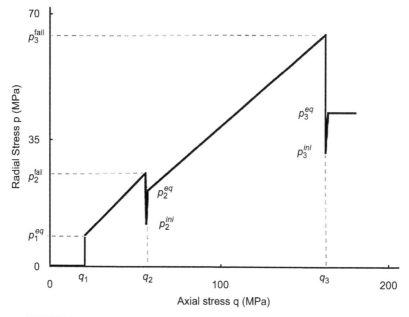

FIGURE 2.23 The average confining pressure developed around the cone.

selected mechanical properties, ie, $c_r = 1.0$ MPa, will not influence the result significantly.

Furthermore, the theoretical average confining pressure developed around the cone and the corresponding axial load in the whole process is illustrated in Fig. 2.23. The sudden

it can be concluded that the theoretical prediction is in good agreement with experimental results.

Compared to the concept proposed in Gaudreau et al. (2004) shown in Fig. 2.12, the yield load of a cone bolt can be estimated as:

$$F_a^{yield} = \pi(R^2 - r^2) \frac{2c_p \cdot \cos \phi_p \cdot \sin(\phi_r + 2i_{min}) \cdot \sin(2i_{min}) + c_r \cdot \cos \phi_r [\cos(\phi_p + 2i_{min}) - \cos \phi_p]}{\sin(2i_{min}) \cdot [\cos i_{min} - \cos(2i_{min})]} \quad (2.41)$$

drop of the confinement pressure represents the stress redistribution caused by subsequent failure.

The theoretical prediction of the axial load on the bolt is approximately 11% lower than the measurement made by Kilic et al. (2003). The main explanation for the discrepancy is most likely that the resistance along the bolt shank is not included in the calculations. Therefore,

In addition, the theoretical elastic displacement can be determined by:

$$u_a^{yield} = \frac{1+v}{E} \cdot \frac{R+r}{2} \cdot \sum \frac{\Delta p^i}{\tan i} \quad (2.42)$$

Where F_a^{yield} and u_a^{yield} are yield points in Fig. 2.12; i_{min} is final bearing angle of the cone; p_i is developed confining pressure with respect to failure angle i.

Therefore, the elastic stiffness, K_e, could be estimated by:

$$K_e = \frac{F_a^{yield}}{u_a^{yield}} \quad (2.43)$$

According to the mechanical model developed in this chapter, both the frictional resistance and the cone geometric resistance should be constant after the yield point, but experimental results (Cai and Champaigne, 2012; Kilic et al., 2003; Gaudreau et al., 2004) suggest that the axial resistance increases slightly with the axial displacement of the bolt, ie, the concept of plastic stiffness proposed by Gaudreau et al. (2004). It is highly probable that the increment of the axial load after the yield point is mainly caused by the volumetric increase of the grouting material during shearing. Accordingly, the axial load after the yield point can be modified by introducing an apparent dilational angle in Eq. (2.31). That is, the theoretical prediction of the axial load corresponding to the plastic stage of Gaudreau's model can be expressed as:

$$q_{fin} = \frac{c_r \cdot \cos \phi_r}{\cos(\phi_r + i_{min}) \cdot \sin i_{min}} + \frac{\tan(\phi_r + i_{min} + i_{dil})}{\tan i_{min}} \cdot p_{fin} \quad (2.44)$$

where i_{dil} is volumetric increase caused dilational angle.

The dashed line in Fig. 2.22 shows the pull-out curve corresponding to Gaudreau's model at a dilational angle of 5 degree.

2.5 The Effect of the Cone's Geometric Face Angles

For a cone bolt with a geometric face angle γ, the axial load needed to cause slipping along the steel–grout interface is in the same format as Eq. (2.26) but using the mechanical properties of the interface, ie,

$$q_{sg} = \frac{c_p^{sg} \cdot \cos \phi_p^{sg}}{\cos(\phi_p^{sg} + \gamma) \cdot \sin \gamma} \quad (2.45)$$

where superscript sg stands for the steel–grout interface.

Let $q_1 = q_{sg}$, then the surface on which failure tends to occur under the condition $p_1^{ini} = 0$ can be determined via:

$$\frac{c_p^{sg} \cdot \cos \phi_p^{sg}}{\cos(\phi_p^{sg} + \gamma) \cdot \sin \gamma} = \frac{2 \cdot c_p \cdot \cos \phi_p}{1 - \sin \phi_p} \quad (2.46)$$

Fig. 2.24 plots γ versus ϕ_p for Eq. (2.46) for given material properties. It suggests that for normal grouting materials, if the cone face angle is great than 73 degree, then anchorage failure will begin as grout shear off as analyzed in previous sections.

If the cone bolt face angle is less than 73 degree, however, the first failure of the bolt occurs at the cone-grout interface. In this case, the axial load induced cone-grout interface failure can be calculated by Eq. (2.47), and subsequent grout failures will be at angles $\gamma/2$, $\gamma/4$ and so on, until a very small angle is reached.

$$q^{sg} = \frac{c_r^{sg} \cdot \cos \phi_r^{sg}}{\cos(\phi_r^{sg} + \gamma) \cdot \sin \gamma} + \frac{\tan(\phi_r^{sg} + \gamma)}{\tan \gamma} \cdot p^{sg} \quad (2.47)$$

Fig. 2.25 plots the theoretical pull-out profiles of cone bolts with different geometric face angles along with the experimental results by Kilic et al. (2003). It suggests that the theoretical prediction developed in this chapter is in good agreement with laboratory test results.

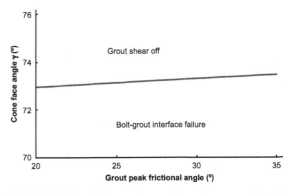

FIGURE 2.24 Failure preference for a given material with $c_p^{sg}/cp = 1 : 10$ and $\phi_p^{sg}/cp = 15$ degree.

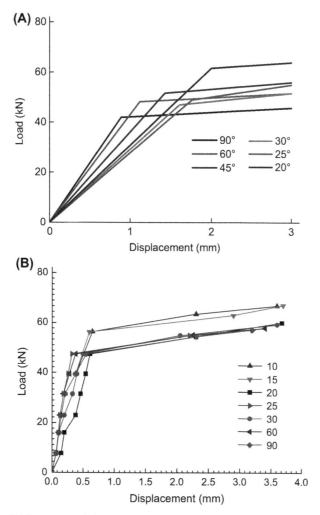

FIGURE 2.25 (A) Experimental data presented by Kilic et al. (2003) and (B) theoretical predictions.

2.6 Conclusions

In this chapter, the interaction between the cone bolt and the surrounding medium is investigated analytically. A simple model for the radial stiffness of a cone bolt subjected to tensile load is first proposed. The deformational process of a cone bolt subjected to tensile load is identified as a series of grouting material being sheared off at a bearing angle that is half of the previous one. For each loading stage, the analytical expression of the pull-out profile is derived based on Mohr–Coulomb material properties. Accordingly, the role of the geometric face angle of the bolt in bolting mechanisms has been studied quantitatively. For a cone bolt with a small bearing angle, the axial resistance has been identified as a combination of friction and cavity expansion. The predicted axial load is compared to experimental data in the literature, and good agreement is demonstrated.

The developed theory explains the roles of the cone geometry, the bolting conditions, and the material properties of the system in load transfer mechanisms. It provides an analytical approach for engineers to predict load—displacement curves of cone bolts and guide the design of their projects.

3. EFFECT OF INTRODUCING AGGREGATE INTO GROUTING MATERIAL

3.1 Introduction

Many researchers have worked theoretically and experimentally on the mechanism of load transfer of fully grouted rockbolts. To date, it is commonly accepted that fully grouted bolts are much more successful in supporting roof strata than other bolting systems.

At the time of writing, approximately 80% of the over 100 million roof bolts installed in US mines, tunnels, and construction projects employ polyester roof bolt-resin. It is estimated that resin consumed in the US each year can encircle the world at the equator approximately three times if in a 22.9-mm diameter cartridge (Blevins and Campoli, 2006).

A fully grouted bolt provides greater shear surface for transmitting the load from the rock to the bolt and vice versa. The grout supplies a mechanism for transferring the load between the rock and the reinforcing element. This redistribution of forces along the bolt is the result of movement in the rock mass, which transfers the load to the bolt via shear resistance in the grout.

This shear resistance within the resin and along the interfaces can be the result of adhesion, friction, and mechanical interlocking, which is a keying effect created when grout fills the irregularities between the bolt and the rock. Therefore, the performance of the reinforcement can be directly enhanced by improvement of the mechanical properties of the grouting material.

Current bolting technology uses a two-part polyester resin cartridge to supply the bolt and borehole with sufficient resin to achieve the desired encapsulated length. The successful performance of the bolt and cartridge system requires that the bolt shreds the plastic cartridge and mixes the separate resin components during installation.

The effectiveness of the installed roof bolts can be compromised by gloving. Gloving refers to the plastic cartridge of a resin capsule encasing a length of bolt, typically with a combination of mixed and unmixed resin filler and catalyst remaining within the cartridge.

This chapter introduces a method to enhance the rockbolting strength by mixing metallic granules into the grouting material, as shown in Fig. 2.26. Experimental results show that the peak load of pull-out tests will increase if some metal granules are mixed into the resin. In addition, the effect of the introduced metallic granules in reducing gloving is also discussed.

3.2 Related Theories

3.2.1 Fully Grouted Bolting

Fully grouted bolting consists of the bolt, grout, and surrounding rock. The relationships between them belong to the continuous mechanically coupled bolt system. A fully grouted bolt is a passive roof support system, which is activated by movement of the surrounding rock. The

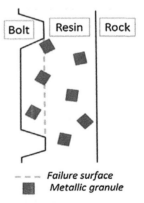

FIGURE 2.26 Schematic of the concept of mixing metallic granules into the resin.

efficiency of load transfer is affected by the mechanical properties of the grout, surface profile of the rockbolt, thickness of the grout annulus, anchorage length, rock properties, confining pressure, and installation procedure.

In a fully grouted rockbolt, the load transfer mechanism depends on the shear stress developed on the bolt—resin and resin—rock interfaces. Peak shear stress and shear stress modulus of the interfaces determine the reaction of the bolt to the strata. Hence, the load transfer is determined by measuring the peak shear stress and system stiffness. In addition, the postfailure behavior of the rockbolting system is also important as it largely determines the total energy absorption of the system.

3.2.2 Failure Mode

Littlejohn (1993) classified various types of axial failure when using grouted bolts as follows: the bolt, grout, rock, bolt—grout interface, and grout—rock interface. The type of axial failure depended on the properties of individual elements. The shear stress at the bolt—grout interface was greater than at the grout—rock interface because of the smaller effective area. If the grout and rock were of similar strengths, failure could occur at the bolt—grout interface. If the surrounding rock was softer then failure could occur at the grout—rock interface.

Based on pull-out tests of cable bolts in the laboratory and in the field, Hyett et al. (1992) have identified two failure modes in cementitious grouted cable bolt. One mode was radial splitting of the concrete cover surrounding the cable, while the other involved shearing of the cable against the concrete. The former concerns the wedge mechanism but it is rarely observed in the resin grouted bolting system. The shearing mechanism involved crushing of the grouting material ahead of the ribs on the bar, eventually making pull-out along a cylindrical friction surface possible. It should also be noted that as the degree of radial confinement increased, the failure mechanism changed from radial fracturing of the cementitious annulus under low confinement, to shearing of the cement flutes and pull-out along a cylindrical friction surface under high confinement.

Recent research work of failure mode analysis suggests that a cylindrical failure surface around the bolt—resin interface is a predominating failure mode in rockbolting (Cao et al., 2013). It occurs for smooth bars and for very closely spaced rebar bolts (like a screw) along the rib tips of the bar. For rebar bolts, experimental observation suggests that if the embedded length is short and the confining material is stiff, parallel shear failure occurs in laboratory pull-out tests. Fig. 2.27 shows a pull-out test bolt of 75-mm embedded length and confined in 8-mm thick-steel tubes.

3.2.3 Dilatancy Behaviors Accompanying Shearing

The discontinuity behavior is often studied under constant normal load (CNL) or constant normal stiffness (CNS) condition. For the

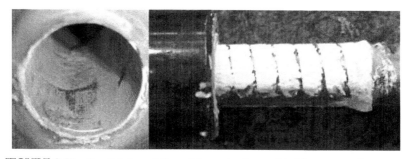

FIGURE 2.27 Parallel shear failure of the resin observed in laboratory pull-out tests.

CNL condition, dilatancy accompanying shearing of the discontinuity surfaces is permitted to occur freely; for example, sliding of an unconstrained block of rock on a slope. Under the CNS condition, however, dilation may be depressed by the surrounding material due to the increased normal stress with shear displacement.

For rockbolting systems, the dilatancy behavior can be better conceptualized under the CNS condition, as shown in Fig. 2.28. That is, dilation of the failure surface may be constrained by the resin annulus and surrounding rock.

The radial stiffness of the rockbolting system can be calculated using the thick-walled cylinder theory, which is widely accepted by theoretical rockbolting research studies (Yazici and Kaiser, 1992). Calculation of the results of some assemblage examples are listed in Table 2.3. The 4.2 mm resin annulus confinement can be used as the lower limit of the radial stiffness of resin grout rockbolting. If the confining material is 8 mm thickness steel tube, the radial stiffness will reach 3.4 GPa/mm. For an infinite medium, the radial stiffness can be found via:

$$K = \frac{E_r}{a(1 + v_r)} \quad (2.48)$$

TABLE 2.3 Radial Stiffness of Commonly Used Experimental Assemblages

Confinement	Redial Stiffness K (GPa/mm)
PVC[a]	0.072
4.2 mm resin annulus only	0.2
Aluminum[a]	0.79
Steel 1[a]	0.77
Steel 2[a]	0.95
300 × 300 mm 40 MPa concrete block	1.10
Steel 3[a]	1.12
Infinite rock mass ($E = 30$ GPa, $v = 0.25$)	2.0
8 mm steel sleeve	3.4

[a] *Reported in Kaiser et al. (1992).*

Where a is hole radius; E_r, v_r are Young's modulus and Poisson's ratio of the rock.

3.2.4 Conceptualization of Introduced Aggregate

Fig. 2.29 conceptualizes introducing aggregate into the resin matrix. If failure of the rockbolting system subjected to axial load occurs

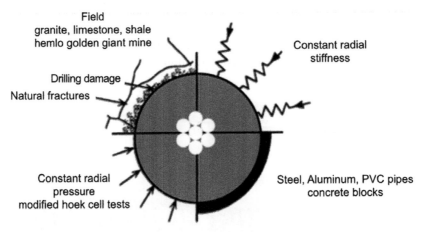

FIGURE 2.28 Schematic of a rock bolting system (Hyett et al., 1992).

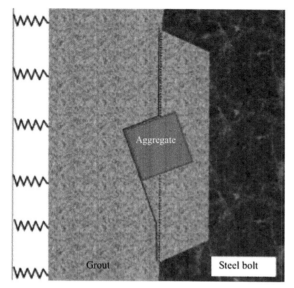

FIGURE 2.29 Alteration of the failure surface due to the introduced metallic granules.

eventually along the rib tips of a rebar bolt, then the aggregate intercepting the failure surface resists the relative slipping of the interface. Due to the interruption, an irregular failure surface is formed with increasing axial loads. The irregularity of the failure surface will cause extra dilation of the interface, increasing the effectiveness of the load transfer of the system.

3.3 Experimental Study

Laboratory pull-out tests were carried out using two kinds of rebar bolts, namely the T2 and T3 bolts, which are popular in the Australian mining industry. The bolt was encapsulated in a steel tube 75-mm long and 8-mm thick using mix and pour resin.

The specifications of the bolts' surface profile configurations can be found in Cao et al. (2013). Two-millimeter diameter steel wire was cut to 2–3-mm long segments and mixed into the resin; with approximately 10 per bolt profile. Fig. 2.30 shows the post-test sheared resin and the introduced metallic granules.

Fig. 2.31 shows the test results. As expected, the load transfer capacity and the total absorbed energy of the bolt are both increased by up to 20% by introducing metallic granules into the resin.

3.4 Potential to Reduce Gloving

Gloving is currently seen as an industrial problem because the gloved and unmixed portions reduce the effective anchor length and adversely affect the reinforcement for the roof strata. Research shows that gloving is a systematic and widespread phenomenon, occurring across the range of resin and/or bolt manufacturers, and in a variety of roof types. It has been found in bolts installed using either handheld pneumatic or continuous miner-mounted hydraulic bolting rigs, under face conditions by operators, and under controlled manufacturers best practice conditions (Campbell and Mould, 2005).

Recent research (Craig, 2012) reported that testing of specific bolt ends of 26–28 mm widths

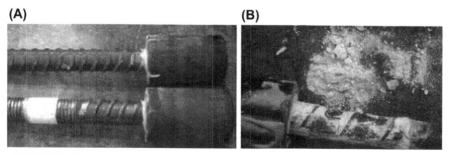

FIGURE 2.30 (A) Pretest of T2 bolting specimen and (B) post-test of T3 bolting specimen.

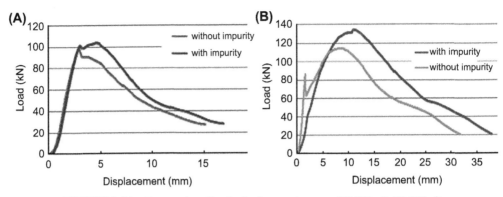

FIGURE 2.31 Test results of load–displacement curves. (A) T2 bolt; (B) T3 bolt.

installed into a hole drilled with a 27-mm bit can significantly reduce gloving, and concluded that gloving could only be significantly reduced by a bolt end that nearly contacted the side of the bolt hole. However, due to installation difficulties, the patent pending bolt cannot be applied using standard Australian bolting rigs.

It is obvious that the plastic film will be ground into pieces if the bolt diameter equaled the bore hole diameter; however, it remains a question whether this can be achieved via other means. Introducing metallic granules into the resin will lead to extra slipping of the plastic film against the granules while the bolt is being installed. This may greatly reduce the extent of the gloving problem because the effect of the introduced metallic granules can be thought as a way to increment bolt diameter without leading to installation difficulties.

3.5 Conclusions

This study provides a new method to increase the load transfer capacity of a fully grouted rockbolting system by introducing metallic granules into the resin. When the granules are mixed into the resin, the failure surface around the interfaces will become irregular for the parallel shear failure mode of the system. This will lead to extra dilation of the failure surface when and after failure occurs. Laboratory pull-out tests were conducted for two kinds of rebar bolts which are commonly used in the Australian mining industry. Results show that both the peak load and total energy absorption can be substantially increased by about 20%. This innovation is also proposed as a possible solution to reduce the gloving of the system; however, more research is required to further examine and quantify this hypothesis.

4. OPTIMIZING SELECTION OF REBAR BOLTS

4.1 Introduction

Rebar bolts have been broadly used as primary reinforcement in civil and mining engineering applications. Steel bolts support the rock mass by decreasing the deformation within the rock. The load transfer capacity of the rockbolting system is the key factor in the reinforcing effect.

The load transfer capacity of the bolt depends on the shear strength of the grout–rock and the grout–bolt interfaces. The grout–rock interface failure rarely occurs in laboratory pull-out tests or in practice. Therefore, the reinforcement effect can largely be attributed to the bonding strength of the grout–bolt interface in a rockbolting system.

It is commonly accepted that the bonding strength of fully grouted rockbolts has three components: cohesion, friction, and mechanical interlock. Singer (1990) showed that there is no adhesion in the grout–bolt interface. The frictional component can be further cataloged into dilational slip, shear failure of grouting material and bolt unscrewing (Hyett et al., 1995). Each of them is related to the confine pressure developed at the interface, which is generated by the interaction of system elements.

Mechanical interlock is a key effect created by the asperity of the grout–bolt interface. Therefore, the rebar bolt surface profile configuration plays an important role in bonding capacity in the rockbolting system. In fact, if the chemical adhesion at the grout–bolt interface is negligible and the frictional resistance can only have its effect after relative movement, then the mechanical interlock must be the predominant factor until rockbolting failure has occurred.

The bolt surface profile configuration is determined by the profile shape, height, and spacing of the rib and the angle of wrap, as shown in Fig. 2.32. Experimental studies suggest that the bolt surface configuration plays an important role in generating shear resistance between the bolt and the surrounding medium (Aydan, 1989; Fabjanczyk and Tarrant, 1992, Ito et al., 2001; Kilic et al., 2002; Aziz et al., 2003; Li, 2012). However, there are no related theories or mechanical models to explain these experimental observations and to identify the role of the bolt profile in load transfer mechanisms.

In traditional rockbolting mechanism studies, the effect of mechanical interlocking is often integrated into the proposed load transfer model in various manners. For example, in the interfacial shear stress (ISS) model, the deformation of surrounding materials is lumped into a zero thickness interface, which is assigned with specific stress–strain or stress–displacement behavior to simulate the mechanical interlocking effect observed in pull-out tests (Li and Stillborg, 1999; Ivanovic et al., 2009). Recently, a trilinear bond-slip model for the grout–bolt interface has been adopted and closed-form solutions for the prediction of the full-range behavior of fully grouted rockbolts under axial loading obtained (Ren et al., 2010; Martin et al., 2011). However, these models have not taken bolt rib geometry into consideration.

To improve the bolt loading capacity by optimizing the rebar surface profile, it is necessary to

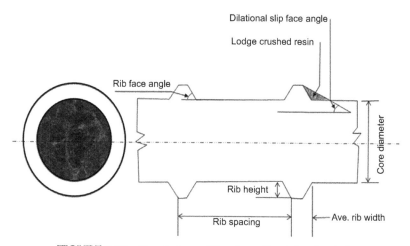

FIGURE 2.32 Terminology of the rib profile of the rebar bolt.

consider the interaction between the rebar bolt profile and the surrounding medium. In this chapter, resin–bolt interface failure is studied and categorized into dilational slip failure and parallel shear failure of the resin based on experimental observations. The analytical expression of the axial loading capacity of each failure mode is derived based on the Mohr–Coulomb's material properties. The results have shown good agreement to experimental data. Accordingly, the role of the rebar bolt profile has been identified with respect to bond failure. As illustrated by the case study, the developed theory offers a tool for engineers to select the optimal bolt in their reinforcement design.

4.2 Failure Modes of Rockbolts Under Axial Loading

Failure mode analysis is the crux of load transfer mechanisms of the rockbolting system. Only resin–bolt interface failure is studied in this chapter as the objective of this research work is to optimize rebar bolt surface profile configuration. Two major failure modes can be identified via pull-out tests, namely parallel shear failure (indicated by the dotted line in Fig. 2.33) and dilational slip failure (indicated by the dashed line in Fig. 2.33).

Parallel shear failure is characterized by a cylindrical failure surface. It occurs for the smooth bar and for very closely spaced rebar bolts (like a screw) along the rib tips of the bar. For rebar bolts, experimental observation suggests that if the embedded length is short and the confining material is stiff, parallel shear failure occurs in laboratory pull-out tests.

Dilational slip failure, on the other hand, is characterized by a conical failure surface. Fig. 2.34 shows dilational slip failure in (A) a pull-out test from concrete block and (B) a CNS test. Experimental observations also suggest that the dilational slip angle is often different to the geometric angle of the rebar profile

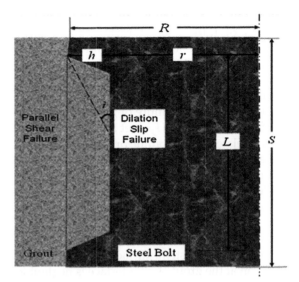

FIGURE 2.33 Failure modes and geometry of one bolt profile subjected to axial loading.

(Fig. 2.32). In this chapter, the geometric angle of the bolt profile is termed bolt rib face angle, to be distinguished from the term slip face angle, which is the real slipping direction of the rebar when dilational slip failure occurs.

4.3 Parallel Shear Failure

4.3.1 *Formulation of Parallel Shear Failure*

Parallel shear failure is a characteristic failure pattern, which occurs for smooth bars along the grout–bolt interface and for very closely spaced rebar bolts (similar to a screw profile) along the profile peaks of the bar. In the latter case, it can be found that the grouting material between the profiles will remain there to form a cylindrical failure surface. The mechanical behavior of the closely spaced rebar bolts subjected to axial loading can be expressed by the following equation:

$$F = \int \tau_{bg} dA_{bg} + \int \tau_g dA_g \quad (2.49)$$

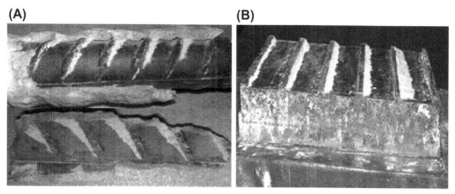

FIGURE 2.34 (A) Dilational slip of the interface in the pull-out test from concrete block and (B) postfailure in the Constant Normal Stiffness (CNS) test.

where F is the resultant force in the axial direction; A is the failure surface area; and τ is the shear resistance at the interface.

The subscripts "bg" and "g" denote the grout–bolt interface and the grout itself, respectively. Once failure occurs, according to the Mohr–Coulomb's failure criterion, the shear resistances can be expressed as:

$$\tau_{bg} = c_{bg} + p \tan \phi_{bg} \text{ and } \tau_g = c_g + p \tan \phi_g \quad (2.50)$$

where c is cohesion of the interface; ϕ is internal friction angle of the interface; and p is confine pressure when failure occurs.

Experimental results suggest that the load capacity of closely spaced rebar bolts is lower compared to ribbed steel bars of the same core diameter (Ito et al., 2001; Kilic et al., 2002).

For rebar bolts, rib width is often much smaller than the rib spacing, ie, $L \gg (S - L)$, such as the bolt shown in Fig. 2.34A; and the strength of the grout–bolt interface is normally much smaller than the resin's. Thus, the strength of the grout–bolt interface can be neglected in Eq. (2.49), and the axial load capacity of one rebar bolt profile for parallel shear failure can be simplified as:

$$F_p = 2\pi RL(c + p \tan \phi) \quad (2.51)$$

where R and L are rib geometric parameters as indicated in Fig. 2.33; c and ϕ are Mohr–Coulomb properties of the resin.

4.3.2 Case Study

To examine the validity of the formulation of parallel shear failure and the importance of failure mode identification, the following case is used to calculate the bond strength and to compare it to experimental data.

Moosavi et al. (2005) conducted a series of laboratory tests to study the effect of confining pressure on the bond capacity of different rebar bolts. A modified triaxial Hoek cell was used to facilitate application of a constant radial confining pressure to the grouted sample while pulling the bolt axially through the cement annulus.

Threaded rockbolts with diameters 22 and 28 mm (called rebar P22 and P28, respectively, in their paper) were tested, as shown in Fig. 2.35A. It can be seen that the profiles of the bar are closely spaced. An enlarged picture is used to obtain the geometric parameters of the rib as they were not provided in the chapter. Measurements indicate that the rib width of the threaded bar is approximately 23% of one rib spacing (Fig. 2.35B), and one rib spacing is approximately 21% of the diameter.

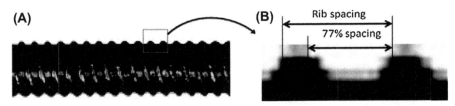

FIGURE 2.35 (A) Deformed bar used in experiment by Moosavi et al. (2005) and (B) measured rib parameter via enlarged picture.

The test results provided in the literature for threaded bars are shown in Fig. 2.36. The dilation diagrams show that the dilation is very small at confining pressures of 3.2, 4.8, and 6.4 MPa for P28 bolts, and also the confining pressure of 3.5 and 5 MPa in cases of P22 bolts. The diagrams also suggest that the failure mode is parallel shear failure of the grout as the residual strength is nearly constant in these cases. As a result, the shear stress in the failure surface can be predicted using Eqs. (2.1) and (2.2).

Before predicting the axial bolt loading capacity, the mechanical properties of the grout and the grout/steel contact surface have to be determined. Type 1 Portland cement mix of the grout used a water to cement ratio of 0.4; the UCS values were provided in the chapter. Assuming the internal friction angle of the grout is

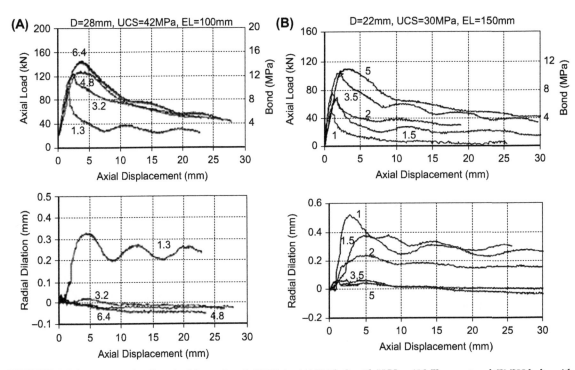

FIGURE 2.36 Results of pull test by Moosavi et al. (2005) for (A) P28 bolt with UCS = 42 MPa grout and (B) P22 bolts with UCS = 30 MPa grout.

33 degree, the cohesion can be estimated as 11.4 and 8.1 MPa for UCS 42 and 30 MPa of the concrete, respectively. In addition, the grout-to-steel adhesion is simply assigned as 2.0 MPa and their frictional angle as 24 degree.

The axial load capacity can be calculated by combining Eqs. (2.49) and (2.50). For example, consider the P28 bolt with grout UCS = 42 MPa, in case of confining pressure $p = 3.2$ MPa, the shear resistance of the grout segment can be determined as: $\tau_g = c_g + p \tan \phi_g = 13.5$ MPa. And the shear resistance in the rib profile segment is: $\tau_{bg} = c_{bg} + p \tan \phi_{bg} = 3.4$ MPa.

From Fig. 2.35B, the grout shear failure surface occupies around 77% of the bolt surface area, and the embedded length = 100 mm. Using Eq. (2.32), the predicted load capacity of the bond is 98 kN.

For all selected cases, the calculations of the axial load were performed and the results summarized in Table 2.4. The average differences between the measured and theoretical axial loads are −16.5% and −10.5% for the P28 and P22 bolts, respectively, indicating agreement.

A sensitivity study has been conducted on different friction angles ranging from 30 to 40 degree (not shown here). The result suggests that the difference in shear strengths is less than 4% in tested confining stress and UCS ranges. So the assumption that $\phi = 33$ degree will not affect the results significantly.

The last row in Table 2.4 is a demonstration of variation in failure modes. Due to the lower confinement of 1.5 MPa, dilation occurs. As a result, failure mode in this test is most likely changed to dilational slip failure, hence formulas used for parallel shear failure are no longer applicable.

4.4 Dilational Slip Failure

In theory, dilational slip is often simulated by adding an introduced dilation angle, i, to the internal friction angle of the slipping surface. The shear strength of the bond can be described by Patton's equation:

$$T = p \tan(\phi + i) \quad (2.52)$$

where i is dilation angle.

It is not easy to determine the dilation angle. In mechanical models of the dilational slip of the interface, the dilation angle can be estimated as a reduction of the geometric angle to form an "apparent dilation angle" to accommodate the data obtained from the experiments (Ladanyi and Archambault, 1977; Yazici and Kaiser, 1992). In this chapter, the dilational slip face angle, i, in rockbolting failure under axial loading is formulated using another approach.

4.4.1 The Governing Equation of Dilational Slip Failure

One bolt rib profile, shown in Fig. 2.33, is the subject of discussion. Before dilational slip failure occurs, the initial radial stress field cannot be affected by the increment of axial loading because they are perpendicular to each other. We assume that the initial confining pressure, p, is universal along the bolting length. That is, the confine pressure will keep its magnitude

TABLE 2.4 Comparison Between Predicted Axial Load Capacity and Experiment Data

Bolts	Confinement (MPa)	Calculated Axial Loads (kN)	Experiment Results (kN)	Difference (%)
P28	3.2	98	118	−16.8
	4.8	107	124	−13.9
	6.4	115	142	−18.9
Average				−16.5
P22	3.5	92	105	−12.7
	5	101	110	−8.2
Average				−10.5
	1.5	79	70	+13.1

until relative axial movement takes place. In this procedure:

$$\sigma_n = p\cos i + \frac{F_d}{A}\sin i \quad (2.53)$$

$$\tau = \frac{F_d}{A}\cos i - p\sin i \quad (2.54)$$

where the conical area of the failure surface $A = \pi(R^2 - r^2)/\sin i$; R and r are rib geometric parameters as shown in Fig. 2.33; σ_n is normal stress of the failure surface.

Once relative movement along the supposed surface occurs, dilational slip failure follows:

$$\tau = c + \sigma_n \cdot \tan\phi \quad (2.55)$$

By combining Eqs. (2.53)–(2.55) and eliminating σ_n, and τ, the axial resultant force which causes dilational slip failure can be calculated:

$$\frac{F_d}{\frac{\pi(R^2-r^2)}{\sin i}} = \frac{c}{\cos i - \sin i \cdot \tan\phi} + \frac{\cos i \cdot \tan\phi + \sin i}{\cos i - \sin i \cdot \tan\phi}p$$

Notice that:

$$\cos(i)\cdot\tan\phi + \sin(i) = \frac{\cos(i)\cdot\sin\phi + \sin(i)\cdot\cos\phi}{\cos\phi}$$
$$= \frac{\sin(i+\phi)}{\cos\phi}$$

$$\cos(i) - \sin(i)\cdot\tan\phi = \frac{\cos\phi\cdot\cos(i) - \sin(i)\cdot\sin\phi}{\cos\phi}$$
$$= \frac{\cos(i+\phi)}{\cos\phi}$$

Then

$$\frac{F_d}{\frac{\pi(R^2-r^2)}{\sin i}} = \frac{\cos\phi}{\cos(i+\phi)}\cdot c + p\cdot\tan(i+\phi)$$

In case of cohesion $c = 0$, this equation reduces to the Patton's equation of inclined discontinuity slipping. In other words, this expression is a general form of Patton's equation for cohesive materials. Rearranging this equation, one gets the axial load in dilational slip failure as:

$$F_d = \pi(R^2 - r^2)\left[\frac{\cos\phi}{\sin(i)\cdot\cos(i+\phi)}\cdot c \right.$$
$$\left. + \frac{\tan(i+\phi)}{\sin(i)}\cdot p\right] \quad (2.56)$$

It can be seen that when dilational slip occurs, the axial force has to overcome two resistances; one is generated by cohesion and another is generated by confining pressure. The magnitude of the axial load will depend on the bolt rib geometric parameters R and r, grout material mechanical properties ϕ and c, redial confinement p, and the dilational slip face angle i.

4.4.2 The Most Vulnerable Slipping Surface

To investigate the role of the bolt rib profile in dilational slip failure of rockbolting, the dilational slip face angle must be acknowledged. Making i the variable in Eq. (2.56), one sees that there is one stationary point at which dilational slip failure commences by a minimum axial load. Hence, this is the most vulnerable slip direction in dilational slip failure of the rockbolting system when subjected to axial load. The stationary point on the curve can be found by $dF_d/di = 0$.

This leads to:

$$\frac{c}{p} = \frac{\sin i - \cos i \sin(i+\phi)\cos(i+\phi)}{\cos\phi\cdot\cos(2i+\phi)} \quad (2.57)$$

It shows that the most vulnerable slip direction depends on the internal friction angle of the grout and the cohesion to confinement ratio, but is irrespective of bolt rib geometry. The mechanical properties of the grout can readily be measured in the laboratory. Therefore, once the installation pressure of the bolt is acknowledged, the most vulnerable dilational slip angle can be calculated. Furthermore, the minimum axial load to commence bond failure is predictable using Eq. (2.56), in which the dilation angle is designated by the most vulnerable dilation angle.

4.4.3 Experimental Study

To examine the theoretical prediction of the axial load at dilational slip failure, a series of short encapsulation pull-out tests were conducted. The confine pressure of the bolt was measured by pulling out a smooth bolt. Then, a rebar bolt commonly used in the Australian mining industry was tested using the same assembly.

As discussed previously, the confining pressure, p, is the initial installation pressure. It can be estimated as 5 MPa for laboratory pull-out assemblies, or $(\sigma_1 + \sigma_2)/2$ in the field, but it is recommended to measure the confine pressure of each engineering application to obtain accurate data.

In the laboratory, the initial installation pressure of the bolting assembly was measured by pulling out a smooth bolt. Firstly, the frictional angle of the grout/smooth steel interface was measured using a shear box that was assigned normal stress from 3 to 7 MPa approximately. Then, the smooth bolt pull-out tests were conducted to measure the residual strength, occurring at axial displacement of around 1 mm. The initial confine pressure of the bolt was then calculated, and it was found that the average initial confine pressure of the pull-out specimen was 5.8 MPa.

A popular rebar bolt (Fig. 2.34A) used in the Australian mining industry was utilized to examine the developed theory. The profile geometric parameters of the bolt, the measured mechanical properties of the resin (same to the smooth bolt pull out) and the calculated initial confining pressure are listed in Table 2.5.

Using Eq. (2.51), the axial load to cause parallel shear failure is calculated as 16.2 kN per bolting profile. Using Eq. (2.57), the most vulnerable dilational slip face angle can be calculated as 26.57 degree. The minimum load to cause failure for one bolting profile is obtained as 8.7 kN using Eq. (2.56). Hence, dilational slip failure is a preferable failure mode in this situation.

TABLE 2.5 Parameters of the Tested Bolting Assembly

T2 BOLT	
Core radius	$r = 10.85$ mm
Average rib height	$h = 1.40$ mm
Rib spacing	$S = 12.5$ mm
Rib tip interval	$L = 10.50$ mm
RESIN PROPERTIES	
Cohesion	$c = 16$ MPa
Internal Frictional angle	$\phi = 35$ degree
BOLTING ASSEMBLY	
Installation pressure	$p = 5.8$ MPa

A series of short encapsulation pull-out tests of the studied rebar bolt were conducted for different embedded lengths. A typical result is shown in Fig. 2.37. The theoretical minimum axial load to commence bond failure of the whole embedded length is estimated by accumulating the axial load of each profile, and denoted by a cross in diagram.

According to the calculation, the theoretical failure load on average is 55% of the peak load. This result is in good agreement with, and also supports, the commonly recognized axial load of bond failure (or, say, elastic stage) in rockbolting mechanism research, such as approximately 50% proposed by Ren et al. (2010).

4.5 Optimization of the Bolt Profile

4.5.1 Formulation of the Optimum Rebar Profile

As discussed previously, the parallel shear failure of the resin and dilational slip failure

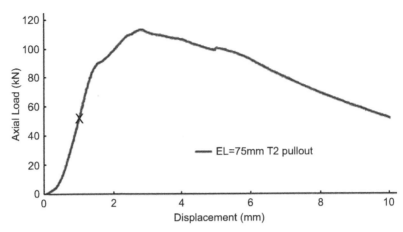

FIGURE 2.37 Pull-out profile using T2 bolts embedded length 75 mm, X = predicted failure load.

are the major failure modes identified in laboratory pull-out tests. The failure load of each failure mode is related to the bolt profile configuration, the grout properties, and the confine pressure of the bolt, and can be predicted using derived equations.

It is logical that the actual rockbolting failure occurs in the failure mode with the smaller loading capacity. From a viewpoint of bolt rib profile design, if the bearable load of each profile is the same in both failure modes, the rib profile configuration of this bolt is optimal with respect to bond failure. That is, the optimal bolt rib profile satisfies $F_p = F_d$.

Substituting Eqs. (2.51) and (2.56) leads to:

$$2\pi RL(c + p\tan\phi) = \frac{\pi(R^2 - r^2)}{\sin i}\left[\frac{\cos\phi}{\cos(i+\phi)}c + p\tan(i+\phi)\right]$$

Substituting Eq. (2.57), the optimal rib geometry can be expressed as:

$$L = \frac{\cos\phi \cdot \cos(i+\phi)}{(1-\cos i)[2\sin i + \sin(2i+2\phi)]} \times \left(h + \frac{rh}{r+h}\right) \quad (2.58)$$

Where $h = R - r$ is the rib height.

This is the equation of the optimal bolt profile design. In the expression, the rib geometry involves the core radius of the bolt, the rib height, and the rib tip interval. The core radius is related to the tensile strength of the bolt; the rib height is concerned with the diameter of the drilled hole and the rib tip interval is the most flexible parameter in steel bolt manufacture.

The expression also indicates that the optimal bolt profile geometry depends on the mechanical properties of the grout and the initial confining pressure because they are related to the dilation angle. In previous optimal bolt profile studies, there's often only one variable that was tested in experiments, such as the rib height or the rib spacing. Results of this study suggest that conclusions drawn from such pull-out tests are only applicable for the same situations as tested in their experiments. There is no universal optimal bolt for all conditions. Since they are all interrelated, bolting installation and grout properties must be taken into account to improve the loading capacity of the bond; this also means that rockbolt selection optimization must be done in a case-by-case manner.

4.5.2 Application to the Metropolitan Colliery Roadway

To evaluate and demonstrate the computation in optimizing the selection of rockbolts presented herein, an underground coal mine roadway is analyzed. Metropolitan Colliery is situated in the Southern Coalfields of the Sydney Basin in NSW, Australia. The mine utilizes a long-wall operation and continuous miner for heading development and long-wall roadways (Jalalifar, 2006). A typical stratigraphical diagram and related material strengths are shown in Fig. 2.38. The roof is mainly sandstone and mudstone, and classified as moderate to strong. The stresses were measured in sandstone with Young's modulus of 16 GPa. From the stress measurements it was found that the maximum horizontal stress = 25 MPa, oriented between parallel with the current heading direction and 30 degree west of the heading direction, the intermediate principle stress = 16 MPa and the minor principle stress = 12.5 MPa vertical.

Table 2.6 columns A—F list the profile geometric parameters of four kinds of rebar bolts commonly used in the Australian mining industry. The core diameters of the bolts are similar, but the surface profiles are quite different. If the primary reinforcement of the roadway is made of four 2.2-m long fully resin grouted rockbolts at 1.0-m interval along the roadway, which bolt should be selected among these four bolts in the preliminary design? Currently, there is no analytical guideline, not even a commonly accepted empirical one, for mine operators to select the appropriate bolt for their application. The following is the computational procedure of optimal bolt selection using the developed theory in this chapter:

1. The immediate roof is mudstone, its stress field can be calculated as $\sigma_1 = 16$ MPa and $\sigma_2 = 12$ MPa using normalization technology.
2. The initial confine pressure of the rockbolts is estimated as 14 MPa.

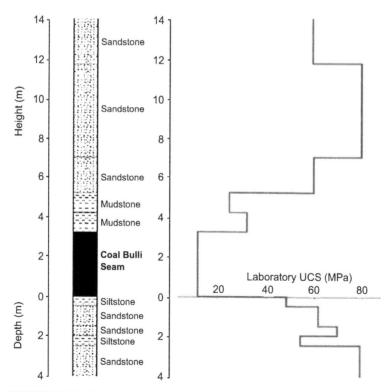

FIGURE 2.38 Modeled geological section and strength profiles (Jalalifar, 2006).

TABLE 2.6 Specifications and Computational Results of Each Studied Bolt

A	B	C	D	E	F	G	H
Bolt	Diagram	R (mm)	r (mm)	L (mm)	S (mm)	F_p/S (N/mm)	F_d/S (N/mm)
T1		11.50	10.85	10.50	12.00	1631	454
T2		12.25	10.85	10.50	12.50	1668	970
T3		12.10	10.85	22.50	25.00	1766	430
T4		11.00	9.80	7.25	9.70	1333	964

3. The mechanical properties of the resin can be measured in the laboratory or obtained from the manufacturer's specification, for example, $c = 16$ MPa and $\phi = 35$ degree.
4. Using Eq. (2.49) and profile parameters, the axial load to cause parallel shear failure can be calculated in load per unit bolting length F_{para}/S (Table 2.6-G).
5. By solving Eq. (2.57), the most vulnerable dilational slip angle can be calculated based on grout properties and *in situ* pressure.
6. Using Eq. (2.56), the minimum axial load per unit bolting length of dilational slip failure can be calculated as F_{dila}/S (Table 2.6-H).

According to the results in Table 2.6, dilational slip failure will occur for all studied bolts in this case as its bearable load is smaller than parallel shear failure's. The T2 bolt and the T4 bolt have high yielding load per unit bolting length, thus they are preferable in primary reinforcement of the roadway. Between them, the core diameters are different, hence the yielding loads of the bolts are different, and this should be considered properly to accommodate the predicted roof deflection.

The yielding load can further be improved by adapting the rib tip interval and profile spacing for this mining condition. For example, the failure loads of T2 and T4 bolt can increase by 38% and 14% per unit bolting length respectively via fine modification of the rib spacing. However, it is difficult for the bolt manufacturer to prepare many different kinds of bolts, and also it is hard to achieve such differentiation according to the current bolt manufacturing procedure.

It should be noted that the selection of T2 or T4 bolts is based on achieving a higher failure resistance of the bolt–resin interface. It is compatible with in situ geoconditions as the roof is classified as moderate to strong, and the displacement of the discontinuity, located in the middle of bolting length, is expected to be modest.

The yielding load of T3 bolting is low; however, experimental data suggest that this bolt is superior in postfailure performance and total energy absorption. The postfailure behavior of rockbolting subjected to axial loading is beyond the scope of this chapter, and will be addressed in the following publications.

The main difference between T1 and T2 bolts is that T1 bolts have a smaller rib height. This design assists to smoothen the flowing of the resin in the bolting installation. In addition, it is also easier to be threaded by the bolt manufacturer.

In summary, each studied bolt has advantages and disadvantages. From the viewpoint of shear resistance against rockbolting failure, the T2 bolt is suitable for the studied roadway support. If the roof is estimated to have good integrity, then the T4 bolt can be selected for economic purposes. In harsh conditions, such as roadway junctions, the T3 bolt should be considered based on the total energy absorption theory. If bolt installation problems, such as resin extrusion, are an issue, the T1 bolt can be considered to be used in modest geoconditions.

The current bolt profile design is admirable in the respect that they still manage to address different purposes, either due to empirical knowledge or simply by chance. This study enlightens the principles behind the design and also provides a tool for engineers to select the optimal bolt in their application.

4.6 Conclusions

In this research work, the interaction between the rebar bolt profile and the surrounding medium is investigated analytically. The interface failure is categorized into dilational slip failure and parallel shear failure of the resin based on experimental observations. For each failure mode, the analytical expression of the axial loading capacity is derived after introducing the Mohr–Coulomb's failure criterion. Accordingly, the role of the rebar bolt profile has been identified quantitatively with respect to bond failure.

Parallel shear failure of the resin occurs for smooth bars or for very closely spaced rebars along the profile peaks and forms a cylindrical failure surface. The predicted failure load is formulated and compared to experimental data in the literature, and showed good agreement.

For dilational slip failure, after distinguishing the slipping face angle from the rib face angle, the equilibrium equation is formulated. It describes the shear slipping occurrence for cohesive materials, and is a general form of Patton's equation.

One major finding of this research work is the introduction of the concept of the most vulnerable slip angle in dilational slip failure, at which the axial load is minimum comparing to other possible failure orientations.

Comparison of the calculated axial load capacities of these two failure modes suggests that dilation slip failure at the most vulnerable slip face angle occurs if the rib spacing is large (rebar bolt profile configurations normally are). A series of laboratory pull-out tests were conducted to examine the developed theory. The initial confine pressure of the bolting assembly was measured by pulling out a smooth bolt. Then, the same setup was used to conduct pull-out tests for a normal rebar bolt. The results show that the calculated failure load is approximately 53% of the peak load, which is supported by other rockbolting mechanism theories.

If the bearable loads of both failure modes are equal, the optimization of the rib profile can be analytically achieved. The optimal bolt profile geometry is related to the mechanical properties of the grout and the confining pressure when failure occurs. Therefore, bolt profile studies must take into account the mechanical properties of the grout and the confine pressure. Furthermore, conclusions drawn from laboratory pull-out tests should be evaluated properly considering both the testing and in situ conditions before any application.

A case study is included to demonstrate the application of the developed theory. Four popular bolts used in the Australian mining industry are supposed to be selected for Metropolitan Colliery roadway support. The advantages and disadvantages of each bolt are discussed analytically for this case.

The developed theory explains the role of the bolt profile in load transfer mechanisms. It provides an analytical approach for engineers in selecting the optimal bolt for their particular application; and also a tool to assist bolt manufacturers in creating new bolt designs.

5. POISSON'S RATIO EFFECT IN PUSH AND PULL TESTING

5.1 Introduction

The introduction of rockbolting to underground mining has significantly improved the stability of openings. Since its introduction, much research has been conducted to improve the understanding of rockbolt performance in both soft and hard rock conditions. Many laboratory and field tests have been undertaken and theories developed. Several laboratory procedures have been used to investigate the load capacity of various rockbolting systems. The adopted method to evaluate the performance of the rockbolts is primarily by using short encapsulation pull or push tests from either concrete or steel pipes. These laboratory tests seem to oversimplify the in situ conditions, but nevertheless, they closely represent the realistic situation.

Short encapsulation push testing is also used in the laboratory, and is often considered interchangeable to examine the effect of various parameters on the mechanical behavior of a bolting system and for investigating the bolt–grout–rock interaction under axial loading. However, the short encapsulation pull test is better because the bolt is extended and deformed similar to in situ conditions. Pushing the bolt out of the tube would contradict the realities of the bolt functioning in situ.

Concerns are often raised about the validity of the short encapsulation push test because it does not reflect the true load transfer capability. The installation and subsequent performance of bolts in situ means they are often placed in tension and in shear. There will be a general reduction in the cross-section as a result of tension. On the contrary, the bolts are in compression in push tests and it leads to an increase of cross-section area.

Rockbolts have a better performance in push tests than pull tests. This is thought to be caused by the Poisson's ratio effect. Aydan (1989) carried out a series of push and pull-out tests to investigate the anchorage mechanism of grouted rockbolts and the effect of various parameters. Two steel bars 13 and 19 mm in diameter were tested. The results showed that the load bearing capacity of bolts was 25% higher in push-out tests than pull-out tests. Aydan suggested that this increase in push test values was attributed to the Poisson ratio effect.

A review paper (Fabjanczyk et al., 1998) of current procedures regarding testing methods used for the assessment of reinforcing tendons concluded that there was no consensus within the mining industry on test procedures, or which tests themselves should be used in the assessment of anchorage. Push tests using grouted cylinders and field pull-out tests were widely used. However, it was believed that with lack of quality control and understanding of some of the procedures, there was a possibility of misleading results being obtained.

Jalalifar (2006) carried out the laboratory tests under both push and pull test conditions. Three kinds of 21.7-mm core diameter bolt, widely used in the Australian mining industry, were tested with an embedded length of 75 mm and confined in steel cylinder. Results showed that the average peak shear stress of the bolts (known as bolts T1, T2, and T3) in push tests was 12.5%, 5.7%, and 7.5% greater than pull-out tests, but the displacement at peak load was greater in the pull test.

To address the limited length of the bolts encapsulated in 75-mm steel sleeve, an additional comparative study was undertaken by Aziz et al. (2006) using 150-mm encapsulation length, with the tests being carried out under both push and pull conditions. Results showed that the average shear stress of the T1, T2, and T3 bolt in a push test was 8.2%, 11.3%, and 11.1% greater than the pull-out tests.

These early experimental studies identified the role of the Poisson's ratio effect in both laboratory pull and push tests; however, there is no analytical work being undertaken to formulate the load transfer capacity of the bolt with respect to the Poisson's ratio effect. Accordingly, this aspect of the analytical evaluation is the subject of discussion in this chapter.

The Poisson's ratio effect in pull and push tests is formulated based on failure mode analysis. Two major failure modes were identified in laboratory push and pull tests, namely parallel shear failure of the resin and dilational slip of the bond. Parallel shear failure is characterized by a cylinder failure surface, which is parallel to the bolt surface. It occurs for smooth bolts or closely spaced rebar bolt. Dilational slip failure is characterized by a conical failure surface. It is often the dominant failure mode for rebar bolts with intermediate to large bolt rib profile space. In the tests, the bond failure is often a combination or alternation of these two major failure modes. The percentage difference of push and pull load has been calculated at bond failure and after bond failure. Results from developed formulas were compared with experimental results, and showed good agreement.

5.2 Poisson's Ratio Effect in Parallel Shear Failure

To investigate the Poisson's ratio effect in push/pull tests, only short encapsulated rockbolts are considered. Mohr–Coulomb's material properties are used to describe the mechanical properties at the failure surface. In addition, the steel bolt is assumed as elastic deformation and the radial stiffness of the confinement is linear.

In push/pull tests, the axial force at the loaded end, F_0, can be expressed as total shear resistance at the interface, that is:

$$F_0 = 2\pi r \int \tau dx \quad (2.59)$$

where r is bolt radius; τ is shear resistance along the bolt. According to material properties, the shear resistance can be expressed as:

$$\tau = c + \sigma_r \tan \phi \quad (2.60)$$

where σ_r is normal stress (radial stress) of the failure surface; ϕ is internal friction angle of the bond; c is bond cohesion.

In laboratory tests, the confining material, such as steel sleeve is normally in an elastic deformation state. Therefore, the normal stress, or confinement pressure of the bolt, can be expressed as:

$$\sigma_r = K u_r \quad (2.61)$$

where K is radial stiffness of confinement; u_r is dilation of the bond.

The radial dilation, u_r, has two components. One of them is caused by the Poisson's ratio effect, that is, it is generated by the expansion/contraction of the bolt cross-section. The other component is caused by other factors, such as irregularity of the slip surface and/or increasing volume due to shearing. It can be thought of as the expansion of the interface rather than the bolt. Therefore, the total radial dilation can be expressed as:

$$u_r = u_{ri} \pm u_{rb} \quad (2.62)$$

where u_{rb} is dilation caused by the Poisson's ratio effect of the steel bolt; u_{ri} is dilation caused by the factors other than the Poisson's ratio effect.

In the expression, the plus sign represents the push test condition and minus sign represents the pull test condition. The two dilational components are assumed independent, that is, the dilation caused by interface deformation is the same magnitude in both pull or push tests.

The Poisson's ratio effect of the steel bolt is described by:

$$\varepsilon_r = v \cdot \varepsilon_a \qquad (2.63)$$

where ε_r is radial strain; ε_a is axial strain; v is Poisson's ratio.

According to the strain definition:

$$\varepsilon_r = u_{rb}/r \text{ and } \varepsilon_a = \Delta L/L = \sigma_a/E \qquad (2.64)$$

where E is Young's modulus of the bolt; σ_a is axial stress of the bolt; ΔL is bolt elongation; L is bolt length.

Putting Eq. (2.64) into Eq. (2.63) and rearranging:

$$u_{rb} = r \cdot v \cdot \sigma_a/E \qquad (2.65)$$

Combining Eqs. (2.59)–(2.62) and (2.65) and rearranging, the axial load at bond failure can be expressed as:

$$F_0 = 2\pi r \left(cL + K \tan \phi \int u_{ri} dx \right.$$
$$\left. \pm \frac{rvK \tan \phi}{E} \frac{h}{\pi r^2} F_0 \right) \qquad (2.66)$$

The minus sign indicates pull test and plus sign indicates push test, hence the axial load in push/pull tests can be solved. A variable h is introduced and is defined by Eq. (2.9). Variable h can be thought of as an equivalent length of fully loaded bolt to produce the same Poisson's ratio effect in the tests:

$$h = \frac{\int \sigma_a dx}{\sigma_{a0}} \qquad (2.67)$$

where σ_{a0} is axial stress of the bolt at the loaded end.

Eq. (2.66) expresses the axial load when shear failure occurs by taking into account the Poisson's ratio effect. If $v = 0$, it reduces to the standard form of the Mohr–Coulomb's failure criterion of the interface.

To investigate the Poisson's ratio effect of the bolt when subjected to axial loading, the percentage difference of the axial load in push and pull tests can be defined as:

$$\Delta\% = \frac{F_{push} - F_{pull}}{F_{pull}} \times 100\% \qquad (2.68)$$

Substituting two expressions solved from Eq. (2.66) into Eq. (2.68) and simplifying, we find:

$$\Delta\% = \frac{4hvK \tan \phi}{E - 2vKh \tan \phi} \times 100\% \qquad (2.69)$$

It shows that the Poisson's ratio effect is related to the mechanical properties of the steel bolt, internal friction angle of the failure surface, radial stiffness of the confinement, and axial load profile of the bolt. The mechanical properties of the bolt can be readily obtained. For smooth bolt, the failure surface is the resin–bolt interface, and the fractional angle can be estimated as 22 degree. For closely spaced rebar bolt, the failure surface includes two segments: one is the resin–bolt contact surface and the other is the resin shear failure surface. A typical value of resin internal friction angle is around 33 degree; hence, the internal friction angle in Eq. (2.69) can be estimated according to the ratio of crest width and spacing in one pitch. The radial stiffness, K, can be calculated using thick-walled cylinder theory, which is widely accepted by rockbolting theoretical research studies. Calculation of the results of some assemblage examples are listed in Table 2.7. The 4.2 mm resin annulus confinement can be used as the lower limit of the radial stiffness of resin grout rockbolting. If the confining material is 8 mm thickness steel tube, the radial stiffness will reach 3.4 GPa/mm. For an infinite medium, the radial stiffness can be found via:

$$K = \frac{E_r}{a(1 + v_r)} \qquad (2.70)$$

TABLE 2.7 Radial Stiffness of Commonly Used Experimental Assemblages

Confinement	Redial Stiffness K (GPa/mm)
PVC[a]	0.072
4.2 mm resin annulus only	0.2
Aluminum[a]	0.79
Steel 1[a]	0.70
Steel 2[a]	0.95
300 × 300 mm 40 MPa concrete block	1.10
Steel 3[a]	1.12
Infinite rock mass ($E = 30$ GPa, $v = 0.25$)	2.0
8 mm steel sleeve	3.4

[a] By Yazici and Kaiser, 1992.

where a is hole radius; E_r, v_r are Young's modulus and Poisson's ratio of the rock.

Once the axial stress profile of the loaded bolt is acknowledged, the Poisson's ratio effect in push/pull tests can be estimated. The axial stress profile in push or pull tests can be found using developed mechanical models in rockbolting mechanisms. The most commonly used models are the linear decay model and the exponential decay model.

5.2.1 Linear Decay Model

In the linear decay model, the shear resistance along the bolt is simply considered to be constant. This is a simple model, but pull-out test results of 12-mm diameter bolt from basalt rock suggested a linear relationship of axial load with embedded length up to 320 mm (Kilic et al., 2002). However, if the confining material is stiff, such as steel tube, the peak load is nonlinear to the embedded length at very short embedment.

As the axial stress is linearly increased along the bolt, hence:

$$h = \frac{\int \sigma_a dx}{\sigma_{a0}} = \frac{L}{2} \quad (2.71)$$

Combining Eqs. (2.69) and (2.71) yields

$$\Delta\% = \frac{2LvK \tan \phi}{E - vKL \tan \phi} \times 100\% \quad (2.72)$$

Fig. 2.39 shows a plot of Eq. (2.72). It suggests that, besides the embedded length, the radial stiffness of the confinement is also a major factor influencing the Poisson's ratio effect in laboratory push/pull tests.

The analytical result indicates that the axial load of rockbolting failure in push tests will increase by 50% compared with pull-out test results in the case of a 22 mm diameter steel bolt embedded in 100 mm steel tube of 8 mm wall thickness. This value is supposed to be overestimating the realistic value, because the shear resistance will no longer be constant in high confinement stiffness. The exponential decay model is more suitable to describe the axial stress of the bolt in high confinement stiffness. In addition, the embedded length in laboratory tests is normally less that 150 mm when the confinement is stiff, otherwise the steel bolts may yield.

5.2.2 Exponential Decay Model

Farmer (1975) developed the exponential decay model of axial stress of rockbolting subjected to axial loading with an elastic anchorage. This model has been widely accepted and further developed by later research (Li and Stillborg, 1999; Ren et al., 2010). In the model, after some approximation, the axial stress profile of the bolt is described by:

$$\sigma_a = \sigma_{a0} \cdot e^{-\alpha \cdot x} \quad (2.73)$$

where $\alpha^2 = 2G/(Ert)$; G is shear modulus of grout material; t is thickness of the resin annulus. Accordingly, Eq. (2.71) can be calculated as:

$$h = \frac{\int \sigma_a dx}{\sigma_{a0}} = \frac{1}{\alpha}\left(1 - e^{-\alpha L}\right) \quad (2.74)$$

Combining Eqs. (2.69) and (2.74), the percentage difference of axial loads in push and pull testing when bond failure occurs is evaluated as:

$$\Delta\% = \frac{4vK(1 - e^{-\alpha L})\tan \phi}{\alpha E - 2vK(1 - e^{-\alpha L})\tan \phi} \times 100\% \quad (2.75)$$

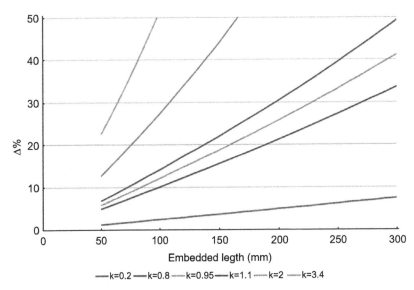

FIGURE 2.39 The percentage difference of the push/pull loads versus embedded length at different radial stiffness (GPa/mm) using linear decay model, in which $E = 200$ GPa, $v = 0.3$ for the steel bolt, and $\tan \phi = 0.40$ for the failure surface.

In the equation, the embedded bolt length is no longer an important parameter, and bond failure is more closely related to the mechanical properties of the confining material. Eq. (2.75) is plotted in Fig. 2.40 using common resin–steel bolt mechanical properties.

Eqs. (2.72) and (2.75) are conclusions drawn from different mechanical models. Eq. (2.72) is

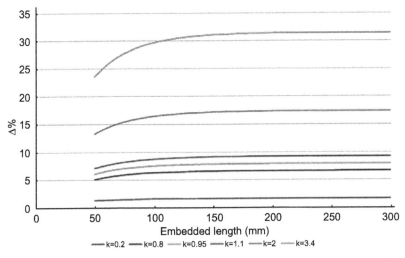

FIGURE 2.40 The percentage difference of the push/pull loads versus embedded length at different radial stiffness (GPa/mm) using exponential decay model, $E = 200$ GPa, $v = 0.3$, $\tan \phi = 0.4$, and $\alpha = 0.03$ mm^{-1}.

suitable for lower confining material and short embedded length to ensure a uniform shear stress along the bolt. Eq. (2.75) can be used for high confinement stiffness with relatively longer embedment, such as steel sleeve confinement or 200$^+$ mm bolt-resin grouted in hard rock. In the intermediate conditions, these two figures can be used in conjunction with each other to form the upper and lower limits of the estimated values.

5.2.3 Poisson's Ratio Effect at Peak Load

To investigate the Poisson's ratio effect at peak load in push/pull tests using Eq. (2.66), it is necessary to construct the axial stress profile of the bolt along the embedded length. Based on the exponential decay theory and considering decoupling of the bond, Li and Stillborg (1999) constructed a model for the shear stress along a fully grouted bolt subjected to axial loading. Fig. 2.41 show the shear stress distribution along a fully grouted rockbolt subjected to an axial load after decoupling occurs. The shear stress profile includes a completely decoupled section with a zero shear stress from the collar to the point x_0, a segment partially decoupled with a residual shear strength s_r till x_1, followed by the residual shear strength linearly increasing to the peak strength s_p from x_1 to x_2, and then exponentially decaying toward the unloaded end of the bolt.

Consequently, the peak load can be expressed as:

$$P_{0max} = \pi d_b s_p \left[\omega \left(L + \frac{d_b}{2\alpha} \ln \omega - m - x_0 \right) + \frac{1}{2} m(1+\omega) + \frac{d_b}{2\alpha}(1-\omega) \right]$$

(2.76)

where $m = x_2 - x_1$; $\omega = s_r/s_p$; d_b = bolt diameter.

$$x_2 = x_0 + \frac{1}{2\omega} \left[\frac{2P_0}{\pi d_b s_p} - \frac{d_b}{\alpha} - m(1+\omega) \right]$$

$$P_0 = \pi d_b \left[s_r(x_1 - x_0) + \frac{1}{2} s_p m(1+\omega) + \frac{d_b}{2\alpha} s_p \left(1 - e^{-\frac{2\alpha}{d_b}(L-x_2)} \right) \right]$$

$$\alpha^2 = \frac{2 G_r G_g}{E_b \left[G_r \ln \frac{d_g}{d_b} + G_g \ln \frac{d_0}{d_g} \right]}$$

Ren et al. (2010) developed a solution for the prediction of the full-range mechanical behavior of fully grouted rockbolts subjected to axial loading. In this approach, a trilinear bond-slip model was used to describe the stress—strain behavior of the failure interface, as shown in Fig. 2.42. The shear stress—strain relationship includes an increased component up to the peak stress at (δ_1, τ_f) followed by a softening branch down to (δ_f, τ_r), and then a horizontal branch representing the nonzero residual frictional strength τ_r after complete debonding. Consequently, Ren et al. defined the full-range behavior of the pull-out profile by five stages. For each stage, closed-form solutions for the bolt axial stress distribution along the bond length were derived.

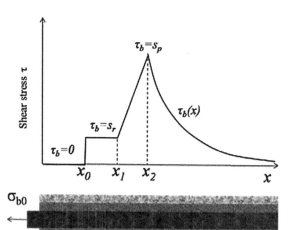

FIGURE 2.41 Shear stress along a fully grouted rockbolt subjected to an axial load (Li and Stillborg, 1999).

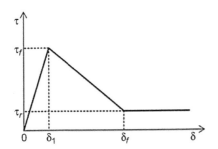

FIGURE 2.42 Trilinear bond-slip model (Ren et al., 2010).

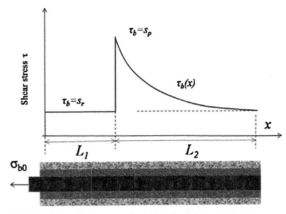

FIGURE 2.43 Simplified shear stress profile used to study the Poisson's ratio effect in push/pull tests.

The ultimate load which is characterized by $\tau(L) = \tau_r$ can be expressed as:

$$P_{ult} = \frac{2\pi r_b \tau_f}{\lambda_2 \sqrt{1-k}} \left[\frac{\lambda_2 \sqrt{1-k}}{\lambda_1} \right. \\ \times \sqrt{\cos\left(a_d \lambda_2 \sqrt{1-k}\right)\left[\cos\left(a_d \lambda_2 \sqrt{1-k}\right) - k\right]} \\ \left. + \sin\left(a_a \lambda_2 \sqrt{1-k}\right) \right] + 2k\pi r_h \tau_f d_u$$

(2.77)

To obtain accurate results, the axial stress profile and peak load of these two models can be used in Eqs. (2.66) and (2.67). However, as an estimation of the Poisson's ratio effect in push/pull tests, a simplified shear stress-shear displacement model is adopted by eliminating the complete debonding and linear increasing segments from Li and Stillborg's model, as shown in Fig. 2.43.

After taking some estimation, the equivalent length can be estimated as:

$$h \approx \frac{L_1}{2} + \frac{1}{\alpha}\left(1 - e^{-\alpha L_2}\right) \quad (2.78)$$

where L_1 is bolt length with residual shear strength; L_2 is bolt length shear strength exponential decaying.

The Eq. (2.78) is in between of Eqs. (2.72) and (2.75). Hence, Figs. 2.40 and 2.41 can be used as the upper and lower boundaries of the Poisson's ratio effect at the peak load, or more generally, postbond failure.

5.3 Poisson's Ratio Effect in Dilational Slip Failure

Conclusions drawn from the previous section are based on the parallel shear failure mode of rockbolting. Another kind of major failure mode identified in laboratory pull-out tests is dilational slip of the bond. The mechanisms in these two failure modes are different. Hence, the influence of the Poisson's ratio effect is expected to be different.

In the theoretical model of dilational slip failure of rockbolting, the deformation at the slip surface is simulated by an increased friction angle. The shear strength of the bond can be described by:

$$\tau = \sigma_r \tan(\phi + i) \quad (2.79)$$

where Dilation angle i is $\tan^{-1}(u_r/u_a)$; u_a is axial displacement.

In theoretical models of rockbolting mechanisms, the dilational angle, i, is often expressed as a reduction of the geometric angle to form an "apparent dilation angle" to accommodate data obtained from the pull-out tests. From the viewpoint of failure mode analysis, the reasons of the reduction of the dilation angle can be explained as the dilational slip failure surface of rockbolting is normally different to the bolt-resin contact surface. This is because the resin shear failure, which is evidenced by the lodged resin in front of the bolt profiles, commonly occurs when the bolt is subjected to axial load. As a result, the dilational slip face angle is smaller than the bolt rib face angle. In addition, the radial dilation can also be depressed if the radial confining pressure is high. For example, Yazici and Kaiser (1992) used the following equation to describe the dilation in the bond strength model:

$$u_r = u_{ult}\left(1 - \frac{\sigma_r}{\sigma_c}\right)^{B/\sigma_c}$$

where u_{ult} is ultimate dilation which is related to the bolt surface geometry. σ_c is compressive strength of grout material. B is introduced reduction parameter.

To investigate the Poisson's ratio effect in push/pull tests using dilational slip model, a small dilation angle δ is introduced to describe dilation caused by the bolt cross-section expansion/contraction, and it is defined as:

$$\tan \delta = u_{rb}/u_a \quad (2.80)$$

Consequently, the shear strength of the bond can be expressed as:

$$\tau = \sigma_r \tan(\phi + i \pm \delta) \quad (2.81)$$

The plus sign indicates push testing and minus sign indicates pull testing. The axial displacement, u_a, can be estimated as the bolt elongation ΔL. Experimental results suggest that this estimation is accurate up to 80% of the peak load (Benmokrane and Chennouf, 1995). Accordingly, the dilation caused by the Poisson's ratio effect can be expressed as the following equation by substituting Eqs. (2.63) and (2.64) into Eq. (2.80):

$$\tan \delta = vr/L \quad (2.82)$$

Thus, the percentage difference of axial load in push/pull tests can be expressed as:

$$\Delta\% \approx \frac{2\tan^2(\phi+i)+2}{\frac{L}{vr}\tan(\phi+i)-\tan^2(\phi+i)-1} \times 100\% \quad (2.83)$$

In the procedure, the influence of the Poisson's ratio effect to the confining pressure is assumed to be small, as the deformation of surrounding materials have been lumped to the dilational angles at the interface in this model.

A sensitivity study on the dilation angle i was conducted (not shown here) and the results suggested that $\Delta\%$ in the Eq. (2.83) were not affected too much by the interfacial dilation angle, hence a constant value of $\phi + i \equiv 45$ degree can be used to evaluate the Poisson's ratio effect in push/pull tests in cases of dilational slip failure of the bond, as shown in Fig. 2.44.

5.4 Comparison of Analytical Results With Experimental Data

Only a few researchers (Jalalifar, 2006; Aziz et al., 2006) that conducted push and pull tests using the same experimental assemblage provided experimental data for further analytical study. To examine the analytical solutions drawn from this chapter, a series of push and pull-out tests were conducted. The confinement was in 8 mm thick steel tube and the embedded lengths were 75, 100, and 150 mm. For accuracy, the same steel bolt was used for each pair of push/pull tests. In addition, both smooth bolt and rebar bolts were used in each embedded length. A typical load-displacement profile is shown in Fig. 2.45, and all experimental results are listed in Table 2.8.

The theoretical prediction is compared with the experimental data obtained from Jalalifar (2006), Aziz et al. (2006), and in this study, as shown in Fig. 2.46. As the confinement is stiff (8-mm steel tube), Eq. (2.75) which is derived using the exponential decay model is used as the upper limit of the percentage difference of axial load in push and pull tests. The lower boundary is generated from the dilational slip model. Fig. 2.46 shows that the theoretical prediction is in agreement with experimental results as most of the experimental data fall into the estimated range.

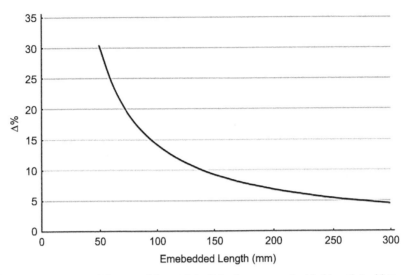

FIGURE 2.44 The percentage difference of the push/pull loads versus embedded length in dilational slip failure.

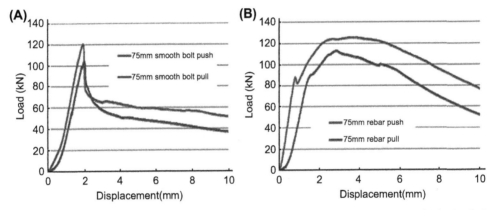

FIGURE 2.45 Push/pull tests profile for 75-mm embedded length of (A) smooth bolt and (B) rebar bolt.

5.5 Conclusions

In this chapter, an analytical approach was conducted to investigate the Poisson's ratio effect in laboratory short encapsulated push/pull tests. The methodology employed is to distinguish the Poisson's ratio effect caused dilation (expansion/contraction of the steel bolt cross-section) from the resin–bolt interface dilation. Two major failure modes of rockbolts, namely parallel shear failure and dilational slip failure, were studied separately, and accordingly equations were developed based on the ISS model and the bond strength model in rockbolting mechanism.

The main conclusions drawn are Eqs. (2.42), (2.75), and (2.83). They can be used for preliminary estimation of the Poisson's effect in short encapsulated push/pull tests for different experimental assemblages. The analytical results suggest that the Poisson's ratio effect in laboratory push/pull tests depends on the mechanical properties of the bolt, properties of

5. POISSON'S RATIO EFFECT IN PUSH AND PULL TESTING

TABLE 2.8 Axial Load Difference in Push/Pull Testing

Bolts and Embedded Length	Average Δ % in Push/Pull Test (%)	
	At Failure	At Peak Load
75-mm rebar bolt T1 (Jalalifar, 2006)	15.6	12.5
75-mm rebar bolt T2 (Jalalifar, 2006)	16.4	5.7
75-mm rebar bolt T3 (Jalalifar, 2006)	25	7.5
75-mm smooth	20	20
75-mm rebar T2		18
100-mm smooth	21	21
100-mm rebar bolt T2	33	24
150-mm rebar bolt T1 (Aziz et al., 2006)	8.2	8.2
150-mm rebar bolt T2 (Aziz et al., 2006)	26	11.3
150-mm rebar bolt T3 (Aziz et al., 2006)	29.2	11.1
150-mm smooth	17	17
150-mm rebar bolt T2	25	15

the failure surface, and the radial stiffness of the confinement.

The main assumptions made in this chapter include elastic deformation of the bolt, linear radial stiffness of the confinement, and identical interfacial deformation in push and pull tests. The first assumption is applicable because the bolt yield is a minor failure mode of rockbolt failure in practice. Consequently, laboratory push/pull tests are often designed to study the bond behavior under bolt yielding load. The last assumption is also reasonable as the deformation of a real bolt is modeled as the sum of the deformation caused by an imaginary bolt whose Poisson's ratio equals to zero and the deformation caused by changing the bolt cross-section area.

For the grout annulus, the longitudinal fractures do not affect the linearity of the radial stiffness because of wedge mechanism. However, hole drilling—caused rock fracture and/or poor resin encapsulation will cause nonlinear behavior in the radial direction, and hence minimize the Poisson's ratio effect of the bolt even at an early stage of axial deformation. In addition, the Poisson's ratio effect in laboratory push/pull tests will be expected to reduce greatly once the radial stiffness enters into the

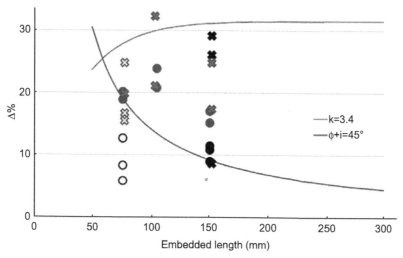

FIGURE 2.46 Comparison of theoretical calculation with experimental data, *cross* = at failure; *dot* = at peak; *yellow* (Jalalifar, 2006); *black* (Aziz et al., 2006); and *red* = this study.

elastoplastic deformation stage. Theoretically, Poisson's ratio effects of the bolt will no longer influence the axial behavior if the confining material enters into a perfect plastic deformation stage in the radial direction.

6. STUDY ON ROCKBOLTING FAILURE MODES

6.1 Introduction

Application of rockbolting as a ground support system has grown rapidly during the past four decades due to a better understanding of load transfer mechanisms and advances made in the bolt system technology. Bolts are used as permanent and temporary support systems in tunneling and mining operations. In surface mining they are used to stabilize slopes and in underground workings they are used for roadway development, shaft sinking, and stoping operations. Rockbolts are installed around openings in mines and tunnels to tie weaker layers to stronger layers above, to prevent sagging and separation and to provide a reinforcement zone in a rock mass that makes greater use of a rocks inherent mass strength to enable it to be self-supporting.

A reinforcement scheme is an arrangement of primary, secondary, and tertiary reinforcement systems in a variety of dimensional and spatial configurations. Some of these may have been installed as pre- or postreinforcement, and may be untensioned, pretensioned, or posttensioned. Winsdor (1997) indicated that three classes of device have evolved: rockbolt (generally less than 3 m), cable bolt (generally in the range from 3 to 15 m), and ground anchor (generally longer than 10 m). All of them comprise four principal components as shown in Fig. 2.47:

While the rock is not generally thought of as being a component of the reinforcement system, it has a marked influence on behavior and must therefore eventually be considered an integral part of the system. For reinforcement with a bolt, the reinforcing element refers to the bolt and the external fixture refers to the face plate and nut. The internal fixture is a medium, such as cementitious grout or resin for bolt encapsulation, or a mechanical action like friction at the bolt interface for frictionally coupled bolts. The internal fixture provides a coupling condition at the interface.

With reference to the component of internal fixture, the reinforcement system has been cataloged into three fundamental types (Windsor and Thompson, 1993):

1. Continuous Mechanically Coupled (CMC) systems,
2. Continuous Frictionally Coupled (CFC) systems and
3. Discreetly Mechanical or Frictionally Coupled (DMFC) system.

According to this classification system, cement and resin grouted bolts belong to the CMC system while split set and Swellex bolts belong to the CFC system. DMFC can be anchored by a slit and wedge mechanism or an expansion shell.

Nowadays, grouted rock anchors have been used extensively in a wide range of geotechnical

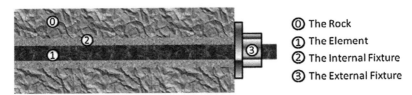

FIGURE 2.47 Four principal components of a reinforcement system (Winsdor, 1997).

and mining applications as temporary or permanent ground supports. A grouted anchor is defined as a structural support comprising a tendon which is inserted into a drilled hole and then grouted. Grouted steel anchors can be made from solid bar or stranded wire cables. Solid bars can be either smooth or deformed with the latter being further classified as either "threadbar" or "rebar." However, the effectiveness of the bolting system requires an in-depth study with closer evaluation of the bolting anchorage system, other than the experimental studies, such as pull-out testing. Load transfer mechanism between rock/resin/medium, modes of the bolt failures, and distribution of the forces generated as a result of the strata deformation are the subject of analytical study discussed in this chapter.

The load transfer concept is central to the understanding of reinforcement system behavior, and the mechanical action of the different devices and their effects on excavation stability. This concept can be visualized as being composed of three basic mechanisms (Stille et al., 1989):

1. Rock movement and load transfer from an unstable zone to the reinforcing element.
2. Transfer of load from the unstable region to a stable interior region via element.
3. Transfer of the reinforcing element load to the stable rock mass.

A fully grouted bolt is a passive roof support system, which is activated by movement of the surrounding rock. The relationships between them belong to the CMC systems. The efficiency of load transfer is affected by the type and properties of the grout, profile of the rockbolt, hole and bolt diameter, anchorage length, rock material, confinement pressure, over and under spinning, and installation procedures. Stress concentration is induced between the roughness of hole wall and the surface profile of the bolt. This localized stress concentration could exceed the strength of the grout and rock resulting in localized crushing that allows additional deformation in the steel.

It is commonly accepted that the fully grouted bolt provides greater shear surface for transmitting the load from rock to bolt and vice versa. In general, only resinous grouts can meet the high strength required for short anchorages. A grouted bolt can transfer greater loads than expanded shell or wedge type anchorages. This may be essential in weaker rock strata where transfer of high loads over a short length borehole may initiate failure at the rock interface.

The grout supplies a mechanism for transferring the load between the rock and reinforcing element. This redistribution of forces along the bolt is the result of movement in the rock mass, which transfers the load to the bolt via shear resistance in the grout. This resistance could be the result of adhesion and/or mechanical interlocking. Adhesion is the actual bonding between grout, steel, and rock, and the mechanical interlocking is a keying effect created when grout fills irregularities between the bolt and the rock.

Singer (1990) demonstrated that there is no adhesion between the grout–bolt and grout–rock interface. Aziz et al. (2003) reported almost no adhesion between the bolt surface and grout only, and Yazici and Kaiser (1992) stressed that the adhesive component was neglected because it cannot be mobilized with frictional strength during the pull-out test.

6.2 Failure Mode of a Two-Phase Material System

By pulling out a steel bar embedded in a concrete column (note that there is no grout material), engineers have been aware that the bond forces radiate out into the surrounding media from the bonding surface of an anchored steel bar. Studies of bonding forces for plain reinforcing bars and rebar show that bond for plain bars is made up of three components (Lutz, 1970): Chemical adhesion, Friction, and Mechanical interaction between concrete and steel.

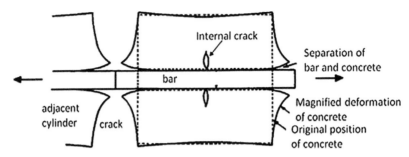

FIGURE 2.48 Deformations of a concrete cylinder with pulled axially embedded plain reinforcing bar (Tepfers, 1979).

Bond strength mainly depends on chemical adhesion and after slip, on friction. There is also some interlocking due to the roughness of the bar surface. The effect of chemical adhesion is small and friction does not occur until there is slip between bar and concrete.

For rebar, slip can occur in two ways: (1) the ribs can split the concrete by wedging action, and (2) the ribs can crush the concrete. When concrete is crushed to a "compacted powder" it becomes lodged in front of the ribs. In addition, even when slip and separation occur, additional transverse cracks and splitting cracks are very probable. Thus, large axial displacement cannot occur without transverse and longitudinal cracking in the surrounding concrete.

Several types of cracks in a concrete cylinder with an axially embedded steel bar were outlined to identify the failure modes of reinforcement (Tepfers, 1979). The breaking of a concrete beam into small columns is called primary cracking, which is the major failure mode. In addition, bond slip and dense minor radial cracks are also presented, as shown in Fig. 2.48.

Tepfers (1979) established an analytical model for the tensile stress distribution causing development of the radial splitting cracks. When pulling out, the interface will result in significant stress concentrations. Due to these accumulated stresses, the debonding process will start and extend inside the specimen along the reinforcing bar. Two types of cracks will occur: cone-shaped cracks and longitudinal splitting cracks, both of which start at the interface as shown in Fig. 2.49. The crack patterns depend on the interface geometry and the properties of the interface and the surrounding concrete; furthermore, these different crack patterns may not form independently from one another, but interact through complicated nonlinear mechanisms.

It is thought that tensile stress is the cause of the splitting cracks. Tepfers (1979) assumed that the radial components of the bond forces can be regarded as hydraulic pressure acting

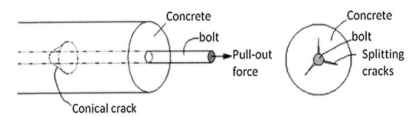

FIGURE 2.49 Internal cone-shaped crack and longitudinal splitting cracks (Tepfers, 1979).

on a thick-walled concrete ring surrounding the reinforcing bar. The shear stress at the interface distributes into the surrounding material by compression under a certain angle and is balanced by tensile stress rings in the concrete, as shown in Fig. 2.50. Accordingly, the radial stress due to bond action on the concrete, which is also regarded as the hydraulic pressure against a thick-walled concrete cylinder, can be calculated out via shear stress of the interface $\sigma_y = \tau \tan \alpha$.

For determination of the resistance against radial cracking, three different stress distributions are applied, referred to as uncracked elastic, partly cracked, and uncracked plastic stage. The cross-section of the deformed concrete beam and terminology in Tepfers' mechanical model are shown in Fig. 2.51.

In the elastic stage (Fig. 2.51B), using thick-walled cylinder theory, the tangential stress can be found:

$$\sigma_t = \frac{\left(\frac{\phi}{2}\right)^2 \cdot \sigma_y}{\left(c_y + \frac{\phi}{2}\right)^2 - \left(\frac{\phi}{2}\right)^2}\left[1 + \frac{\left(c_y + \frac{\phi}{2}\right)^2}{r^2}\right] \quad (2.84)$$

When the bond reaches the plastic stage, the cylinder will not break until the stresses in the tangential direction at every part of the cylinder have reached the ultimate tensile strength. The tangential stress in the cylinder can be expressed as $\sigma_t = \phi \cdot \sigma_y / 2c_y$ (Fig. 2.51D). In the intermediate stage of these two cases, the ring has internal partial cracks where the circumferential stresses have reached the ultimate tensile strength of concrete. The bond force is now transferred through

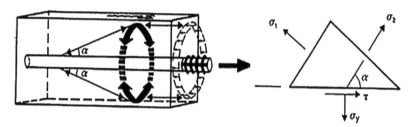

FIGURE 2.50 Radial components of the bond forces are balanced against tensile stress rings and bond stresses in the concrete adjacent to the bar (Tepfers, 1979).

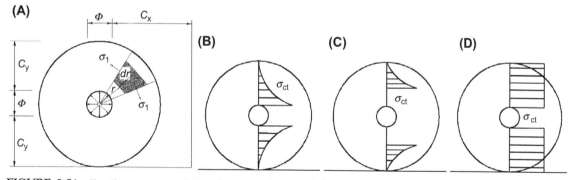

FIGURE 2.51 Tensile zone in a reinforced concrete beam and the variations in tensile stress in different conditions. (A) Tensile zone; (B) elastic deformation; (C) partial crack; (D) completely plastic.

the concrete teeth between the internal cracks to the uncracked part of the ring. Tepfers (1979) found that when the radius of cracked zone $r_e = 0.486(c_y + \phi/2)$ the bond force capacity of the concrete ring reaches its maximum value (Fig. 2.51C).

This analysis is based on the specific bond failure mode, ie, cone cracks and radial cracks. In the rockbolting system, however, it is not always the case due to grouting material which dominates the failure of the bond. The thick-wall theory and associated methodology of elastic and plastic analysis used in Tepfers (1979) paper are admirable and employed in later research work on the fully grouted rockbolt composed of three phase material with two interfaces.

6.3 Failure Modes of Cable Bolting System

Yazici and Kaiser (1992) developed a conceptual model for fully grouted cable bolts (Fig. 2.52) called the bond strength model (BSM).

According to the theory, the bond strength is mainly frictional and hence depends on the pressure build-up at the interface which in turn depends on the dilational movement against the confining grout or rock. The cable bolt surface was simplified to be zigzag (twisting of the cable is ignored), thus a bilinear dilation-dependent joint strength concept can be applied as $\tau = \sigma \tan(i_0 + \phi)$. For small angle, the bond strength can be expressed in terms of friction and dilation angle:

$$\tau = \sigma \tan\left(i_0\left[1 - \left(\frac{\sigma}{\sigma_c}\right)^\beta\right] + \phi\right) \quad (2.85)$$

where τ is shear or bond stress; σ is radial stress at the bolt—grout interface; i_0 is dilation angle at the bolt—grout interface given by surface geometry; β is reduction coefficient of dilation angle; σ_c is compressive strength of grout; ϕ is friction angle between the steel and grout.

The BSM involves four main components: axial displacement, lateral displacement, confining pressure, and bond strength. In

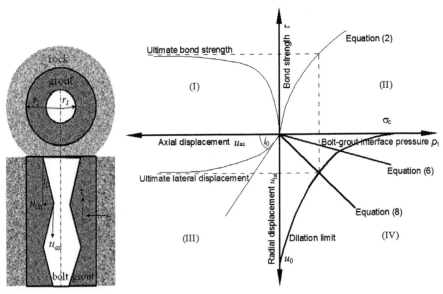

FIGURE 2.52 Schematic diagram reflecting the geometry of a rough cable bolt and relationship of each component in BSM. *After Yazici, S., Kaiser, P.K., 1992. Bond strength of grouted cable bolts. International Journal of Rock Mechanics and Mining Sciences and Geomechanics Abstracts 29 (3), 279—292.*

Fig. 2.52, the schematic diagram illustrates these interrelated components in four quadrants:

1. The first quadrant shows the variation of bond strength with axial displacement. It represents the pull-out test graph.
2. The second quadrant relates to the confining pressure at the bolt–grout interface to the bond strength using Eq. (2.85).
3. The third quadrant shows the relation between axial and lateral displacements. Since apparent dilation angle is not constant, the relation is nonlinear and asymptotically approaches an ultimate lateral displacement.
4. The dilation acts outward on the grout column and creates the interface pressure as illustrated by the fourth quadrant. The straight lines showed that the grout may split under the dilational pressure.

In the fourth quadrant of the BSM, the dilatational behavior of grout is: (1) elastic; (2) fully split; or (3) a transition zone of partially split with an elastic portion. In the elastic grout expansion (Fig. 2.53A), the radial displacement at the bolt–grout interface can be derived from the plane strain thick-walled cylinder equations:

$$u_{1g} = \frac{(1+\nu_g)(1-2\nu_g)}{E_g} \frac{p_1 r_1^2 - p_2 r_2^2}{r_2^2 - r_1^2} r_1$$
$$+ \frac{(1+\nu_g)}{E_g} \frac{(p_1 - p_2) r_1^2 r_2^2}{r_2^2 - r_1^2} \frac{1}{r_1} \quad (2.86)$$

$$u_{2g} = \frac{(1+\nu_g)(1-2\nu_g)}{E_g} \frac{p_1 r_1^2 - p_2 r_2^2}{r_2^2 - r_1^2} r_2$$
$$+ \frac{(1+\nu_g)}{E_g} \frac{(p_1 - p_2) r_1^2 r_2^2}{r_2^2 - r_1^2} \frac{1}{r_2} \quad (2.87)$$

The radial displacement of the rock, induced by an internal pressure in a circular hole of radius in an infinite medium, is given by:

$$u_{2g} = p_2 r_2 / [(1+\nu_r) E_r] \quad (2.88)$$

Combining these three equations, the displacement at the bolt–grout interface can be expressed in terms of the internal pressure in the form:

$$u_{1g} = \left[\frac{(1+\nu_g)(1-2\nu_g)}{E_g} \frac{r_1^2 - \frac{r_2^2}{X}}{r_2^2 - r_1^2} r_1 \right.$$
$$\left. + \frac{(1+\nu_g)}{E_g} \frac{(1-\frac{1}{X}) r_1^2 r_2^2}{r_2^2 - r_1^2} \frac{1}{r_1} \right] p_1 \quad (2.89)$$

where $X = \left\{ \frac{1+\nu_r}{E_r} + \frac{1+\nu_g}{E_g(r_2^2 - r_1^2)} \left[(1-2\nu_g) r_2^2 + r_1^2 \right] \right\} \Big/ \left\{ \frac{1+\nu_g}{E_g(r_2^2 - r_1^2)} \left[(1-2\nu_g) r_1^2 + r_1^2 \right] \right\}$

If the tangential stress exceeds the tensile strength of the grout, the grout will fully split and the tangential stress in the grout column becomes zero. This changes the thick-walled grout cylinder to a wedge-shaped geometry. The new

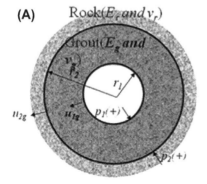
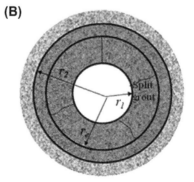

FIGURE 2.53 Conceptual cross-section through (A) a grouted bolt and (B) partially split grout column.

state of stress can be found according to the wedge theory as $p_1 r_1 = p_2 r_2$. Consequently, the difference of the displacements between the boundaries of the split grout column can be calculated:

$$u_{1g} - u_{2g} = \frac{1 - \nu_g^2}{E_g} p_1 r_1 \ln \frac{r_2}{r_1} \quad (2.90)$$

Substituting u_{2g} with Eq. (2.88), the displacement at the bolt–grout interface for the totally split grout cylinder is:

$$u_{1g} = r_1 \left(\frac{1 + \nu_r}{E_r} + \frac{1 - \nu_g^2}{E_g} \ln \frac{r_2}{r_1} \right) p_1 \quad (2.91)$$

In the transition zone of these two cases (Fig. 2.53B), the interface pressure is obtained via elastoplastic behavior:

$$p_1 = \sigma_T (r_2^2 - r_e^2) \bigg/ \left\{ \frac{r_1}{r_e} \left[r_e^2 + r_2^2 \left(1 - \frac{2}{X} \right) \right] \right\} \quad (2.92)$$

where r_e is radius of the cracked zone, as shown in Fig. 2.53-2; σ_T is tensile strength of the grout; X is obtained via Eq. (2.89) using $r_1 = r_e$.

Thus, the dilation for partially split grout is found by algebraically adding the displacements for split and intact grout:

$$u_{1g} \Bigg\{ \frac{r_1}{r_e} \left[\frac{(1 + \nu_g)(1 - 2\nu_g)}{E_g} \frac{\left(r_e^2 - \frac{r_e^2}{X}\right)}{r_2^2 - r_e^2} r_e \right.$$
$$\left. + \frac{(1 + \nu_g)}{E_g} \frac{(1 - \frac{1}{X}) r_e^2 r_2^2}{r_2^2 - r_e^2} \frac{1}{r_e} \right] + r_1 \left(\frac{1 + \nu_r}{E_r} \right.$$
$$\left. + \frac{1 - \nu_g^2}{E_g} \ln \frac{r_e}{r_1} \right) \Bigg\} p_1 \quad (2.93)$$

This equation is only applicable for $r_1 < r_e < r_2$, and u_{1g} is not a linear function of p_1 because the length of crack r_e is also a function of p_1. A closed-form solution could not be found and, hence, u_{1g} is determined iteratively starting from $r_e = r_1$.

To complete BSM, the dilation limit must be determined. While failure occurs, the area of the grout teeth in contact with the bolt decreases but the stress acting on an individual tooth increases leading eventually to complex modes of failure. An empirical model was chosen to describe it as:

$$u_{1g} = u_0 \left(1 - \frac{p_1}{\sigma_c} \right)^{B/\sigma_c} \quad (2.94)$$

where u_0 = maximum dilation ≈ teeth height; σ_c is compressive strength of grout, and B is a constant which can be determined from pull-out data.

Hyett et al. (1992, 1995, 1996) carried out a series of laboratory and field pull-out tests to investigate the major factors influencing the bond capacity of grouted cable bolts. All tests were conducted on 15.9 mm diameter seven-strand cable bolts grouted using type 10 Portland cement pastes. Their results indicate that cable bolt capacity most critically depends on the cement properties, embedment length, and radial confinement. They found that cable bolt capacity increased with embedment length although not in direct proportionality. Furthermore, higher capacities were obtained under conditions of higher radial confinement.

From pull-out tests, two failure modes have been observed. One mode involves radial splitting of the concrete cover surrounding the cable, and the other shearing of the cable against the concrete. The radial splitting mechanism is induced by the wedging action between the lugs of the bar and the concrete. This exerts an outward pressure on the inside of the concrete annulus that is balanced by the induced tensile circumferential stress within the annulus. However, if the tensile strength of the cement is exceeded, radial splitting will occur and the circumferential stress in the concrete annulus will be reduced to zero, as will the associated reaction force at the steel–concrete interface, resulting in failure. The shearing mechanism involves crushing of the concrete ahead of the ribs on the bar, eventually making pull out along a cylindrical frictional surface possible. Thus, it can be concluded that as the degree of radial confinement increased the failure mechanism changed from radial fracturing and lateral displacement of the grout annulus under low confinement, to shear of the cement flutes and pull out along a cylindrical frictional surface under high confinement.

The successive stages in the failure during a pull test were summarized schematically as shown in Fig. 2.54 (Hyett et al., 1992). In the essentially linear response (stage 1), as the experimental initial stiffness is significantly less than that predicted from elastic solutions, Hyett argued that the adhesional bond between the cable and the cement is negligible because (1) the cement paste is porous, and (2) the bond is not continuous but instead comprises a series of point contacts. Consequently, the mechanical interlock and frictional resistance is related with the initial linear response during a pull test, although partial adhesion probably involves additional components. From stage 2, the failure mechanism is dependent on the radial confining pressure. The stress drop may correspond to radial fracturing of the grout annulus

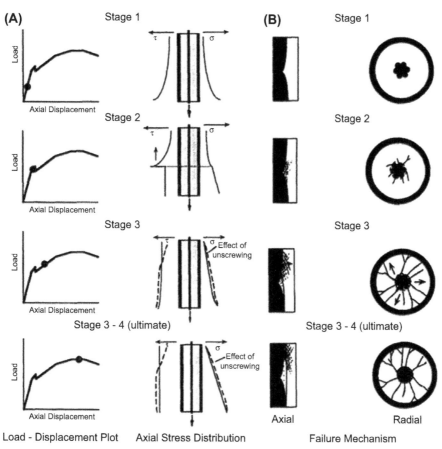

FIGURE 2.54 Successive stages in the failure during a pull test (Hyett et al., 1992).

and/or shear failure through the grout flutes. From then on, as cable displacement increases, the radial confining pressure is controlled by the potential for greater geometric mismatch between the cable and cement flutes. How far the individual wedges that now comprise the grout annulus can be pushed aside is determined by the radial stiffness of the confining medium. When the radial stiffness is low the favorable failure mechanism is lateral displacement of the wedges; when it is high, dilation is suppressed and failure is more likely to occur by shear of the grout flutes and pull-out along a cylindrical frictional surface.

To construct a mechanical model for bond failure of fully grouted cable bolts, another series of pull tests were conducted (Hyett et al., 1995), in which the confining pressure at the outside of the cement annulus was maintained constant using a modified Hoek cell. The data were used to develop a frictional-dilational model for cable bolt failure in a mathematical form which is amenable to implementation in numerical programs.

In this model, the bond strength is frictional, so it depends on the pressure generated at the cable–grout interface, p_1, which in turn depends on the reaction force generated at the borehole wall caused by dilation during bond failure. The frictional resistance can be cataloged into:

1. For dilational slip:

$$f = A \cdot p_1 \cdot \tan(\phi_{g-s} + i) \quad (2.95)$$

2. For nondilational unscrewing:

$$f = (A \cdot p_1 \cdot \tan \phi_{g-s})/\sin \alpha + Q \quad (2.96)$$

3. For shear failure of the cement flutes:

$$f = A(c + p_1 \tan \phi_g) \quad (2.97)$$

In which I is dilation angle; A is interface contact area; ϕ_{g-s} is sliding friction between grout and steel; ϕ_g is internal angle of friction for grout; c is grout cohesion; and α is cable twisting angle.

Micrographs reveal that shearing of the grout flutes only occurred within 75 mm of the exit point. The only viable explanation is that, along the majority of the test section, failure involves unscrewing of the cable from the cement annulus. To include the unscrewing effect, Q is introduced as the component of the pull-out force required to untwist the free length of cable. Based on work considerations, the formula is given by:

$$Q = 4\pi^2 \, C u_a / \left[l^2 (u_a + L_f) \right] \quad (2.98)$$

In which C = torsional rigidity of cable; l = pitch length; u_a = pull-out length; and L_f = free length of the cable between test and anchor sections.

After 50 mm of axial displacement, the radial dilations measured at the midpoint of the test section are from approximately 0.15 mm for 1 MPa radial confining pressure, to 0.02 mm for 15 MPa. Since the dilation angles are small ($i < 0.2$ degree), the pull force component related to dilational slip may be ignored. Thus, the axial pull-out force, F_a, may be approximately written as:

$$F_a = \frac{L_s}{L} \cdot A(c + p_1 \tan \phi_g) + \left(1 - \frac{L_s}{L}\right) \times \frac{A \cdot p_1 \cdot \tan \phi_{g-s}}{\sin \alpha} + Q \quad (2.99)$$

In which L is embedment length and L_s is sheared length of the grout flutes. However, as the shear failure length is undeterminable, an average coefficient of friction ϕ' over the whole test section is introduced, then

$$F_a = A \cdot p_1 \cdot \tan(\phi') + Q \quad (2.100)$$

The average coefficient of friction angle can be evaluated as the slope of the linear portion of plot $(F_a - Q)/A$ against confine pressure, which in turn is independent of confining pressure.

In the cable–grout interface, the pressure dependent closure is assumed to be hyperbolic, and then the total dilation due to splitting may be written as:

Radial splitting dilation:
$$v_r = v_{r0} - \frac{p_1 \cdot v_{r0}}{K_{r0} \cdot v_{r0} + p_1}$$

where v_{r0} is the dilation generated by splitting when $p_1 = 0$, and K_{r0} represents the radial stiffness (MPa/mm) of the cable–grout interface immediately following splitting. Appropriate values for K_{r0} and v_{r0} can be determined from the radial displacement–axial displacement plots.

The cable is not rigid; radial contraction of the cable due to the application of p_1 is considered and evaluated as p_1/K_{rc}. Therefore:

Radial splitting dilation:
$$v_r = v_{r0} - \frac{p_1 \cdot v_{r0}}{K_{r0} \cdot v_{r0} + p_1} - \frac{p_1}{K_{rc}}$$

Based on this, a simple mechanical model to characterize the radial deformability of the cable–grout interface is assumed after the cable has been pulled by an amount u_{r1}:

$$u_{r1} = \frac{k_1}{p_1}(u_a - 1) + v_{r0} - \frac{p_1 \cdot v_{r0}}{K_{r0} \cdot v_{r0} + p_1} - \frac{p_1}{K_{rc}} \quad (2.101)$$

where u_a is axial displacement and k_1 is empirical constant determined by best fit.

Combining Eqs. (2.100) and (2.101), a differential formulation for the deformability of the cable joint interface during bond failure, ie, tangent stiffness matrix, can be obtained:

$$\begin{bmatrix} dF_a \\ dp_1 \end{bmatrix} = \begin{bmatrix} K_1^i & K_2^i \\ K_3^i & K_4^i \end{bmatrix} \begin{bmatrix} du_{r1} \\ du_a \end{bmatrix} \quad (2.102)$$

The behavior of grout annulus is discussed in three scenarios based on the assumption that it has fully split after 1 mm of axial pull. Thereafter, the cement annulus will be unable to support a tensile tangential stress. That is, the fracture is free to open or close depending on confining pressure p_2 and dilation u_{r1}.

While the tangential stresses are compressive, the grout annulus will behave identically to an intact hollow cylinder, and the plane strain elastic solution for a thick-walled hollow cylinder can be applied. For the case when the radial fractures are fully open, a series of individual grout wedges are formed. The solution to the stresses and dilations are identical with the BSM (Yazici and Kaiser, 1992).

If the radial fractures are partially open, ie, the outer annulus is in compression but the inner annulus is in tension, the tangential stress at the common boundary must be zero. Thus, the radius for which fractures are open can be solved:

$$r_c = \frac{p_2 r_2^2 - r_2\sqrt{p_2^2 r_2^2 - p_1^2 r_1^2}}{r_1 p_1} \quad (2.103)$$

Consequently, the radial displacement equation and stiffness matrix can be formulated.

6.4 Interfacial Shear Failure of Rockbolt

6.4.1 Failure Modes of Rockbolt

It is known that the ultimate failure of rockbolts may occur: (1) in the bolt, (2) in the grout, (3) in the rock, (4) at the bolt–grout interface, (5) at the grout–rock or steel tube interface, and (6) a combination of these failure modes (Ren et al., 2010). Under the debonding failure, zero thickness interface represents the materials adjacent to the critical surface where debonding occurs. The deformation of the surrounding rock or grout is often negligible, ie, all deformations in the surrounding grout and rock outside the critical interface are lumped in the interface. As a result, the bolt can be assumed to be under uniaxial tension and the bolt–grout interface layer under interfacial shear deformation only. If debonding occurs at the grout–rock interface, the idealized model is still applicable by treating the bolt and grout together as a "hybrid bolt" under uniaxial tension.

For the steel bar bolting system, failure takes place along the weakest interface unless the bolt itself yields. The product of the hole diameter and the bond strength of the grout–rock interface is greater than the product of the bolt diameter and the bond strength of the bolt–grout interface. Hence, failure may occur by the bolt pulling out, as is sometimes observed in the case of smooth rebars. Such failure of grouted bolts can be prevented by shaping the bolt surface. However, failure may also initiate within the grout annulus or at the grout–rock interface, owing to impaired grout strength development or poor adhesion of grout to the borehole wall.

A series of push- and pull-out tests (Aydan, 1989) were carried out to investigate the anchorage mechanism of grouted rockbolts and the effect of various parameters, such as the ratio of the bolt to borehole diameter and the behavior of the bolt to grout interface under triaxial stress. The results showed that the load bearing capacity of bolts was 25% higher in push-out tests than pull-out tests. Similar tests (Jalalifar, 2006) resulted in a load bearing capacity increase at around 10% in favor of push tests. This increase in push test values was attributed to the Poisson's ratio effect (the radial stress is of compressive character in the push-out case while it tends to become tensile in the pull-out tests).

Ayden's investigation (1989) also showed that an increase in bearing capacity was attributable to the normal compressive stress resulting from the geometric dilation of the surface. It suggested that shearing might occur along one of the surface of weakness in the rockbolt system (grout–rock interface or bolt–grout interface), and classified the failure modes in the push/pull tests as follows:

1. Failure along the bolt–grout interface. This occurred in every test on bars with a smooth surface and deformed bars installed in a large borehole.
2. Failure along the grout–rock interface. This occurred in deformed bars installed only in smaller diameter boreholes.
3. Failure by splitting of grout and rock annulus.

6.4.2 Analysis of the Behavior of Rockbolt

Li et al. (1999) developed an analytical model for predicting the behavior of rockbolts under three different conditions: (1) for bolts in pull-out tests; (2) for bolts installed in a uniformly deformed rock mass; and (3) for bolts subjected to the opening of rock joints.

The development of these models was based on the description of the mechanical coupling at the interface between the bolt and grout medium for the grouted tests. Based on the exponential decay theory and decoupling of the bond, they constructed a model for the shear stress along a fully grouted bolt as shown in Fig. 2.55.

Before decoupling occurs at the interface for fully grouted rockbolts, the attenuation of the shear stress is expressed as:

$$\tau_b = \frac{\alpha}{2}\sigma_{b0} e^{-2\alpha \frac{x}{d_b}} \quad (2.104)$$

where $\alpha^2 = \dfrac{2 G_r G_g}{E_b \left[G_r \ln \frac{d_g}{d_b} + G_g \ln \frac{d_0}{d_g} \right]}$; d_b is bolt diameter; d_g is borehole diameter; E is Young's modulus; G is shear modulus and d_0 is diameter of rock outside which influence of the bolt disappears.

After decoupling occurs, for equilibrium of the bolt, the applied load P_0 should equal the total shear force at the bolt interface, ie,

$$P_0 = \pi d_b \int_{x_0}^{L} \tau_b dx = \pi d_b \left[s_r(x_1 - x_0) \right.$$
$$\left. + \frac{1}{2} s_p \Delta (1 + \omega) + \frac{d_b}{2\alpha} s_p \left(1 - e^{-\frac{2\alpha}{d_b}(L - x_2)} \right) \right]$$
$$(2.105)$$

And the maximum applied load can be expressed as:

$$P_{0-\max} = \pi d_b s_p \left[\omega \left(L + \frac{d_b}{2\alpha} \ln \omega - \Delta - x_0 \right) \right.$$
$$\left. + \frac{1}{2}\Delta(1 + \omega) + \frac{d_b}{2\alpha}(1 - \omega) \right]$$
$$(2.106)$$

where L is length of the bolt; $\Delta = x_2 - x_1$; $\omega = s_r/s_p$; $x_2 = x_0 + \frac{1}{2\omega}\left[\frac{2P_0}{\pi d_b s_p} - \frac{d_b}{\alpha} - \Delta(1+\omega)\right]$.

Ground anchor is also one common reinforcing method in civil engineering. It can make

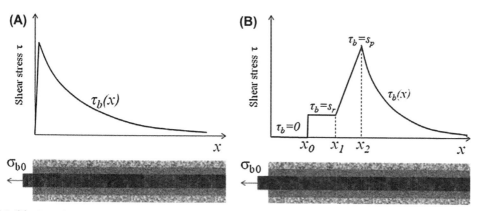

FIGURE 2.55 Distribution of shear stress along a fully grouted rockbolt subjected to an axial load. (A) Before decoupling occurs; (B) after decoupling occurs. *After Li, C., Stillborg, B., 1999. Analytical models for rockbolts. International Journal of Rock Mechanics and Mining Sciences and Geomechanics Abstracts 36 (8), 1013–1029.*

effective use of the soil potential and enhance its self-stability. According to the modes of load transfer, anchorage can be divided into three types: tension, pressure, and shearing. Tension anchor is more commonly used and its reinforcement mechanism is to transmit the supporting force from anchor to stable stratum through bonding resistance. There are three failure modes of the tension anchor, that is, (1) anchor breaking, (2) anchor and grout body bonding failure, and (3) the anchorage body and soil shear failure. The former two failure modes hardly occur in practice so the main task of soil anchorage design is to determine the side resistance distribution between the anchorage body and surrounding soil to avoid the last failure mode. Side resistance distribution of the anchorage segment is conventionally assumed to be uniform. However, some recent investigations have shown that the side resistance is not uniform but has a peak in the front part and then decreases gradually and finally approaches zero. Based on this, a model of the tensile anchor using shear displacement method (Li and Stillborg, 1999) was developed and is shown in Fig. 2.56.

In Xiao and Chen's work, the softening feature of the soil was considered. The shear stress–strain relationship of soil surrounding the anchorage body was simplified into a trilines model consisting of an elastic phase, an elastoplastic phase, and a residual phase (Fig. 2.56). The shear stress–strain relationship can be expressed as

$$\begin{cases} \tau/G & \text{Elastic phase} \\ \gamma = \tau_1/G + (\tau - \tau_1)/K & \text{Elastoplastic phase} \\ \tau_1/G + (\tau_2 - \tau_1)/K & \text{Residual phase} \end{cases}$$
(2.107)

According to elastic theory, the shear stress and strain of soil surrounding pile are

$$\tau = \tau_0 r_0 / r \quad \text{and} \quad \gamma = ds/dr \qquad (2.108)$$

where r is distance from soil to pile center; r_0 is radius of anchorage; τ_0 is shear stress of anchorage surface; and s is soil displacement. Consequently, at a depth z, the soil displacement at the anchor interface can be obtained as

$$\begin{cases} \dfrac{\tau_0 r_0}{G} \ln \dfrac{r_m}{r_0} & \text{Elastic phase} \\ s = \dfrac{\tau_1 r_0}{G} \ln \dfrac{r_m}{r_0} + \dfrac{\tau_1 r_0}{K} \ln \dfrac{\tau_0}{\tau_1} & \text{Elastoplastic phase} \\ \text{uncertainty} & \text{Residual phase} \end{cases}$$
(2.109)

where r_m is the soil radius surrounding anchorage body that shear displacement can be

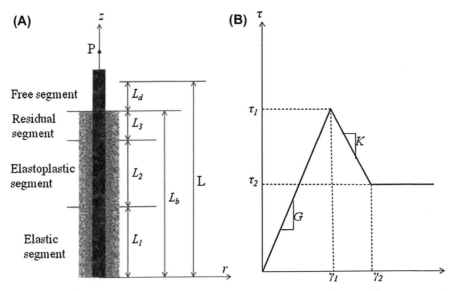

FIGURE 2.56 (A) Schematic diagram of tension type anchor and (B) relationship between shear stress and strain of soils. *After Li, C., Stillborg, B., 1999. Analytical models for rockbolts. International Journal of Rock Mechanics and Mining Sciences and Geomechanics Abstracts 36 (8), 1013–1029.*

ignored. The shear displacement, s, is also a function of depth z. According to the definition, the shear displacements in the elastic phase and elastoplastic phase are

$$\begin{cases} \text{Elastic phase:} & s = \dfrac{\tau_0(z)r_0}{G}\ln\dfrac{r_m}{r_0} & 0 \leq z \leq L_b \\[6pt] & s = \dfrac{\tau(z)r_0}{G}\ln\dfrac{r_m}{r_0} & 0 \leq z \leq L_1 \\[6pt] \text{Elastoplastic phase:} & s = \dfrac{\tau_1(z)r_0}{G}\ln\dfrac{r_m}{r_0} + \dfrac{\tau(z)r_0}{K}\ln\dfrac{\tau(z)}{\tau_1} & L_1 \leq z \leq L_b \end{cases} \quad (2.110)$$

The governing equation is still

$$\dfrac{d^2 s}{dz^2} - \dfrac{2}{r_0 E}\tau = 0 \quad (2.111)$$

When only elastic deformation exists, this governing equation can be solved. When the surrounding soil enters into the elastoplastic phase, it is doubtful if the close form solution can be obtained since the equation is a transcendental equation. Nevertheless, Xiao and Chen provided a method to potentially predict the full-range behavior of the anchorage.

Ren et al. (2010) developed a closed-form solution for the prediction of the full-range mechanical behavior of fully or partially grouted rockbolts under tension. In this solution, a trilinear bond-slip model is used to accurately model the interfacial debonding mechanism between the grout and the bolt. The full-range behavior consists of five consecutive stages:

elastic stage, elastic-softening stage, elastic-softening-debonding stage, softening-debonding stage, and total debonding stage. For each stage, closed-form solutions for the load–displacement relationship, ISS distribution, and bolt axial stress distribution along the bond length were derived. The ultimate load and the effective anchor length were also obtained. Their analytical model was calibrated with two pull-out experimental studies. The predicted load–displacement curves as well as the distributions of the ISS and the bolt axial stress are in close agreement with test results.

This study adopts a trilinear bond-slip model, similar with previous description (Fig. 2.56B), in which an ascending branch up to the peak stress at (δ_1, τ_f) is followed by a softening branch down to (δ_f, τ_r), and then a horizontal branch representing the nonzero residual frictional strength τ_r after complete debonding.

Let k be the ratio of the residual strength τ_r to the peak stress τ_f so the shear stress can be expressed as:

$$\tau(\delta) = \begin{cases} \dfrac{\tau_f}{\delta_1} & \text{for } 0 \le \delta \le \delta_1 \\ \dfrac{k\tau_f(\delta-\delta_1) + \tau_f(\delta_f-\delta)}{\delta_f-\delta_1} & \text{for } \delta_1 \le \delta \le \delta_f \\ k\tau_f & \text{for } \delta_f \le \delta \end{cases}$$

(2.112)

Based on force equilibrium, the governing equation of the grouted rockbolt and the axial stress in the bolt are

$$\frac{d^2\delta}{dx^2} - \lambda^2 \frac{\delta_f}{\tau_f}\tau(\delta) = 0 \qquad (2.113)$$

where $\lambda^2 = \frac{2\tau_f}{\delta_f E_b r_b}$.

The governing equation can be solved once $\tau(\delta)$ is defined. Fig. 2.57 illustrates the evolution of ISS distribution and corresponding load–displacement curve. In the elastic stage, the solutions of the governing equations are

$$\delta = \frac{\delta_1 P \lambda_1 \cosh(\lambda_1 x)}{2\pi r_b \tau_f \sinh(\lambda_1 L)}; \quad \tau = \frac{P\lambda_1 \cosh(\lambda_1 x)}{2\pi r_b \sinh(\lambda_1 L)};$$

$$\sigma_b = \frac{P \sinh(\lambda_1 x)}{\pi r_b^2 \tau_f \sinh(\lambda_1 L)}$$

(2.114)

where $\lambda_1^2 = \frac{\delta_f}{\delta_1}, \lambda^2 = \frac{2\tau_f}{\delta_1 E_b r_b}$.

The slip at the loaded end, ie, $x = L$, is defined as the displacement of the rockbolt and denoted as Δ. The following load-displacement expression can then be obtained

$$P = \frac{2\pi r_b \tau_f \tanh(\lambda_1 L)}{\delta_1 \lambda_1}\Delta \qquad (2.115)$$

The elastic stage ends when the shear stress reaches the bond shear strength τ_f at a slip of δ_1 at $x = L$. Setting $\Delta = \delta_1$, the load at the initiation of interface softening is found to be

$$P_{\text{sof}} = \frac{2\pi r_b \tau_f \tanh(\lambda_1 L)}{\lambda_1} \qquad (2.116)$$

As the pull-out force increases, softening commences at the loaded end and the peak shear stress is transferred toward the embedded end (state II). With the development of the softening length a, the load P continues to increase because more interface is mobilized to resist the pull-out force. At the end of this stage, P reaches the debonding load P_{deb}. The following differential equations for the elastic-softening stage can be obtained:

$$\frac{d^2\delta}{dx^2} = \begin{cases} \lambda_1^2 \delta & 0 \le \delta \le \delta_1 \\ \lambda_2^2(\delta_f - k\delta_1) - (1-k)\lambda_2^2 \delta & \delta_1 \le \delta \le \delta_f \end{cases}$$

(2.117)

where $\lambda_2^2 = \frac{\delta_f}{\delta_f-\delta_1}, \lambda^2 = \frac{2\tau_f}{(\delta_f-\delta_1)E_b r_b}$.

The boundary conditions are: $\delta_b = 0$ at $x = 0$; $\delta = \delta_1$ or $\tau = \tau_f$ at $x = L - a$; and $\delta_b = \frac{P}{\pi r_b^2}$ at $x = L$.

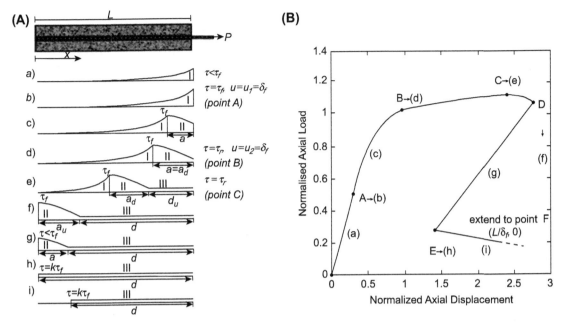

FIGURE 2.57 (A) Evolution of interfacial shear stress distribution and (B) propagation of debonding and typical full-range theoretical nondimensional load–displacement curve. (a and b) Elastic stage; (c and d) elastic-softening stage; (e and f) elastic-softening-debonding stage; (g) softening-debonding stage; (h and i) debonding stage. I, II, and III represent elastic, softening, and debonding stress states, respectively. *After Ren, F.F., Yang, Z.J., Chen, J.F., Chen, W.W., 2010. An analytical analysis of the full-range behaviour of grouted rockbolts based on a tri-linear bond-slip model. Construction and Building Materials 24, 361–370.*

The solution for the elastic region of the interface with $0 \leq x \leq L - a$ is

$$\delta = \frac{\delta_1 \cosh(\lambda_1 x)}{\cosh[\lambda_1(L-a)]}; \tau = \frac{\tau_f \cosh(\lambda_1 x)}{\cosh[\lambda_1(L-a)]};$$

$$\delta_b = \frac{2\tau_f \sinh(\lambda_1 x)}{r_b \lambda_1 \cosh[\lambda_1(L-a)]}$$

(2.118)

The solution for the softening region with $L - a \leq x \leq L$ is

And

$$P = \frac{2\pi r_b \tau_f}{\lambda_2 \sqrt{1-k}} \left[\frac{\lambda_2 \sqrt{1-k}}{\lambda_1} \cos\left[a\lambda_2 \sqrt{1-k}\right] \right.$$
$$\left. \tanh[\lambda_1(L-a)] + \sin\left[a\lambda_2 \sqrt{1-k}\right] \right]$$

(2.120)

$$\begin{cases} \delta = (\delta_f - \delta_1) \left[\frac{\lambda_2 \sin\left[\lambda_2(x-L+a)\sqrt{1-k}\right] \tanh[\lambda_1(L-a)]}{\lambda_2 \sqrt{1-k}} - \frac{\cos\left[\lambda_2(x-L+a)\sqrt{1-k}\right]}{1-k} + \frac{\delta_f - k\delta_2}{(1-k)(\delta_f - \delta_1)} \right] \\ \tau = -\tau_f \left[\frac{\lambda_2 \sqrt{1-k}}{\lambda_1} \sin\left[\lambda_2(x-L+a)\sqrt{1-k}\right] \tanh[\lambda_1(L-a)] - \cos\left[\lambda_2(x-L+a)\sqrt{1-k}\right] \right] \\ \sigma_b = \frac{2\tau_f}{\lambda_2 r_b \sqrt{1-k}} \left[\frac{\lambda_2 \sqrt{1-k}}{\lambda_1} \cos\left[\lambda_2(x-L+a)\sqrt{1-k}\right] \tanh[\lambda_1(L-a)] + \sin\left[\lambda_2(x-L+a)\sqrt{1-k}\right] \right] \end{cases}$$

(2.119)

$$\Delta = (\delta_f - \delta_1)\left[\frac{\lambda_2 \sin[a\lambda_2\sqrt{1-k}]\tanh[\lambda_1(L-a)]}{\lambda_1\sqrt{1-k}}\right.$$
$$\left. - \frac{\cos[a\lambda_2\sqrt{1-k}]}{1-k} + \frac{\delta_f - k\delta_1}{(1-k)(\delta_f - \delta_1)}\right]$$
(2.121)

Debonding initiates at the loaded end when τ reduces to τ_r at $x = L$. Substituting $\tau = \tau_r$ and $x = L$ into shear equation leads to

$$\frac{\lambda_2\sqrt{1-k}}{\lambda_1}\sin\left[\lambda_2 a\sqrt{1-k}\right]\tanh\left[\lambda_1(L-a)\right]$$
$$- \cos\left[\lambda_2 a\sqrt{1-k}\right] = -k$$
(2.122)

debonding propagates, the peak shear stress continues to move toward the embedded end. Thus there are three possible stress states within the bond length: the elastic state (state I), the softening state (state II), and the debonding state (state III) (Fig. 2.57E). The debonded length is denoted by d and the solution for the elastic-softening zones, ie, Eq. (2.36), is still valid if L is replaced by $(L - d)$. The differential equation for the debonding zone can be obtained by substituting Eq. (2.112) into Eq. (2.111). The solution for the debonding region with $L - d \leq x \leq L$ is

$$\begin{cases} \delta = \frac{\delta_f}{2\lambda_1\lambda_2\sqrt{1-k}}\left[\lambda_1\left[\lambda_2\sqrt{1-k}\left(2 + kd^2\lambda^2 + k\lambda^2(L-d)^2 - 2kd\lambda^2(L-x)\right)\right.\right. \\ \qquad\left.\left. - 2\lambda^2(L-x-d)\sin\left(\lambda_2 a_d\sqrt{1-k}\right)\right] + 2\lambda^2\lambda_2\sqrt{1-k}(L-x-d)\cos\left(\lambda_2 a_d\sqrt{1-k}\right)\tanh[\lambda_1(a_d+d-L)]\right] \\ \tau = k\tau_f \\ \delta_b = \frac{2\tau_f}{r_b}\left[k(d-L+x) + \frac{\sin\left(\lambda_2 a_d\sqrt{1-k}\right)}{\lambda_2\sqrt{1-k}} - \frac{\cos\left(\lambda_2 a_d\sqrt{1-k}\right)\tanh\left[\lambda_1(a_d+d-L)\right]}{\lambda_1}\right] \end{cases}$$
(2.125)

Thus the softening length a at the initiation of debonding at the loaded end, denoted as a_d, can be solved as:

$$a_d = \frac{1}{\lambda_2\sqrt{1-k}}\sin^{-1}$$
$$\times \left[\frac{\sqrt{\delta_f - \delta_1}\left[\sqrt{\delta_f - k\delta_1 - k^2(\delta_f - \delta_1)} - k\sqrt{\delta_1(1-k)}\right]}{\delta_f - k\delta_1}\right]$$
(2.123)

Debonding load P_{deb} can be found as

$$P_{deb} = \frac{2\pi r_b \tau_f}{\lambda_2\sqrt{1-k}}\left[\frac{\lambda_2\sqrt{1-k}}{\lambda_1}\cos\left[a_d\lambda_2\sqrt{1-k}\right]\right.$$
$$\left. + \sin\left[a_d\lambda_2\sqrt{1-k}\right]\right]$$
(2.124)

Once the shear stress decreases to τ_r at $x = L$, debonding initiates at the loaded end. As

and

$$P = \frac{2\pi r_b \tau_f}{\lambda_2\sqrt{1-k}}\left[\frac{\lambda_2\sqrt{1-k}}{\lambda_1}\cos\left(a_d\lambda_2\sqrt{1-k}\right)\right.$$
$$\left. \times \tanh[\lambda_1(L-d-a_d)] + \sin\left(a_d\lambda_2\sqrt{1-k}\right)\right]$$
$$+ 2k\pi r_b \tau_f d$$
(2.126)

$$\Delta = \frac{\delta_f}{2\lambda_1\lambda_2\sqrt{1-k}}\left[\lambda_1\left[\lambda_2\sqrt{1-k}\left(2 + kd^2\lambda^2\right.\right.\right.$$
$$\left.\left. + 2\lambda^2 d\sin\left(\lambda_2 a_d\sqrt{1-k}\right)\right)\right]$$
$$- 2\lambda^2\lambda_2\sqrt{1-k}\cos\left(\lambda_2 a_d\sqrt{1-k}\right)$$
$$\left. \times \tanh\left[\lambda_1(a_d + d - L)\right]\right]$$
(2.127)

P reaches its maximum value when the derivative is zero with respect to d (denoted as d_{ult}),

$$d_{ult} = L - a_d - \frac{1}{\lambda_1}\tanh^{-1}\sqrt{\frac{\cos(a_d\lambda_2\sqrt{1-k})-k}{\cos(a_d\lambda_2\sqrt{1-k})}}$$

(2.128)

This analysis shows that the full-range mechanical behavior of rockbolts under tension consists of five distinct stages. The important points are point A (P_1, u_1) corresponding to the initiation of interface softening; point B (P_2, u_2) corresponding to the initiation of debonding; and point C (P_3, u_3) corresponding to the ultimate load. These three points may be identified from an experimental load−displacement curve, and used to calibrate the parameters in the trilinear bond-slip model.

6.5 Conclusion

Rockbolts are widely used in mining and tunneling engineering to support underground excavation or to stabilize a jointed rock mass. In this work, up-to-date failure modes of several kinds of rockbolting reinforcement system were presented and discussed. The stress state in a concrete beam due to bond forces from a steel bar is analyzed. Upon the specific bond failure modes, ie, cone cracks and radial cracks, the stresses were calculated for an elastic stage, a plastic stage, and an elastic stage with internal cracks.

For fully grouted cable bolts, the BSM explains observations from laboratory pull-out tests by predicting the elastic, partially split, and fully split grout behavior. The load level during splitting of grout is a function of the grout confinement by the rock, grout stiffness, and grout strength. The grout column of cable bolts confined by relatively soft or disturbed rock, with soft or low tensile strength grout will be susceptible to grout splitting which in turn leads to a reduction in ultimate bond strength.

Hyett et al. (1992, 1995, 1996) emphasized that the failure involves unscrewing of the cable from the cement annulus. This type of failure mechanism is due to the helical form and low torsional rigidity of a seven-wire strand, and it distinguishes the mechanical behavior of cable bond failure from a solid deformed bar.

Finally, a closed-form solution for predicting the full-range behavior of rockbolts under tension based on a trilinear bond-slip model is presented. Formulations for the shear slip and shear stress on the grout−bolt interface, the load-displacement relations, and the axial stress in the bolt, have been derived for each of the five distinct loading stages. The control parameters in the solution can be calibrated from pull-out test data. It offers a theoretical basis of the rockbolt behavior under tension, and provides practical application. Once the bond-slip model is calibrated using the analytical solution from pull-out tests, it can be used in the modeling of complex engineering problems.

7. STEEL BOLT PROFILE INFLUENCE ON BOLT LOAD TRANSFER

7.1 Introduction

The bolt load transfer capacity is governed by the shear strengths developed between the rock−grout and the grout−bolt interface. It is commonly accepted that the bonding strength has three components: cohesion, friction, and interlock. Singer (1990) demonstrated that there is no adhesion between the grout to bolt and interface but in most reported cases, there is a very small adhesion between grout−rock and grout−bolt as shown in Fig. 2.58 (Aziz and Webb, 2003).

The frictional components can be cataloged into dilation slip, shear failure of surrounding medium, and torsional unscrewing of bolt (Hyett et al., 1995). Each of the components depend on

FIGURE 2.58 Resin–bolt separation after postencapsulation. *After Aziz, N., Pratt, D., Williams, R., 2003. Double shear testing of bolts. In: Proceedings 4th Underground Coal Operator's Conference, Wollongong, Australia.*

the stress generated at the bolt–grout interface, which in turn depends on the internal reaction forces of the whole system, which has three phase materials with two interfaces.

It is commonly accepted that the mechanical interlock component plays an important role in load transfer capacity of the rockbolting system. It in turn is influenced by the bolt profile configuration. The profile configuration is defined by the rib profile shape, profile height, angle of wrap, and spacing between the ribs, as shown in Fig. 2.59.

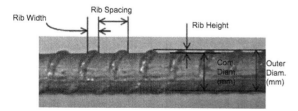

FIGURE 2.59 Steel bolt rib profile configuration.

Research work has shown that smooth bar has a very low load transfer capacity compared with ribbed steel bar (Aydan, 1989; Kilic et al., 2002). In another extreme, if a bolt has too many ribs, such as thread bar, its load bearing capacity is also small (Kilic et al., 2002; Ito et al., 2001). In fact, a thread bar can be thought of as a smooth bar with a larger diameter.

In traditional rockbolting load transfer mechanism analysis, the effect of mechanical interlocking is often integrated into the analytical model using various ways without reference to the rib geometric configuration. Farmer (1975) has shown an exponential decay of the axial stress along the bolt in case of perfect bonding and elastic deformation. Yazici and Kaiser (1992) proposed a BSM to predict the ultimate load transfer capacity of fully grouted cable bolt. In this model the mechanical interlock is simulated as zigzag surface of the cable, which generates dilation or radial movement when debonding occurs. For the same problem, Hyett et al. (1995) emphasized the "unscrewing" effect during the deformation of the cable bolt and introduced an untwisting component into the axial force formula to quantify the interaction of the bolt and the grouting material. In the so-called ISS model, the deformation of surrounding materials is lumped into a zero thickness interface, which is assigned with specific stress–strain behavior to simulate the mechanical interlocking observed in pull-out tests. For instance, Li and Stillborg (1999) developed an ISS model for predicting the behavior of rockbolts in pull-out tests, in uniformly deformed rock mass and when subjected to opened joints. They considered the elastic, linear softening, residual, and debonding stages of the interface coupling. Ivanovic and Neilson (2009) developed a lumped parameter model with varying shear-load failure properties along the fixed anchor length to analyze the bolt behavior under static or dynamic load. Closed-form solutions have also been obtained for the prediction of full-range behavior of fully grouted rockbolting under axial load (Ren et al., 2010; Martin et al., 2011). A trilinear bond-slip model with residual strength at the grout–bolt interface was adopted and consequently five consecutive deformation stages of the interface were identified to describe the interaction of each component in the rockbolting system.

It should be noted that the geometric configuration of the bolt rib has been ignored in these models, instead they included their functions of the bolt profiles into calculations to model the influence. One can argue that "mechanical interlocking" is actually a general term to describe the complex interactions of each component within the rockbolting system. If its function is being described in one way or another, then the structure can be eliminated. This is true on a macroscopic scale, such as the full bolt length; however, during a "local pull-out" of a small bolt section the deformation may only concern a small number of bolt profiles. In addition, the resin's mechanical properties and volume may have a close relationship with rib profile to achieve optimum design. Fig. 2.60 shows a comparison of ISS between rebar and conceptualized smooth bar when subjected to a same axial load within elastic range.

There is some literature that includes the rib profile as part of the research work. In studies of the reinforced concrete beams, civil engineers found that during the pull-out tests the ribbed bar can crush the concrete in certain patterns consisting of primary cracks, radial cracks, and cone cracks (Tepfers, 1979, Fig. 2.61). For bars with a rib face angle varying between 40 degree and 105 degree, it is likely to produce approximately identical behavior of the concrete during the pull-out tests. If the face angle is less than 30 degree, then the bonding action is different. After the concrete is crushed to a compacted powder, it becomes lodged in front of the ribs and effectively produces ribs with face angle of 30–40 degree. In an analytical model Tepfers (1979) assumed that the radial components of the bond forces can be regarded as a hydraulic pressure acting on a thick-walled concrete ring surrounding the reinforcing bar. The shear stress at the interface distributes into the surrounding material by compression under a certain angle (angle of the cone-shaped crack) and is balanced

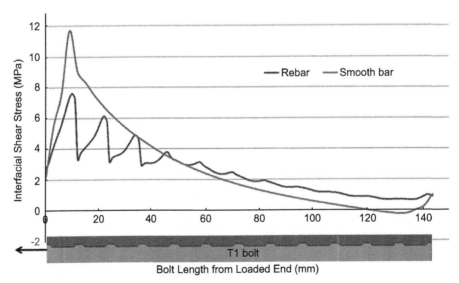

FIGURE 2.60 FLAC axisymmetric model to compare the shear stress distribution for a rebar and a smooth bar.

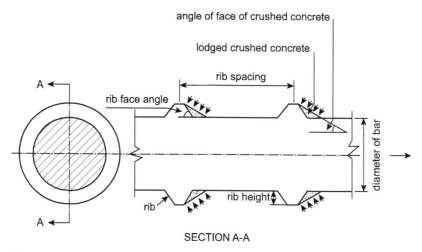

FIGURE 2.61 The geometry of a deformed reinforcing bar and the mechanical interaction between the bar and the concrete. *After Tepfers, R., 1979. Cracking of concrete cover along anchored deformed reinforcing bars. Magazine of Concrete Research 31 (106), 3—12.*

by tensile stress rings in the concrete. However, in this model the rib geometry is ignored.

Fabjanczyk and Tarrant (1992) investigated the load transfer mechanism in push-out tests. They found that bolts with a lower profile height had smaller stiffness (Fig. 2.62) and concluded that the load transfer was a function of parameters, such as hole geometry, resin properties, and bar surface configuration.

Kilic et al. (2002) studied cone-shaped lugs of cement grouted steel bolt by pull-out tests. Single, double, and triple conical-lugged bars with different face angles were tested and the experimental results showed that the conical-lugged rockbolt provide better anchorage strength. Ito et al. (2001) used an X-ray CT scanner to visualize the patterns of failure in pull-out tests. The tests were conducted on four types of bolts grouted into an artificial rock with cement paste. The results show strong influence of bolt types on the load—displacement deformational behavior (Fig. 2.63).

Blumel (1996) reported on the influence of profile spacing on load transfer capacity of the bolt. Pull-out testing of equal diameter bolts

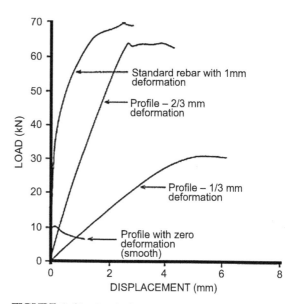

FIGURE 2.62 Load—displacement curve for rebar with various amounts of bar deformation removed. *After Fabjanczyk, M.W., Tarrant, G.C., 1992. Load transfer mechanisms in reinforcing tendons. In: Proceedings of the 11th International Conference on Ground Control in Mining, The University of Wollongong, pp. 1—8.*

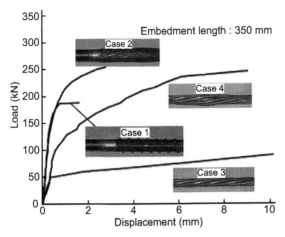

FIGURE 2.63 Load—displacement curves of four cases in the pull-out tests. *After Ito, F., Nakahara, F., Kawano, R., Kang, S.S., Obara, Y., 2001. Visualization of failure in a pull-out test of cable bolts using X-ray CT. Journal of Construction and Building Materials 15, 263—270.*

with different profile spacing was carried out on bolts of 13.7, 27.4, and 54.8 mm profile spacing. Blumel found out that widening of the spacing between the profiles enhanced the load transfer capacity of the bolting system installation. Later, Blumel et al. (1997) reported on the finite element modeling of the bolts with different profile spacings. Their study supported the experimental laboratory findings clearly demonstrating that higher stresses with more significant peaks being developed in the bolt with wider-spaced ribs as compared to the small rib distance. Aziz et al. (2003) studied on profile configurations to include push testing of bolts installed in cylindrical steel tubes, 75-mm long and 27-mm internal diameter. The tests were made using chemical resin instead of cement. Moosavi et al. (2005) also studied the profile configurations in cementitious grout, leading to similar conclusions.

Since these early works, there has been no further analytical or numerical work being undertaken to advance the load transfer capacity of the bolts with respect to profile configuration. Accordingly, this aspect of the topic is currently being evaluated both analytically and numerically, which is the subject of discussion in this chapter.

7.2 Methodology and Governing Equations

A rockbolting system with three phase materials is studied. A unit rib which includes all bar surface features is shown in Fig. 2.64.

The rockbolt problem is reduced to an axisymmetric problem in three dimensions. When the bolt is subjected to an axial force, the unit bolt will experience a net axial load. This load is then transferred into the resin via rib profile. The stress distribution within the resin can be calculated according to solutions of a

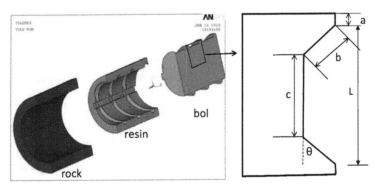

FIGURE 2.64 Three units of rockbolting system and rib parameters of a unit bolt used in this study.

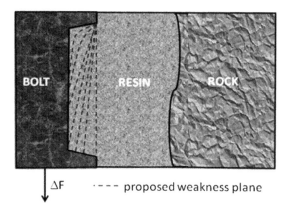

FIGURE 2.65 A schematic drawing of a single spacing and proposed weakness planes.

half-space or full-space problem. In the resin, the weakness surfaces are assumed as shown in Fig. 2.65. For each proposed weakness surface, critical load is calculated according to a nominated failure criterion. By comparing the sustainable load of each surface, the weakest surface, which is most likely to initialize the bond slip, will be found.

It is essential to test a large number of weakness planes to establish the most probable failure surface. It is also necessary to confirm the theoretical calculation with experiments; to observe the initiation of resin failure in the laboratory is rather difficult. Past studies used flattened surface of the real bolt to examine the resin shear failure under CNS conditions in laboratory, as shown in Fig. 2.66A. This experimental technique is now being introduced in this study, whereby the axisymmetrical rockbolting problem is reduced to a plane stress problem so that the theoretical solutions can be verified from controlled experiments. The experiment is designed as a bolt profile on a flat plane to mimic a flattened bolt surface. The failure mode of resin covered on this flattened plate which contains all surface features of the bolt is easily to identify, as shown in Fig. 2.66B.

Stress analysis on this flattened plate is a half-space problem. Initiated from the Lord Kelvin's problem, Boussinesq derived fundamental solutions for various loads on infinite or semiinfinite elastic media (Poulos and Davis, 1974). While loading an infinite strip on the surface of a semi-infinite mass (Fig. 2.67), the stress tensor anywhere within the media can be calculated as a function of the load, position, and material properties.

For a uniform normal load, the stress tensor can be calculated using the Boussinesq equations:

$$\sigma_z = p[\alpha + \sin \alpha \cos(\alpha + 2\delta)]/\pi \qquad (2.129)$$

$$\sigma_z = p[\alpha - \sin \alpha \cos(\alpha + 2\delta)]/\pi \qquad (2.130)$$

$$\sigma_y = 2pv\alpha/\pi \qquad (2.131)$$

$$\tau_{xz} = p[\sin \alpha \sin(\alpha + 2\delta)]/\pi \qquad (2.132)$$

For the uniform shear load shown in Fig. 2.67, the stress distribution can also be calculated via Cerutti's equations:

$$\sigma_z = q[\sin \alpha \sin(\alpha + 2\delta)]/\pi \qquad (2.133)$$

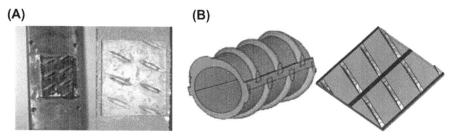

FIGURE 2.66 (A) Flattened real bolt and (B) designed experimental plate with all bolt rib features.

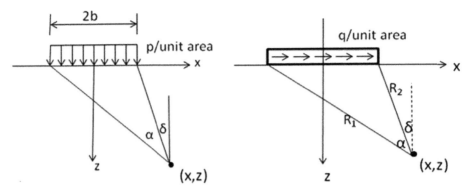

FIGURE 2.67 Stress tensor can be found at any position given by x and z coordinates within the semiinfinite elastic medium loaded by a uniformly distributed loads (p and q).

$$\sigma_x = \frac{q}{\pi}\left[\ln\frac{R_1^2}{R_2^2} - \sin\alpha\sin(\alpha + 2\delta)\right] \quad (2.134)$$

$$\tau_{xz} = q[\alpha - \sin\alpha\cos(\alpha + 2\delta)]/\pi \quad (2.135)$$

7.3 Modeling of Fully Grouted Bolt Profiles

To correlate the load transfer mechanism with the bolt rib configuration, a single spacing between two bolt profiles is modeled and examined. When the bolt is loaded, the load is transferred to the resin as shown in Fig. 2.68. The direction of these loads depends on the bolt profile, while their magnitudes depend on both the bolt profile and the material properties. To investigate the type of resin failure that will occur, several weakness planes are considered. As a first trial, a plane of weakness spanning between the bolt profile tips is identified as shown in Fig. 2.68.

During bolt loading, the distributed loads within the bolt can be represented as shear forces and normal forces. Assume that the initial bonding shear forces $S1$, $S3$, $S4$, and $S5$ between the bolt and the resin at the interface are small and the tensile bond between the bolt and resin is also small, then (1) $S1 = S3 = S4 = S5 \approx 0$; (2) N (tension) ≈ 0.

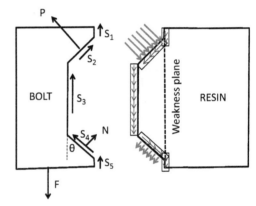

FIGURE 2.68 Load transfer between the steel bolt and the fully encapsulated resin.

Under these assumptions, the free body diagram of the bolt can be thus simplified as shown in Fig. 2.69, where only one normal force and one shear force to the inclined bolt profile remains. In most cases these stress components play a major role in stress distribution within the resin.

Three failure criteria can be used in the calculations. These are the shear failure, the Mohr–Coulomb and the distortional energy (Von Misses) failure criterion. From the three methods or models the Mohr–Coulomb criterion of failure is chosen in this chapter. Therefore, the normal and shear stresses to the chosen plane of weakness must be calculated.

For static equilibrium of the bolt, $\sum F_x = 0$ and $\sum F_y = 0$.

Assume stress is evenly distributed along the bolt profile side, then:

$$p = F \sin \theta / b, \quad s_2 = q = F \cos \theta / b \quad (2.136)$$

Where F is net axial force on unit bolt profile (for 2-D case); p is normal load on bolt boundary b; s_2 is shear load on bolt boundary b.

7.3.1 Stresses on the Proposed Failure Plane Due to Normal Load

The resin segment corresponding to the bolt unit is sketched in Fig. 2.70. Load p is the normal load expressed in Eq. (2.136). Solid line PQ is the assumed plane of failure connecting two corners of the bolt. Point A represents an arbitrary point on the plane of weakness and the variable h indicates its distance from point P. For convenience, line PQ is rotated around x-axis and coordinate system is reset as shown in Fig. 2.70.

According to rib profile parameters of the steel bolt defined in Fig. 2.64, where $PQ = L$ and $PB = b$, we get:

$$L = c + 2b \cos \theta \quad (2.137)$$

$$\alpha + \delta + \theta = \pi/2 \quad (2.138)$$

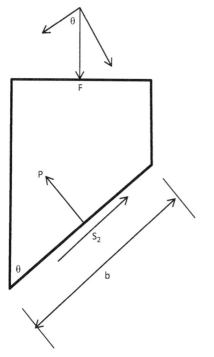

FIGURE 2.69 Free body diagram of the bolt after approximation.

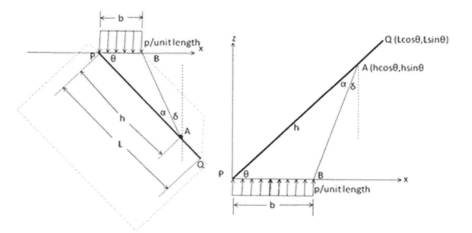

FIGURE 2.70 Resin section under normal load p, and coordinate system after rotation.

$$\sin \alpha = \frac{b \sin \theta}{\sqrt{b^2 + h^2 - 2b \cos \theta}} \quad (2.139)$$

$$\tan \delta = \frac{h \cos \theta - b}{h \sin \theta} \quad (2.140)$$

$$R_1 = AP = h \quad (2.141)$$

$$R_2 = AB = \sqrt{b^2 + h^2 - 2b \cos \theta} \quad (2.142)$$

where b, c, h, L, and θ are rib geometrical parameters shown in Fig. 2.64; and α, δ, R_1, and R_2 are positioning parameters shown in Fig. 2.70.

Both Boussinesq's and Cerutti's equations are for half-space problems. However, the contact of bolt and resin is not a half space, rather with irregularities due to rib profile. The sensitivity studies (Cao et al., 2011) dealing with this issue was presented previously. Another difference is that the resin and the rock are two materials with a common boundary. Study of rock and resin stiffness and resin failure near the bolt indicates that this effect is very small. In practice, the resin–rock boundary is invariably rifled due to drilling with wing bit, as shown in Fig. 2.70. The analysis can be simplified by extending the resin boundary to infinity as failure along the resin–rock interface is unlikely to occur.

The shear and normal stresses along the plane of proposed weakness are calculated using the stress tensor calculated by Boussinesq Eqs. (2.129)–(2.132). For the final solutions the stresses are calculated parallel and perpendicular to the weakness/failure plane. The stress tensor is thus transformed to a coordinate system parallel to the plane of weakness/failure using the following equations:

$$\sigma_x = \frac{1}{2}(\sigma_{x'} + \sigma_{z'}) + \frac{1}{2}(\sigma_{z'} - \sigma_{x'})\cos 2\theta$$
$$- \tau_{x'z'} \sin 2\theta \quad (2.143)$$

$$\sigma_z = \frac{1}{2}(\sigma_{x'} + \sigma_{z'}) - \frac{1}{2}(\sigma_{z'} - \sigma_{x'})\cos 2\theta$$
$$+ \tau_{x'z'} \sin 2\theta \quad (2.144)$$

$$\tau = \frac{1}{2}(\sigma_{z'} - \sigma_{x'})\sin 2\theta + \tau_{x'z'} \cos 2\theta \quad (2.145)$$

Thus the normal and shear stress to the failure plane would be:

$$\sigma_n = \frac{1}{2}(\sigma_x + \sigma_z) + \frac{1}{2}(\sigma_z - \sigma_x)\cos 2\theta$$
$$- \tau_{xz} \sin 2\theta \quad (2.146)$$

$$\tau = \frac{1}{2}(\sigma_z - \sigma_x)\sin 2\theta + \tau_{xz} \cos 2\theta \quad (2.147)$$

The calculations of normal stress, $\int \sigma_n dh$, and shear stress, $\int \tau dh$, caused by load p, on the supposed weakness plane yield:

$$\int \sigma_n dh = \int \left(\frac{\sigma_x + \sigma_z}{2} + \frac{\sigma_z - \sigma_x}{2}\cos 2\theta \right.$$
$$\left. - \tau_{xz} \sin 2\theta \right) dh$$

Substituting the three expressions from Eqs. (2.129), (2.130), and (2.132), we obtain:

$$\frac{\sigma_x + \sigma_z}{2} = \frac{\alpha}{\pi}p; \quad \frac{\sigma_z - \sigma_x}{2}$$
$$= \frac{p}{\pi}\sin \alpha \cos(\alpha + 2\delta);$$

$$\tau_{xz} = \frac{p}{\pi}\sin \alpha \sin(\alpha + 2\delta)$$

$$\int \sigma_n dh = \int \left[\frac{\alpha}{\pi}p + \frac{p}{\pi}\sin \alpha \cos(\alpha + 2\delta)\right.$$
$$\left. - \frac{p}{\pi}\sin \alpha \sin(\alpha + 2\delta)\right]dh$$

$$= \frac{p}{\pi}\int(\alpha - \sin \alpha \cos \alpha)dh$$
$$(2.148)$$

In a similar manner:

$$\int \tau dh = \frac{p}{\pi} \int \sin^2 \alpha dh \quad (2.149)$$

7.3.2 Stresses on the Proposed Failure Plane Due to Shear Load

The stress calculations due to the shear load q are similar to the normal load calculations presented in the last section (Fig. 2.71).

7.3.3 Superposition of Stress Tensor on the Failure Surface

Through superposition, Eqs. (2.149)–(2.151) are added and yield the final expression of stress tensor on the plane of weakness:

$$\int \sigma_n dh = \frac{p}{\pi} \int [\alpha - \sin\alpha \cos\alpha] dh$$

$$+ \frac{q}{\pi} \int \left[2\sin^2\theta \ln\frac{R_1}{R_2} + \sin^2\alpha \right.$$

$$\left. - \alpha \sin 2\theta \right] dh$$

$$\int \tau dh = \frac{p}{\pi} \int \sin^2\alpha dh + \frac{q}{\pi} \int \left[\sin\alpha \cos\alpha \right.$$

$$\left. - \sin 2\theta \ln\frac{R_1}{R_2} + \alpha \cos 2\theta \right] dh$$

Substituting p and q from Eq. (2.136), yields

$$\int \sigma_n dh = \frac{F}{\pi b} \int \left[-\alpha \sin\theta \cos 2\theta \right.$$

$$- \sin\theta \sin\alpha \cos\alpha + \cos\theta \sin^2\alpha$$

$$\left. + \sin\theta \sin 2\theta \ln\frac{R_1}{R_2} \right] dh$$

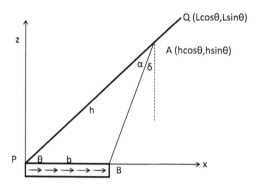

FIGURE 2.71 Shear load s_2 and the assumed plane of failure.

Calculation of normal and shear stresses due to s_2 at the weakness plane:

$$\int \sigma_n dh = \int \left(\frac{\sigma_x + \sigma_z}{2} + \frac{\sigma_z - \sigma_x}{2} \cos 2\theta \right.$$

$$\left. - \tau_{xz} \sin 2\theta \right) dh$$

Substituting the three expressions from Eqs. (2.133)–(2.135) will yield:

$$\frac{\sigma_x + \sigma_z}{2} = \frac{q}{\pi} \ln\frac{R_1}{R_2}; \quad \frac{\sigma_z - \sigma_x}{2}$$

$$= \frac{q}{\pi} \left[\sin\alpha \sin(\alpha + 2\delta) - \ln\frac{R_1}{R_2} \right];$$

$$\tau_{xz} = \frac{q}{\pi} \left[\alpha - \sin\alpha \sin(\alpha + 2\delta) \right]$$

$$\int \sigma_n dh = \frac{q}{\pi} \int \ln\frac{R_1}{R_2} dh + \frac{q}{\pi} \cos 2\theta \int \left(\sin\alpha \sin(\alpha + 2\delta) - \ln\frac{R_1}{R_2} \right) dh - \frac{q}{\pi} \sin 2\theta \int [\alpha - \sin\alpha \sin(\alpha + 2\delta)] dh$$

$$= \frac{q}{\pi} \int \left[2\sin^2\theta \ln\frac{R_1}{R_2} + \sin^2\alpha - \alpha \sin 2\theta \right] dh$$

(2.150)

$$\int \tau dh = \frac{q}{\pi} \int \left\{ \sin 2\theta \left[\sin\alpha \sin(\alpha + 2\delta) - \ln\frac{R_1}{R_2} \right] + \cos 2\theta [\alpha - \sin\alpha \sin(\alpha + 2\delta)] \right\} dh$$

$$= \frac{q}{\pi} \int \left[\sin\alpha \cos\alpha - \sin 2\theta \ln\frac{R_1}{R_2} + \alpha \cos 2\theta \right] dh$$

(2.151)

$$\int \tau dh = \frac{F}{\pi b} \int \left[\sin\theta \sin^2\alpha + \cos\theta \sin\alpha \cos\alpha \right.$$
$$- \sin 2\theta \cos\theta \ln\frac{R_1}{R_2}$$
$$\left. + \alpha \cos\theta \cos 2\theta \right] dh$$

The procedure of simplification of these two expressions is presented in the Appendix. Closed-form solutions can be obtained:

$$\int \sigma_n dh = \frac{F}{\pi b} 2 \left[b \sin^2\theta \cos 2\theta \ln\sqrt{\frac{L^2 - 2bL\cos\theta + b^2}{L^2 + 2bL\cos\theta + b^2}} \right.$$
$$- L\sin\theta \cos 2\theta \left(\frac{\pi}{2} - \theta - \tan^{-1}\frac{L\cos\theta - b}{L\sin\theta}\right)$$
$$\left. - L\sin\theta \sin 2\theta \ln\frac{\sqrt{L^2 - 2bL\cos\theta + b^2}}{L} \right]$$

$$\int \tau dh = \frac{F}{\pi b} \left[\left(\frac{\pi}{2} - \theta + \tan^{-1}\frac{L - b\cos\theta}{b\sin\theta}\right) b + \left(\frac{\pi}{2} - \theta\right) \right.$$
$$- \tan^{-1}\frac{L\cos\theta - b}{L\sin\theta} \right) L\cos\theta \cos 2\theta$$
$$+ L\sin 2\theta \cos\theta \ln\frac{\sqrt{L^2 - 2bL\cos\theta + b^2}}{L}$$
$$\left. + \frac{b\sin 4\theta}{8} \ln\frac{L^2 + 2bL\cos\theta + b^2}{L^2 - 2bL\cos\theta + b^2} \right]$$

7.3.4 Mohr–Coulomb Failure Along the Weakness Plane

Two combined stress fields are considered within the resin. The first one is the initial preloading stress tensor at the failure surface T_0 and the second is the bolt load induced stress tensor T. Thus, the failure criterion (f) is expressed as net resistant force that can be summed together:

$$f = T_0 + T$$
$$= (c_w + \mu\sigma_{n0} - \tau_0)L + \left(\int \mu\sigma_n dh - \int \tau dh\right)$$
$$(2.152)$$

Where $L = c + 2b\cos\theta$ is failure length; h is distance from any chosen point along the plane of weakness from 0 to L; $\mu = \tan\phi$, where ϕ is an internal frictional angle of the resin; σ_{n0} is initial normal stress; c_w is cohesion; τ_0 is initial shear stress; τ is shear stress introduced by axial load on the bolt; b, c, and θ are bolt profile parameters.

Each component along the plane of weakness is shown in Fig. 2.72. Failure criteria expression can be written as

$$f = (c_w + \mu\sigma_{n0} - \tau_0)L + \mu\int \sigma_n dh - \int \tau dh$$
$$= T_0 + \frac{\mu F}{\pi b}\left[b\sin^2\theta \cos 2\theta \ln\sqrt{\frac{L^2 - 2bL\cos\theta + b^2}{L^2 + 2bL\cos\theta + b^2}} \right.$$
$$- L\sin\theta \cos 2\theta \left(\frac{\pi}{2} - \theta - \tan^{-1}\frac{L\cos\theta - b}{L\sin\theta}\right)$$
$$\left. - L\sin\theta \sin 2\theta \ln\frac{\sqrt{L^2 - 2bL\cos\theta + b^2}}{L} \right]$$
$$- \frac{F}{\pi b}\left[\left(\frac{\pi}{2} - \theta + \tan^{-1}\frac{L - b\cos\theta}{b\sin\theta}\right) b \right.$$
$$+ \left(\frac{\pi}{2} - \theta - \tan^{-1}\frac{L\cos\theta - b}{L\sin\theta}\right) L\cos\theta \cos 2\theta$$
$$+ L\sin 2\theta \cos\theta \ln\frac{\sqrt{L^2 - 2bL\cos\theta + b^2}}{L}$$
$$\left. + \frac{b\sin 4\theta}{8} \ln\frac{L^2 + 2bL\cos\theta + b^2}{L^2 - 2bL\cos\theta + b^2} \right]$$

7.4 Application Example

Consider a variable profile shape, with θ varying from 45 to 90 degree as shown in Fig. 2.73. Let the failure length $L = 10$ mm, $a = 1$ mm, and rib height $(b\sin\theta) = 1.5$ mm. As c changes from 7 mm to 10 mm, then θ will change from 45 to 90 degree. For small θ, the failure is likely to occur elsewhere, such as at the bolt–resin interface. As θ reaches maximum, the failure may occur at the proposed plane of weakness.

7. STEEL BOLT PROFILE INFLUENCE ON BOLT LOAD TRANSFER

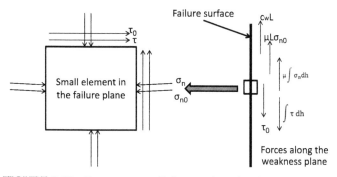

FIGURE 2.72 Forces on a small element along the plane of weakness.

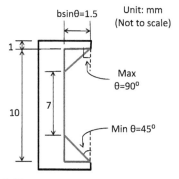

FIGURE 2.73 An example of bolt with different face angle with color lines indicating boundaries of the bolt profile.

considered to be small as most of the resin along the bolt cures after pretensioning the bolt.

Therefore, for the initial conditions: $\sigma_{n0} = 0$ MPa and $\tau_0 = 0$ MPa.

Now, let resin cohesion $c = 16$ MPa and the angle of internal friction $\Phi = 35$ degree. Then

$$T_0 = L(c_w + \mu\sigma_{n0} - \tau_0)$$
$$= 10(16 + \tan35° \times 0 - 0)$$
$$= 160 \text{ N (per mm of profile thickness)}$$

Use the failure criteria formula:

$$f = T_0 + \frac{\mu F}{\pi b}\left[b\sin^2\theta\cos 2\theta \ln\sqrt{\frac{L^2 - 2bL\cos\theta + b^2}{L^2 + 2bL\cos\theta + b^2}} - L\sin\theta\cos 2\theta\left(\frac{\pi}{2} - \theta - \tan^{-1}\frac{L\cos\theta - b}{L\sin\theta}\right)\right.$$
$$\left. - L\sin\theta\sin 2\theta \ln\frac{\sqrt{L^2 - 2bL\cos\theta + b^2}}{L}\right] - \frac{F}{\pi b}\left[\left(\frac{\pi}{2} - \theta + \tan^{-1}\frac{L - b\cos\theta}{b\sin\theta}\right)b\right.$$
$$\left. + \left(\frac{\pi}{2} - \theta - \tan^{-1}\frac{L\cos\theta - b}{L\sin\theta}\right)L\cos\theta\cos 2\theta + L\sin 2\theta\cos\theta\ln\frac{\sqrt{L^2 - 2bL\cos\theta + b^2}}{L} + \frac{b\sin 4\theta}{8}\ln\frac{L^2 + 2bL\cos\theta + b^2}{L^2 - 2bL\cos\theta + b^2}\right]$$
$$= 160 - F \cdot G(\theta)$$

Due to the bolt installation procedure, the initial preloading stress within the resin (T_0) is

where F is the pull-out force acting onto 1 mm thick half profile and $G(\theta)$ is the "influence

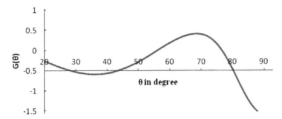

FIGURE 2.74 Influence factor $G(\theta)$ versus the rib slope angle θ.

factor" of the rib face angle θ. The graph in Fig. 2.74 shows the calculated influence factor versus the rib face angle θ for this geometry.

If the influence factor is negative, the failure cannot occur on the nominated plane of weakness no matter how large the applied pull-out force is. If the influence factor is positive, then failure may occur on the nominated plane of weakness.

From Fig. 2.74, it can be concluded that:

1. If the $55° < \theta < 78°$, $G(\theta) \geq 0$. In this case, as $f = 160 - F \times G(\theta)$ and when the pull-out force F is large enough, it will overcome the cohesion of 160 N and failure is possible.
2. The derived failure criterion indicates that if $\theta < 55°$ or $>78°$, failure will never occur on the proposed plane no matter how large is the axial force of the bolt.
3. When $G(\theta) \geq 0$, the maximum $G(\theta) = 0.4$. It means that the minimum pull-out force to cause the resin failure for the geometry given is

$$F_c = \frac{160}{0.4}$$
$$= 400 \text{ N(per mm of profile thickness)}$$

This will occur when the rib face angle θ is about 70 degree.

The derived mathematical equations used in this example can also be used to calculate resin failure for different variables. For example, using the same equations, the axial load can also be calculated for different rib height and different rib height and face angle or varying rib spacing, shown in Fig. 2.75. In this situation, the conditions apply for the same proposed failure surface and the same failure criterion.

Failure criterion can also be altered. In the example expressed in this chapter, similar solutions have been obtained using shear failure criterion or Von Misses criterion. While using Von Misses criterion, closed-form solution is not reachable but step-based calculation was employed.

In a similar manner, closed-form solution can also be obtained in any designated plane. The results can then be compared to get a better understanding of failure mechanism in the rockbolting system. For curved weakness surfaces, closed-form solutions are difficult to access except for some arc surfaces located in specific positions. For a general curve, step-based calculation has to be employed.

7.5 Conclusions

From the solutions presented in this study, it can be concluded that the mechanical properties of the resin are the key factor in the failure mechanism of rockbolting system. No definition to describe the mechanical behavior of the resin in this study is presented. That is based on the notion that, if the crushed resin becomes compacted powder and stays in front of the bolt rib, then they can be through as part of the bolt profile, especially during the shear failure

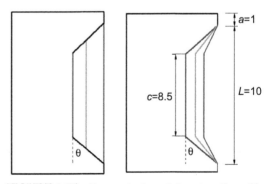

FIGURE 2.75 Demonstration of changing rib profile.

initiation. If this hypothesis is close to practice, then the crushed powder has less effect on the shear failure of the remaining resin. In this situation, the resin acts as an elastic material at the initial stage of failure in spite of progressive plastic deformation at the contact of the resin and the steel bolt.

Identification of initial resin failure surface is the central point of this study. Analytical results should be compared with computer simulations and experimental observations. Current numerical simulation software, to the best knowledge of authors, often require their user to predefine the potential failure surface. Consequently, the problem of optimum design of rib profile cannot be simulated directly.

Identification of resin crack initiation in pullout tests is difficult. However, there are indirect ways to test the cracks initiation and failure mechanism. To obtain experimental verification, the rib profile problem studied in this chapter has been transformed from an axisymmetric problem to a plane stress problem. The experimental results, together with comparison of analytical studies, will be the subject of follow-up studies.

8. TENSILE STRESS MOBILIZATION ALONG A ROCKBOLT

8.1 Introduction

There has been much theoretical and experimental research undertaken examining the effectiveness of fully grouted rockbolt since the 1970s (Farmer, 1975; Hagan, 2004; Mahony and Hagan, 2006). Some of this work has been directed at understanding the axial and shear behavior between the rock and grout when tension is applied on the rockbolt (Farmer, 1975; Li and Stillborg, 1999). Signer et al. (1997) measured the in situ loading of rockbolts to assist the evaluation and selection of roof bolt. Both axial and bending forces were measured by strain gauges at multiple locations along the length of fully grouted bolts during different stages of mining.

However, underground investigations showed that a high proportion of reinforcing elements can fail in shear (Haile, 1999). Researchers tried to use the strain gauges to measure the behavior of axial and bending when the rockbolt was subjected to shearing load. McHugh and Signer (1999) conducted the single shear tests to measure the distribution of axial and bending strain along the rockbolt with different axial load. In their tests, strain gauges were installed in a groove machined along the length of a rockbolt. However, it was demonstrated that the slot reduced axial bolt strength by approximately 10%. Another problem with this test is the strain gauge signal is often terminated before reaching the peak load (McHugh and Signer, 1999). Double shear tests were carried out by Jalalifar (2006) to study the contribution of rockbolt to the bolted joint shear strength. Numerical modeling was also adopted to investigate the axial stress distribution. However, the results were not verified by strain gauges.

Factors that contribute to bolt performance can be complex and variable. This makes analytical and numerical approaches difficult. A method based on bolt load measurement can be useful to supplement these methods to obtain a better understanding of support and rock interaction. Many methods have been adopted to study the reinforcing effect of rockbolt but the mechanism and effect of bolt action in jointed rock mass are still not sufficiently clear.

In this chapter, the double shear tests were conducted to examine the tensile stress generated along a rockbolt when subjected to shear loading. Strain gauges were installed along the rockbolt to measure the axial stress distribution. Two load cells were mounted, one each at either end of the rockbolt, to measure the normal load generated on the rock blocks.

TABLE 2.9 Mechanical Properties of Materials

Mechanical Properties	Block	Grout	Rockbolt
Compressive strength (MPa)	53.7	77	
Young's modulus (GPa)	29.7	13.2	200
Friction angle (degree)	32	–	
Yield force (kN)	–	–	78
Ultimate force (kN)	–	–	97

8.2 Adhesion Strength Tests

8.2.1 Double Shear Test Setup

Double shear tests were conducted using an instrumented 16 mm rebar grouted into 24 mm holes in a simulated rock sample. The samples were cast as separate blocks using a cement-based material. After 28 days of curing, tests were undertaken to determine the strength and modulus of the material, and the results are summarized in Table 2.9. The dimensions of each block were 300 × 300 × 200 mm³. Each test sample comprised three blocks joined by a rockbolt, as shown in Fig. 2.76. The vertical shear load, normal load, and shear displacement were monitored by load cell and LVDT.

An Avery hydraulic universal press with a capacity of 3600 kN was used in the shear tests. Two steel plates and four steel bars were used to constrain the two end blocks as shown in Fig. 2.76. This differs from a fully enclosed test sample arrangement in which it is assumed the rock block is under a perfect confining pressure to simulate an infinite rock mass. However, under this situation, the rock is assumed to be rigid and nondeformable, thereby making it difficult to understand the interaction between the rock and rockbolt.

To ensure the testing was conducted under pure shear loading, three steel plates, each of 20-mm thickness, were placed underneath the blocks to ensure even load distribution and avoid sample failure before the rockbolt yielded. The tests were conducted at different pretension loads of 0, 20, and 50 kN.

8.2.2 Strain Gauge Installation

Five pairs of diametrically opposed strain gauges were mounted along the length of the rockbolt to determine changes in axial strain on both sides of the shear plane. The strain gauges were positioned as close as possible to the shear plane. The distance from each pair of strain gauges to the joint plane were −70, −30, +30, +70, and +140 mm for the pairs numbered 1−5, respectively, in Fig. 2.77.

To maximize the axial strain profiles, the strain gauges when installed in the sample were aligned perpendicular to the shear load direction. This allowed analysis of the strains occurring on opposite sides of the rockbolt with expected zones of tension and compression.

The strain gauge was attached on the surface of rockbolt based on industry recommended

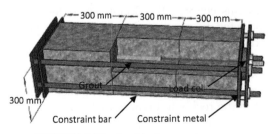

FIGURE 2.76 Double shear testing setup.

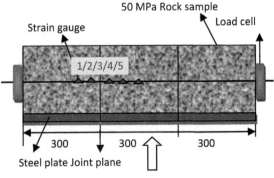

FIGURE 2.77 The double shear testing scheme.

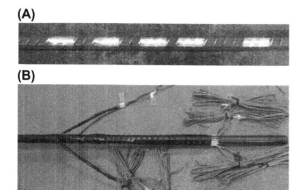

FIGURE 2.78 Rockbolt and the strain gauges. (A) Ground and polish sections of a rockbolt prior to application of strain gauges; (B) view of final instrumented strain-gauged rockbolt.

procedure. The type of strain gauge is FLA-3-11-3 L, which is suitable for 16 mm diameter rebar. The strain gauge had a nominal resistance of 120 Ω and gauge factor of 2.12% ± 1%.

Because the length of each strain gauge was longer than the rib spacing on the rockbolt, the surface was first ground and then polished to achieve a smooth finish. Fig. 2.78A shows the location of the polished profiles Fig. 2.78B shows a fully instrumented strain-gauged rockbolt.

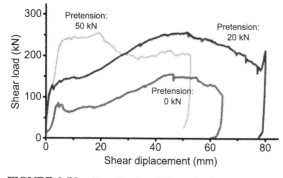

FIGURE 2.79 Shear load and shear displacement curve in double shear test.

8.3 Results Analysis

8.3.1 Shear Load and Normal Load

Fig. 2.79 shows a graph of shear displacement against shear load under the three different pretensions.

Three phases of behavior can be observed in the graph: rapid loading, slow loading, and decreased loading, as shown in Fig. 2.79. During the first phase, the shear load increased rapidly, indicating high initial stiffness of the system, until the elastic yield point was reached. In this phase there was elastic deformation of the rockbolt with only small shear displacement occurred. In the second phase after the elastic limit had been reached, plastic deformation occurred. Interestingly there was resistance to further shear displacement with load slowly increasing. It is likely that this is due to the rockbolt but also due to the strength of the test sample. In the final phase beyond the ultimate load there was a rapid decrease in load with little resistance to shear displacement after failure of the rockbolt and consequently loss of confinement of the test samples.

The elastic limit of the rockbolt in the first phase increased with pretension as shown in Fig. 2.79. In addition, the higher pretension increased the stiffness of the system. Pretensioning of the rockbolt also contributed to the elastic limit being achieved at lower levels of shear deformation as the rockbolt had already undergone some elastic deformation. A small reduction in shear load can also be observed in Fig. 2.79 due to rock fracture around the rockbolt.

Fig. 2.80 shows the variation in normal load during shearing. When the middle block was pushed up, the two end plates restricted movement of the rockbolt. Hence normal load progressively increased with shear displacement.

With no pretension applied on the rockbolt, the final normal load reached 70 kN. This is of a similar magnitude as the value measured by strain gauge pair #1 of 76 kN as shown in Table 2.12.

2. ROCKBOLTING

TABLE 2.10 Measured Strain and Deflection of Instrumented Rockbolt

Strain Gauge No.	Deflection (mm)	Strain (10^{-3})	Distance From Joint Plane (mm)
1-L	4.27	2.601 (+)	−70
1-U	4.27	2.564 (−)	−70
2-L	4.29	5.965 (+)	−30
2-U	4.29	5.843 (−)	−30
3-L	13.27	5.620 (−)	+30
3-U	13.27	5.790 (+)	+30
4-L	23.05	4.651 (−)	+70
4-U	23.05	4.819 (+)	+70
5-L	37.05	2.263 (−)	+140
5-U	37.05	2.312 (+)	+140

Note. U indicates upper-strain gauge and L indicates lower-strain gauge.

8.3.2 The Strain Along the Deformed Bolt

The deflection and axial strains of the instrumented rockbolt measured in the double shear test are shown in Table 2.10.

The original length of rockbolt was 1158 mm; the rockbolt was found to have elongated as a result of shear loading as shown in Fig. 2.81.

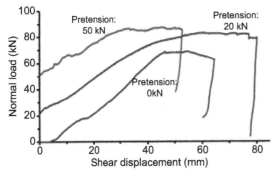

FIGURE 2.80 Variation in normal load with shear displacement.

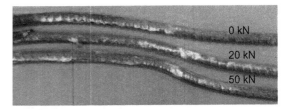

FIGURE 2.81 Deformed rockbolt for different levels of pretension.

The elongation after testing is shown in Table 2.11. The rockbolt elongation resulted in an increase in strain, which will further contribute to the rockbolt reaching the elastic limit.

Rockbolt experienced compression and tension at the same time on either side of maximum axial force line as shown in Fig. 2.82. There was no bending at the location with the maximum axial force, and it was the transitional point from compression to tension upside of rockbolt. Vice versa for the downside of rockbolt; the stress is transferred from tension to compression.

8.3.3 Axial Stress Distribution Along the Rockbolt

Axial stress is calculated through Eq. (2.153) and the distribution curve is shown in Fig. 2.82. Table 2.12 shows the variation in axial strain, stress, and force along the axial stress, and force could be determined by the Eqs. (2.154) and (2.155).

$$\varepsilon_{axial} = (\varepsilon_D + \varepsilon_U)/2 \qquad (2.153)$$

TABLE 2.11 Rockbolt Elongation

Pretension (kN)	Length After Testing (mm)	Percentage Elongation (%)
0	1163	0.43
20	1168	0.86
50	1173	1.29

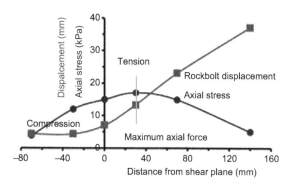

FIGURE 2.82 Rockbolt deformation and axial stress distribution with zero pretension.

TABLE 2.12 Axial Stress and Axial Force

Strain Gauge Pair	Strain (10^{-3})	Stress (kPa)	Force (kN)
1	0.019	3.8	76.4
2	0.061	12.2	245
3	0.085	17.0	342
4	0.084	16.8	338
5	0.025	5.0	100.5

where $\varepsilon_{\text{axial}}$ is axial strain (mm); ε_D is strain down the bolt; and ε_U is strain up the bolt.

$$\sigma = \varepsilon E \quad (2.154)$$

$$F = \frac{1}{4}\pi d^2 \sigma \quad (2.155)$$

Where ε is axial strain (10^{-3}); σ is axial stress (kPa); E is elastic modulus (GPa); d is the rockbolt diameter (mm); and F is normal load (kN).

8.4 Discussion

Aziz et al. (2003) indicated that rock sample strength is critical to the performance of rockbolt. In this testing, a 50-MPa rock sample strength was sufficient to test 16-mm diameter rockbolts. However, the confining condition is different with infinite rock masses. During testing, the rock sample cracked in regular position as shown in Fig. 2.83. Once the rock cracked, it can no longer provide a pure shear condition. Shear failure becomes bending failure due to the sample crushing around the intersection of joint and bolts. Even though more shear load applied, the energy released through rock fracture. Therefore, it is not easy to break the rockbolt in this condition. In addition, the reserved 10-mm shear displacement is not enough for the rockbolt failure.

As shown in Fig. 2.83, a void appeared in the central rock block. This caused differential movement between the monitored displacement on the rock and the actual rockbolt deflection. The interaction of rock and rockbolt resulted in different magnitudes of shear and bend. In weak rock, it is easy to crush the surrounding rock. However, in stronger rocks it is inclined to brittle failure with straight cracks and in such cases, the rockbolt may provide less resistance to shear loading.

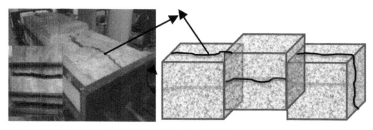

FIGURE 2.83 Rock failure type.

The pretension applied on the rockbolt plays a significant role in responding to shear loading. Pretensioning causes the rockbolt to be in a state of elastic elongation before shearing begins. The force is transferred to the grout through bond or shear stresses at the bolt–grout interface causing differential bolt extension and grout shear along the bolt.

The tensile force will tend to increase resistance to shear displacement. And the shear force leads to larger tension along the bolt in turn. Shear loading of the bolts results in tensional (positive) and compressional (negative) strain at the bending site.

There are two factors that are of importance in evaluating the effectiveness of different type of rockbolt in shear.

1. The rockbolt must withstand large shear and tensile forces.
2. The rockbolt shear resistance ability is relied on the tensile elongation. The rockbolt should also be able to withstand tensile deformation under shear loading.

8.5 Conclusions

Double shear testing is a useful method in assessing rockbolt shear behavior in resisting jointed mass movement. It is representative of rockbolt shear capacity in the behavior of tensile force mobilization. With higher pretension applied, the normal load is progressively increased. It is concluded that the shear performance is dependent on the tensile elongation capacity. The tensile stress and force distribution were derived through strain gauge data. Furthermore, the interaction between rock and rockbolt with greater pretension created higher magnitude of tension and eventual bending.

References

Abrams, D.A., 1913. Tests of bond between concrete and steel. In: Bulletin no. 71, 1913. University of Illinois Engineering Experiment Station, Urbana.
Ansell, A., 2006. Dynamic testing of steel for a new type of energy absorbing rockbolt. Journal of Constructional Steel Research 62 (5), 501–512.
ASTM A615/A615M–09a, September 1996. Standard Specification for Deformed and Plain Billet-steel for Concrete Reinforcement. American Society for Testing and Materials, West Conshohocken, Pennsylvania.
Aydan, O., 1989. The Stabilisation of Rock Engineering Structures by Rock Bolts, Geotechnical Engineering (Ph.D. thesis). Nagoya University, Nagoya.
Aziz, N., Jalalifar, H., Concalves, J., 2006. Bolt surface configurations and load transfer mechanism. In: Proc. 7th Underground Coal Operators Conference, Wollongong, pp. 236–244.
Aziz, N., Pratt, D., Williams, R., 2003. Double shear testing of bolts. In: Proceedings 4th Underground Coal Operator's Conference, Wollongong, Australia.
Aziz, N., Webb, B., 2003. Study of load transfer capacity of bolts using short encapsulation push test. In: Proc. 4th Underground Coal Operators Conference, pp. 72–80.
Benmokrane, B.A., Chennouf, A., 1995. Laboratory evaluation of cement-based grouts and grouted rock anchors. International Journal of Rock Mechanics and Mining Sciences and Geomechanics Abstracts 32 (7), 633–642.
Blevins, T.C., Campoli, A.A., 2006. The origin and history of U.S. mine resin. In: Proceeding of 25th International Conference on Ground Control, Morgantown, WV, Aug 1–3, pp. 409–411.
Blumel, M., 1996. Performance of grouted rockbolts in squeezing rock. In: Proceedings EUROCK'96, Predictions and Performance in Rock Mechanics an Rock Engineering. Balkema, Rotterdam, pp. 885–891.
Blumel, M., Schweger, H.F., Golser, H., 1997. Effect of rib geometry on the mechanical behaviour of grouted rockbolts. In: World Tunnelling Congress, 23rd General Assembly of the International Tunnelling Ass. Wien. 6 p.
Cai, M., Champaigne, D., 2012. Influence of bolt-grout bonding on MCB cone bolt performance. International Journal of Rock Mechanics and Mining Sciences and Geomechanics Abstracts 49, 165–175.
Cairns, J., Jones, K., 1995. Influence of rib geometry on strength of lapped joints: an experimental and analytical study. Magazine of Concrete Research 47 (172), 253–262.
Campbell, R., Mould, R.J., 2005. Impacts of gloving and unmixed resin in fully encapsulated roof bolts on geotechnical design assumptions and strata control in coal mines. International Journal of Coal Geology 64 (1–2), 116–125.
Cao, C., 2012. Bolt Profile Configuration and Load Transfer Capacity Optimization (Ph.D. thesis). University of Wollongong, Wollongong, Australia.
Cao, C., Nemcik, J., Aziz, N., 2011. Improvement of rockbolt profiles using analytical and numerical methods. In: Proc 12th Underground Coal Operators Conference, Wollongong, pp. 236–244.

REFERENCES

Cao, C., Nemcik, J., Aziz, N., Ren, T., 2013. Analytical study of steel bolt profile and its influence on bolt load transfer. International Journal of Rock Mechanics and Mining Sciences and Geomechanics Abstracts 60, 188–195.

Choi, O.C., Lee, W.S., 2002. Interfacial bond analysis of deformed bars to concrete. ACI Structural Journal 99 (6), 750–756.

Clark, A.P., 1946. Comparative bond efficiency of deformed concrete reinforcing bars. Journal of Research of the National Bureau of Standards 37, 399–407.

Clark, A.P., 1949. Bond of concrete reinforcing bars. Journal, American Concrete Institute; Proceedings 46 (11), 161–184.

Craig, P., 2012. Addressing resin loss and gloving issues at a mine with coal roof. In: 12th Coal Operators' Conference. UoW, pp. 120–128.

Darwin, D., Graham, E.K., 1993. Effect of deformation height and spacing on bond strength of reinforcing bars. ACI Structural Journal 90 (6), 646–657.

Fabjanczyk, M.W., Tarrant, G.C., 1992. Load transfer mechanisms in reinforcing tendons. In: Proceedings of the 11th International Conference on Ground Control in Mining. The University of Wollongong, pp. 1–8.

Fabjanczyk, M., Hurt, K., Hindmarsh, D., 1998. Optimisation of roof bolt performance. In: Proc of International Conference on Geomechanics Ground Control in Mining and Underground Construction, vol. 1, pp. 413–424.

Farmer, I.W., 1975. Stress distribution along a resin grouted rock anchor. International Journal of Rock Mechanics and Mining Sciences and Geomechanics Abstracts 12 (11), 347–351.

Gaudreau, D.M., Aubertin, S., Simon, R., 2004. Performance assessment of tendon support systems submitted to dynamic loading. In: Ground Support in Mining and Underground Construction. Balkema, London, pp. 299–312.

Goto, Y., 1971. Cracks formed in concrete around deformed tension bars. ACI Journal 68 (4), 244–251.

Hagan, P., 2004. Variation in the load transfer of fully encapsulated rockbolts. In: Proceedings 23rd International Conference on Ground Control in Mining, Morgantown, WV, USA.

Haile, A.T., 1999. Observation of the dynamic support performance of South African tunnel support systems. In: Proceedings Rock Support and Reinforcement Practice in Mining. Balkema, Rotterdam.

Hamad, B.S., 1995. Comparative bond strength of coated and uncoated bars with different rib geometries. ACI Material Journal 92 (6), 579–590.

He, M., Gong, W., Wang, J., Qi, P., Tao, Z., Du, S., Peng, Y., 2014. Development of a novel energy-absorbing bolt with extraordinarily large elongation and constant resistance. International Journal of Rock Mechanics and Mining Sciences and Geomechanics Abstracts 67, 29–42.

Hong, S., Park, S.K., 2012. Uniaxial bond stress-slip relationship of reinforcing bars in concrete. Advances in Materials Science and Engineering 2012, 1–12, 328570.

Hyett, A.J., Bawden, W.F., Reichert, R.D., 1992. The effect of rock mass confinement on the bond strength of fully grouted cable bolts. International Journal of Rock Mechanics and Mining Sciences and Geomechanics Abstracts 29 (5), 503–524.

Hyett, A.J., Bawden, W.F., Macsporran, G.R., Moosavi, M., 1995. A constitutive law for bond failure of fully-grouted cable bolts using a modified Hoek cell. International Journal of Rock Mechanics and Mining Sciences and Geomechanics Abstracts 32 (1), 11–36.

Hyett, A.J., Mossavi, M., Bawden, W., 1996. Load distribution along fully grouted bolts, with emphasis on cable bolt reinforcement. International Journal for Numerical and Analytical Methods in Geomechanics 20, 517–544.

Idun, E.K., Darwin, D., 1999. Bond of epoxy-coated reinforcement: coefficient of friction and rib face angle. ACI Structural Journal 90 (4), 773–782.

Ito, F., Nakahara, F., Kawano, R., Kang, S.S., Obara, Y., 2001. Visualization of failure in a pull-out test of cable bolts using X-ray CT. Journal of Construction and Building Materials 15, 263–270.

Ivanovic, A., Neilson, R.D., 2009. Modelling of debonding along the fixed anchor length. International Journal of Rock Mechanics and Mining Sciences and Geomechanics Abstracts 46 (4), 699–707.

Jalalifar, H., 2006. A New Approach in Determining the Load Transfer Mechanism in Fully Grouted Bolts (Ph.D. thesis), (chapter 9). University of Wollongong, Australia.

Jeng, F.S., Huang, T.H., 1999. The holding mechanism of under-reamed rockbolts in soft rock. International Journal of Rock Mechanics and Mining Sciences and Geomechanics Abstracts 36 (6), 761–775.

Kaiser, P.K., Tannant, D.D., McCreath, D.R., 1996. Canadian Rockburst Support Hand-book. Geomechanics Research Centre, Sudbury, Ont, Canada (Laurentian University).

Kaiser, P.K., Yazici, S., Nose, J., 1992. Effect of stress change on the bond strength of fully grouted cables. International Journal of Rock Mechanics and Mining Sciences and Geomechanics Abstracts 29 (3), 293–306.

Kilic, A., Yasar, E., Gelik, A.G., 2002. Effect of grout properties on the pull out load capacity of fully grouted rockbolt. Tunnelling and Underground Space Technology 17, 355–362.

Kilic, A., Yasar, E., Atis, C.D., 2003. Effect of bar shape on the pull out capacity of fully grouted rockbolts. Tunnelling and Underground Space Technology 18, 1–6.

Kokubu, M., Okamura, H., 1997. Studies on usage of large deformed bar. Concrete Library of the Japan Society of Civil Engineers 43, 19–29.

Ladanyi, B., Archambault, G., 1977. Shear strength and deformability of filled joints. In: Proc. Int. Symp. on the Geotechnics of Structurally Complex Formations, Capri, vol. 1, pp. 317–326.

Li, C.C., 2010. A new energy-absorbing bolt for rock support in high stress rock masses. International Journal of Rock Mechanics and Mining Sciences and Geomechanics Abstracts 47 (3), 396–404.

Li, C., 2012. Performance of D-bolts under static loading. Rock Mechanics and Rock Engineering 45 (2), 183–192.

Li, C., Stillborg, B., 1999. Analytical models for rockbolts. International Journal of Rock Mechanics and Mining Sciences and Geomechanics Abstracts 36 (8), 1013–1029.

Littlejohn, S., 1993. Rock reinforcement-technology, testing, design and evaluation. In: Hudson, J.A. (Ed.), Comprehensive Rock Engineering Principals, Practice and Projects, vol. 4, pp. 413–451.

Lutz, L.A., 1970. Analysis of stresses in concrete near a reinforcing bar due to bond and transverse cracking. Journal Proceedings 67 (10), 778–787.

Lutz, L.A., Gergely, P., 1967. The mechanics of bond and slip of deformed bars in concrete. Journal Proceedings 64 (11), 711–721.

Lutz, L.A., Gergely, P., Winter, G., 1966. The Mechanics of Bond and Slip of Deformed Reinforcing Bars in Concrete. Structural Engineering Report No. 324. Cornell University.

Mahony, L., Hagan, P., 2006. A laboratory facility to study the behaviour of reinforced elements subjected to shear. In: Proceedings 7th Underground Coal Operators' Conference, Wollongong, NSW, Australia.

Malan, D.F., Basson, F.R.P., 1998. Ultra-deep mining: the increased potential for squeezing conditions. Journal of the South African Institute of Mining and Metallurgy 98, 353–364.

Martin, L.B., Tijani, M., Hadj-Hassen, F., 2011. A new analytical solution to the mechanical behaviour of fully grouted rockbolts subject to pull-out tests. Construction and Building Materials 25, 749–755.

McHugh, E., Signer, S.D., 1999. Roof bolt response to shear stress: laboratory analysis. In: Proceedings 18th International Conference on Ground Control in Mining, Morgantown, WV, USA.

Menzel, C.A., 1939. Some factors influencing results of pull-out bond tests. ACI Journal of Proceeding 35, 517–544.

Moosavi, M., Jafari, A., Khosravi, A., 2005. Bond of cement reinforcing bars under constant radial pressure. Cement and Concrete Composites 103–109. Elsevier.

Murata, J., Kawai, A., 1984. Studies on bond strength of deformed bar by pullout test. Concrete Library of the Japan Society of Civil Engineers 1 (348), 113–122.

Poulos, H., Davis, E., 1974. Elastic Solutions for Rock Mechanics. John Willey & Sons Inc, New York. TA710 P67 624.1513 73–17171, Printed in New York.

Rehm, G., 1957. The fundamental law of bond. In: RILEM Symposium on Bond and Crack Formation in Reinforced Concrete, Stockholm, vol. 2, pp. 491–498.

Ren, F.F., Yang, Z.J., Chen, J.F., Chen, W.W., 2010. An analytical analysis of the full-range behaviour of grouted rockbolts based on a tri-linear bond-slip model. Construction and Building Materials 24, 361–370.

Roberts, M.K.C., Brummer, R.K., 1988. Support requirements in rock burst conditions. Journal of the South African Institute of Mining and Metallurgy 88 (3), 97–104.

Signer, P.P., Cox, D., Johnston, J., 1997. A method for the selecion of rock support based on loading measurements. In: Proceedings 16th International Conference on Ground Control in Mining, Morgantown, WV, USA.

Singer, S.P., 1990. Field Verification of Load Transfer Mechanics of Fully Grouted Roof Bolts. US Bureau of Mines (9301).

Skorbogatov, S.M., Edwards, A.D., 1979. The influence of the geometry of deformed steel bars on their bond strength in concrete. Proceedings of the Institution of Civil Engineers 67 (2), 327–339.

Soretz, S., Holzenbein, H., 1979. Influence of rib dimensions of reinforcing bars on bond and bendability. ACI Journal Proceedings 76 (1), 111–125.

Stille, H., Holmberge, M., Nord, G., 1989. Support of weak rock with grouted bolts and shotcrete. International Journal of Rock Mechanics and Mining Sciences and Geomechanics Abstracts 26 (1), 99–113.

St-Pierre, L., Hassani, F.P., Radziszewski, P.H., Ouellet, J., 2009. Development of a dynamic model for a cone bolt. International Journal of Rock Mechanics and Mining Sciences and Geomechanics Abstracts 46 (1), 107–114.

Tannant, D.D., Brummer, R.K., Yi, X., 1995. Rockbolt behaviour under dynamic loading: field tests and modeling. International Journal of Rock Mechanics and Mining Sciences and Geomechanics Abstracts 32, 537–550.

Tannant, D.D., Buss, B.W., 1994. Yielding rockbolt anchors for high convergence or rockburst conditions. In: Proceedings of the 47th Canadian Geotechnical Conference.

Tepfers, R., 1979. Cracking of concrete cover along anchored deformed reinforcing bars. Magazine of Concrete Research 31 (106), 3–12.

Thompson, A.G., Player, J.R., Villaescusa, E., 2004. Simulation and analysis of dynamically loaded reinforcement systems. In: Ground Support in Mining and Underground Construction: Proceedings of the 5th International Symposium on Ground Support, Perth, Australia, pp. 341–355.

Turner, M.H., 2002. Seismically active mines-to buy or not to buy. In: Operators Conference. Australian Institute of Mining and Metallurgy, Townsville.

Winsdor, C.R., 1997. Rock reinforcement systems. International Journal of Rock Mechanics and Mining Sciences and Geomechanics Abstracts 34 (6), 919–951.

Windsor, C.R., Thompson, A.G., 1993. Rock reinforcement-technology, testing, design and evaluation. In: Hudson, J.A. (Ed.), Comprehensive Rock Engineering Principles, Practice and Projects, vol. 4, pp. 451–484.

Wu, C., Chen, G., Jeffery, S., Volz, R.K., Brow, M., Koenigstein, L., 2012. Local bond strength of vitreous enamel coated rebar to concrete. Construction and Building Materials 35, 428–439.

Yazici, S., Kaiser, P.K., 1992. Bond strength of grouted cable bolts. International Journal of Rock Mechanics and Mining Sciences and Geomechanics Abstracts 29 (3), 279–292.

Further Reading

Ansell, A., 2005. Laboratory testing of a new type of energy absorbing rockbolt. Tunnelling and Underground Space Technology 20, 291–300.

Patton, F.D., 1996. Multiple modes of shear failure in rock. In: Proc 1st Congr Int Soc Rock Mech, Lisbon, vol. 1, pp. 509–513.

Xiao, S.J., Chen, C.F., 2008. Mechanical mechanism analysis of tension type anchor based on shear displacement method. Journal of Central South University of Technology 15, 106–110.

CHAPTER 3

Grouted Cable

1. LOAD TRANSFER MECHANISM OF FULLY GROUTED CABLE

1.1 Introduction

A cable is a flexible tendon consisting of a quantity of wound wires that are grouted in boreholes at certain spacings to provide ground reinforcement of excavations (Hutchinson and Diederichs, 1996). They were first introduced into the mining industry in the 1960s (Thorne and Muller, 1964) and since the early 1970s have been used in both hard rock and coal mining operations. Over time, cables have become the dominant form of ground support particularly in highly stressed ground conditions.

Originally, cables were only used as a temporary reinforcement element. This was because many earlier cables were made from discarded steel ropes which had very poor load transfer properties as a consequence of their smooth surface profile. Over subsequent years a number of modifications have been made to the basic plain-strand cable, such as buttoned strand (Schmuck, 1979), double plane strand (Matthews et al., 1983), epoxy-coated strand (Dorsten et al., 1984), Fiberglass Cable (FCB) (Mah, 1990), birdcaged strand (Hutchins et al., 1990), bulbed strand (Garford, 1990), and nutcaged strand cables (Hyett et al., 1993). These changes to the cable surface geometry have been undertaken in an effort to improve the load transfer efficiency and anchorage capacity that has resulted in the more widespread use of cables for permanent reinforcement.

Despite these developments in design, failure of cable reinforcement systems still occurs. Rupture of the cable strands rarely occurs as it requires the shear resistance between the cable strand and the grouted surface of the strand being larger than the cable's maximum tensile capacity (Mitri and Rajaie, 1992). Potvin et al. (1989) stated that it is more likely for a cable to fail at either of the cable/grout or grout/rock interfaces but more likely the cable/grout interface, which is a function of the load transfer between the cable and rock mass.

To evaluate load transfer efficiency, both peak shear stress capacity and system stiffness need to be determined. Although values for these can be estimated, most researchers tend to use the load versus displacement curves obtained from laboratory tests to study and compare the load transfer characteristics of cables. Thomas (2012) proposed the load transfer index to evaluate the cable load transfer efficiency.

Hartman and Hebblewhite (2003) stated there are three sets of factors that have an impact on the cable load transfer, including the reinforcing element, rock mass, and loading conditions. The following sections outline results of the effect of relevant parameters on cable load transfer together with the evolution in design of testing facilities showing the development in understanding the load transfer mechanism of cables with respect to axial loading.

1.2 Load Transfer Behavior of Cables

1.2.1 Split-Pull/Push Tests

The earliest "split-pull" testing equipment as shown in Fig. 3.1 was designed by Fuller and Cox (1975) and used in a study of the load transfer mechanism of cables. In this design, steel split pipes were used to represent the rock mass and provide confinement to the grouting material surrounding the cable. Within this facility, although the rotating behavior of cables was constrained, the steel tube provided a level of confinement that was markedly different from that of a rock mass as evident by the stress-strain relationship. The consequence of this was very high peak loads being achieved, much greater than was achieved in field measurements. The facility was used to evaluate the effect of surface geometry on the performance of cables and they found the shape and conditions of the cable had a critical impact on the load transfer. Any protrusion, such as surface rust on the wire strands improved the load transfer whereby the location of each protrusion would influence the characteristic of the residual load but it had little impact on the peak load. Following that, the effect of wire indentations on the performance of cables was reported by Cox and Fuller (1977). They found that the indentations have a positive effect on load transfer for mill-finished wires. However, for rusted wires, indentations reduced the effective rusted surface area and thereby the change in load transfer capacity. Thus, it was suggested, the wire surface should be slightly rusted and nonindented. They also showed that high grout strength had a positive impact on load transfer.

Goris and Conway (1987) went on to use a similar test rig design to investigate the impact of epoxy coating, showing that epoxy-coated cables had a larger bearing capacity compared with conventional cables. In addition, with respect to the position of steel buttons along the cable in a grout column, they found that an increase in the distance between the button and joint significantly enhanced the performance of the cables. Finally, the impact of birdcage node location with respect to rock fractures was investigated. Here, if the node was located at the pipe discontinuity the bearing capacity of cables was nearly 31% higher than that of strands in which the antinode was located at the discontinuity, indicating the birdcage node, if located near a rock joint, would degrade the load transfer

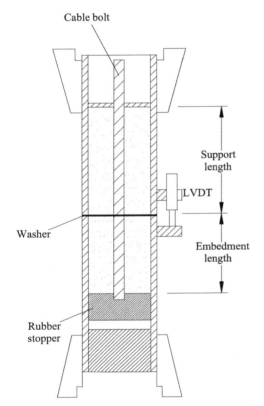

FIGURE 3.1 Split-pulling rig. *After Fuller, P.G., Cox, R.H.T., 1975. Mechanics of load transfer from steel tendons to cement based grout. In: Proceedings Fifth Australasian Conference on the Mechanics of Structures and Materials, Melbourne, Australia, pp. 189–203.*

efficiency. However, this effect is only suitable for single birdcage cables whereas double birdcage cables were less sensitive to the location of nodes or antinodes with respect to rock joints (Goris, 1991).

Comprehensive experiments were carried out by Goris (1990) to investigate the axial performance of cables. The ultimate bearing capacity of plain cables improved linearly with embedment length from 203.2 to 812.8 mm. The bearing capacity was also found to increase with the presence of two wound cables, high curing temperatures, low water-cement ratios, and sand-cement grouts. Furthermore, the load transfer of cables was not influenced by breather tube size and the existence of a breather tube so long as the breather tube was fully filled with grout.

Strata Control (1990) paid particular attention to the effect of bulb density of twin-strand Garford bulb cables. A linear relation was found between the cable bearing capacity and bulb density with bulb density frequency ranging from 3.9 to 6 bulbs per strand per meter, indicating load transfer increased with the bulb frequency and number.

Although the "split-pull" test provided much useful information on the different kinds of cables, the design was defective in the extra confinement created near the pulling threads given by the screw gripping assembly that tended to overestimate the measured pull-out load. To overcome this issue, Reichert (1991) designed the "split-push" test, which was a modification of the traditional pulling test. In this arrangement, the grout column and pipe were pushed off from the cable as indicated on the right in Fig. 3.2, rather than being "pulled" as in the conventional sense shown on the left.

Using this test arrangement aluminum, PVC and steel pipes were used to model the effect of different radial confinements or stiffness. It was reported that with this "split-push" test equipment, Hyett et al. conducted tests on various kinds of modified cables. In 1993, the impact of

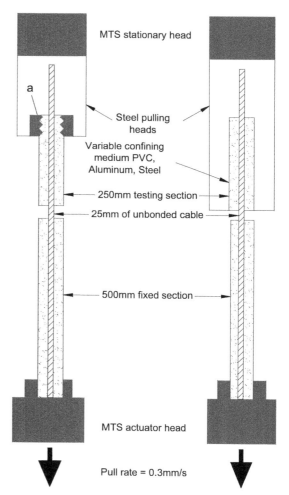

FIGURE 3.2 Comparison of conventional and modified test. *After Reichert, R.D., Bawden, W.F., Hyett, A.J., 1992. Evaluation of design bond strength for fully grouted cables. CIM Bulletin 85 (922), 110–118.*

nutcase geometry was evaluated and they found that the nutcase cables were less sensitive to high water to cement ratios and lower confinement. As for the nutcase size the larger the nutcase, the larger the nutcase cable stiffness (Hyett et al., 1993). In later follow-up work 1994, they also studied the impact of low radial confinement and low water to cement ratio on 25-mm Garford bulb cables. In this case, radial confining pressure had little impact on the load transfer capacity of this cable. As for the effect of the water to cement ratio, it was found that the load carrying capacity remained largely unchanged even at a high water to cement ratio of 0.5 (Hyett and Bawden, 1994).

The larger capacities were achieved with higher radial confinement. In addition, bearing capacity of cables increased with embedment length though not in direct proportion. Finally, cable capacity increased by 50–75% when using stiffer grouts having low water-cement ratios of less than 0.40.

Further improvements to the design were made by Macsporran (1993) when he pointed out that previous research on the performance of cables was based on constant normal stiffness conditions. This was as a consequence of the use of metal pipes, concrete blocks, or actual rock to provide confinement to the cable. However, load transfer under constant normal pressure conditions had not been studied. To this end he incorporated a modified Hoek cell (MHC) to apply constant radial pressure as shown in Fig. 3.3.

The MHC was integrated into the "split-push" testing rig. Using this design, the effect of confining pressure and water to cement ratio on cable performance was studied. A direct link was reported between the carrying capacity of the cable and confining pressure whereby carrying capacity increased with confinement. As for the impact of water to cement ratio, a low ratio resulted in larger bond capacity.

1.2.2 Single Embedment Pull Test

Stillborg (1984) had taken a different approach using concrete blocks to represent the rock mass and provide confinement to the cable, carrying out both short and long single embedment length pull-out tests. This approach has several advantages; firstly, concrete blocks more closely model the properties of a rock mass compared to metal tubes. Furthermore, both the borehole roughness and radial stiffness

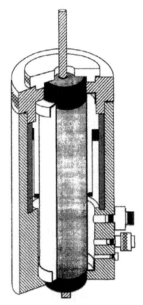

FIGURE 3.3 Cutaway section of a modified Hoek cell. *After Macsporran, G.R., 1993. An Empirical Investigation into the Effects of Mine Induced Stress Change on Standard Cable Bolt Capacity (Master thesis). Queen's University, Kingston.*

of concrete blocks better simulate that of boreholes within a rock mass. It was found that bearing capacity of the cable was not directly related to embedment length. Also the surface properties of the cable, curing conditions, and grout type all influenced load transfer to a large extent. But the design had the disadvantage in that a length of cable was left free, which allowed rotation or unraveling of the cable under load.

Farah and Aref (1986b) used a similar method to study axial behavior of cables using a fast loading rate to simulate dynamic loading environments. They compared the effects of a mortar mix of sand, water, and cement as the grout material against a concrete mix of mortar plus an aggregate. They found the cables grouted with concrete had a larger ultimate bonding strength and ductility as well as higher load at bonding failure compared to mortar-based grouts. These are desirable properties in dynamic loading environments.

Hassani and Rajaie (1990) conducted tests to study the effect of shotcrete as aggregate on load transfer of cables, finding that when using grouts having shotcrete, the peak strength of cables was larger than that of cables with traditional grout materials. Furthermore, large residual bonding strength could be attained. However, the bearing capacity and bonding stiffness decreased.

Mah (1990) just used Schedule 80 pipe to confine the grouted FCB strand and conducted tests to understand the performance of FCB used in hard rock mines. After the experiment, the author indicated that three parameters including hand-mix time, embedment length as well as water to cement ratio were the most important for FCB performance.

Similar single embedment tests were reported by Hassani et al. (1992). In their research, PVC and steel pipes were used to represent rock mass with different stiffness. They found that rock mass with larger stiffness tended to generate more confinement, enhancing the cable load transfer capacity.

Benmokrane et al. (1992) used a concrete cylinder having a diameter of 200 mm to represent the rock mass and conducted tests on two kinds of reinforcing tendons including a seven-wire cable and a deformed bar, which is shown in Fig. 3.4.

They studied the effect of grout type on bond stress-slip relationship, using six kinds of grouts. It was found that there was a difference in the load transfer mechanism between the steel bars and cables, which was later verified by Ito et al. (2001). Further work by Benmokrane et al. (1995), based on a theoretical analysis approach, he developed a model for the behavior of the rock tendons, the results of which are shown in Fig. 3.5.

This model is trilinear in nature and can be represented by Eq. (3.1):

$$\tau = ms + n \quad (3.1)$$

where τ is shear stress on the tendon–grout interface, and s is slip between the tendon and grouts.

The coefficients of m and n for three phases in the bond stress versus slip curve were depicted as:

In the case when $0 \leq s \leq s_1$

$$m = m_1 = \frac{\tau_1}{s_1} \text{ and } n = 0 \quad (3.2)$$

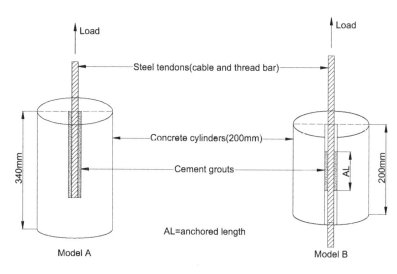

FIGURE 3.4 Concrete cylinders reinforced with two different tendons. *After Benmokrane, B., Chennouf, A., Ballivy, G., 1992. Study of bond strength behaviour of steel cables and bars anchored with different cement grouts. In: Proceedings Rock Support in Mining and Underground Construction: Proceedings of the International Symposium on Rock Support, Sudbury, Canada, pp. 293–301.*

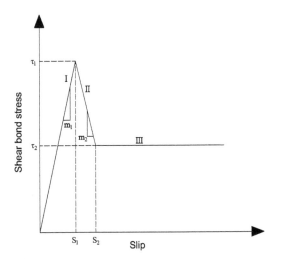

FIGURE 3.5 Trilinear bond stress-slippage model. *After Benmokrane, B., Chennouf, A., Mitri, H.S., 1995. Laboratory evaluation of cement-based grouts and grouted rock anchors. International Journal of Rock Mechanics and Mining Sciences 32 (7), 633–642.*

When $s_1 \leq s \leq s_2$

$$m = m_2 = \frac{\tau_1 - \tau_2}{s_1 - s_2} \text{ and } n = \frac{\tau_2 s_1 - \tau_1 s_2}{s_1 - s_2} \quad (3.3)$$

Finally, when $s_2 \leq s$

$$m = 0 \text{ and } n = \tau_2 \quad (3.4)$$

During the first phase of loading (indicated as I in Fig. 3.5), there is a linear relation between the bonding stress and slip. Immediately following the ultimate load (II), debonding takes place resulting in a reduction in stress. In the third and final phase (III) some level of residual stress was achieved due to interface friction.

Thompson and Windsor (1995) studied the impact of pretension on load transfer in an axial direction, finding that pretension did not seem to improve load transfer performance. This finding was later confirmed by Mirabile et al. (2010).

Martin et al. (1996) carried out tests using resin-grouted cables to study the effect of surface buttons on load transfer. It was found that those cables that had buttons had a much greater stiffness. The effect of borehole size on Garford bulb cables was also evaluated. When the borehole diameter was within the range of 25.4–35 mm, there was little change in load transfer efficiency. However, when testing from 42 to 106 mm there was a degradation in performance. Similar findings were reported by Mosse-Robinson and Sharrock (2010) who found that a smaller borehole diameter resulted in larger load transfer capacity for bulb cables.

Hyett et al. (1996) conducted 75 pull-out tests to study the effect of bulb spacing on performance of Garford bulb cables. According to their research, at shorter bulb spacing and longer embedment length as well as higher radial stiffness of the confining medium, axial load increased as did bond stiffness, which is illustrated in the graph in Fig. 3.6.

The effect of stress change on axial performance of plain and bulb cables in hard and soft

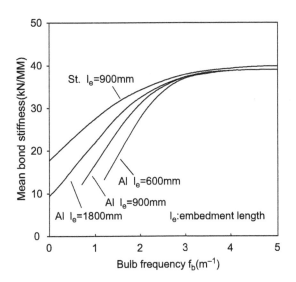

FIGURE 3.6 Effect of bulb spacing on bonding stiffness. *After Hyett, A.J., Moosavi, M., Bawden, W.F., 1996. Load distribution along fully grouted bolts, with emphasis on cable bolt reinforcement. International Journal for Numerical and Analytical Methods in Geomechanics 20 (7), 517–544.*

rock mass was studied by Prasad (1997). He reported that the load transfer capacity of the plain cables in weaker rock was found to be more sensitive to both isotropic and anisotropic stress change compared with stronger rock. However, the effect of stress change on the bearing capacity of Garford bulb cables was negligible.

Tadolini et al. (2012) studied the indentation geometry impact on the behavior of PC strands, finding the indentations had an important effect on cable load transfer whereby both cable bearing capacity and stiffness increased with indentation depth. They concluded that indentations on cable wires enhanced the mechanical interlock at the interface between the cables and grouts, which in turn improved load transfer efficiency.

1.2.3 Double-Embedment Pull Test

The double-embedment pulling test illustrated in Fig. 3.7 was initially proposed by Hutchins et al. (1990) to investigate the load transfer features of birdcage cables. This newly devised testing method differed from previous pull test methods in that it enabled the study of the effect of embedment lengths on either side of the discontinuity. However, this type of test could not properly simulate underground conditions, particularly the grout/rock interface since the tube used in the test was specially threaded internally to prevent failing along this interface.

The effect of debonding was studied whereby parts of the surface of the cable were painted. In the case of standard cables, the ultimate pull-out load decreased significantly, which was also verified by Satola (1999). This reduction in load was not repeated with birdcage cables although there was a reduction in system stiffness. There was also little effect of node location relative to the discontinuity on load transfer.

Later, Satola and Aromaa (2004) designed a new double pipe test system to investigate the impact of epoxy and zinc coatings on the performance of cables. In this arrangement, the embedment length was increased nearly tenfold to 2.0 m. It was found that the corrosion protection mechanisms of epoxy coating and zinc galvanizing on the cable surface increased the cable ultimate load capacity and stiffness to a large extent.

1.2.4 Laboratory Short Encapsulation Pull Test

Thomas (2012) reported on a modified laboratory short encapsulation pull test (LSEPT) as originally reported by Clifford et al. (2001) and shown in Fig. 3.8. Here a thick-walled steel cylinder is used to provide confinement to a sandstone core in which the cable is grouted. Incorporated between the two halves of the test cell is an antirotation device that prevents the cable from rotating or unraveling during a test.

A total of 14 different types of grouted cables were tested with the aim of evaluating the effect of cable design and borehole diameter on load transfer. As expected, there was a marked difference in load transfer with the different cable designs. The impact of borehole diameter on

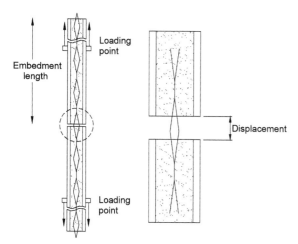

FIGURE 3.7 Double-embedment length test set up. *After Hutchins, W.R., Bywater, S., Thompson, A.G., Windsor, C.R., 1990. A versatile grouted cable bolt dowel reinforcing system for rock. In: Proceedings the AusIMM Proceedings, pp. 25—29.*

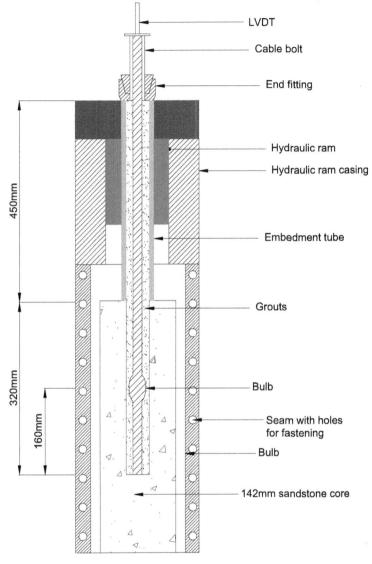

FIGURE 3.8 Schematic diagram of modified laboratory short encapsulation pull test. *After Thomas, R., 2012. The load transfer properties of post-groutable cable bolts used in the Australian Coal Industry. In: Proceedings 31st International Conference on Ground Control in Mining, Morgantown, USA, pp. 1–10.*

cables was not consistent. For bulbed and nutcaged cables, the load transfer efficiency increased with borehole diameter from 45 to 71 mm. Whereas for plain-strand cables, the load transfer efficiency decreased as borehole diameter increased from 28.5 to 61 mm.

1.3 Discussion

Previous studies found that load transfer mainly relies on the shear resistance at the cable/grout interface, and this resistance is provided by three basic mechanisms: chemical adhesion, mechanical interlock, and friction (Gambarova, 1981). However, the influence of chemical adhesion is only temporary since a small displacement of approximately 0.2 mm can damage this adhesive bond (Fuller and Cox, 1975). Hence in most instances the latter two mechanisms dominate (Stillborg, 1984). Mechanical interlock can be enhanced by the relative movement between the cable and cement grout, compressing the grout within the borehole and generating extra normal pressure at the cable/grout interface. As for friction, it occurs along the cable/grout interface as a result of shear resistance, preventing the cable from slipping, which is the most important part in determining the load transfer behavior. This also explains the reason why the load transfer efficiency of modified cables is much greater than that of conventional cables. In the case of modified cable designs, structures, such as the bulb on the strand, especially when they are filled with grout, increases the geometric mismatch and normal pressure at the cable/grout interface, resulting in the much higher load transfer capacity.

1.4 Conclusions

Many different testing procedures and equipment have been developed and used to determine the influence of a wide range of parameters on the load transfer behavior of fully grouted cables in the axial direction. Based on previous research, the impacts of critical parameters including the rock mass confinement, the cable surface geometry, water to cement ratio, embedment length, and so forth on the axial strength of cables are well understood. However, there is a lack of knowledge on the load transfer behavior of the wide range of cables currently available for use in ground control particularly based on a common testing methodology. A project is underway with support of the Australia Coal Association through their research funding organization, the Australian Coal Association Research Program (ACARP) to devise a testing facility. This project focuses on studying the axial performance of some particular types of cables and assessing the impacts of corresponding factors on them. The design of the new testing facility, which is shown in Fig. 3.9, is based on the recommendation design principles outlined in the British Standard with modifications that accommodate the requirements of the wider range of cable designs used in Australia.

The main objectives of this project mainly include the following aspects:

To design and establish a robust axial test rig for kinds of fully grouted cables; To evaluate the behavior of cables including the twin strand, PC strand, indented PC strand, sumo bolt, and TG bolt.

1. To assess the effects of soft and medium rock confinement on those types of cables.
2. To study the impacts of normal and larger boreholes on corresponding cables.
3. To investigate the influences of normal and higher grout strength on the load transfer of cables.

2. THEORETICAL ANALYSIS OF LOAD TRANSFER MECHANICS

2.1 Introduction

Fully grouted cables have been used in the mining industry for more than 40 years. With the incremental tendon length, they are effective in reaching into the depth of rock mass and reinforcing large rock blocks together to increase the rock internal strength and prevent bed separation from occurring. Although there is a drastic improvement in the amount of cable utilization in underground mines, some accidents near the cabling area still happen; for instance, the falling of large rock blocks, which is led by the failure of cable supports, especially the relative movement on the cable-grout interface (Hutchinson and

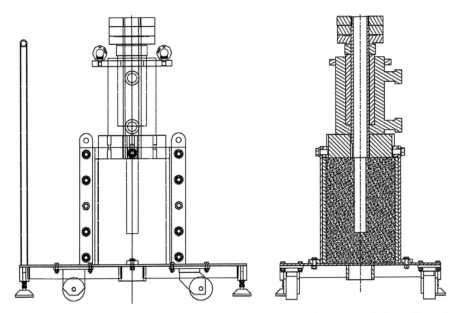

FIGURE 3.9 Schematic of the design of a proposed new laboratory short encapsulation pull test facility.

Diederichs, 1996). This is related to the generic area of a load transfer issue and demanding modifications on the design of cables.

The engineering design of cable reinforcing is primarily dependent on load–displacement curves of strands, which are obtained from pull-out experiments on grouted cables. Since cables were introduced into mining, numerous laboratory and field tests have been conducted. Cox and Fuller (1977) conducted "split-pipe pull" tests on cables, suggesting that light rust on the strand could improve the bearing capacity of cables. Stillborg (1984) performed long embedment length pull tests, discovering that an increased embedded length resulted in a larger pull force. Beyond that, Maloney et al. (1992) undertook field tests and concluded that stress change evoked by mining resulted in bad performance of traditional strands. According to Satola and Aromaa (2003), who carried out double-embedment length pull tests, the corrosion coating on the strand increased the bearing capacity. This was followed by Thomas (2012), who accomplished LSEPTs on nutcaged cables with results that the nutcage geometry improved the stiffness of cables up to two orders of magnitudes. These tests are successful in understanding the behavior of grouted cables and helping engineers to modify relevant parameters in practice to improve the load transfer efficiency of strands.

However, compared with physical experiments, limited research has been carried out on the theoretical analysis to explain the load transfer mechanics between rock mass and cables. This chapter aims to present a review of previous theoretical studies in this area. Relevant analytical models and their basic deducing process are given. Following that, some parametric studies that relied on those models are referred to. The authors also provide the shortcomings and applied range of each model. Based on these shortcomings, the reasons for the different performance of plain and modified cables as well as two different bond failure modes, are ultimately concluded.

2.2 Load Transfer Mechanics

2.2.1 Computational Formula

A conceptual model named the bond strength model (BSM) was proposed by Yazici and Kaiser (1992) to study load transfer mechanics of fully grouted cables. The surface of a cable was simplified to be a zigzag geometry like a rough joint (Fig. 3.10).

Under this assumption, when a cable is pulled out, the bond failure is a friction-dilation procedure. Then, the joint strength concept (Patton, 1966) was adopted to achieve the bond strength, expressed as:

$$\tau = \sigma \tan\left(i_0[1 - \sigma/\sigma_{Cg}]^\beta + \varphi\right) \quad (3.5)$$

Major elements including axial displacements, lateral displacements, pressure on cable-grout interface and bond strength were connected in the BSM shown in Fig. 3.11.

There are four quadrants in Fig. 3.11. The first quadrant showed the pull-out test graph. The second quadrant presented the relationship between bond strength and pressure on the cable-grout interface. Variation of lateral displacements with axial displacements increasing was showed in quadrant 3. Finally, lateral displacement or dilation created pressure on the cable-grout interface as presented in the fourth quadrant. To better illustrate this relationship, based on the fact that grout might split due to the pressure, the status of the grout cylinder was classified as elastic, fully split, and partially split with an elastic portion.

When the grout cylinder was elastic (Fig. 3.12A), thick-walled cylinder equations under plane strain state were used to acquire the relationship between displacement and pressure on the cable-grout interface, which was:

$$u_{1g} = M' p_1 \quad (3.6)$$

$$M' = \frac{(1+v_g)(1-2v_g)}{E_g} \frac{r_1^2 - r_2^2/X}{r_2^2 - r_1^2} r_1$$

$$+ \frac{(1+v_g)}{E_g} \frac{(1-1/X)r_1^2 r_2^2}{r_2^2 - r_1^2} \frac{1}{r_1} \quad (3.7)$$

If the grout was fully split (Fig. 3.12B), wedge theory was used and the relation was changed as:

$$u_{1g} = M'' p_1 \quad (3.8)$$

$$M'' = r_1\left((1+v_r)/E_r + \left(1-v_g^2\right)/E_g \cdot \ln(r_2/r_1)\right) \quad (3.9)$$

When the grout cylinder was partially split, the grout was classified into two parts. They

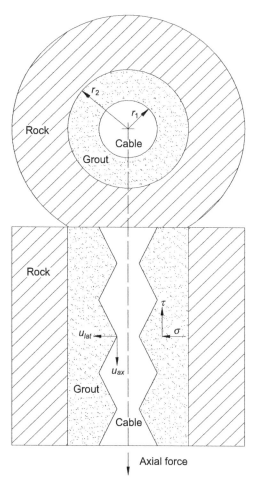

FIGURE 3.10 Geometry diagram of a rough cable. *After Yazici, S., Kaiser, P.K., 1992. Bond strength of grouted cable bolts. International Journal of Rock Mechanics and Mining Sciences Geomechanics Abstracts 29 (3), 279–292.*

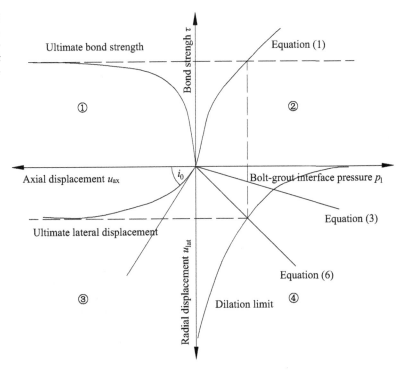

FIGURE 3.11 Schematic of bond strength model. *After Yazici, S., Kaiser, P.K., 1992. Bond strength of grouted cable bolts. International Journal of Rock Mechanics and Mining Sciences Geomechanics Abstracts 29 (3), 279–292.*

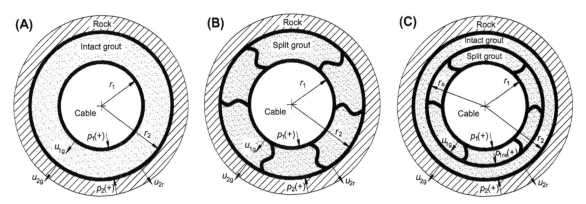

FIGURE 3.12 Grout column in different conditions. (A) Elastic grout column; (B) fully spilt grout column; (C) partially spilt grout column. *After Yazici, S., Kaiser, P.K., 1992. Bond strength of grouted cable bolts. International Journal of Rock Mechanics and Mining Sciences Geomechanics Abstracts 29 (3), 279–292.*

proposed an imaginary crack tip boundary with radius of r_e shown in Fig. 3.12C. In the inner part of this boundary ($r_1 < r < r_e$), grout was fully split while in the outer part of this boundary ($r_e < r < r_2$), the grout remained intact. Therefore, the interface displacement for partially split grout was computed through adding displacements for fully split and intact grout and expressed as:

$$u_{1g} = M''' p_1 \qquad (3.10)$$

$$M''' = r_1 M'_e / r_e + r_1 \Big((1 + v_r)/E_r \\ + \Big(1 - v_g^2\Big) \ln(r_e/r_1) \Big/ E_g \Big) \qquad (3.11)$$

The equations including (3.6), (3.8), and (3.10) described the stress path in quadrant 4 of Fig. 3.11. To connect this quadrant with the ultimate bond strength, the "dilation limit," which was the limiting lateral displacement during the pull-out process, was determined. An empirical equation was then adopted by them to illustrate this limit:

$$u_{lat} = u_0 (1 - p_1/\sigma_{Cg})^{B/\sigma_{Cg}} \qquad (3.12)$$

Then, this BSM was verified with physical experiments. According to the comparison, the BSM was effective in studying the load transfer mechanics of cables. Therefore, they conducted analytical parametric research with this model, discovering that the bearing capacity of cables improved with high rock-to-grout stiffness ratio and large grout strength. In addition, it was found that the load transfer capacity of cables decreased with borehole size increasing. However, this is not consistent with results obtained by Martin et al. (1996) whose research shows that borehole size has no effect on the load transfer capacity of cables.

From BSM, it can be seen that the bond strength of cables is generated by friction between strand and grout, indicating that any pressure variation on the cable-grout interface will influence the bond strength significantly.

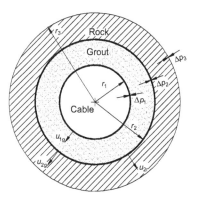

FIGURE 3.13 Cross-section of a grouted cable under stress change. *After Kaiser, P.K., Yazici, S., Nosé, J.P. 1992. Effect of stress change on the bond strength of fully grouted cables. International Journal of Rock Mechanics and Mining Sciences & Geomechanics Abstracts 29 (3), 293–306.*

Therefore, Kaiser et al. (1992) implemented stress change into BSM to study the effect of stress change on load transfer of cables.

A cross-section of a grouted cable under stress change Δp_3 in rock is shown in Fig. 3.13. This stress change induced stress variation on grout-rock and cable-grout interfaces which are Δp_1 and Δp_2, respectively. According to the thick-walled cylinder theory, when the grout column was elastic, the displacement $u_{1g\Delta}$ on cable-grout interface caused by stress change was obtained:

$$u_{1g\Delta} = m' \Delta p_3 \qquad (3.13)$$

$$m' = -(1 + v_g)(2 - 2v_g) Q r_2^2 r_1 / \big(E_g (r_2^2 - r_1^2)\big) \qquad (3.14)$$

Adding this displacement with the dilation displacement which was given in original BSM, they acquired the displacement on cable-grout interface under the elastic grout state as:

$$u_{1g} = M' p_1 + m' \Delta p_3 \qquad (3.15)$$

With similar methods, the interface displacement when grout was fully split was also achieved:

$$u_{1g} = M'' p_1 + m'' \Delta p_3 \qquad (3.16)$$

$$m''' = -2r_2(1 - v_r^2/E_r) \quad (3.17)$$

As for the partially split grout, the interface displacement was calculated as:

$$u_{1g} = M''' p_1 + m''' \Delta p_3 \quad (3.18)$$

$$m''' = -(1 + v_g)(2 - 2v_g)Qr_2^2 r_e/(E_g(r_2^2 - r_e^2)) \quad (3.19)$$

By substituting new established displacements into the BSM, the bond strength of cables under stress change was acquired. The calculating results were confirmed well with laboratory tests, showing that the BSM was also capable of studying the effect of stress change on bearing capacity of cables. According to the results, it was indicated that a stress increase improved the bond strength while a stress decrease reduced the strength; this impact was more distinct, especially within relatively soft rock mass.

The BSM is successful in predicting the bond strength of cables and the effect of relevant parameters on it; however, this model is only suitable for cases where the cable does not rotate during the pull-out process. Beyond that, rotation of a standard strand in the grout column is likely to happen, especially when the confining pressure is low. In this situation, the BSM shows its weakness. Therefore, based on the results obtained from the laboratory for nonrotating and rotating pull-out tests, Diederichs et al. (1993) modified the BSM by determining two dilation limits with the same equation of (3.12). One was used for the nonrotating case while the other was suitable for the cable untwisting situation. According to comparison with experimental data, the modified BSM has confidence in predicting the bond strength of cables under those two corresponding conditions. Furthermore, with calibration, this modified model is capable of determining the shear strength at different axial slippages of the cable and thereby can be used to simulate the pull-out tests. As a result of these advantages, this modified model has been merged into one numerical modeling code named CABLEBND which was designed to compute the bond strength of cables.

After that, Benmokrane et al. (1995) proposed a trilinear model to illustrate the relationship between bond stress and slip on the cable-grout interface, which is shown in Fig. 3.14.

The universal formula for this model was expressed as:

$$\tau = ms + n \quad (3.20)$$

When $0 \leq s \leq s_1$

$$m = m_1 = \tau_1/s_1 \text{ and } n = 0 \quad (3.21)$$

When $s_1 \leq s \leq s_2$

$$m = m_2 = \frac{\tau_1 - \tau_2}{s_1 - s_2} \text{ and } n = \frac{\tau_2 s_1 - \tau_1 s_2}{s_1 - s_2} \quad (3.22)$$

When $s \geq s_2$

$$m = 0 \text{ and } n = \tau_2 \quad (3.23)$$

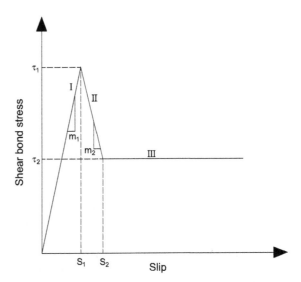

FIGURE 3.14 Trilinear bond stress-slippage model. *After Benmokrane, B., Chennouf, A., Mitri, H.S., 1995. Laboratory evaluation of cement-based grouts and grouted rock anchors. International Journal of Rock Mechanics and Mining Sciences 32 (7), 633–642.*

With this model, the pull tests of cables were explained as three stages. In this first stage, the bond strength had a linear relation with slippage while debonding occurred in the second stage. Lastly, the cable maintained its residual stress by friction on the cable-grout interface. Therefore, it indicated that friction was the most significant component in determining the load transfer of grouted cables. This constitutive model is satisfactory in explaining the pull-out process of cables but it cannot be adopted to compute the bond stress for different kinds of cables since the parameters given by them are obtained from axial tests on plain cables.

Although the previous models are able to study the load transfer between rock mass and grouted cables, they are only applicable for constant normal stiffness (CNS) conditions in which a cable is pulled out from confining pipes, concrete, or actual rock blocks. As for the constant normal load (CNL) environment, Hyett and Bawden (1996) and Hyett et al. (1995b) proposed another constitutive model to study the load transfer mechanics. The pressure on the cable-grout interface was expressed with the lateral and axial displacement on the interface:

$$dp_1 = \frac{(k_1/p_1)du_{ax} - du_{1g}}{(k_1/p_1^2)(u_{ax}-1) + K_0 v_0^2 / (K_0 v_0 + p_1)^2 + 1/K_{rc}} \quad (3.24)$$

As for the pull-out force, it was composed of the friction resistance along the grouted length and the untwisting force along the free length, expressed as:

$$dF_{ax} = \partial F_{ax}/\partial p_1|_{u_{ax}} \cdot dp_1 + \partial F_{ax}/\partial u_{ax}|_{p_1} \cdot du_{ax} \quad (3.25)$$

Substituting Eqs. (3.24) into (3.25), a differential formulation which represented the cable-grout interface deformation was acquired:

$$\begin{bmatrix} dF_{ax} \\ dp_1 \end{bmatrix} = \begin{bmatrix} K_1^i & K_2^i \\ K_3^i & K_4^i \end{bmatrix} \begin{bmatrix} du_{1g} \\ du_{ax} \end{bmatrix} \quad (3.26)$$

Since the status of grout annulus had an important role in determining the load transfer process, those authors also classified it as three different situations including tightly closed (elastic), fully open (split), and partially closed (partially split). After that, the cable-grout interface pressure and displacements on grout-rock interface could be expressed as:

$$\begin{bmatrix} dp_1 \\ du_{2g} \end{bmatrix} = \begin{bmatrix} K_1^a & K_2^a \\ K_3^a & K_4^a \end{bmatrix} \begin{bmatrix} dp_2 \\ du_{1g} \end{bmatrix} \quad (3.27)$$

Through combining Eqs. (3.26) with (3.27), the constitutive law between the cable pull-out force and the axial displacements was expressed as:

$$dF_{ax} = \frac{K_1^i \cdot K_1^a}{K_3^i - K_2^a} dp_2 + \frac{K_2^i \cdot K_3^i - K_2^i \cdot K_2^a - K_4^i \cdot K_1^i}{K_3^i - K_2^a} du_{ax} \quad (3.28)$$

Using Eq. (3.28), the pull-out test process of cables under CNL setting was simulated. Physical experiments were also conducted so as to confirm the accuracy of this analytical model. During the experiment, an MHC was designed to provide constant radial pressure on seven-wire cables. Based on the comparison, this constitutive model simulated the pull-out tests of cables under CNL circumstances relatively well. It should be noted that even though this model was deduced under CNL conditions, it was extended to CNS situations under the assumption that the borehole wall had a linear relation with the grout-rock interface radial displacement, which was also verified by later experiments (Moosavi, 1997). Therefore, this model is appropriate for any boundary conditions including both CNL and CNS. With this superiority, the algorithm has already been coupled to the software of Phase 2, enabling simulating of the seven-wire strand geometry to be conducted. In addition to that, Moosavi and Grayeli (2006) incorporated it into the initial discontinuous deformation analysis (DDA) program, assisting researchers to realistically reinforce the blocky rock mass by means of fully grouted cables. Nevertheless, this model is barely designed for conventional cables since the pull-out force in it is composed of friction and untwisting force.

For modified cables, the untwisting force hardly occurs because of the appearance of deformed structures along the cable length. In view of this matter, Moosavi (1999) adopted another approach to study the load—displacement reaction of modified cables under CNL environment. He considered that the shear stress on the cable-grout interface ascended with the axial displacement increased until the bounding shear strength. But that ascending rate or Keff was not constant and gradually decreased based on the initial shear stiffness, which was defined as:

$$K_{eff} = K_s(1 - \tau/\tau_m) \quad (3.29)$$

Then the relation between shear stress and axial displacement of cables was expressed in an incremental form:

$$\Delta\tau = K_s(1 - \tau/\tau_m)\Delta u_a \quad (3.30)$$

With Eq. (3.30), the shear stress was computed incrementally with the axial displacement rising. Since the pull-out load had a certain relationship with this shear stress, it was stated as:

$$F_{ax} = \tau A_s \quad (3.31)$$

2.2.2 Test Verification

The constitutive model between the pull-out load and the axial displacement for modified cables was achieved. By using this model, also extended to the CNS situation, the author simulated the pull-out process of the Garford bulb and nutcase cables under both CNL and CNS environments. Close simulation of physical experiments was acquired (Fig. 3.15), indicating that this model was capable of studying the load transfer mechanics of modified cables under any boundary conditions.

Based on this model, Moosavi et al. (2002) investigated the impact of bulb spacing on the ultimate bearing capacity of Garford bulb cables, finding that decreasing the bulb spacing improved the bearing capacity as well as stiffness. This is in accordance with conclusions achieved from long-pipe tests on 25-mm Garford bulb cables with different bulb spacing conducted by Hyett et al. (1996).

2.3 Discussion

To better understand the mechanics of load transfer between rock mass and fully grouted cables, some researchers conducted theoretical analysis in this area. Although the process of transferring load is replied on chemical adhesion, mechanical interlock and friction on the bolt—grout interface (Littlejohn and Bruce, 1975), the component of adhesion is always neglected in analytical models. This is because, firstly, the adhesive strength is easily removed after only a few millimeters of relative displacement of cables (Fuller and Cox, 1975). In addition to that, the gradual reduction of adhesion is difficult to be simulated with appropriate models (Moosavi, 1999). Therefore, the cable-grout assembly can be analyzed as a nonadhesive system. Since the helical geometry of the strand and ambient grout produce a rock—joint interface, when an installed cable is activated by rock mass movement and suffers axial displacement, grout will shift in the adverse direction. Those features show similarities with shear behavior of rock joints. Consequently, researchers regarded the grouted cable system as columniform joints and adopted theories in rock joints to investigate the load transfer within cable reinforcing systems. However, even though the cable-grout interface is similar to the rough joint surface, distinctions between them still exist. Primarily, the free length of a strand is likely to rotate, which seldom occurs in rock joints. Besides, when shear movement happens, roughness on rock joints may crack on two sides

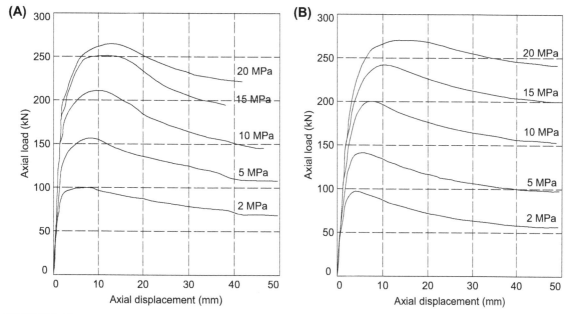

FIGURE 3.15 Pull-out curves for modified cables obtained from physical tests and analytical modeling. (A) Pull-out curves for modified cables; (B) analytical modeling. *After Moosavi, M., 1999. A comprehensive comparison between bond failure mechanism in rock bolts and cable bolts. In: Proceedings the Congress of the International Society for Rock Mechanics, pp. 1463–1466.*

of joints while the grout column is the unique one which will be cracked in the cable reinforcing case.

2.4 Conclusions

Based on the previous analytical models, it is found that the bearing capacity and stiffness of modified cables is apparently larger than that of conventional strands. One important reason for that is when a modified cable is pulled out from a borehole, the deformed geometry on the strand compresses the surrounding grout extremely, resulting in more dilation force. The reaction force of this increasing dilation will, in turn, grip the cable radially, allowing more load to be transferred from unstable rock mass to stable rock mass within the strand. It is also found that standard cables are more sensitive to the circumjacent conditions, such as the high water to cement ratio and stress change, indicating that modified cables will be more suitable for wicked environments like soft rock mass quality. Beyond that, the load transfer mechanics of seven-wire and modified cables lead to two different bond failure modes. For standard cables, the untwisting of cables is the major type because of the helical geometry and low torsional rigidity of cables. However, for modified cables, bond mainly fails by grout shearing on the cable-grout interface, which is attributed to the presence of specific geometries, such as bulbs, preventing modified cables from rotating and making strands shear on the cable-grout interface like deformed bars.

Theoretical models are proved to be effective in analyzing the load transfer process within cable reinforcing systems and improving the design of cables. However, compared with physical and field tests, limited number of analytical research has been done and there is still scope for enrichment with this medium. Previous calculations explained the cable-grout interface response when cables are subjected to pure axial tension and the constituent of cable shear displacement induced by bending or relative movement of rock mass strata is not considered. Therefore, a suitable model, which includes the shear displacement of a cable, should be created to study the load transfer behavior of fully grouted cables under combined axial and shear loading environment.

3. IMPACTING FACTORS ON THE DESIGN FOR CABLES

3.1 Introduction

The use of fully grouted cables as an additional form of rock reinforcement system was introduced into the underground mining industry some decades ago (Thorne and Muller, 1964). With the combined advantages of higher load-bearing capacity compared with conventional rock bolts, their longer length, and their ability to anchor further into the rock mass preventing bed separation from occurring along planes of weakness, cables are increasingly being used in the more arduous ground conditions, such as at intersections in underground roadways in coal mines, the backs of stopes in hard rock mines and permanent openings in tunnels and other civil projects (Hutchinson and Diederichs, 1996). With this increasing use and application, new manufacturers are entering the supply market such that there is now a plethora of cables on the market of different dimensions and design that will consequently all perform differently. The issue faced by end-users is how to compare these different models and match the performance to different requirements.

To be able to differentiate between the failure modes of the various rock reinforcement systems, Windsor (1997) classified the reinforcement systems into three fundamental types; namely continuously mechanically coupled (CMC) systems, continuously frictionally coupled systems, and discretely mechanically or frictionally coupled systems. Using this classification method, the fully grouted cables belong to the CMC systems. A substantial amount of research has been reported on the axial behavior of cables over the past 30 years. Among the methods used, the simplest is the single embedment length test. In this case, one section of the cable is installed in the confining medium, such as a rock or concrete test sample while the other section is left protruding from the sample to which a force is applied. Due to its simple design, many researchers adopted this method to study the performance of cables. For instance, Stillborg (1984) tested the effect of embedment length on the bearing capacity of cables, finding that the relationship between them is not directly linear. Farah and Aref (1986a,b) investigated the dynamic characteristics of cables, saying that the concrete specimen is more desirable in the dynamic loading environments rather than the cement paste specimen. Mosse-Robinson and Sharrock (2010) evaluated how the large borehole diameter influenced the pulling load versus displacement relationship of cables, indicating that the smaller borehole size results in a relatively lower load transfer capacity. However, a common problem in each of these tests is the tendency of the free section of the cable tendon to rotate due to its special helical surface geometry and consequently the measured pulling force is lower than the true value (Hyett et al., 1992a,b).

To overcome this shortcoming, two alternate approaches to testing methods have been devised. The "split-pipe pull test" was first reported by Fuller and Cox (1975). In this method, a cable is grouted into two adjoining lengths of mild steel tube. One length of tube is used to represent the rock material while a second longer length of tube provides confinement preventing rotation during loading. A washer can be placed between the two lengths to simulate a discontinuity. This approach has been used by researchers to evaluate the influences of many parameters on the load transfer behavior of cables, including surface rusts (Cox and Fuller, 1977), birdcaged geometry (Goris and Conway, 1987), grout curing temperature (Goris, 1990), nutcaged geometry (Hyett et al., 1993), bulb size (Hyett et al., 1995a,b), and so forth. A modification of this method was reported by Hutchins et al. (1990) who developed the double-embedment length test (DELT). Here the grouted cable is confined by two steel tubes but each of similar length. The advantage is that the load transfer of tendon itself can be studied on either side of the discontinuity and because of this, the DELT found widespread use (Strata Control, 1990; Renwick, 1992; Satola and Aromaa, 2003, 2004; Satola, 2007) and was incorporated into the British Standard (BS7861-1, 2007).

Several deficiencies were also found with this approach. One of the more important being the metal tube that represents the confining medium is largely different in behavior compared to rock, the former being much stiffer, resulting in much larger pull-out forces being recorded. What's more, the internal surface of the metal tube is nearly always threaded to ensure good contact between the grout and tube. However, this in turn prevents any study of the failure mode at the grout/rock interface. To overcome this issue the LSEPT was proposed by Clifford et al. (2001). This method is similar to that used in anchorage testing of rock bolts (Hagan, 2003). The method uses a cylindrical cored sample placed within a biaxial cell to confine the cable within the embedment length section. This testing method has been adopted in the UK and included in the current British Standard, BS7861-2 (2009) (Reynolds, 2006; Bigby, 2004).

A study by Thomas (2012) reported several shortcomings with the current LSEPT as it was mainly devised for cables typically in use in the UK, but not suited to the modified bulb designs used elsewhere.

For the purpose of better understanding the load transfer behavior and failure modes of fully grouted cables used in the Australia underground coal mining industry, Hagan et al. (2014) reviewed the LSEPT design as part of an ACARP-funded project. The purpose is to develop a more robust axial test rig capable of testing a wide range of cable designs. The results of a critical analysis of the design process including the important parts of the equipment suggested for use in testing are explained in this paper. Furthermore, the testing procedure of cables with this equipment is outlined.

3.2 Laboratory Design

3.2.1 Components of the Test Rig

The proposed modified axial test rig is based on the LSEPT with changes that overcome its major disadvantages as mentioned earlier. In order that the cementitious grout can be easily poured into the borehole and obtain a good bond with the cable, the equipment is aligned in a vertical direction, as shown in Fig. 3.16.

The test unit is comprised of the followings elements: embedment length section in concrete; bearing plate; anchor length section in a steel tube; and terminating device. A double-acting hollow cylinder, specially designed electrically actuated hydraulic power pack and accompanying load cell are used to apply a load near the midpoint of the embedded cable adjacent to the bearing plate, while a linear variable differential transformer (LVDT) is used to measure displacement. Detailed information regarding the important parts within this equipment is given in the following sections.

3.2.2 Dimensions of the Test Sample

The lower section of the test rig is the most important in this test arrangement. Within this section the cable is embedded in a rock-like

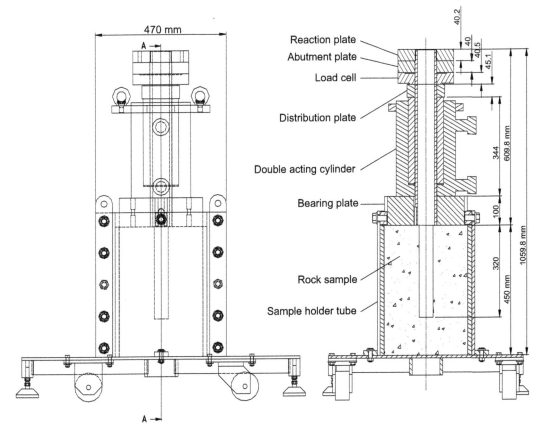

FIGURE 3.16 Front and side views with dimensions of the various components in the test rig.

material. While the BS7861-2, 2009 standard calls for a cylinder of sandstone rock sample, the much larger diameter of 300 mm in this modified test is more than double the size sample in the standard. Obtaining rock samples of size for each test can be difficult compounded by needing to be homogeneous and similar in material properties. Hence the test samples are cast using a cementitous grout of consistent strength.

During the testing of a cable, the friction-dilation mechanism between the cable and surrounding material is activated leading to stresses being induced in the material. These stresses can therefore eventually result in dilation and radial cracking dependent on the size of the sample and hence impact on the load transfer performance (Chen et al., 2014a,b). Previous work by Rajaie (1990) found a correlation between test sample diameter and pull-out force of a cable as shown in Fig. 3.17. He observed there was a minimum diameter of test sample of approximately 250 mm above which there was further increase in pull-out force, that is, the size of the sample no longer influenced the performance of the cable.

This implies that for the size of the test sample to have a minimal effect on the performance of a cable, the diameter of a test sample needs to be at least 250 mm. However, this result was for the case of a 15.2-mm plain-strand cable, whereas

3. IMPACTING FACTORS ON THE DESIGN FOR CABLES   171

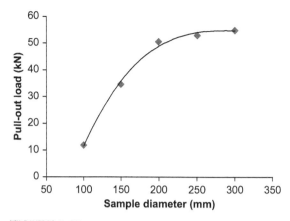

FIGURE 3.17 Variations in pull-out force or bearing capacity of a plain-strand cable. *After Rajaie, H., 1990. Experimental and Numerical Investigations of Cable Bolt Support Systems (Ph.D. thesis (Unpublished)). McGill University, Montreal.*

for many cables in use today, this minimum size may not be applicable especially for the case of high capacity modified cables due to the increasing dilation behavior on cable/grout interface, which has already been verified by Holden and Hagan (2014). It is, therefore, appropriate to determine whether there is any further influence of test sample diameter on the pull-out force of modified cables.

Cylindrical test samples with five different diameters ranging from 150 to 508 mm are cast using a cement-based material, examples of which are shown in Fig. 3.18. The samples were cast with holes having a diameter of 42 mm and length of 280 mm for installation of a cable. A sumo strand cable manufactured by Jennmar Australia was selected as it has one of the largest load transfer capacity in Australian underground coal mining. In this way, the results obtained would be no less applicable to other lower load transfer cables. The bulb with a diameter of 36.5 mm was located midway on the borehole. The cable was grouted in the hole using a polyester resin.

A hydraulic cylinder was used to provide the axial pull-out force to the cable during each test in which force and displacement were recorded with a typical load/displacement characteristic graph shown in Fig. 3.19.

It can be seen from Fig. 3.20 that the maximum pulling force of the Sumo strand increases with the test sample diameter in a similar fashion to that observed by Rajaie (1990). A major difference in this case is that pull-out load continued to increase beyond 250 mm. Indeed, it is only after 500 mm had been reached that it is shown to have plateaued but the undulation

FIGURE 3.18 Range of test samples of differing diameters.

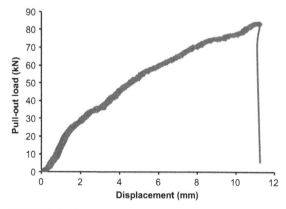

FIGURE 3.19 Characteristics of sumo strand confined by the sample with a diameter of 305 mm.

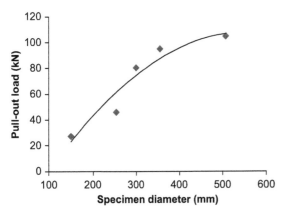

FIGURE 3.20 Influence of concrete sample size on the bearing capacity of the sumo strand.

can occur with some cables, as it tends to unwind when under load. Accordingly, a steel tube is used in this design to hold the concrete sample. Moreover, this tube is split in the middle, as shown in Fig. 3.21. Consequently, after the test, the sample holder tube can be removed and the concrete sample can be cut to check the failure status within the cable reinforcing area.

3.2.3 Anchor Length Section

An anchor tube is set up within the anchor length section, which is used to grip the cable and prevent the cable from unwinding during the pull-out process. The anchor tube shown in Fig. 3.22 has an internal diameter of 50 mm, which is sufficient to contain all types of cables. In addition, a high strength cementitious grout with low viscosity is used to fill the gap between the internal surface of the anchor tube and the cable. It should be noted that the internal surface of this anchor tube is threaded with a pitch of 2 mm and a depth of 1 mm in deep to ensure good bonding contact between the anchor tube and the cable.

point is likely to be as low as 330 mm. Therefore, the sample holder tube should be designed to have a diameter of approximately 330 mm.

Each of the test configurations was replicated three times with the results shown in Fig. 3.20.

As for the construction of the test sample holder tube, Thomas (2012) stated that the design of the biaxial cell used in the British Standard could not deal with the rotation that

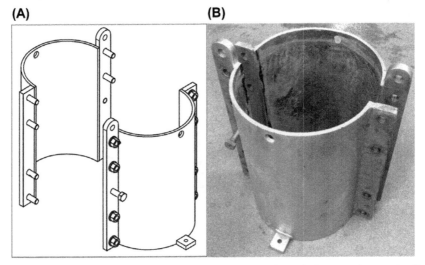

FIGURE 3.21 Concept design of split tube and steel confinement tube. (A) Concept design of split tube; (B) steel confinement tube.

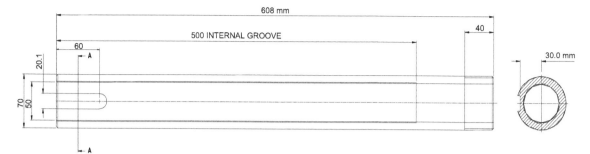

FIGURE 3.22 Anchor tube with fully threaded internal surface.

The hydraulic cylinder is installed along the anchor tube while the abutment plate, load cell, and the reaction plate are placed at the top end as shown in Fig. 3.22.

3.2.4 Bearing Plate

The bearing plate is an important element in the test unit. Its purpose is to distribute the axial pull-out load from the hydraulic cylinder evenly across the full surface of the test sample. More importantly, the hole size in the middle of the bearing plate has a crucial role in determining the failure mode of this cabling system. However, there is only limited research regarding the hole diameter effect (Hagan, 2004). Generally, it is common to see that in some tests the internal diameter of the bearing plate is of similar size as the rock tendon outside diameter. As a result, the grout annulus within the borehole is constrained by the bearing plate and failure must occur at the bolt/grout interface. This arrangement does not simulate reality where there is no confinement around the cable at the surface of a discontinuity where the ground separates. In fact, there is no restriction around the borehole when a rock joint develops between the embedment and anchor length in reality (Thomas, 2012), hence the bearing plate used in the laboratory should have a larger internal hole diameter compared to the borehole size.

On the other hand, Khan (1994) found that if the internal hole of the bearing plate was too large (over 100 mm bigger than the cable diameter), it would have an adverse impact on the results as is shown in Fig. 3.23.

Considering these problems, a bearing plate with an internal diameter of 70.4 mm is designed as shown Fig. 3.24. Since the largest borehole within all tests is limited to 50 mm, there is a minimum difference of 20 mm between the borehole diameter and the internal hole diameter of the bearing plate, allowing the grout column and concrete surrounding the borehole to be liberated. By this means, different failure modes of the cabling system that commonly occur in the field can be studied properly (Jeremic and Delaire, 1983).

3.2.5 Antirotation Devices

The rotation behavior of cables under load can untwist itself from the confining medium, resulting in a lower pull-out force. This can be prevented successfully within the embedment length by locking the thick-wall steel tube that confines the concrete sample to the anchor tube via keyways and a locking key as shown in Fig. 3.25. As the bearing plate is prevented from rotation with respect to the sample holder tube by two keyways, the whole system within the anchor length section can move along the axial pull-out direction but is constrained from rotation.

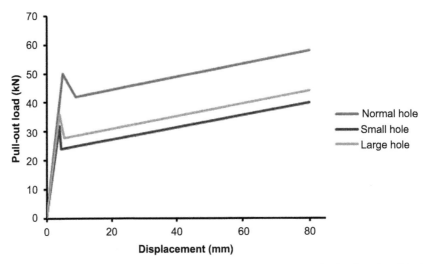

FIGURE 3.23 Influence of bearing plate hole diameter on pulling load. *After Khan (1994).*

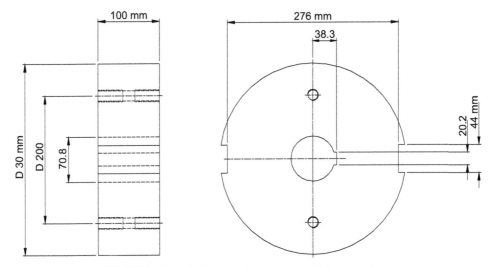

FIGURE 3.24 Outline and dimensions of the bearing plate.

3.3 Testing Procedure

3.3.1 Concrete Sample Preparation

To cast the test samples, a cylindrical mold made from cardboards with a length of 450 mm and an outside diameter of 310 mm is prepared as shown in Fig. 3.26A. The cement-based material is poured into the molds to create the test samples, as shown in Fig. 3.26B.

It should be noted in the center of each mold is placed a plastic pipe around which is wound a plastic tube of 5-mm thickness with a pitch or lay length of 20 mm, which is shown in Fig. 3.27A. This tube is adhered on the basement

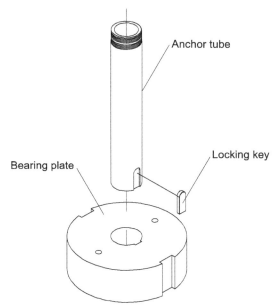

FIGURE 3.25 Location of keyways and the associated locking key.

board. Its purpose is to create a rifled borehole within the sample as shown in Fig. 3.27B. After the material has set for 24 h, both the middle tube and the outer cardboard mold are removed.

All test samples are cured fully submerged for a minimum of 28 days before use.

3.3.2 Cable and Anchor Tube Installation

The diameter of the borehole pipe in each test sample matches the required diameter for the particular cable to be tested. The cable is grouted in the test sample and anchor tube using a low viscosity grout (Stratabinder) that is commonly used in bottom-up grouting of cables in boreholes.

To ensure no relative slippage between the cable and the anchor tube, three measures are taken. First, the mix of grout used in the anchor tube has a slightly higher strength to attain a stronger bond. Second, a bulb in the cable is located midway in the anchor tube to enhance the mechanical interlock between the cable and grouts. Finally, an ending plate or barrel and wedge system is used to secure the top of the cable.

3.3.3 Setting up of the Hydraulic and Monitoring System

The bearing plate is placed on the concrete sample and coupled with the sample holder tube as well as the anchor tube with the locking

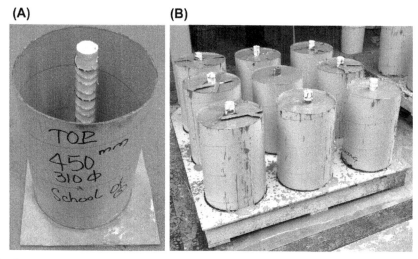

FIGURE 3.26 Concrete sample mold and casting process of concrete samples. (A) Concrete sample mold; (B) casting process of concrete samples.

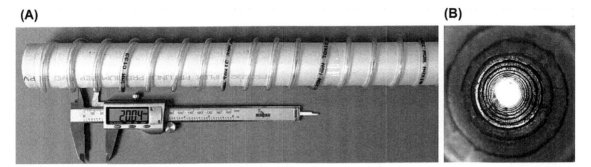

FIGURE 3.27 Spiral plastic tube wound around outside of the pipe and the corresponding rifled borehole. (A) spiral plastic tube wound around outside of the pipe; (B) corresponding rifled borehole.

nut together with the locking key, respectively. The hydraulic cylinder is seated on the bearing plate followed by the distribution plate, load cell, abutment plate, and finally the reaction plate are fixed along the anchor tube one on top of the other. A displacement transducer is attached on the testing frame to monitor and record displacement during the pull-out process. This arrangement of the different elements is shown in Fig. 3.28.

Before a test begins, the load cell and displacement transducer are calibrated. Once the test begins, an increasing pressure is applied by the hydraulic pump to the cylinder. At the same time, the axial displacement advances at a constant rate of 0.3 mm/s under constant displacement control. The cable is loaded until failure occurs within the embedment length section. During this process, the NI LabVIEW system records data from the pressure and displacement transducers at a sampling interval of 0.001 s. A typical pulling load variation curve during the pull-out process is shown in Fig. 3.29.

3.4 Conclusions

There have been numerous studies undertaken in the past to evaluate the axial performance of fully grouted cables used in underground mines. Until recently, the most reliable testing method is the LSEPT as it overcomes some of the shortcomings of previously developed methods; however, several issues have been found. These include lack of an antirotational capability and a relatively small

FIGURE 3.28 The assembled modified laboratory short encapsulation pull test cable testing unit.

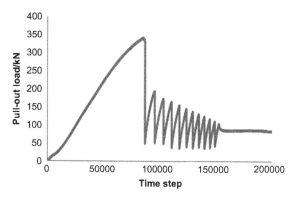

FIGURE 3.29 Typical result of the variation in load with time during a test.

test sample diameter of 142 mm, a size that is within the range of influence on pull-out load and size of hole in the bearing plate.

After consideration of these issues, a modified axial test unit based on the LSEPT method has been designed and constructed at the School of Mining at the University of New South Wales (UNSW) in Australia. The unit is capable of testing the wide range of different cable designs on the market that currently exceeds 12. With this test unit, the load transfer process of cables from surrounding concrete to the cable tendon can be studied. Furthermore, the failure modes of the cabling system, especially the relative slippage at both the cable/grout and grout/rock interfaces, can be investigated.

4. MECHANICAL PROPERTIES OF CEMENTITIOUS GROUT

4.1 Introduction

The stability of underground excavations, including drifts, open stope backs, permanent openings, and so forth, is always the most significant issue with which mining engineers are concerned because any instability behavior is likely to result in a loss of life and property. In view of this problem, most excavations in underground coal mines are always stabilized to guarantee the completeness of surrounding rock masses. In fact, Hoek and Brown (1980) suggested that "The principal objective in the design of underground excavation support is to help the rock mass to support itself." To realize this aim, fully grouted cables are widely used in underground coal mines to sustain the dead weight of separated rock mass, reinforcing the bedded rock masses together.

According to the fundamental classification of rock reinforcement system defined by Windsor (1997), fully grouted cables are involved in the CMC scheme. Therefore, the axial performance of fully grouted cables is mainly relied on the load transfer between surrounding rock masses and bolts via the grout annulus (Fabjanczyk and Tarrant, 1992). Although there are many different kinds of grouts used in underground coal mines for rock reinforcement tendons, such as polyester resins, cementitious grouts, and shotcrete, the cement-based grouts are more popular for cable usage. In this case, cementitious grout properties have a significant effect in determining the load-bearing capacity of cables. Initially, Stillborg (1984) conducted pull-out tests on 15.2-mm plain strand, using ordinary Portland cement with two different water/cement (w/c) ratios, namely 0.3 and 0.4. The results showed that both the ultimate as well as the residual pulling load were relatively larger when lower w/c ratio was adopted. However, it should also be mentioned that the chemical adhesion provided from those two different grouts on the cable was the same, representing the pulling force at the general bond failure was not influenced distinctly by different w/c ratios. Similar results were also verified by later researchers under both static (Goris, 1990; Hassani and Rajaie, 1990, 1992; Diederichs et al., 1993; Chen and Mitri, 2005) and dynamic loading conditions (Farah and Aref, 1986a,b), while the only difference is that Hassani et al. (1992) and Chen and Mitri (2005) studied the effect of w/c ratio on the load-bearing capacity of cables at different embedment lengths. In the same period, considering the Portland cement grouts were commonly used in underground cabling systems, Hyett et al. (1992a,b) undertook a comprehensive study to evaluate the physical and mechanical properties of this grout. The w/c ratio ratios were ranged from 0.70 to 0.35 and corresponding parameters including density, Uniaxial compressive strength (UCS), Young's modulus, and so on were given. This kind of research provides an essential element to optimize the cable installation, which is much beneficial for operators in the field. Later, Benmokrane et al. (1992) mixed six cement-based grouts, aiming at studying the effect of grout type on the shear stress and slip relationship for a seven-wire plain strand. It was found that the stiffness of cable reinforcement system was apparently influenced by the grout type. Furthermore, the grout strength had an important effect on the load transfer of the cabling system, especially when large displacements occurred. Based on the research mentioned earlier, it was concluded that the grout strength had a critical impact in determining the performance of plain cables. However, whether this effect was the same for modified cables had not been decided yet. Thus, Hyett et al. (1993) conducted this kind of research, showing that the w/c ratio of grouts had a less important effect in determining the load transfer behavior of nutcase cables. Similar conclusions were also applicable to Garford bulb cables (Hyett and Bawden, 1994), indicating that compared with plain strands, the modified cables could still generate large radial dilation or confinement during the pull-out process due to their specific surface geometries even when high w/c ratios were used. As a consequence, it was recommended that modified cables could be selected for serious environments; for example, when poor grout quality was obtained in the field (Hutchinson and Diederichs, 1996). Compared with axial tests on cementitious grouts, Moosavi and Bawden (2003) conducted direct shear tests on Portland cement grouts with different w/c ratios, finding that thicker grouts as well as larger normal confinement resulted in larger shear strength of grouts, which is beneficial for designing the fully grouted cabling systems. In recent years, using single embedment length pull-out tests, Mosse-Robinson and Sharrock (2010) evaluated the effect of grout property on the pulling strength of single-bulbed cables for large boreholes. According to their experimental program, the lower w/c ratio or the higher grout strength could lead to a larger shear resistance on the cable/grout interface. This enhanced frictional resistance might exceed the axial tensile strength of cable itself, especially for the modified cables. Finally, the failure mode of cables would be changed from slippage on the cable/grout interface to the rupture of the cable tendon itself.

At present, with the advantage of high strength and good flowability, the Stratabinder HP grout produced by the Minova Company, which belongs to one kind of bottom-up cementitious grouts, is widely accepted and used in cable reinforcement in the Australian underground coal mining industry. To better understand the mechanical properties of the Stratabinder HP grout and provide some basis for the cable installation in the field, a vast of unconfined compressive tests were conducted on grout samples with different w/c ratios. Both cylindrical and cubic grout samples were cast and cured in this research project, following the International Society for Rock Mechanics (ISRM) and Australian Standard, respectively. Based on that, results obtained from different samples were calculated and compared. Beyond that, important parameters including UCS, Young's modulus as well as Poisson's ratio were recorded. Finally, the effects of w/c ratio as well as sample size on the mechanical behavior of the Stratabinder HP grouts were analyzed.

4.2 Experimental Program

4.2.1 Preparation of Samples

So as to measure the mechanical properties of grouts, reasonable samples should be firstly prepared. In fact, there are different standards used at present. For example, the ISRM Standard recommends that a cylindrical sample with a length/diameter (L/D) ratio ranged from 2.5 to 3 is appropriate for testing (ISRM, 2007) while the Australian Standard prefers to use a cubic sample, especially for the concrete testing (SAA and SAI Global, 2005). Considering this, both cylindrical and cubic samples were cast, as shown in Figs. 3.30 and 3.31, respectively.

The cylindrical samples have an average length of 140 mm and a meaning diameter of 54 mm. Thus, the L/D ratio of those samples is around 2.60, which is shown in Table 3.1. As for cubic samples, they have an average edge length of 50 mm, which are listed in Table 3.2. Four different w/c ratios ranged from 0.35 to 0.45 are used to cast samples and all samples were cured within moist environments for at least 28 days.

4.2.2 Testing Process

Unconfined compressive strength tests were performed on grout samples. A rigid MTS load frame with the maximum capacity of 1500 kN was adopted to impose the axial displacement, as shown in Fig. 3.32. It should be mentioned that the axial and circular strain gangues were attached on cylindrical samples to measure the Poisson's ratio. Displacement control was selected in the whole testing process and the compressing rate was kept constant at

FIGURE 3.30   Cylindrical samples based on the International Society for Rock Mechanics Standard.

FIGURE 3.31   Cubic samples followed the Australian Standard.

TABLE 3.1 Dimensions of Cylindrical Samples

Number	w/c Ratio	Average Length (mm)	Average Diameter (mm)	L/D Ratio	Replication Times
1	0.35	141.20	53.56	2.64	5
2	0.38	141.38	53.45	2.64	5
3	0.42	139.65	53.47	2.61	5
4	0.45	140.55	53.61	2.62	5

TABLE 3.2 Dimensions of Cylindrical Samples

Number	w/c Ratio	Average Length (mm)	Average Width (mm)	Average Height (mm)	Replication Times
1	0.35	50.39	50.07	50.08	5
2	0.38	50.29	50.18	50.26	5
3	0.42	50.37	50.19	50.25	5
4	0.45	50.33	50.18	50.24	5

0.003 mm/s. One load cell and a displacement transducer were used to measure the reaction force as well as the advancing displacement during the process. Each test was replicated to five times and the best three were selected to represent the final results.

4.3 Results of Experiments

4.3.1 Processing Procedures

After all tests, the axial stress and strain for both cylindrical and cubic samples were calculated. Following that, the axial stress—strain relationship curves of samples can be given. It should be noted that for cylindrical samples, within the axial stress versus strain curves, the meaning slope of the more-or-less straight part is regarded as the Young's modulus. The Poisson's ratio is obtained with the following equation:

$$v = - \text{slope of axial stress} - \text{strain curve/slope of diametric stress} - \text{strain curve}$$

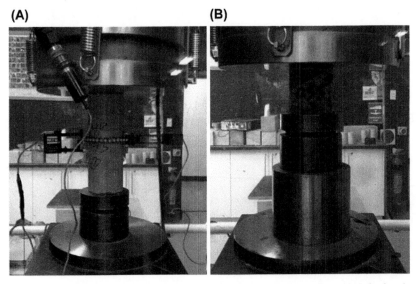

FIGURE 3.32 Uniaxial compressive strength tests on cylindrical and cubic samples. (A) Cylindrical sample; (B) cubic sample.

4.3.2 Stress–Strain Relationships of Samples

4.3.2.1 W/C RATIO OF 0.35

With the method mentioned earlier, when the w/c ratio is 0.35, the stress–strain relationships of both cylindrical and cubic samples are shown in Fig. 3.33. To be more specific, the corresponding values are given in Tables 3.3 and 3.4, respectively.

4.3.2.2 W/C RATIO OF 0.38

Similarly, the stress versus strain curves of those two kinds of samples with a w/c ratio of 0.38 are shown in Fig. 3.34. To be more specific, the corresponding values are illustrated in Tables 3.5 and 3.6 separately.

4.3.2.3 W/C RATIO OF 0.42

When the w/c ratio is increased to 0.42, the stress–strain relationships of both cylindrical and cubic samples are plotted in Fig. 3.35. In the meantime, Tables 3.7 and 3.8 state the homologous values.

4.3.2.4 W/C RATIO OF 0.45

Lastly, the stress versus strain curves for both two types of samples with a w/c ratio of 0.45 are given in Fig. 3.36. Specific values can be found in Tables 3.9 and 3.10 accordingly.

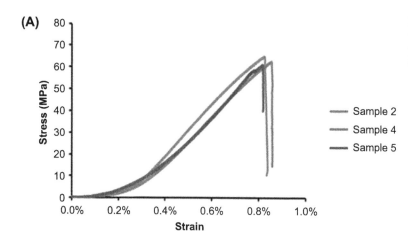
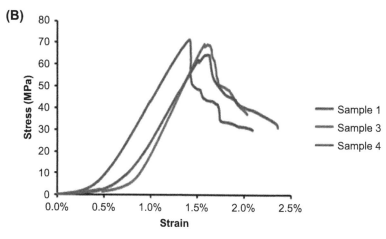

FIGURE 3.33 Stress–strain relationships of cylindrical and cubic samples. (A) Cylindrical sample; (B) cubic sample.

TABLE 3.3 Stress–Strain Values of Cylindrical Samples

List	Force (kN)	Strength (MPa)	Average	Young's Modulus (GPa)	Average	Poisson's Ratio	Average
2	145.56	65.14	63.10	11.97	11.82	0.20	0.23
4	140.14	62.81		11.40		0.23	
5	138.74	61.35		12.09		0.25	

TABLE 3.4 Stress–Strain Values of Cubic Samples

List	Area (mm^2)	Force (kN)	Strength (MPa)	Average (MPa)
1	2512.47	179.76	71.55	68.60
3	2526.52	175.83	69.60	
4	2526.57	163.38	64.67	

4.3.3 Failure Modes of Grout Samples

4.3.3.1 CYLINDRICAL SAMPLES

During the UCS testing process, it is found that cylindrical samples always fail along an inclined plane, which is shown in Fig. 3.37. However, it should be noted that the inclination angle of the failure plane is a bit different for cylindrical samples with different strengths, as illustrated in Fig. 3.38.

Sometimes, failure may also occur near one section of cylindrical samples and the whole grouts within that area are kicked off (Fig. 3.39). However, this is probably because the surface on one side of samples is not smooth enough and stress concentration exists within the unsmooth area during the compressive process. As a consequence, this stress concentration in turn breaks the grout in that region.

4.3.3.2 CUBIC SAMPLES

Compared with cylindrical samples, it is more interesting to see that cubic samples are likely to have a different failure mode. Figs. 3.40 and 3.41 show typical cubic samples during and after the UCS test. It can be seen that with the loading force increasing gradually, failure appears along a horizontal plane within the sample. And this kind of failure mode accounts for a large proportion.

Beyond that, the shear failure of cubic samples along an inclined plane is also found, as stated in Fig. 3.42. However, compared with the failure mode shown in Fig. 3.41, this type of failure behavior has a much smaller percentage.

4.4 Analysis and Discussions

4.4.1 Effect of W/C Ratio

For the w/c ratio ranged from 0.35 to 0.45, the strength variation curves of both cylindrical and cubic samples are shown in Fig. 3.43 and Table 3.11. It can be seen that for cylindrical samples, the strength decreases from 63.10 to 43.43 MPa gradually while a similar trend also arises on cubic samples with the strength reducing from 68.60 to 49.82 MPa. To be more specific, as indicated in Fig. 3.44, the Young's modulus of cylindrical samples is likely to drop linearly from 11.82 to 8.70 GPa as the w/c ratio increasing from 0.35 to 0.45, representing that the smaller the w/c ratio is, the stiffer the material of the mixed Stratabinder HP grouts will be. However, as for the Poisson's ratio, there is no distinct relationship between it and w/c ratio since the Poisson's ratios of all samples are around 0.25.

4.4.2 Effect of Sample Size

The effect of sample size on the UCS of Stratabinder HP grouts can also be analyzed from

4. MECHANICAL PROPERTIES OF CEMENTITIOUS GROUT

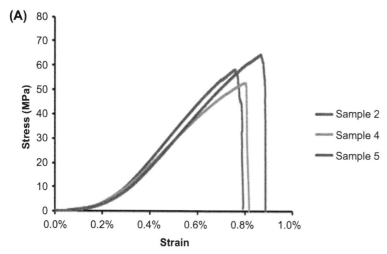

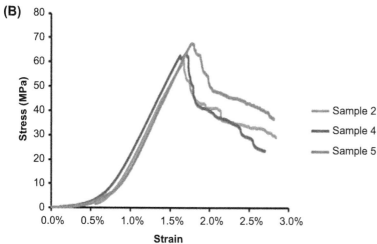

FIGURE 3.34 Stress–strain relationships of cylindrical and cubic samples. (A) Cylindrical sample; (B) cubic sample.

TABLE 3.5 Stress–Strain Values of Cylindrical Samples

List	Force (kN)	Strength (MPa)	Average	Young's Modulus (GPa)	Average	Poisson's Ratio	Average
2	130.14	58.46	58.63	11.57	10.84	0.14	0.20
4	117.74	52.87		9.95		0.23	
5	145.47	64.57		11.01		0.24	

TABLE 3.6 Stress–Strain Values of Cubic Samples

List	Area (mm²)	Force (kN)	Strength (MPa)	Average (MPa)
2	2531.09	153.62	60.69	64.13
4	2527.58	161.56	63.92	
5	2518.03	170.70	67.79	

Fig. 3.43. It is found that for different w/c ratios, all cylindrical samples have smaller strengths compared with cubic samples. The reason for this is probably because those two types of samples have apparent differences in the profile. Furthermore, cylindrical samples are slenderer than cubic blocks, increasing the instability of this type of concrete material. However, it should also be noted that the UCS differences between cylindrical samples and cubic samples are relatively small, indicating that either of those

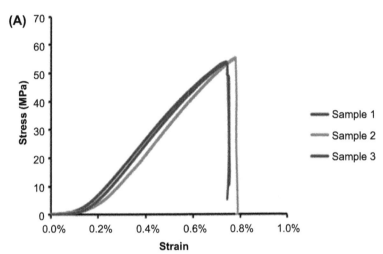
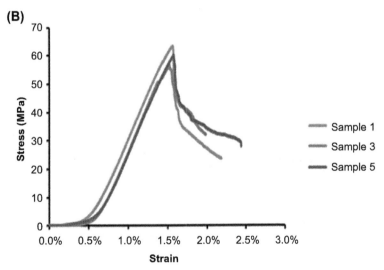

FIGURE 3.35 Stress–strain relationships of cylindrical and cubic samples. (A) Cylindrical sample; (B) cubic sample.

TABLE 3.7 Stress–Strain Values of Cylindrical Samples

List	Force (kN)	Strength (MPa)	Average	Young's Modulus (GPa)	Average	Poisson's Ratio	Average
1	120.18	53.88	54.25	9.42	9.94	0.18	0.21
2	122.90	55.37		10.40		0.28	
3	119.25	53.51		10.01		0.18	

TABLE 3.8 Stress–Strain Values of Cubic Samples

List	Area (mm^2)	Force (kN)	Strength (Mpa)	Average (MPa)
1	2525.55	160.62	63.60	60.20
3	2527.58	143.40	56.73	
5	2533.59	152.73	60.28	

samples can be selected to measure the UCS values for grout materials.

Compared with peak strength, the sample size has a more important effect in determining the residual behavior of grout materials. From Fig. 3.33 to Fig. 3.36, it is found that the stress of all cylindrical samples has a sudden drop after the peak load and the residual load is almost

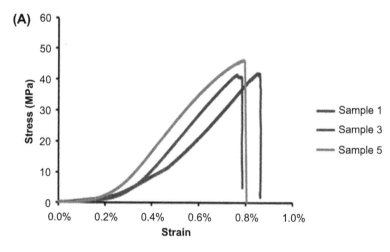

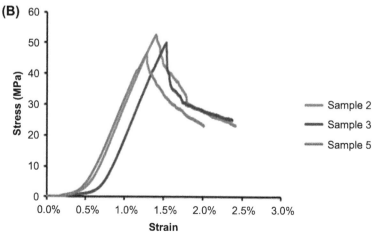

FIGURE 3.36A Stress–strain relationships of cylindrical and cubic samples. (A) Cylindrical sample; (B) cubic sample.

TABLE 3.9 Stress–Strain Values of Cylindrical Samples

List	Force (kN)	Strength (MPa)	Average	Young's Modulus (GPa)	Average	Poisson's Ratio	Average
1	96.14	42.18	43.43	8.45	8.70	0.38	0.34
3	93.14	41.71		9.17		0.26	
5	105.65	46.39		8.49		0.37	

TABLE 3.10 Stress–Strain Values of Cubic Samples

List	Area (mm²)	Force (kN)	Strength (MPa)	Average (MPa)
2	2525.55	133.01	52.66	49.82
3	2523.55	126.45	50.11	
5	2533.59	118.30	46.69	

close to zero. This is due to the reason that before the peak load, there is only relatively small strain (around 0.80%) occurred. After the peak load, the energy accumulated in cylindrical samples releases all of a sudden and brittle failure turns up. For cubic samples, it can be seen that the corresponding strain when peak load happens is near 1.50%, which is much larger than that of cylindrical samples. During the compressive process, the energy is releasing all the time and ductile failure appears when the peak load is reached. Beyond that peak point, the stress of cubic samples decreases gradually and the residual load is remained around 30 MPa.

4.5 Conclusions

A series of unconfined compressive tests were performed on the Stratabinder HP grout which is one kind of bottom-up grout commonly used in

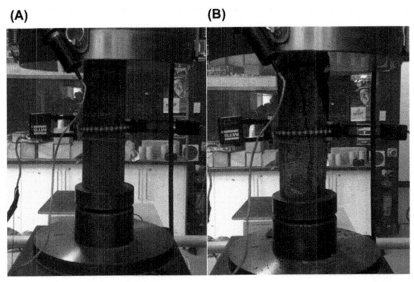

FIGURE 3.37 A cylindrical sample before and after uniaxial compressive strength test. (A) Before test; (B) after test.

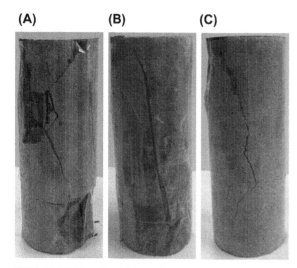

FIGURE 3.38 Shear failure distributions along the cylindrical sample. (A) Along the inclined plane; (B) replication test; (C) partial shear failure.

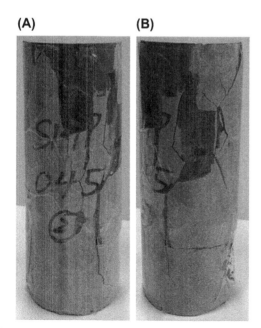

FIGURE 3.39 Failure distributions along the cylindrical sample. (A) Around the top section; (B) replication test.

the Australian coal mining industry. Following two different standards, both cylindrical and cubic samples were cast and tested with four different w/c ratios. During the tests, important mechanical parameters including the peak strength, Young's modulus, residual load, and so forth were recorded. After tests, it is found that:

With the w/c ratio increasing, the UCS values of both cylindrical and cubic samples decrease gradually. Specifically, the peak strength of cylindrical samples is always smaller than that of cubic samples at all w/c ratios. But the differences between them are relatively small, indicating either cylindrical or cubic samples can be selected to measure the peak strength for grout materials. Beyond that, the Young's modulus of cylindrical samples is likely to decline linearly as the w/c ratio increases, while the Poisson's ratio is found to keep steady when the w/c ratio is varying.

Cylindrical samples are more likely to fail along one inclined plane within the samples while cubic samples always have failure along a horizontal plane. Furthermore, the brittle failure behavior always happens in cylindrical samples and the residual load is almost close to zero due to the sudden break of samples. In contrast, ductile failure is common to see in cubic samples and there is always a gradually decreasing residual load occurring.

5. ANCHORAGE PERFORMANCE TEST OF CABLES

5.1 Introduction

A research project was undertaken to develop a new laboratory-scale test apparatus to study the axial performance of fully grouted cables that are used in the Australian underground coal mining industry. Work by Thomas (2012) and others in recent years found there were a number of deficiencies in the testing

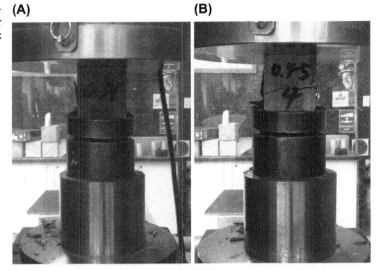

FIGURE 3.40 Failure along a horizontal plane within the cubic sample under different w/c ratios. (A) w/c ratio of 0.38; (B) w/c ratio of 0.45.

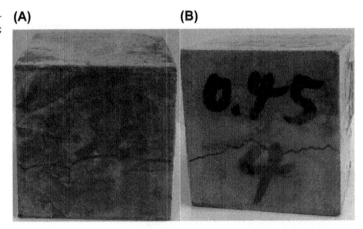

FIGURE 3.41 Failure of cubic samples under different w/c ratios. (A) w/c ratio of 0.38; (B) w/c ratio of 0.45.

FIGURE 3.42 Shear failure of cubic samples under different w/c ratios. (A) w/c ratio of 0.35; (B) w/c ratio of 0.38.

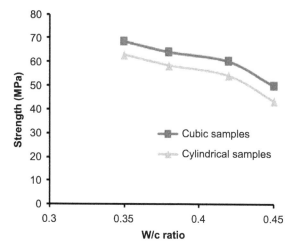

FIGURE 3.43 Effect of w/c ratio on the strength of samples.

methodologies making it difficult to determine and compare the performance of the dozen or more different types of cables on the market. This has meant that geotechnical practitioners have been hamstrung in optimizing the design of ground support systems in underground environments that are increasingly more arduous.

A cable is a flexible tendon consisting of a quantity of wound wires that are grouted in boreholes at defined distances between holes to provide ground reinforcement of excavations (Hutchinson and Diederichs, 1996). They were initially introduced into the underground hard rock mining industry in the 1960s (Thorne and Muller, 1964) and since the early 1970s have been brought to coal mining operations.

TABLE 3.11 Testing Results of Both Cylindrical and Cubic Samples

Number	w/c Ratio	Cubic Samples	Cylindrical Samples	Young's Modulus	Poisson's Ratio
1	0.35	68.60	63.10	11.82	0.23
2	0.38	64.13	58.63	10.84	0.20
3	0.42	60.20	54.25	9.94	0.21
4	0.45	49.82	43.43	8.70	0.34

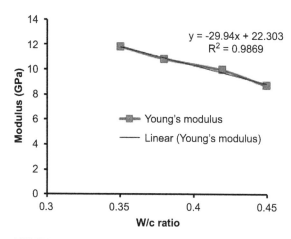

FIGURE 3.44 Effect of w/c ratio on Young's modulus.

Originally, cables were only used as a temporary reinforcement element. One reason for this being that many of the earlier cables were made from discarded steel ropes that had very poor load transfer properties due to their smooth surface profile, lacking the equivalent to ribs found on rockbolts. Over subsequent years a number of modifications have been made to the basic plain-strand cable, such as buttoned strand (Schmuck, 1979), double plain strand (Matthews et al., 1983), epoxy-coated strand (Dorsten et al., 1984), fiberglass cable (Mah, 1990), birdcaged strand (Hutchins et al., 1990), bulbed strand (Garford, 1990), and nutcaged strand cables (Hyett et al., 1993). These changes to the cable surface geometry have been undertaken in an effort to improve the load transfer efficiency and anchorage capacity that has resulted in the more widespread use of cables for permanent reinforcement.

Load transfer is a process resulting from differential movement within a rock mass, such as that caused by bed separation or movement along joint planes. It was originally defined as the mechanism by which force is generated and sustained in a supporting tendon as a consequence of strata deformation (Fabjanczyk and Tarrant, 1992). Later, Windsor (1997) modified this concept and indicated it is composed of three fundamental mechanisms: rock movement, which induces the transfer of load from unstable rock mass to the reinforcing tendon; transfer of load from unstable rock mass to interior stable rock mass by means of the reinforcing tendon; transfer of load from the reinforcing element tendon to the interior stable rock mass.

A laboratory-scale axial-loading cable testing facility capable of assessing the anchorage performance of the entire range of cables is required. One of the major benefits of a laboratory test facility is that it allows repeated tests under controlled conditions with certainty of what variables have been changed between each test so that results can be compared. This ability to control the test environment thereby provides more certainty in any analysis of test results and subsequent conclusions that are made about anchorage behavior.

In this context, there are questions as to what difference, if any, there is in the maximum load-bearing capacity, stiffness, postfailure load-bearing capacity, of rock type, and so forth. Efforts have been made in the past to develop a suitable test methodology and facility to examine these differences. Overall, they have been successful taking into account the materials available at that time.

Nevertheless, as cabling technology has evolved over the past 60 years to meet the new challenges as mining conditions change and demands increase in terms of performance, there has also been a need for changes in testing methods. This is no better illustrated than the current British Standard for cables. The foundation for this test was developed over a decade ago and was based on the cabling practice in the UK at that time. With subsequent new cables, such as plain-strand cables, work reported by Thomas (2012) found deficiencies with the LSEPT developed by RMT Ltd., such as twisting of the cable as it unwinds under load. Similarly, in the 1970s and 1980s with the use of 25 mm diameter cables, the testing facilities had much lower load capacities than that needed today, with 31 mm diameter modified cables, for example.

Having said that, there are limitations of a laboratory test unit and this is mainly due to scale. There are constraints on the range of test parameters chiefly to do with modeling the response of the rock mass and the length of encapsulation. These affect size and hence mass of test samples that can be safely produced and handled in a general laboratory setting as well as cost to conduct a test. Larger-scale tests are possible to simulate in-field conditions but this requires funding to build, operate, and maintain a dedicated large laboratory facility.

With respect to rock mass, varying approaches have been taken over the years, such as the adoption of steel tubes and biaxial cells to overcome the effects these can have on masking performance. The new testing facility is a hybrid design incorporating a near doubling in the diameter of the test rock sample placed within a rigid fixed frame. This design combines both confinement as would be seen in a rock mass in the field but while allowing interaction to occur between the cable, grout, and rock mass.

In terms of length of encapsulation and aside from double-embedment tests, which have their own limitations, most test facilities provide for a limited encapsulation length and importantly a short range of displacement. The total length of cable encapsulation in the test sample is 320 mm, similar to the RMT Ltd. test facility.

The new test facility allows for a much longer controlled pull-out length. Previous facilities have typically displayed a capability of controlling the load/displacement behavior up to around 40–50 mm with plain-strand cables. The current facility has doubled this to 100 mm.

The longer length of displacement control allows for a more extensive assessment of the behavior in the postfailure region. Thomas (2012) reported there were indications of significant differences between the various types of cables beyond the point of peak anchorage load. For example, in testing modified bulb cables with high anchorage loads, little data was measured in the postfailure region resulting in total measured displacements of typically less than 30 mm. It is postulated that this might have been due to the large uncontrolled release of the strain energy on anchorage failure. Displacement control in the current facility allows for load measurement over the entire range of displacement up to 100 mm independent of cable type.

One advantage of displacement control and the extended range of measured displacement is that it allows a study of postfailure behavior. This could range from a cable having:

1. Little or no load-bearing capacity postfailure as was evident in some instances with the use of modified bulb cable.
2. Some level of resistance and hence support, such as plain-strand cables.
3. The phenomena described as "slip/lock" mechanism. This has previously been observed in some tests with indented wire cables and 25-mm diameter bulbed cables but to date has not been shown to occur with larger diameter modified cables.

5.2 Method

The new test unit is comprised of the followings elements: embedment length section in concrete; bearing plate; anchor length section in a steel tube; and terminating device. A double-acting hollow cylinder, specially designed electrically actuated hydraulic power pack, and accompanying load cell are used to apply and measure the load near the midpoint of the embedded cable adjacent to the bearing plate, while an LVDT is used to measure displacement. Detailed information regarding new cable test facility is provided by Chen et al. (2014a,b). In order that the cementitious grout can be easily poured into the borehole and obtain a good bond with the cable, the equipment is aligned in a vertical direction, as shown in Fig. 3.16. The fully assembled test unit is shown in Fig. 3.45.

A test program was undertaken to examine the behavior of two types of cables. The range of test parameters is shown in Table 3.12. To reduce variability, a commercially available cementitious material was used to create the weak or low strength test samples and the strong test samples. The tests samples for each of the two strengths were cast from the one batch of material.

5.3 Test Results

5.3.1 Types of Failure Modes

Two types of failure modes were observed with the cables depending on the testing conditions, these being:

1. Failure at the cable/grout interface as the cable was drawn through the grout that remained bound to the surrounding rock as shown in Fig. 3.46.
2. Failure at the grout/rock interface, such as the MW9 cable when a "plug" consisting of the cable and grout was extruded from the borehole as shown in Fig. 3.47.

This is in line with the observations by Jeremic and Delaire (1983) and Hutchinson and Diederichs (1996) who noted there are essentially five different failure types of cabling, namely bond failure at the cable/grout interface, failure within the grout column, relative slippage at

FIGURE 3.45 Line-up of prepared cables samples and fully assembled cable test facility prior to a test. (A) Line-up of prepared cables samples; (B) fully assembled cable test facility prior to a test.

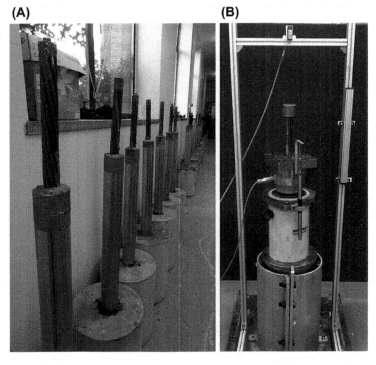

the grout/rock interface, failure within surrounding rock mass, and rupture of the cable tendon itself. Potvin et al. (1989) pointed out that it is more likely for a cable to fail at either of the cable/grout or grout/rock interfaces but more likely the cable/grout interface which is a function of the load transfer between the cable and rock mass.

The postfailure behavior was continuously monitored under controlled conditions over a total displacement of 100 mm. There were significant differences in postfailure behavior between the two cable types.

5.3.2 Superstrand Cable

In the case of the Superstrand cable, the average peak loads were 112 and 265 kN in the weak and strong test samples, respectively, in a rifled borehole at the standard recommended borehole diameter. This was achieved at a much lower displacement in the weak test sample compared to the strong sample as indicated in Fig. 3.48A. In both cases, there was an appreciable and sustained level of load-bearing capacity after the maximum load was achieved as shown in Fig. 3.48B.

TABLE 3.12 Range of Modified Parameters in the Test Program

Parameter	Level
Cable types	Megabolt MW9 and Jennmar plain Superstrand 9 mm
Strength of test samples	10 MPa (weak strength) and 68 MPa (strong strength)
Borehole size	As per manufacturer's recommendation
Grout strength	Constant at 60 MPa (Stratabinder)
Hole rifling	Rifled hole

FIGURE 3.46 An example of failure of the cable at the cable/grout interface.

FIGURE 3.47 An example of failure of the cable at the grout/rock interface.

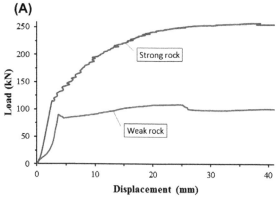

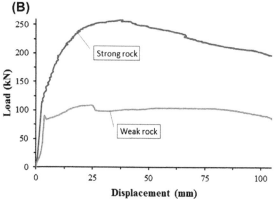

FIGURE 3.48 Load/displacement curves for a cable for the initial 40 mm and the full 100-mm displacement. (A) The initial 40-mm displacement; (B) the full 100-mm displacement.

After 100-mm displacement in the weak test sample, the cable was still able to maintain a load-bearing capacity of nearly 80 kN. Interestingly, there was little reduction in bearing capacity in the postfailure load being consistently maintained over the entire range of displacement. This compares to behavior in the strong rock where there was a near 25% reduction in load-bearing capacity. Even so, the load-bearing capacity after 100-mm displacement in the strong rock was quite high at 200 kN.

5.3.3 MW9 Modified Bulb Cable

In earlier test work reported by Thomas (2012) undertaken by RMT Ltd in the UK, measurements were truncated soon after the peak load was reached. This was typically within a total displacement of 10–15 mm as shown in Fig. 3.49. Little if any postfailure behavior of the cables could be measured in each case.

By comparison, when using this new test facility design, the postfailure behavior for the MW9 cable was measure well beyond the peak load. As shown in Fig. 3.50, the postfailure behavior over a displacement range of 100 mm was measured.

The average peak loads were much higher than the Superstrand at 380 and 208 kN in the strong and weak test samples, respectively, in a rifled borehole at the standard recommended borehole diameter.

Immediately postfailure, there were reductions of 20 and 100 kN, respectively, in load-bearing capacity in the weak and strong sample. Beyond the initial displacement of about

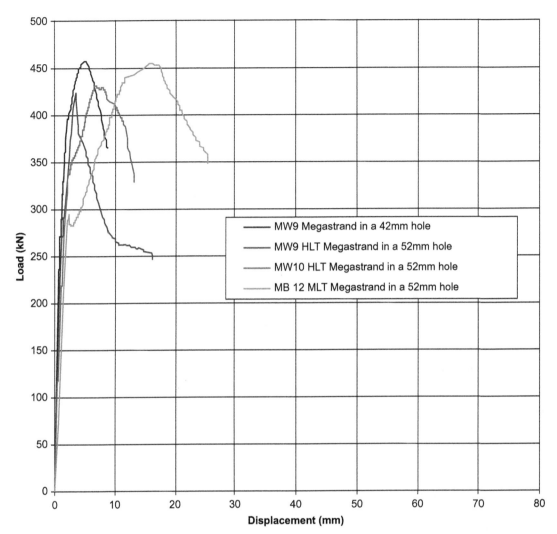

FIGURE 3.49 Load/displacement curves for the bulbed cables indicating truncation of readings soon after peak load reached with little indication of postfailure behavior. *After Thomas, R., 2012. The load transfer properties of post-groutable cable bolts used in the Australian Coal Industry. In: Proceedings 31st International Conference on Ground Control in Mining, Morgantown, USA, pp. 1–10.*

30–40 mm there was a near-uniform reduction in load signifying little effective bond strength or resistance between the cable and grout. There was a mirror image in behavior between the strong and weak rock for the MW9 cable with a near equal stiffness of −3.7 kN/mm as reflected in the constant gradient observed in the postfailure load-bearing capacity to approximately 100 kN. By the 100-mm mark in soft rock, there was little remaining load-bearing capacity whereas in strong rock, this range was to nearly 125 mm.

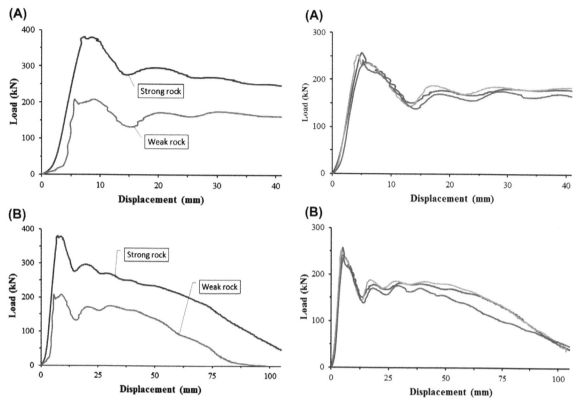

FIGURE 3.50 Load/displacement curves for MW9 cable in strong and weak rocks for the initial 40 mm and the full 100-mm displacement. (A) The initial 40-mm displacement; (B) the full 100-mm displacement.

FIGURE 3.51 Similarity in the behavior between the three test replications using an MW9 cable for the initial 40 mm and the full 100-mm displacement. (A) The initial 40-mm displacement; (B) the full 100-mm displacement.

There was evidence of slip/lock mechanism with the MW9 modified cable using the new test facility within a narrow range of displacement of 15–30 mm as had been previously observed by Thomas (2012) with the Hilti spirally indented cable. This behavior was observed in both the strong and weak rocks as shown in Fig. 3.50 and replicated in each of the three tests shown in Fig. 3.51A.

5.3.4 Effect of Rock Strength

Differences between the test samples strengths of 10 and 60 MPa were found to have some influence particularly on the postfailure behavior. There was a near doubling in the maximum load-bearing capacity observed with both cable types with an increase in rock strength.

The average peak loads were 112 and 265 kN for the weak and strong test samples, respectively, for the Superstrand while the MW9 achieved much higher values of 208 and 380 kN under standard comparable conditions. Interestingly while the initial stiffness of both cables were similar at around 51 and 69 kN/mm for the weak and strong test samples, respectively, there was over a fivefold difference in the overall displacement to peak load being on average 6 mm for the MW9 and 33 mm for the Superstrand.

5.4 Conclusions

Results of the test program have shown the new laboratory-scale axial-loading cable testing facility has the capability of consistently determining the postfailure behavior over a much wider range of displacement of 100 mm.

Two cables at either end of the design spectrum were tested under control conditions, the Superstrand and MW9 cables at two different rock strengths. The tests were undertaken using a cementitious material having a consistent manufactured rifled borehole and consistent grout strength of 60 MPa.

The maximum anchorage load was 112 kN for the Superstrand cable and 208 kN for the MW9 cable or an 86% difference. Whereas in the stronger sample, this difference was much less at only 43% with 265 kN attained for the Superstrand (representing a 137% increase) and 380 kN (83% increase) for the MW9 bolt, an overall difference in this case of 43%. Hence, the Superstrand was more sensitive to changes in rock strength.

There were distinct differences in the stiffness of the cables both before and after maximum anchorage load. In both instances of testing, the initial stiffness was similar at approximately 51 and 69 kN/mm in the weak and strong rock, respectively. However, stiffness began to reduce above a load of 100 kN with nearly 33-mm total displacement to the maximum load using the Superstrand. In the case of the MW9 cable, stiffness remained constant with only around 6-mm displacement to maximum load, hence the latter provides a much stiffer resistance to load.

Postfailure, the Superstrand showed very little reduction in load-bearing capacity over the entire displacement range of 100 mm but more so in the weak rock and a 25% reduction from 265 kN to a still significant load capacity of 200 kN in the strong rock or a stiffness of −1.1 kN/mm. In the weak rock, failure occurred at the grout/rock interface resulting in a "plug" of grout and cable being extruded from the borehole. For the MW9 cable, what resembled the characteristics of the slip/lock mechanism seemed to have been activated postfailure for up to a total 30–40-mm displacement in both the weak and strong rocks. Beyond this there was a similar stiffness measured in both the weak and strong rock of 3.8 kN/mm.

6. AXIAL PERFORMANCE OF A FULLY GROUTED MODIFIED CABLE

6.1 Introduction

Fully grouted cables have been used in the underground mining industry for many decades. Initially, plain strands manufactured by twisting seven flexible steel wires together were commonly used (Thorne and Muller, 1964). Laboratory and field tests lead to a better understanding of the transfer process between grouted cables and the surrounding rock mass, which is beneficial to improve the reinforcing performance of cables in engineering practice (Fuller and Cox, 1975; Cox and Fuller, 1977; Schmuck, 1979; Stillborg, 1984). Nevertheless, as in many cases the cable was sourced from discarded steel with a smooth surface along the strand, plain cables did not always meet performance expectations (Windsor, 1992). Over time, it was realized that cable surface geometry plays an important role in determining anchorage performance. Consequently, modified forms of cables were developed. For instance, twin-strands which were made by using spacers to combine two plain cables together improved bearing capacity twofold (Matthews et al., 1983). However, this required a large borehole to be drilled. Epoxy-coated strands had remarkable performance in practice but were difficult to install without damaging the epoxy layer (Dorsten et al., 1984). In the 1990s, there was a rapid development in the cable design. A typical example was birdcaged cables. They were initially suggested by Hutchins et al. (1990) via unwrapping traditional seven-wire strands and generating a number of nodes together with antinodes along the strand. But it was difficult to make birdcaged cables that could be coiled for transporting.

Garford (1990) produced bulbed cables by forcing together two parts of one standard strand, causing the wires to separate from each other. Hyett and Bawden (1994) found there was a limit to this as extra-large bulb size (more than 40 mm) would lead to poor load transfer behavior. Furthermore, they found the bulb spacing has a significant effect on axial performance. An Ultrastrand cable was proposed by Renwick (1992) via assembling metal spacers at a constant interval along the middle wire, though again, spacer spacing was critical to performance. Hyett et al. (1993) reported installation of a hexagonal nut on the middle wire to make nutcaged cables. Mah (1990) proposed a cuttable strand manufactured from fiberglass, named as fiberglass cables. Satola (2007) pointed out that galvanized strands had pronounced better performance than plain strands. Recently, Tadolini et al. (2012) proposed PC-strand tendons by adding regular indentation along peripheral wires. However, it should be mentioned that the indentation geometry has an apparent effect on bearing capacity of strands. It can be seen that most of these different types of cables are manufactured based on standard seven-wire strands, which are classified as modified cables.

In the Australian mining industry, the MW9 cables were developed by the Megabolt Company and are in wide use. This type of modified cables has a nominal diameter of 31 mm and has fully bulbed geometry with a diameter of 36 mm and an average bulb spacing of 500 mm (Megabolt, 2012). To understand the load transfer performance of MW9 cables, Bigby (2004) conducted a series of axial pull tests. However, the testing method used followed a British Standard, using steel pipes to simulate rock mass to confine cables (BS7861-2 1997). As a consequence, the results obtained are much stiffer than reality. Bigby and Reynolds (2005) developed the LSEPT for flexible tendons, conducting experiments on Megastrands. To be more specific, a biaxial cell was used to provide the confining pressure to a sandstone core sample, creating a CNL environment around the installed cable. However, this is not a true reflection of loading surroundings in the field since the underground confining pressure is variable not only along the length of cables but also within the service life of grouted tendons. Considering this problem, Reynolds (2006) modified the traditional LSEPT using a steel shell to restrict the sandstone core, creating a CNS circumstance. Similar to that, Thomas (2012) selected a split steel tube to serve as the confining medium. Both of those two methods successfully simulated the changeable loading environment around grouted cables. Nevertheless, neither approach has considered the size effect of the sandstone samples on performance of flexible tendons. In consideration of this problem, a new axial test apparatus was established. Then, various pull-out tests were performed with MW9 cables to better understand axial performance. Finally, several parameters including embedment length, test material strength, borehole size, and so forth were evaluated.

6.2 Building a New LSEPT Apparatus

Generally, axial pull tests on fully grouted cables can be classified as either an unconstrained or constrained test. In unconstrained tests, the cable is likely to rotate due to their special helical surface geometry, resulting in a poor pulling response (Hagan et al., 2014). Based on this consideration, it is necessary to build a nonrotating test rig. A new LSEPT apparatus was designed and constructed in the School of Mining Engineering, UNSW Australia, which is depicted in Fig. 3.16. This equipment is composed of two main parts, which are the embedment section and anchor segment.

6.2.1 Embedment Section

The lower part or the embedment section is the most important part of this new test rig. Usually, a sandstone core with a diameter of 142 mm is used to confine the grouted cable. However, a recent study has found that the size of the test material sample has a notable impact on the bearing capacity of a cable. In addition, the bond stress on the cable/grout interface cannot be transferred if the test sample is not sufficiently large (Holden and Hagan, 2014). Therefore, to determine the appropriate sample size, two series of pulling tests were executed on sumo strands, which have almost the largest load transfer capacity of all types of cables used in the Australian mining industry (Brown et al., 2013). Firstly, sumo strands were pulled out from artificial rock cylinders with the diameter ranging from 150 to 500 mm. Detailed information regarding the sample preparation and pull-out process has been described previously by Chen et al. (2014a,b). After testing, the relationship between the peak capacity along the cable/grout interface and sample size can be acquired, as displayed in Fig. 3.52.

It can be found that generally there is a bilinear relationship between the bond strength of the cable/grout interface and sample diameter. When the sample diameter is in the range of 150 and 350 mm, the peak capacity of cable/grout interface rises linearly to 95 kN, and the bond strength still increases but with a much lower slope.

Taking the in situ stress impact into consideration, a second series of tests was implemented on test samples confined within a steel cylinder. Following previous work, the cylinder used has a thickness of 10 mm. During the pull-out process, dilation of the cable/grout interface propagated radially. As a consequence, the confining pressure on the sample increased until the dilation limit was reached. After testing, the variation of peak capacity with sample diameter is shown in Fig. 3.53.

It shows that generally, there is again a bilinear relationship between the pull-out load and sample size. Specifically, within a limited size of 300 mm, the peak capacity rises linearly up to 110 kN. However, beyond 300 mm, the peak capacity along the cable/grout interface is likely to remain unchanged. It should also be mentioned that the maximum bearing capacity of cable/grout interface in the confined situation is always higher than without confinement, as noticed in Table 3.13. As a result, it is mainly reliant on the friction-dilation mechanism along the cable/grout interface (Chen et al., 2014a,b). When the sample is constrained by the steel cylinder, more confining pressure can be provided

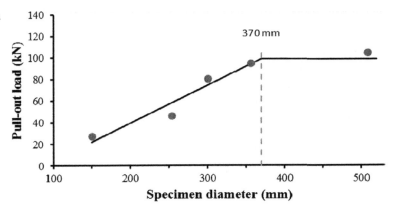

FIGURE 3.52 Tests conducted in unconfined samples.

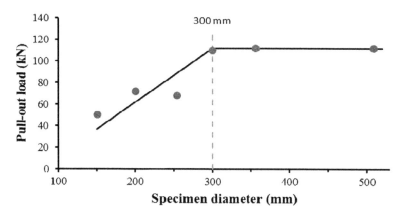

FIGURE 3.53 Tests conducted in confined samples.

TABLE 3.13 Bond Strength of Cable/Grout Interface

Sample Size (mm)	Peak Load Without Confinement (kN)	Peak Load Under Confinement (kN)
150	27.3	50.5
200	–	72.4
254	46.4	68.3
300	80.5	110.2
356	95.0	112.2
508	104.9	112.0

on the cable, enhancing the interface shear strength.

It is analyzed that under confined situation, once the test sample has a diameter of at least 300 mm, stable peak capacity of the cable/grout interface can be achieved. It was decided to set the diameter of the sample within the embedment section to 300 mm.

6.2.2 Anchor Section

The upper part in Fig. 3.16 represents the rock anchor in the underground reinforcement situation. To be more precise, an anchor tube with a length of 608 mm is used to restrict the grouted cable. Attention should be paid to the fact that this anchor tube is threaded internally with a constant pitch of 2 mm and depth of 1 mm. Consequently, after cementitious grouts are poured into the anchor tube, the bond among the cable, grout, and the anchor tube can be enhanced to a great extent. Another benefit gained from this design is that the barrel and wedge system is not necessary. In this case, the effect of different barrels and wedges on the final result can be avoided.

A load cell with a capacity of 1000 kN is used to measure the pull-out force. At the same time, two different displacement transducers were used to monitor pull displacements. One LVDT was attached on the hydraulic cylinder, measuring the relative movement between the sample and the cable, and the other laser monitor fastened to the testing frame to record the relative movement between the cable and ground, which is shown in Fig. 3.54

6.3 Results Analysis

Once the test apparatus has been set up, a number of pull tests were conducted using MW9 cables. Parameters included embedment length, test material strength, and borehole size were studied. Each test was replicated five times and the best three were selected and averaged as the final result.

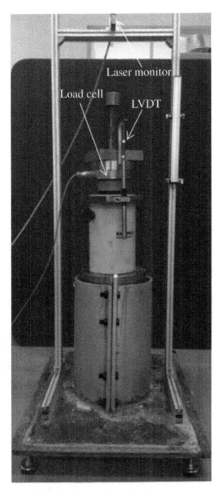

FIGURE 3.54 Front view of the test rig showing the displacement transducers.

6.3.1 Influence of Embedment Length

The embedment length has a critical impact on the load transfer performance of cables. A minimum length of 320 mm was used in the tests mainly as it was previous found that if the tested length is not long enough, scatter of pulling results usually occur (Hyett et al., 1992a,b). Beyond 320 mm, an interval of 20 mm was selected with the longest tested length 380 mm. A commercial cementitious grout with a UCS of 6 MPa, was formed to cast the test samples. In the center of the test sample, a standard rifled borehole with a diameter of 42 mm was reserved for the cable. Stratabinder grout with a UCS of 54 MPa was used to bond the cable to surrounding test material.

The failure of this cabling system always occurred along the cable/grout interface within the embedment section, as depicted in Fig. 3.55. The reason for this is during the loading process, the shear stress on the cable/grout interface was in excess of the interface's shear strength.

The load transfer performance of MW9 cables with different embedment lengths is displayed in Fig. 3.56. It can be seen that the embedment length has almost no effect in determining the initial stiffness of the bolted system. Nevertheless, the peak force does vary. To be more specific, within a limited range of 360 mm, the peak force increases linearly with embedment length. Once the embedment length is longer than that, the maximum load transfer capacity is stable. Further attention should be paid that the embedment length has a prominent influence on the residual behavior of MW9 cables. With an

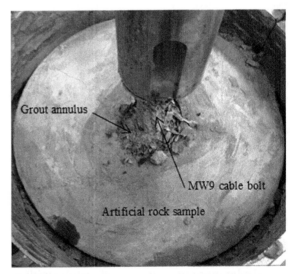

FIGURE 3.55 Shear failure along the cable/grout interface.

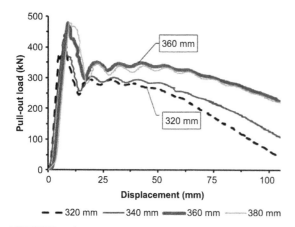

FIGURE 3.56 Performance of an MW9 cable at different embedment lengths.

embedment length of 320 mm, and after the peak load has been reached, there is some oscillation in measured pull force until the pull displacement reaches 60 mm. With the embedment length of 380 mm, this oscillation phenomenon is observed over the full pull-out stroke of 100 mm. This is determined by the slip-lock mechanism along the cable/grout interface. In detail, the bulbed geometry along the cable reacts with the surrounding confining grout annulus periodically until the front of the bulb is pulled within 10 mm far away from the borehole collar (Fig. 3.57).

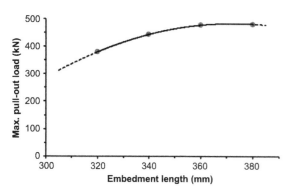

FIGURE 3.57 Impact of embedment length on max pull-out load of an MW9 cable.

6.3.2 Influence of Test Sample Strength

To study the effect of sample strength on load transfer performance of MW9 cables, another grout material with a much lower strength of 8.8 MPa was used to simulate weak rock conditions. The performance was compared and displayed in Fig. 3.58. Detailed information regarding the average peak load and the initial stiffness are shown in Table 3.14.

According to Fig. 3.58, the sample strength has a pronounced influence on the maximum bearing capacity. Furthermore, the postpeak behavior is largely different. When cables were pulled from strong material, after the peak load, oscillation of the pulling capacity was

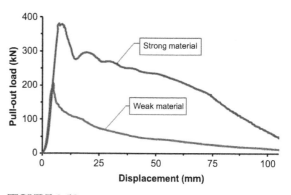

FIGURE 3.58 Performance comparison of MW9 cable in strong and weak materials.

TABLE 3.14 Testing Results of Test Material Strength Effect

Sample Type	Peak Load (kN)	Average Peak Load (kN)	Initial Stiffness (kN/mm)
Strong material	381	380	69.4
	379		
	381		
Weak material	208	208	52.2
	210		
	206		

apparent. While the cables were installed in weak material, the pulling capacity drops monotonically after the peak force. The reason for this is induced by the different failure behavior. Specifically, for an MW9 cable installed in weak test material, failure appears along the grout/rock interface, as shown in Fig. 3.59.

6.3.3 Influence of Borehole Size

It is not uncommon in the field to see larger boreholes used for cables. Because cables are flexible, reinforcing tendons usually have long lengths, making it difficult to install cables into small boreholes. For the purpose of evaluating the large borehole effect on the performance of cables under weak rock mass conditions, holes with a diameter of 52 mm were cast in the middle of the test sample for further testing. A comparison between results acquired using a standard or recommended diameter borehole and a 10 mm oversized borehole is displayed in Fig. 3.60. More detailed results about the peak load and the initial stiffness are tabulated in Table 3.15.

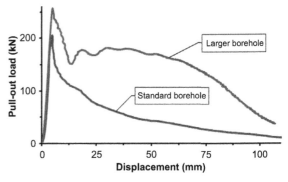

FIGURE 3.60 Comparison of performance between standard and oversized boreholes with an MW9 cable.

TABLE 3.15 Testing Results of Borehole Size Effect

Sample Type	Borehole Size	Peak Load (kN)	Average Peak Load (kN)	Initial Stiffness (kN/mm)
Weak material	Standard	—	208	52.2
	Large	257	250	51.0
		252		
		242		

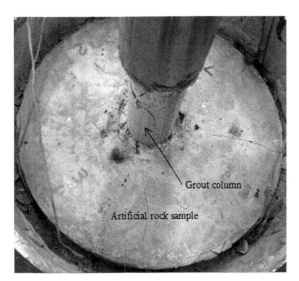

FIGURE 3.59 Cable, grout, and sample extruded en masse from the borehole in weak material.

It is clear that using a relatively large borehole improved the axial performance of an MW9 cable in weak material. The main reason for this improved performance is the grout used in the borehole. As mentioned in the previous section, failure of the MW9 reinforced system for a standard borehole situation occurs at the grout/rock interface. By contrast, with a lager borehole, failure occurs along the cable/grout interface, as shown in Fig. 3.61. As a result, both the grout annulus and test material can be regarded as the intact confining medium. The grout used had a strength of 54 MPa, being much larger than the test material of 8.8 MPa. It is reasonable to say that the bigger the grout annulus, the larger the confinement force will be. Finally, when large boreholes are used within soft material conditions, the load transfer performance of cables will be improved.

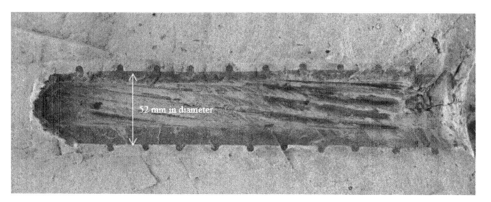

FIGURE 3.61 Failure evident at the cable/grout interface.

6.4 Conclusions

A series of tests using a new LSEPT were outlined in this paper. The test apparatus comprises an embedment section and an anchor section. To optimize the sample size within the embedment section, a number of tests were conducted using Sumo strands with test samples in unconfined and confined states. It was found that the bond strength of the cable increased with sample diameter up to 370 mm, thereafter remained unchanged for unconfined samples. With confined samples, width limit was reached at 300 mm. As a result, a standard diameter of 300 mm was used.

Various pull-out tests were executed on a type of modified cable, namely an MW9 cable. Three parameters were studied. First, the embedment length was evaluated. With an increase in the encapsulated length from 320 mm to 360 mm, the maximum bearing capacity of cables increased almost linearly. While over 360 mm, the peak capacity remained constant. In addition, the embedment length has an effect on the residual behavior with oscillation of the pulling load discernible at 360 mm. The load transfer behavior of MW9 cables confined within weak and strong material is largely different. When the MW9 cable was pulled out from weak material, failure always occurred along the grout/rock interface. Finally, the borehole size influenced the axial performance of MW9 cables in weak material. A large borehole was more effective in increasing both the peak and residual bearing capacity of cables. Therefore, if weak is encountered in the field, it is recommended to use a large borehole for modified cables.

7. SAMPLE DIMENSIONS ON ASSESSING CABLE LOADING CAPACITY

7.1 Introduction

The application of cable systems has advanced rapidly in recent years due to better understanding of the load transfer carrying capacity mechanisms and the advances made in cable system technology. Cables are used as part of temporary and permanent support systems in both civil tunneling and mining operations throughout the world. In mining they are used for slope stability applications in surface mining and a variety of ground support purposes in underground operations, such as stopping, roadway development, and shaft sinking. Cables are used to prevent the movement between discontinuity planes by transferring load across the discontinuity when relative strata layer movement takes place with separation.

The most common type of failure mechanism identified in the field is failure at the cable-grout interface (Hyett et al., 1996; Singh et al., 2001). This type of failure is common due to insufficient frictional resistance between the cable strand ridges and the grout material usually due to poor ground conditions and/or poor quality control at installation which leads to weak shear bond strength at the interface. This will often result in premature failure of the system before the steel capacity is mobilized. Due to the vast majority of failures being identified at the cable-grout interface, it can be concluded that a standardized testing methodology should focus on failures at the cable-grout interface (Rajaie, 1990; Hutchinson and Diederichs, 1996).

As reported by Hagan et al. (2014), a range of testing methods has been developed over the years including the double embedment and the LSEPT. The latter overcomes many of the deficiencies in the earlier tests. An issue with the LSEPT method highlighted by Thomas (2012) is the use of a small diameter test sample of approximately 142 mm placed within a pressurized Hoek cell arrangement and its inability to withstand the torsional loads generated during a pull test. Rajaie (1990) reported a study on the anchorage strength of cable and the effect of the diameter of the test sample. Nearly 300 pull tests were conducted using test samples in an unconfined state to define the characteristic and behavior of the cable element using conventional grout and grout-aggregate. The cable used was a plain-strand cable with a diameter of approximately 15 mm in test samples having a constant embedment length and borehole diameter. Tests were conducted in test rock samples having diameters ranging between 100 and 300 mm. As the results in Fig. 3.62 show, the load-carrying capacity of the cable varied with sample diameters up to 200 mm beyond which there was no change. This

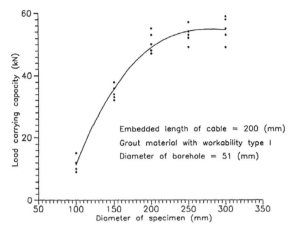

FIGURE 3.62 Variation in load-carrying capacity with sample diameter (Rajaie, 1990)

phenomenon was due to the stress generated with the test sample as a result of the load transfer between the cable, grout, and rock. Rajaie recommended that pull-out tests be standardized to test samples having a diameter of 250 mm.

Since that time, there have been a number of significant developments in the design of cables including the availability of modified bulbed cables having much greater load-bearing capacity then the plain-strand cables as used by Rajaie. These higher capacity cables are likely to generate much higher stresses in the surrounding rock mass when tested to full capacity. Subsequent work reported by Holden and Hagan (2014) repeated the work by Rajaie using a high capacity cable. They report the pull-out load continued to increase beyond the 200 mm diameter limit as shown in Fig. 3.63.

The approach used for cable embedment and test sample confinement has also changed. In tests, such as the double-embedment tests, the cable is grouted in a small bore steel tube. Confinement within the tube creates a constant stiffness-testing environment that inhibits any dilation effect otherwise induced by the cable when under load due to load transfer. As Thomas (2012) noted, this arrangement does

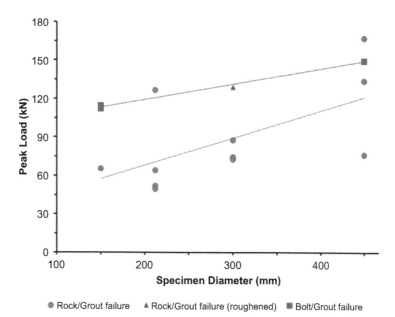

FIGURE 3.63 Variation in peak load with composite medium diameter (Holden and Hagan, 2014).

not allow assessment of the rock-to-grout interface. In the LSEPT test method, the cable is grouted within a sample of rock that itself is confined within a biaxial cell pressurized to 10 MPa. The use of a biaxial cell creates a constant stress environment that is not necessarily a true and consistent reflection of the underground environment in that (1) very little is known in regard to the in situ magnitude of borehole closure in underground coal mines and (2) it is almost certainly a dynamic variable that will vary both along the length of the cable and during the "life" of the cable (Thomas, 2012). To overcome this issue, the LSEPT was modified in the test program of Thomas with the sample instead grouted in a thick-walled steel cylinder.

7.2 Methodology

In line with the developments in testing methodology, the work of Holden and Hagan (2014) was repeated, but in this case instead of the test sample being unconfined, the samples were confined within a steel cylinder for each sample diameter. To ensure a more consistent mode of failure, the borehole in which the cable was grouted was rifled to provide better and more consistent bonding between the grout and test sample.

7.2.1 Sample Preparation

For testing, the test samples were made from a cement-based material cast in molds ranging in diameter from 150 to 500 mm with an overall length of 320 mm and borehole diameter of 42 mm. An indented sumo strand cable, manufactured by Jennmar Australia, was selected as it is a high load transfer capacity cable commonly used in the underground coal mining industry. Fig. 3.64 shows the bulb design sumo strand cable as used in the tests.

The sumo strand cable was chosen as it represents the worst-case scenario in terms of the high loads generated and hence high stresses induced within the test sample as a result of load transfer.

FIGURE 3.64 Sumo strand cable.

Hence the findings would be equally applicable to lower capacity cables. Preparation involved the following:

1. Preparing molds (including a rifling mold) and pouring of mortar to create test samples.
2. Initial curing for 24 h at which time the rifling mold and the casting mold were removed and the test sample allowed to cure for the remainder of the 28-day period under fully saturated conditions.
3. Grouting a cable into the test sample using a polyester resin.

The test samples were cast in molds made from thick-walled cardboard cylinders with a height of 320 mm and diameters ranging between 150 and 350 mm. The cardboard molds were glued to the base of a wooden board as shown in Fig. 3.65A using industrial silicone to ensure the molds retained its round shape and prevented any leakage during the cement pouring stage.

The manufactured rifled boreholes were prepared from a hollow PVC tube around which were wrapped 3-mm electrical wire at a pitch of 36 mm as shown in Fig. 3.65B. The purpose of the wire was to create the rifling effect in the borehole wall that would better promote interlock with the infill resin simulating the load transfer mechanism between resin and rock.

The cement-based product used to prepare the sample had an ultimate compressive strength of 32 MPa. Once the silicone dried and set, cement was poured into the molds as shown in Fig. 3.66. A mechanical vibrator was used to remove air bubbles in the cement during the mixing stage.

Within 24 h of pouring the mortar, both the middle PVC tube and the outer cardboard

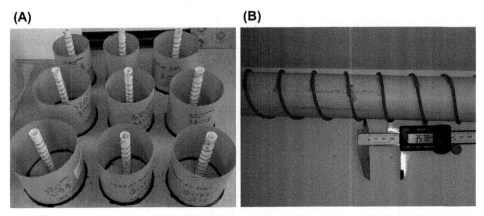

FIGURE 3.65 Prepared molds prior to pour cement and PVC tube used to create borehole with rifling effect. (A) Prepared molds prior to pour cement; (B) PVC tube used to create borehole.

FIGURE 3.66 Post the pouring of cement into the casting molds.

mold were removed, leaving a test sample with the rifled borehole as shown in Fig. 3.67.

After removing the outer cardboard casting mold and the rifling PVC tube, the samples were left to cure fully submerged for 28 days in either a large holding tub or plastic bags as shown in Fig. 3.68.

After curing of the rock cylinder samples, a cable was grouted into each rock cylinder using a slow-set resin with a setting time of 20–25 min. The resin and oil based catalyst were mixed for 13 min. An electric mixer was employed to combine the two components to ensure a

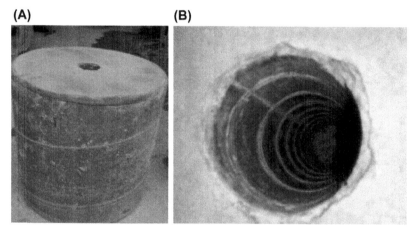

FIGURE 3.67 Test sample with rifling borehole and rifling effect. (A) Test sample with rifling borehole; (B) rifling effect.

FIGURE 3.68 Test samples fully submerged in water for curing.

208   3. GROUTED CABLE

FIGURE 3.69   Cable embedded in test sample using a slow set resin.

thorough and even distribution of catalyst throughout the resin, which is imperative in achieving the ultimate strength of the cured resin. The mixed resin was poured into the boreholes to a height 50 mm below the collar of the borehole. This allowed room for displacement of the resin after the cable was installed into the borehole. Excess resin was removed from around the rim of the borehole. The resin was left to cure for a day before it was used for testing. An example of the cured resin with cable and rock is shown in Fig. 3.69.

7.2.2 Test Arrangement

The setup arrangement for the tests is illustrated in Fig. 3.70. The cable was grouted in the test sample shown in the lower section of Fig. 3.70 and load applied to the cable using hollow hydraulic cylinder acting against a steel plate located on the top surface of the test

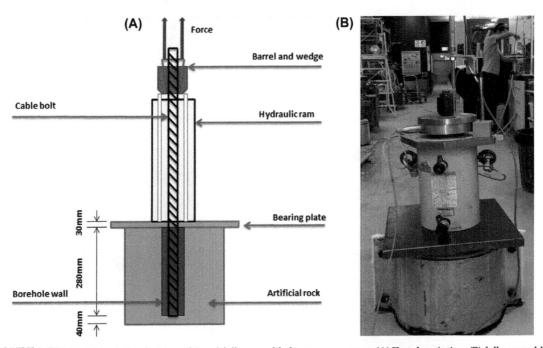

FIGURE 3.70   Test description (not to scale) and fully assembled test arrangement. (A) Test description; (B) fully assembled test arrangement.

sample. The level of applied load was measured using a pressure transducer and load cell, and the displacement was measured using an LVDT.

Prior to each test, the test sample was placed within a split steel cylinder as shown in Fig. 3.71 and the narrow gap or annulus between the sample and cylinder backfilled with cement. A 15 mm gap was left between the faceplates that were bolted together to join the two halves of the steel cylinder. This gap was filled with foam to prevent spillage of cement during backfilling. To ensure a consistent level of contact between the test sample, cement, and steel cylinder that might otherwise alter the maximum pull-out load; preconfinement was applied to the steel cylinder and test sample by tensioning the bolts with a micrometer torque wrench.

7.2.3 Test Variables

Tests were undertaken to determine the size effect of the test sample on pull-out load under:

(1) unconfined conditions; (2) confined conditions with zero torque; (3) confined conditions with 40 Nm; and (4) confined conditions with 80 Nm.

7.3 Test Results

7.3.1 Unconfined Conditions

The peak load carrying capacities of each size of test sample in the unconfined state is plotted in Fig. 3.72. Three test replications were undertaken at each level of sample diameter.

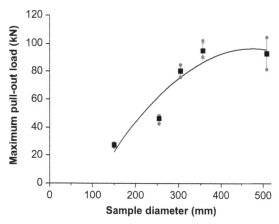

FIGURE 3.72 Variation in maximum pull-out load with test sample diameter in the unconfined state.

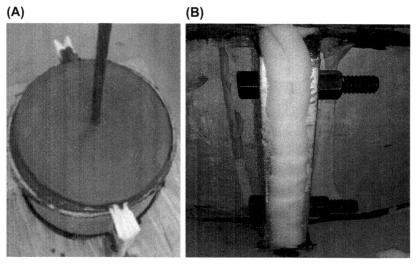

FIGURE 3.71 Test sample placed in assembled steel cylinder and gap filled with foam and bolts on side of cylinder. (A) Test sample placed in assembled steel cylinder; (B) gap filled with foam and bolts on side of cylinder.

There was a near threefold variation in maximum pull-out load from approximately 30 kN with the smallest size of sample up to over 100 kN achieved in the largest of test samples. This reaffirms the earlier findings by Rajaie (1990) and Holden and Hagan (2014) of the sensitivity of changes in pull-out being dependent on sample size in any laboratory determination of cable performance.

The maximum pull-out load increased with size of the test sample until some threshold value was reached beyond which there was no further increase in load. This is in line with the observation made that in the unconfined condition, the smaller size samples tended to dilate creating two or often three fracture planes as shown in Fig. 3.73. The larger test samples generally remained intact after testing. The general trend is similar to that reported by Rajaie (1990) except the threshold value of sample diameter was in this case in the order of 400 mm, much larger than the diameter of 250 mm as recommended by Rajaie.

7.3.2 Confined With Zero Torque

In this series, the test samples were all placed in a steel cylinder that was intended to provide confinement to the sample similar to that experienced by a rock mass surrounding a cable in situ. The maximum pull-out load in this confined condition of the test samples is plotted in Fig. 3.74. In this case no torque was applied to the bolts joining the two halves of the steel cylinder.

Similar to the unconfined tests, the maximum pull-out load was sensitive to changes in sample diameter. The main effect of confinement was that it reduced the threshold size above which there was little change in load. In this case, the threshold diameter was found to be in the order of 330 mm, down from the 400 mm observed in the unconfined tests.

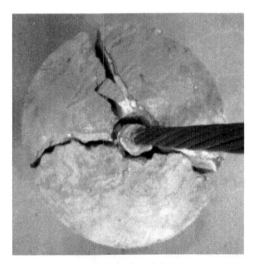

FIGURE 3.73 Typical failure mode of test samples with two or three fractures.

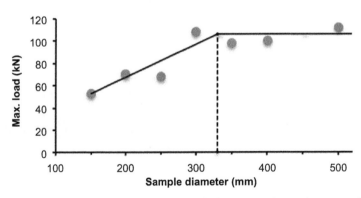

FIGURE 3.74 Variation in maximum load with test sample diameter with sample contained in steel cylinder.

7.3.3 Confinement With a Bolt Torque of 40 Nm

In this series, the samples were again placed in a steel cylinder, but in this case the joining bolts were tightened with a torque wrench to a torque level of 40 Nm are shown in Fig. 3.75. A similar result was achieved with a reduction in the threshold diameter with confinement in the steel cylinder.

7.3.4 Confinement With a Bolt Torque of 80 Nm

The variation in load with sample diameter in the case of tightening the joining bolts to a torque level of 80 Nm is shown in Fig. 3.76. A doubling in the level of bolt torque of the bolt did not appear to have any significant effect in reducing the threshold diameter.

7.4 Discussion

The results from the unconfined test combined with the three different scenarios of confinement are plotted in Fig. 3.77. The graph shows a definite upward shift in the maximum pull-out load attained with the confined test samples especially for test samples with diameters less than 300 mm. The difference, however, reduces as the diameter approaches 400 mm. Interestingly, the pull-out load is insensitive to the level of confinement, at least over the range of confinements investigated in this test program

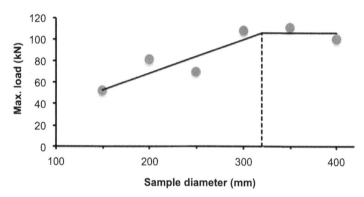

FIGURE 3.75 Variation in maximum load with test sample diameter at a bolt torque of 40 Nm.

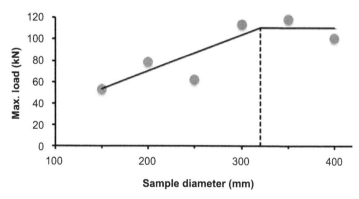

FIGURE 3.76 Variation in maximum load with test sample diameter at a bolt torque of 80 Nm.

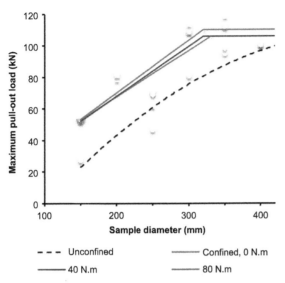

FIGURE 3.77 Comparison of the effect of varying levels of confinement on the maximum pull-out load for different sample diameters.

in terms of both the maximum pull-out load of approximately 100 kN and threshold sample diameter of 300–330 mm as there is very little apparent difference.

Overall, it can be concluded that consistent pull-out test results can be obtained with a test sample that is in the confined condition with a diameter of at least 300 mm. Moreover, while the maximum pull-out load varies little with the amount of confinement, a tangible amount of confinement provided by tightening the joining bolts to the same low level of torque of 40 Nm will provide a standardized testing environment and hence is more likely to provide more consistent results.

7.5 Conclusions

The results obtained from performance testing a high capacity modified bulb sumo cable embedded in test samples in an unconfined state found a similar trend to that reported by Rajaie (1990) in that there is an increase in the maximum pull-out load with diameter of the test sample up to some limiting or threshold diameter. Beyond this threshold diameter there is little change in the load of the cable. Rajaie stated the threshold to be around 200 mm in tests that used a low capacity plain-strand cable, whereas with the high capacity cable used in this project, the threshold is nearly double at 400 mm. Confinement of the test sample by placing it in a rigid steel cylinder was found to reduce this threshold diameter to around 300 mm. Interestingly, over the range of confinement levels studied, the performance of the cable was essentially insensitive to the actual level of confinement. To provide a standardized testing environment for the range of cables now available it is recommended that the test sample in which the cable is embedded is placed within a split steel cylinder of at least 300 mm diameter and the bolts that join the two halves of the cylinder be tightened to a torque of 40 Nm to ensure a consistent level of confinement.

References

Benmokrane, B., Chennouf, A., Ballivy, G., 1992. Study of bond strength behaviour of steel cables and bars anchored with different cement grouts. In: Proceedings Rock Support in Mining and Underground Construction: Proceedings of the International Symposium on Rock Support, Sudbury, Canada, pp. 293–301.

Benmokrane, B., Chennouf, A., Mitri, H.S., 1995. Laboratory evaluation of cement-based grouts and grouted rock anchors. International Journal of Rock Mechanics and Mining Sciences 32 (7), 633–642.

Bigby, D., 2004. Coal Mine Roadway Support System Handbook. Rock Mechanics Technology Limited, Staffordshire, pp. 1–67.

Bigby, D., Reynolds, C., 2005. Development of the laboratory short encapsulation pull test for a revised British standard on rock reinforcement components used in coal mining. In: Proceedings 24th International Conference on Ground Control in Mining, Morgantown, USA, pp. 313–322.

Brown, R.E., Garcia, A., Hurt, K., 2013. Split Cylinder Pull Tests on 3 Samples of the SUMO Cable Bolt Manufactured by Jennmar, pp. 1–5.

REFERENCES

BS7861-1, 2007. Strata Reinforcement Support System Components Used in Coal Mines – Part 1: Specification for Rockbolting, pp. 1–44.

BS7861-2, 1997. Strata Reinforcement Support System Components Used in Coal Mines – Specification for Birdcaged Cablebolting, pp. 1–24.

BS7861-2, 2009. Strata Reinforcement Support Systems Components Used in Coal Mines – Part 2: Specification for Flexible Systems for Roof Reinforcement, pp. 1–48.

Chen, J., Hagan, C.P., Saydam, S., Zhang, S., 2014a. Consideration of the factors impacting on the design of the laboratory short encapsulation test for cable bolts. In: Proceedings 8th Asian Rock Mechanics Symposium-Rock Mechanics for Global Issues-Natural Disasters, Environment and Energy, Sapporo, Japan, pp. 1–9.

Chen, J., Mitri, H.S., 2005. Shear bond characteristics in grouted cable bolts. In: Proceedings 24th International Conference on Ground Control in Mining, Morgantown, USA, pp. 342–348.

Chen, J., Saydam, S., Hagan, P.C., 2014b. The load transfer mechanics of fully grouted cable bolts: a theoretical analysis. In: Proceedings EuRock 2014, ISRM European Regional Symposium, Vigo, Spain, pp. 1045–1050.

Clifford, B., Kent, L.K., Altounyan, P., Bigby, D., 2001. Systems used in coal mining development in long tendon reinforcement. In: Proceedings 20th Internatioanl Conference on Ground Control in Mining, Morgantown, USA, pp. 235–241.

Cox, R.H.T., Fuller, P.G., 1977. Load Transfer Behaviour between Steel Reinforcement and Cement Based Grout. Division of Applied Geomechanics, CSIRO, Australia, pp. 1–12.

Diederichs, M.S., Pieterse, E.S., Nosé, J.S., Kaiser, P.K., 1993. A model for evaluating cable bolt bond strength: an update. In: Proceedings ISRM International Symposium-Eurock 93, Lisboa, Portugal, pp. 83–90.

Dorsten, V., Hunt, F.F., Kent, P.H., 1984. Epoxy coated seven-wire strands for prestressed concrete. Prestressed Concrete Institute 29 (4), 120–129.

Fabjanczyk, M.W., Tarrant, G.C., 1992. Load transfer mechanisms in reinforcing tendons. In: Proceedings 11th International Conference on Ground Control in Mining. University of Wollongong, Wollongong, Australia, pp. 212–219.

Farah, A., Aref, K., 1986a. Design considerations of fully grouted cable bolts. In: Proceedings 88th Annual General Meeting aof CIM, Montreal, Canada, pp. 1–18.

Farah, A., Aref, K., 1986b. An investigation of the dynamic characteristics of cable bolts. In: Proceedings of 27th U.S. Symposium on Rock Mechanics. American Rock Mechanics Association, Tuscaloosa, pp. 423–428.

Fuller, P.G., Cox, R.H.T., 1975. Mechanics of load transfer from steel tendons to cement based grout. In: Grundy, P., Stevens, L.K. (Eds.), Proceedings Fifth Australasian Conference on the Mechanics of Structures and Materials, Melbourne, Australia, pp. 189–203.

Gambarova, P.G., 1981. On aggregate interlock mechanism in reinforced concrete plates with extensive cracking. IABSE Reports of the Working Commissions 34, 99–120.

Garford, 1990. An Improved, Economical Method for Rock Stabilisation. Garford, Pty Ltd, Perth, pp. 1–4.

Goris, J.M., 1990. Laboratory Evaluation of Cable Bolt Supports (In Two Parts) 1. Evaluation of Supports Using Conventional Cables. US Bureau of Mines, United States, pp. 1–23.

Goris, J.M., 1991. Laboratory Evaluation of Cable Bolt Supports (In Two Parts) 2. Evaluation of Supports Using Conventional Cables With Steel Buttons, Birdcage Cables, and Epoxy-Coated Cables. US Bureau of Mines, United States, pp. 1–14.

Goris, J.M., Conway, J.P., 1987. Grouted flexible tendons and scaling investigations. In: Proceedings 13th World Mining Congress, Stockholm, Sweden, pp. 783–792.

Hagan, C.P., 2004. Variation in the load transfer of fully encapsulated rockbolts. In: Proceedings 23rd International Conference on Ground Control in Mining, Morgantown, USA, pp. 242–249.

Hagan, P.C., 2003. The effect of resin annulus on anchorage performance of fully encapsulated rockbolts. In: Proceedings ISRM 2003-Technology Roadmap for Rock Mechanics, Gauteng, South Africa, pp. 447–450.

Hagan, P.C., Chen, J., Saydam, S., 2014. The load transfer mechanism of fully grouted cable bolts under laboratory tests. In: Proceedings 14th Coal Operators' Conference, Wollongong, Australia, pp. 137–146.

Hartman, W., Hebblewhite, B., 2003. Understanding the performance of rock reinforcements under shear loading through laboratory testing. In: Proceedings 1st Australasian Ground Control in Mining Conference. Australasian Mining Rock Mechanics Society, Sydney, Australia, pp. 151–160.

Hassani, F.P., Mitri, H.S., Khan, U.H., Rajaie, H., 1992. Experimental and numerical studies of the cable bolt support systems. In: Proceedings Rock Support in Mining and Underground Construction : Proceedings of the International Symposium on Rock Support, Sudbury, Canada, pp. 411–417.

Hassani, F.P., Rajaie, H., 1990. Investigation into the optimization of a shotcrete cable bolt support system. In: Proceedings 14th Congress of the Council of Mining and Metallurgical Institution, Edinburgh, United Kingdom, pp. 119–129.

Hoek, E., Brown, E.T., 1980. Underground Excavations in Rock. Institution of Mining and Metallurgy, London, pp. 244–245.

Holden, M., Hagan, P.C., 2014. The size effect of rock sample used in anchorage performance testing of cable bolts. In: Aziz, N., Kininmonth, B., Nemcik, J., Black, D., Hoelle, J., Cunbulat, I. (Eds.), Proceedings Coal

Operators' Conference. University of Wollongong, Wollongong, Australia, pp. 128–136.

Hutchins, W.R., Bywater, S., Thompson, A.G., Windsor, C.R., 1990. A versatile grouted cable bolt dowel reinforcing system for rock. In: Proceedings the AusIMM Proceedings, pp. 25–29.

Hutchinson, D.J., Diederichs, M.S., 1996. Cablebolting in Underground Mines. BiTech Publishers Ltd, Richmond, pp. 1–2.

Hyett, A.J., Bawden, W.F., 1994. The 25 mm Garford Bulb Anchor for Cable Bolt Reinforcement Part 1: Laboratory Results. Queen's University, Ontario, pp. 1–16.

Hyett, A.J., Bawden, W.F., Coulson, A.L., 1992a. Physical and mechanical properties of normal Portland cement pertaining to fully grouted cable bolts. In: Proceedings Rock Support in Mining and Underground Construction : Proceedings of the International Symposium on Rock Support, Sudbury, Canada, pp. 341–348.

Hyett, A.J., Bawden, W.F., Hedrick, N., Blackall, J., 1995a. A laboratory evaluation of the 25 mm Garford bulb anchor for cable bolt reinforcement. CIM Bulletin 88 (992), 54–59.

Hyett, A.J., Bawden, W.F., Macsporran, G.R., Moosavi, M., 1995b. A constitutive law for bond failure of fully-grouted cable bolts using a modified hoek cell. International Journal for Numerical and Analytical Methods in Geomechanics Abstracts 32 (1), 11–36.

Hyett, A.J., Bawden, W.F., Powers, R., Rocque, P., 1993. The nutcase cable bolt. In: Proceedings Innovative Mine Design for the 21st Century : Proceedings of the International Congress on Mine Design, Kingston, Canada, pp. 409–419.

Hyett, A.J., Bawden, W.F., Reichert, R.D., 1992b. The effect of rock mass confinement on the bond strength of fully grouted cable bolts. International Journal of Rock Mechanics and Mining Sciences 29 (5), 503–524.

Hyett, A.J., Bawden, W.F., 1996. The Effect of Bulb Frequency on the Behaviour of Fully Grouted Garford Bulb Cable Bolts. Garford PTY Ltd., Canada, pp. 1–54.

Hyett, A.J., Moosavi, M., Bawden, W.F., 1996. Load distribution along fully grouted bolts, with emphasis on cable bolt reinforcement. International Journal for Numerical and Analytical Methods in Geomechanics 20 (7), 517–544.

ISRM, 2007. The Complete ISRM Suggested Methods for Rock Characterization, Testing and Monitoring: 1974–2006, pp. 138–140.

Ito, F., Nakahara, F., Kawano, R., Kang, S.S., Obara, Y., 2001. Visualization of failure in a pull-out test of cable bolts using X-ray CT. Construction and Building Materials 15 (5–6), 263–270.

Jeremic, M.L., Delaire, G.J.P., 1983. Failure mechanics of cable bolt systems. CIM Bulletin 76 (856), 66–71.

Kaiser, P.K., Yazici, S., Nosé, J.P., 1992. Effect of stress change on the bond strength of fully grouted cables. International Journal of Rock Mechanics and Mining Sciences & Geomechanics Abstracts 29 (3), 293–306.

Khan, U., 1994. Laboratory Investigation of Steel Cables and Composite Material Tendons for Ground Support (Ph.D. thesis (Unpublished)). McGill University, Montreal.

Littlejohn, G.S., Bruce, D.A., 1975. Rock Anchors-State of the Art. Foundation Publications LTD., Essex.

Macsporran, G.R., 1993. An Empirical Investigation into the Effects of Mine Induced Stress Change on Standard Cable Bolt Capacity (Master thesis). Queen's University, Kingston.

Mah, G.P., 1990. The Development of a Fibreglass Cable Bolt (Master thesis (Unpublished)). The University of British Columbia, Vancouver.

Maloney, S., Fearon, R., Nose, J., Kaiser, P.K., 1992. Investigations into the effect of stress change on support capacity. In: Proceedings Rock Support in Mining and Underground Construction: Proceedings of the International Symposium on Rock Support, Rotterdam, Netherland, pp. 367–376.

Martin, L., Girard, J.M., Curtin, R., 1996. Laboratory Pull Tests of Resin-grouted Cable Bolts. National Institute for Occupational Safety and Health, Spokane, pp. 455–468.

Matthews, S.M., Tillmann, V.H., Worotnicki, G., 1983. A modified cable bolt system for the support of underground openings. In: Proceedings AusIMM Annual Conference, Broken Hill, Australia, pp. 243–255.

Megabolt, 2012. Pretensioned & Grouted Secondary Support, pp. 1–4.

Mirabile, B., Poland, R., Campoli, A., 2010. Application of tensioned cable bolts for supplemental support. In: Proceedings 29th Internatioanl Conference on Ground Control in Mining, Morgantown, USA, pp. 1–6.

Mitri, H.S., Rajaie, H., 1992. Shear bond stresses along cable bolts. In: Proceedings 10th International Conference on Ground Control in Mining, Morgantown, USA, pp. 90–95.

Moosavi, M., 1997. Load Distribution Along Fully Grouted Cable Bolts Based on Constitutive Models Obtained from Modified Hoek Cells (Ph.D. thesis). Queen's University, Kingston.

Moosavi, M., 1999. A comprehensive comparison between bond failure mechanism in rock bolts and cable bolts. In: Proceedings the Congress of the International Society for Rock Mechanics, pp. 1463–1466.

Moosavi, M., Bawden, W.F., 2003. Shear strength of Portland cement grout. Cement & Concrete Composites 25, 729–735.

Moosavi, M., Bawden, W.F., Hyett, A.J., 2002. Mechanism of bond failure and load distribution along fully grouted cable-bolts. Mining Technology 111 (1), 1–12.

Moosavi, M., Grayeli, R., 2006. A model for cable bolt-rock mass interaction; integration with discontinuous deformation analysis (DDA) algorithm. International Journal of Rock Mechanics and Mining Sciences 43 (4), 661–670.

Mosse-Robinson, S., Sharrock, G., 2010. Laboratory experiments to quantify the pull-out strength of single strand cable bolts for large boreholes. In: Hagan, P., Saydam, S. (Eds.), Proceedings Second Australasian Ground Control in Mining Conference 2010. Australasian Institution of Mining and Metallurgy, Sydney, Australia, pp. 201–209.

Patton, F.D., 1966. Multiple modes of shear failure in rock. In: Proceedings 1st ISRM Congress, pp. 509–513.

Potvin, Y., Hudyma, M., Miller, H.D.S., 1989. Design guidelines for open stope support. CIM Bulletin 52 (926), 53–62.

Prasad, K., 1997. Effect of Isotropic and Anisotropic Stress Changes on the Capacity of Fully Grouted Cable Bolts (Master thesis). Queen's University, Kingston.

Rajaie, H., 1990. Experimental and Numerical Investigations of Cable Bolt Support Systems (Ph.D. thesis (Unpublished)). McGill University, Montreal.

Reichert, R.D., 1991. A Laboratory and Field Investigation of the Major Factors Influencing Bond Capacity of Grouted Cable Bolts (Master thesis). Queen's University, Kingston.

Reichert, R.D., Bawden, W.F., Hyett, A.J., 1992. Evaluation of design bond strength for fully grouted cables. CIM Bulletin 85 (922), 110–118.

Renwick, M.T., 1992. Ultrastrand - an innovative cable bolt. In: Proceedings Fifth Underground Operators' Conference. A.A. Balkema, pp. 151–156.

Reynolds, C., 2006. Testing and Standards for Rock Reinforcement Consumables. Rock Mechanics Technology Limited, Norwich, pp. 1–25.

SAA and SAI Global, 2005. Chemical Admixtures for Concrete, Mortar and Grout Part 2: Methods of Sampling and Testing Admixtures for Concrete, Mortar and Grout, pp. 1–41.

Satola, I.S., 1999. Testing of the yielding cable bolt achieved by debonding. In: Proceedings Congress of the International Society for Rock Mechanics, pp. 1467–1470.

Satola, I.S., 2007. The Axial Load-Displacement Behaviour of Steel Strands Used in Rock Reinforcement (Ph.D. thesis). Helsinki University of Technology, Helsinki.

Satola, I.S., Aromaa, J., 2003. Corrosion of rock bolts and the effect of corrosion protection on the axial behaviour of cable bolts. In: Proceedings ISRM 2003-Technology Roadmap for Rock Mechanics, Gauteng, South Africa, pp. 1039–1042.

Satola, I.S., Aromaa, J., 2004. The corrosion of rock bolts and cable bolts. In: Villaescusa, Potvin (Eds.), Proceedings of Ground Support in Mining and Underground Construction, Perth, Australia, pp. 521–528.

Schmuck, C.H., 1979. Cable bolting at the Homestake gold mine. Mining Engineering 31 (12), 1677–1681.

Singh, R., Mandal, P.K., Singh, A.K., Singh, T.N., 2001. Cable-bolting-based semi-mechanised depillaring of a thick coal seam. International Journal of Rock Mechanics and Mining Sciences 38 (2001), 245–257.

Stillborg, B., 1984. Experimental Investigation of Steel Cables for Rock Reinforcement in Hard Rock (Ph.D. thesis). Högskolan I Luleå University, Luleå.

Strata, Control, 1990. Testing of Garford Bulb Tendons. Strata Control Technology Pty. Limited, Wollongong, pp. 1–5.

Tadolini, S.C., Tinsley, J., McDonnell, J.P., 2012. The next generation of cable bolts for improved ground support. In: Proceedings 31st Internatioanl Conference on Ground Control in Mining, Morgantwon, USA, pp. 1–7.

Thomas, R., 2012. The load transfer properties of post-groutable cable bolts used in the Australian Coal Industry. In: Proceedings 31st International Conference on Ground Control in Mining, Morgantown, USA, pp. 1–10.

Thompson, A.G., Windsor, C.R., 1995. Tensioned cable bolt reinforcement-an integrated case study. In: Balkema, A.A. (Ed.), Proceedings 8th ISRM Congress. International Society for Rock Mechanics, Tokyo, pp. 679–682.

Thorne, L.J., Muller, D.S., 1964. Pre-stressed Roof Support in Underground Engine Chambers at Free State Geduld Mines Ltd. Trans. Assn Mine Mngrs South Africa, pp. 412–428.

Windsor, C.R., 1992. Invited lecture. Cable bolting for underground and surface excavations. In: Proceedings Rock Support in Mining and Underground Construction : Proceedings of the International Symposium on Rock Support, Sudbury, Canada, pp. 349–366.

Windsor, C.R., 1997. Rock reinforcement systems. International Journal of Rock Mechanics and Mining Sciences 34 (6), 919–951.

Yazici, S., Kaiser, P.K., 1992. Bond strength of grouted cable bolts. International Journal of Rock Mechanics and Mining Sciences Geomechanics Abstracts 29 (3), 279–292.

Further Reading

Hyett, A.J., Bawden, W.F., 1996. The Effect of Bulb Frequency on the Behaviour of Fully Grouted Garford Bulb Cable Bolts. Queen's University, Ontario, pp. 1–34.

Littlejohn, G.S., Bruce, D.A., 1977. Rock anchors-design and quality control. In: Proceedings 16th US Rock Mechanics Symposium, pp. 77–88.

CHAPTER 4

Tunnel Engineering

1. CONSTRUCTION OPTIMIZATION FOR A SOFT ROCK TUNNEL

1.1 Introduction

With the rapid development of the Chinese economy and investment increases in infrastructure construction, many tunnel projects have developed rapidly, and the construction technology and the mechanization levels of the tunnel have been significantly improved. The development of tunnel engineering has been toward deep-buried and long distance tunnels in China, and in this situation it is inevitable that tunnel construction will come across high stress, soft, and broken surrounding rock or affect the complex geological environments (Fu, 2003; Tao et al., 2006).

According to the literature, when tunnels experience complicated and difficult conditions, some large deformation and failure phenomena often occur in tunnel projects, such as cross-section shrinking, lining cracking, shotcrete spalling, steel arch distortion, and so forth (Kang et al., 2001; Liu et al., 2006, 2008). For example, Muzhailing tunnel, Laoyeling tunnel, and Laotougou tunnel in Baozhong railway, and Jiazhuqing tunnel through coal-bearing strata in Neikun railway had similar cases of large deformation destructive phenomena, which had brought great difficulties to tunnel construction and supporting design. The more serious large deformation destructive phenomena had occurred in Erlang Mountain tunnel and Huayin Mountain tunnel, which were in the soft rock construction; there are also many typical engineering examples, such as Japan Ena Mountain highway tunnel, Tauern railway tunnel, Arlberg tunnel, and so forth. At present, the theoretical and experimental studies on large deformation for soft rock tunnels have been conducted in many countries, and a number of countermeasures have been adopted in the tunnel construction process, but due to the differences of engineering-geological conditions in different places, there are still not mature theoretical methods and construction techniques.

For studies on large deformation of soft rock tunnel, there has been an important change from the passive support of early Terzaghi theory to the active support of New Austrian Tunnelling Method (NATM). Scholars in China have put forward the axial variation theory (Yu, 1982), the loose circle theory (Dong et al., 1994), the united support theory (Li and Li, 2008), the shotcrete with arc-plate support theory (Zhang et al., 2003), and so on. In view of the large deformation problem and the difficult support problem for the soft rock tunnel, Chinese scholar M.C. He focused on the determination of deformation mechanism and the conversion technology of deformation mechanism from complex to single types for a soft rock tunnel. He put forward the engineering mechanics theory and the support methods of soft rock tunnels by a comprehensive integrated analysis method with engineering geology and large deformation mechanics. The theory and the method had been applied in coal, water conservancy, and other industries; technical and economic benefits had been achieved as a result.

Based on the theory and method of engineering mechanics of soft rock, combined with the practical tunnel engineering, the problems of release pressure excavation and reasonable support were analyzed under both high stress and large deformation conditions, which was of important theoretical significance and practical value for guiding similar practice.

1.2 Engineering Background

1.2.1 Tunnel Design

The soft rock tunnel was 20.05-km long with an elliptical cross section. The excavation height and the excavation width of the tunnel was about 8.8 and 7.8 m, respectively; the excavation area was about 58 m². The original support design of the tunnel was the reinforced concrete lining with 500-mm thick (Fig. 4.1).

1.2.2 Climatic Conditions

The tunnel altitude was 2447—2663 m, and the highest peak was covered with snow all year round. The climate was cold and dry, the winter and the spring were affected by sand and wind, and the average annual precipitation was about 600 mm, which is a typical northern arid climate.

1.2.3 Rock Lithology

The soft rock tunnel engineering passed through the sedimentary rock, metamorphic rock and igneous rock, and its strata lithology was complex, which mainly consisted of mudstone, sandstone, shale, slate, andesite, diorite, phyllite, and so on. The mudstone was of the swell characteristic when watered. The surrounding rock of the tunnel was crushing, low strength, poor cementation, and scattered body-like structure and belonged to rock grade V ~ VI.

1.2.4 Geological Structure

The tunnel passed through four large regional faults of F_4, F_5, F_6, and F_7, which formed a large extrusion tectonic belt with many regional fractures and developmental joint fissures, and therefore the tunnel engineering was in the complicated geological conditions. As shown in Fig. 4.2, the strike of each fault was NWW, and the angle between the strikes of these faults and the tunnel axis was about from 50 to 57 degrees (Zhao et al., 2009).

1.2.5 Geostress

The maximum depth of the tunnel ridge area was 1100 m and belonged to the high-stress area. The horizontal stress was 15—25 MPa and the vertical stress was 9.15—20.5 MPa. The maximum principal stress direction of the macro stress field was N 21.2° E, and the angle between the maximum principal stress direction and the tunnel axis was 38.2 degrees.

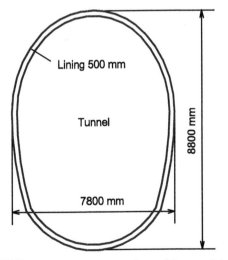

FIGURE 4.1 The cross-section shape of the tunnel design.

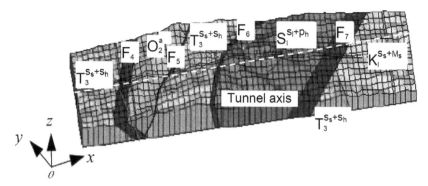

FIGURE 4.2 The strike of each fault along the tunnel axis (Zhao et al., 2009).

1.2.6 Deformation Failure Characteristics

After the primary support, the tunnel displayed the phenomena of concrete-lining cracking and large deformation failure. The subsidence of the tunnel arch was 567–1015 mm, and the deformation rate was 30–35 mm/d. The horizontal deformation of the sidewall, which was on 1.5-m elevation above the floor, was 700–1034 mm.

Due to passing through the different lithology and geological position, the tunnel produced the following deformation and failure characteristics:

FIGURE 4.3 Concrete-lining crack and damage of the tunnel.

1. When the excavation length of the tunnel reached to 416 m in the mud gravel zone of F_7 fault, the primary support at the 250 m position displayed concrete-lining cracking and damage (Fig. 4.3), and the steel arch supports with No. 20 I-beam all twisted in the range of 30 m, and the steel arch supports on the tunnel vault, which were in the range of 2.0 m of both sides, were partly cut down (Fig. 4.4). The subsidence of the tunnel arch was 177–241 mm, and the horizontal shrinkage of both side walls was 267–645 mm (Fig. 4.5).
2. When the excavation length of the tunnel reached 220 m in the phyllite slate, the primary support of the tunnel produced large deformation and vault collapse, and the

FIGURE 4.4 No. 20 I-beam steel distortion and cracking of the tunnel.

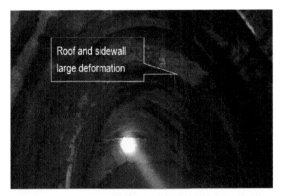

FIGURE 4.5   Roof and sidewall large deformation of the tunnel.

FIGURE 4.6   Large deformation of steel beam and the surrounding rock.

collapse height was 4–8 m. According to the observation, in seven days the horizontal shrinkage of the sidewall was 408–447 mm after excavation, the average shrinkage amount was 58–64 mm every day, and therefore the steel crossbars in the arch foot, which were with 6 m length and 219 mm diameter, were pressed into a triangular shape or distorted shape with the No. 20 I-beam steel arch support in the vault also twisted (Fig. 4.6).

1.3 Deformation Mechanism and Its Conversion Technology

1.3.1 Deformation Mechanism of Soft Rock Tunnel

According to some investigations including geological structure, lithology, rock mechanical properties, and supporting conditions and deformation failure characteristics, it was found that the surrounding rock of the tunnel was composed of compound features of high stress, jointed, and swell soft rock (HJS).

The deformation mechanism of the surrounding rock of the tunnel showed $I_A II_{BD} III_{BB} III_{CB} III_{DB}$. The meaning of each symbol was as follows: I_A was for the swell mechanism when watered; II_B was for the gravity mechanism; II_D was for the engineering deviatoric stress mechanism; III_{BB} was for the weak layer oblique type; III_{CB} was for the beddings oblique type; and III_{DB} was for the joint oblique type.

Since the deformation mechanism of HJS type was not a single type but the complex deformation mechanism, which had a variety of deformation mechanisms with complication and syndrome feature, it was difficult to support the soft rock tunnel. Therefore, to effectively carry out the support of the soft rock tunnel, a single type of support method was ineffective, and the appropriate engineering countermeasure, just like a doctor prescribing medicine according to the disease, should be taken.

1.3.2 Conversion Technology of Composite Deformation Mechanism

The key support technology of an HJS soft rock tunnel was the effective transformation technology from the complex deformation mechanism to a single type. According to the geological characteristics and the engineering conditions of the soft rock tunnel, the following conversion technologies were chosen, as shown in Fig. 4.7.

Step 1: Section step-by-step pressure release enlarging excavation.

Step 1: $I_A\ II_{BD}\ III_{BB}\ III_{CB}\ III_{DB} \longrightarrow II_{BD}\ III_{BB}\ III_{CB}\ III_{DB}$
Step 2: $\longrightarrow II_{BD}$
Step 3: $\longrightarrow II_B$

FIGURE 4.7 Steps of conversion technologies of the composite deformation mechanism.

For the excavation of HJS type soft rock tunnel, the original full-face excavation by steps method was adjusted to the pilot tunnel pressure release and enlarging excavation, the latter method could release part of the excessive pressure of the surrounding rock, weakened the partial stress influence of the surrounding rock, and made the stress force of the surrounding rock and the supporting force uniform. Thus the transformation from $I_A II_{BD} III_{BB} III_{CB} III_{DB}$ to $II_{BD} III_{BB} III_{CB} III_{DB}$ preliminarily was realized.

Step 2: Bolting with wire mesh combined with steel-fiber shotcrete.

Steel meshes and steel-fiber shotcrete were used to constrain the crushing surrounding rock, by adding three-dimension bolting reinforcement to weaken the effects of the oblique bedding and joints, the transformation from $II_{BD} III_{BB} III_{CB} III_{DB}$ to II_{BD} was realized preliminarily.

Step 3: Small-diameter anchor supporting and the conducted grouting technology.

The small-diameter anchor supporting and the grouting conducted could eliminate the impact of the weak interlayer of the soft rock tunnel roof, could take advantage of the strength of the deep surrounding rock, and could improve the stress state of the surrounding rock in a loose zone. The rigid-gap flexible layer supporting technology could be implemented to the soft rock tunnel. The reinforced concrete lining of the secondary support was used to improve the overall strength and stiffness of the soft rock tunnel so as to resist the excessive deformation energy which was beyond the primary support withstanding. Through the technologies mentioned earlier, one could realize the transformation from II_{BD} to II_B.

1.4 Numerical Analysis and Field Test of Conversion Technologies

1.4.1 Numerical Analysis of Different Excavation Methods

1.4.1.1 COMPUTATIONAL MODEL AND ITS PARAMETERS

The computational model was built by using FLAC^{2D}. According to Saint Venant's principle, the calculation model was 22-m wide and 24-m high, which was divided into 8448 elements. The vertical stress applied to the model was 20.5 MPa, and the horizontal stress was 16 MPa. The horizontal displacements of both lateral boundaries of the model were restricted, the top of the model was applied the vertical load to simulate the overlying rock loads, and its bottom was fixed. The material of the model was supposed to meet the Mohr–Coulomb strength criterion, and the physics and mechanics parameters selected were listed in Table 4.1.

1.4.1.2 NUMERICAL SIMULATION SCHEMES

1.4.1.2.1 PILOT-TUNNELING PRESSURE RELEASE ENLARGING EXCAVATION First, the initial stress equilibrium was calculated, and then the displacement field and the velocity field were set into zero. Second, the pilot tunnel was excavated by three steps and the displacements of the vault and the side wall of the pilot tunnel

TABLE 4.1 Physics and Mechanics Parameter of the Model

Name	Density (kg/m^3)	Friction Angle (degrees)	Cohesion (MPa)	Elasticity Modulus (GPa)	Poisson Ratio
Phyllite	2400	25	50	1.60	0.31

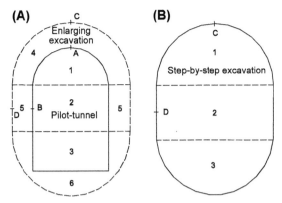

FIGURE 4.8 Excavation sequence and monitoring points. (A) Pilot tunnel excavation and (B) step-by-step excavation.

were monitored. Finally, the designed section was enlarged by three steps, and the displacements of the vault and the side wall of the tunnel were monitored. The excavation sequence (1, 2, 3, 4, 5, and 6) and the monitoring points (points A, B, C, and D) were shown in Fig. 4.8A.

1.4.1.2.2 SECTION STEP-BY-STEP EXCAVATION
First, the initial stress equilibrium was calculated, and then the displacement field and the velocity field were set to zero. Second, the tunnel was excavated by three steps and the displacements of the vault and the side wall of the tunnel were monitored. The excavation sequence (1, 2, and 3) and the monitoring points (points C and D) were shown in Fig. 4.8B.

1.4.1.3 SIMULATION RESULTS AND ANALYSIS

1.4.1.3.1 DISPLACEMENT VECTOR FIELD As seen from Fig. 4.9, after the pilot-tunneling and enlarging excavation, the maximum displacement of the surrounding rock was 301 mm, and the peripheral displacements distribution of the surrounding rock were more uniform. On the other hand, after the tunnel step-by-step excavation, the maximum displacement of the surrounding rock was 530 mm, and the displacements of the surrounding rock in the arch and both side walls were not uniform, where the large deformation and failure phenomenon were prone to occur.

1.4.1.3.2 MONITORING DISPLACEMENT ANALYSIS The displacement curves of

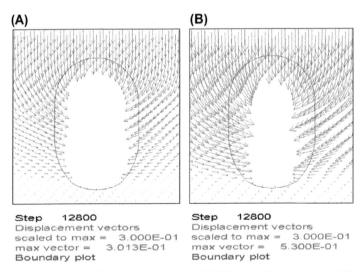

FIGURE 4.9 Displacement vector field of the surrounding rock. (A) Pilot tunnel excavation and (B) step-by-step excavation.

1.4.2 Field Test of Different Excavation Methods

According to the field monitoring data of the soft rock tunnel, the displacement convergence rate was 10–20 mm/d in the first four days of the primary support when the pilot tunnel and enlarging excavation was adopted and did not consider the effect of the secondary lining support, and in the back four days the displacement convergence rate was 10 mm/d or less, the ultimate convergence displacement was 250–350 mm.

When the step-by-step excavation was adopted, the displacement convergence rate was greater than that of the former, the eventually cumulative convergence displacement, which was 700–800 mm, also was greater than that of the former. Therefore the pilot tunnel and enlarging excavation could release the deformation energy of the surrounding rock significantly, and improve the safety factor and the technical benefits of the secondary lining.

The monitoring displacement curves of a certain section of side wall in the primary support were shown in Fig. 4.11. A_1, A_2, B_1, and B_2 were the monitoring points of the pilot tunnel and enlarging excavation and the step-by-step

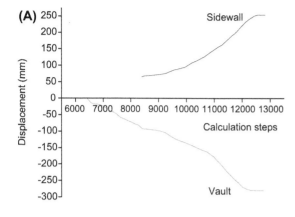
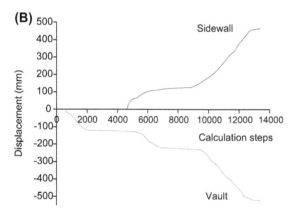

FIGURE 4.10 Displacement curves of the monitoring points. (A) Pilot tunnel excavation and (B) step-by-step excavation.

monitoring points in the vault and the side walls were shown in Fig. 4.10. After pilot tunnel and enlarging excavation, the maximum displacement of the vault and the side wall was 291 and 286 mm, respectively. On the other hand, after the tunnel step-by-step excavation, the maximum displacement of the vault and the side wall was 503 and 485 mm, respectively. Therefore, the displacement of the latter was significantly greater than that of the former, so the effect of the pilot tunnel and enlarging excavation was obvious.

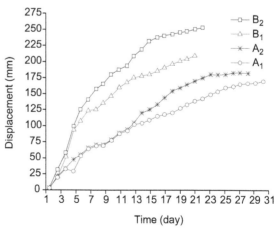

FIGURE 4.11 The monitoring displacement curves of the tunnel with different excavations.

excavation, respectively. The deformation of A_1 and A_2 was smaller than that of B_1 and B_2, which also validated the numerical calculation results.

1.4.3 Parameters Optimization of the Primary Support

To reasonably select the primary support parameters, the parameter optimization tests of the primary support were carried out on site. Table 4.2 presented the primary support parameters of two trial sections.

In the field test, taking the convergence displacement of the primary support in the arch foot and the appearance of failure as the main evaluation factors, the effects of different support parameters were judged.

Analysis of steel-fiber shotcrete cracking: the steel support of No. 20 I-beam was cracked and was twisted seriously and needed to be replaced. While the steel support of H175 crack occurred mainly in the ring, which had no large distortions, the tunnel had no chip-off falling.

After a month or so of convergence monitoring, the typical test sections of the displacement and displacement rate curves were shown in Figs. 4.12 and 4.13, respectively.

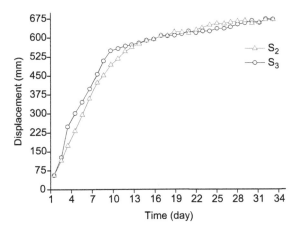

FIGURE 4.12 Displacement-time curves of S_2 and S_3 trial sections.

As shown in Figs. 4.12 and 4.13, in S_3 trial section after the tunnel excavation of the upper and lower section, the convergence rate of the side wall was about 50–100 mm/d, while in S_2 trial section that was 60 mm/d, the cumulative deformation in S_3 and S_2 trial section were both about 670 mm. The deformation trend of the displacement curves in the two trial sections were basically the same, and the difference was not obvious. By comparison, it was found that the

TABLE 4.2 Primary Support Parameters of Two Trial Sections

Number	Location	Advanced Support	I-beam Steel	Steel Mesh	Shotcrete	Bolting	Small-Diameter Anchor
S_2	YDK175+440 to YDK175+458	The diameter was 42 mm, the length was 4–6 m, and the spacing was 25 cm of the leading small-conduct pipe.	The spacing of the H175 steel bracket was 67 cm.	The diameter of the double-layer steel mesh was 8 mm, and the grid was 20 cm × 20 cm.	The thickness of the C_{30} steel-fiber-shotcrete was 25 cm.	The diameter was 32 mm, the length at the arch was 400 cm, the length at the wall was 600 cm, and the spacing was 80 cm of the grouting pipe.	The diameter was 15.2 mm and the length was 6 m of the anchor.
S_3	YDK175+458 to YDK175+467		The spacing of the H175 steel bracket was 80 cm.		The thickness of the C_{20} steel-fiber-shotcrete was 25 cm.		

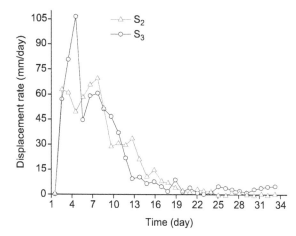

FIGURE 4.13 Displacement rate–time curves of S_2 and S_3 trial sections.

effect of 80 cm span steel support was no significantly fewer than that of 67 cm span steel support.

Based on the analysis in this section and the effect of the field construction, the rigid-gap flexible layer support technology was adopted in the soft rock tunnel, the support parameters of S_3 trial section were optimized, which had brought the good technical and economic effects.

1.5 Conclusions

Based on the geological structure, lithology, rock mechanical properties, and the engineering conditions, the deformation mechanism of the surrounding rock of the soft rock tunnel was determined.

The conversion technologies of the soft rock tunnel from the complex deformation mechanism into a single type were put forward, and the conversion technologies were validated by numerical analysis and the field tests.

Based on the transformation technologies mentioned earlier, the rigid-gap flexible layer supporting technology was carried out in the field and brought the good technical and economic effects.

2. WATER INRUSH CHARACTERISTICS OF ROADWAY EXCAVATION

2.1 Introduction

Confined water in the Ordovician limestone is one of the hidden troubles that threaten safe production of mines in north China, which is characterized by high water pressure and a large amount of water. With the continuous increase of the depth, the mining is subject to different Ordovician threats of flooding. Especially in the case of fault structures development, mining disturbance has the potential to cause fault activation, and due to aquifer seepage, the mining face can cause water inrush. In cases of water inrush, the mine's normal production will be affected, and the workers' lives will be endangered by such a disaster, which can lead to the mine closing down (Wu et al., 2007).

Both domestic and foreign scholars have conducted research on the mechanism of water inrush in coal mines with a number of achievements obtained. Hu et al. (2007) studied the deformation and failure process in the mining floor over the confined water, the water level variation, and the influence of faults to the mine water inrush. Kong et al. (2007) conducted a risk analysis for a mining field of water inrush using the theory of seepage instability. Liu et al. (2008) did the numerical analysis of the floor failure depth under different mine pressure with Rock Failure Process Analysis (RFPA). Zhu et al. (2008a) studied the water inrush process of the mining floor under the combined effect of the mining stress and the confined water pressure. Yin (2009) put forward the water inrush mode of the seam floor, and did the water inrush mechanism analysis according to the different geological conditions. Han et al. (2009) analyzed the water inrush mechanism of the mining face near the fault. Liu et al. (2010) completed the simulation analysis of the whole water inrush process from the seam floor under the mining disturbance with RFPA. Wildemeersch et al. (2010) performed an uncertainty analysis based on both the hydraulic head and water discharge rate observation using the Hybrid Finite Element Mixing Cell method. Wu et al. (2011) identified the most dangerous time when water inrush to a coal mine is possible, based on the variable parameter rheology-seepage coupling model. Li et al. (2011) analyzed the fault activation law of the mining floor based on a practical water inrush example. Tang et al. (2011) obtained the expression of critical water pressure for the collapse pillar water inrush using the thick plate theory of elastic mechanics, and so on.

Although various researches have been conducted on the water inrush mechanism in coal mines, the roadway water inrush modes and their evolution process have been rarely reported by far. Therefore, concerning the roadway excavation in the Danhou coal mine, it is of great importance to analyze the water inrush modes and their evolution characteristics of the roadway excavation.

2.2 Engineering Situation

The Danhou coal mine is located in Yuxian County, Zhangjiakou City, Hebei Province, China. Its annual production capacity designed is 1.50 Mt/a. The No. 3 mining area is located in the Yuxian basin, with complex hydrogeological conditions, and the normal production of the mining area is seriously constrained by Ordovician limestone aquifer.

The hydrogeological type of No. 3 mining area is a complex karst water-filled deposit, which is mainly a floor water inrush from the Ordovician limestone aquifer. The Ordovician limestone aquifer is located in the seam floor, with full recharge, well permeability, large water gushing and rapid recovery the water level, and the complexity conditions. The hydrostatic pressure of the confined aquifer is usually 2.00–4.12 MPa.

Between the seam and the Ordovician limestone, a layer of impermeable mudstone is better developed. However, the deposition thickness of this layer changes from the seam as 0.47–35.43 m.

There, developed faults in the No. 3 mining area are mainly normal faults. The total number of these faults is 19, all of which are cut deep to the limestone and may conduct the Ordovician limestone aquifer on condition that mining construction expose these faults.

2.3 Fault Activation Conditions and Water Inrush Modes

2.3.1 Fault Activation Mechanical Model

The so-called fault activation is that the fault plane produces plastic failure and come into a hydraulic conductivity state from the nonhydraulic conductivity state.

Assuming that a normal fault is located in front of the coal roadway, with the dip angle α, the loading over the roadway can be simplified with a uniformly distributed stress $H\gamma$, where H is the tunnel depth and γ is the average density of the rock on the roadway. The fault activation mechanical analysis model is shown in Fig. 4.14.

Assume that the footwall and the hanging wall on both sides of the fault are elastic

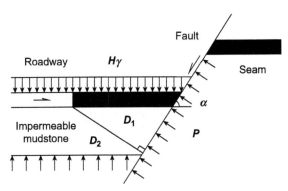

FIGURE 4.14 Fault activation mechanical analysis model.

material. Under the mining disturbance stress and the confined aquifer water pressure, when the shear force of the fault plane is greater than the maximum shear capacity of the fault material, the fault plane can shear slip and come into the plastic failure state. So, the fault can be regarded as being activated.

2.3.2 Fault Activation Mechanics Analysis

Taking a small unit from the fault for the mechanical analysis as shown in Fig. 4.15, assume that the horizontal stress and the vertical stress of the unit are σ_3 and σ_1 respectively. The cohesion of the fault plane is c and the internal friction angle is φ. When the fault is in the activation critical state and the unit is under vertical stress and horizontal stress conditions, the fault plane can produce the normal stress σ_a and the shear stress τ_a. Meanwhile, the failure plane of the unit will produce the shear stress τ_f to resist being activated, assuming that the confined water pressure on the fault plane is P.

According to the mechanical equilibrium relations, the normal stress σ_a and the shear stress τ_a, acting on the unit are as follows:

$$\sigma_a = \frac{\sigma_1 + \sigma_3}{2} + \frac{\sigma_1 - \sigma_3}{2}\cos 2\alpha - P \quad (4.1)$$

$$\tau_a = \frac{\sigma_1 - \sigma_3}{2}\sin 2\alpha \quad (4.2)$$

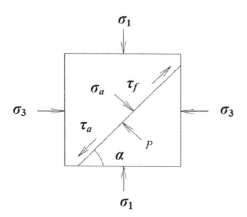

FIGURE 4.15 Unit mechanical analysis model.

Assuming that the strength of the unit material obeys the Mohr–Coulomb yield criterion, the shear stress τ_f to resist being activated is written as Eq. (4.3).

$$\tau_f = c + \sigma_a \tan \varphi \quad (4.3)$$

Then the fault activation condition is $\tau_a \geq \tau_f$. When the fault is in the critical activation state, combining Eq. (4.1)–(4.3), Eq. (4.4) can be obtained as follows:

$$P = \frac{\sigma_1 + \sigma_3}{2} + \frac{\sigma_1 - \sigma_3}{2}\cos 2\alpha \\ + \frac{2c - (\sigma_1 - \sigma_3)\sin 2\alpha}{2\tan \varphi} \quad (4.4)$$

According to the theory of mining pressure, the vertical stress σ_1 and horizontal stress σ_3 have the following relationships with the tunnel depth H:

$$\sigma_1 = H\gamma \quad (4.5)$$

$$\sigma_3 = K\sigma_3 = KH\gamma \quad (4.6)$$

where K is the lateral pressure coefficient.

If the critical water pressure value of the roadway floor as the fault being activated is P_c, then Eq. (4.7) can be obtained from Eq. (4.4).

$$P_c = \frac{\gamma H(1+K)}{2} + \frac{\gamma H(1-K)}{2}\cos 2\alpha \\ + \frac{2c - \gamma H(1-K)\sin 2\alpha}{2\tan \varphi} \quad (4.7)$$

Eq. (4.7) is the expression of the critical water pressure as the fault being activated.

2.3.3 Roadway Water Inrush Mode Analysis

Roadway water inrush must meet two conditions: (1) the fault is activated; (2) the fractured channel of water inrush is formed, where (1) is the prerequisite and (2) is the necessary condition.

After the fault is activated, the confined water level can rise above the hanging wall along the fault failure plane. With the roadway excavation approaching to the fault, the plastic failure depth of the roadway floor will increase and the front seam protection thickness will decrease synchronously. Then, according to the spatial location and the chronological order, there are three modes of water inrush: inrush from the floor, inrush from the working face, and inrush from both the floor and the working face.

As shown in Fig. 4.14, α is the fault dip angle, assuming that D_1 and D_2 are the minimum horizontal distance and the minimum normal distance from the roadway working face to the fault, respectively. On the other hand, assuming that the seam effective strength within D_1 is S_1, and the impermeable rock effective strength within D_2 is S_2, then if the water inrush from the floor and the working face occur together during the roadway excavation, the following equation can be obtained:

$$S_1 D_1 = S_2 D_2 \qquad (4.8)$$

According to Fig. 4.14, Eq. (4.9) can be expressed as follows:

$$D_2 = D_1 \sin \alpha \qquad (4.9)$$

So the critical angle α_c of the fault can be obtained from Eq. (4.9).

$$\alpha_c = \arcsin \frac{S_1}{S_2} \qquad (4.10)$$

It can be concluded from Eq. (4.10) that when the fault dip angle $\alpha > \alpha_c$, $S_1 D_1 > S_2 D_2$, the water inrush is from the roadway working face and conversely, when the fault dip angle $\alpha < \alpha_c$, $S_1 D_1 < S_2 D_2$, the water inrush is from the roadway floor.

2.4 Computational Model and Simulation Analysis Programs

2.4.1 Building the Calculation Model

On the basis of the roadway water inrushes in the Danhou coal mine, the tunnel surrounding rock and the fault spatial location, the roadway excavation engineering-geological model and three-dimensional numerical simulation model were built. As shown in Fig. 4.16, the model consists of sandstone, seam, mudstone, and Ordovician limestone aquifer. The calculation

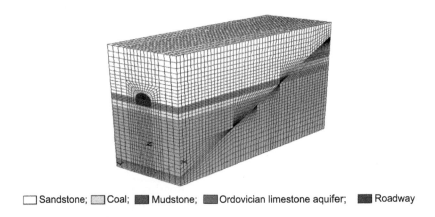

☐ Sandstone; ☐ Coal; ■ Mudstone; ■ Ordovician limestone aquifer; ■ Roadway

FIGURE 4.16 Computational model and its meshes.

conditions in Fig. 4.16 are as follows: the fault dip angle is 30 degrees; the fault displacement is 5.0 m; and the impermeable layer thickness of the roadway floor is 2.0 m. The computational model was 55.36-m long, 20-m wide, and 40-m high in the x-, y-, and z-axis. The roadway excavation dimension was 4.6-m wide and 4.0-m high. The model was divided into 57,496 zones and 61,709 grid-points.

2.4.2 Boundary Conditions and Water Inrush Criteria

The model was restricted by horizontal movement on four sides with its bottom fixed and its top free. The destruction of rock material was assumed to obey the Mohr–Coulomb strength criterion. A uniformly distributed load 6.0 MPa was applied on the model top to simulate the weight load on the roadway.

The model was surrounded by impermeable boundary on four sides, but the permeable boundary was assumed on the model top and at its bottom. A confined water pressure was fixed at the model bottom.

The fault was simulated with a contact plane element, and the hanging wall and the footwall could slip along the fault plane. According to field investigation, exploration data, and indoor and outdoor test results, the physical mechanics parameters and the hydrodynamics parameters of the rocks were listed in Table 4.3.

Due to the two conditions of roadway water inrush in the calculation model, the water inrush evaluation criterion is that the roadway surrounding rock comes into plastic failure and forms the conduct water channel after the fault is activated.

2.4.3 Simulation Analysis Programs

First, considering the confined water influence, the initial stress field balance calculation of the model was conducted, and after that the displacement field and the velocity field are cleared.

Second, the excavation process of the roadway was simulated step by step, and the initial excavation length of the roadway at each step was 5.0 m. To reveal the water inrush evolution characteristics, when the roadway was near the fault, the excavation length of the roadway at each step reduced to 1.0 m.

Finally, under the conditions of different water pressure, impermeable rock thickness, fault displacement, and fault dip angles, the roadway water inrush modes and their evolution characteristics were comparatively analyzed. The comparative analysis programs were listed in Table 4.4.

2.5 Water Inrush Characteristics Under Different Conditions

2.5.1 During the Process of Roadway Excavation

For simulation program No. 1, when the confined water pressure acting on the roadway

TABLE 4.3 The Physical Mechanics and the Hydrodynamics Parameters

Name	Density (kg/m³)	Bulk Modulus (GPa)	Shear Modulus (GPa)	Cohesion (MPa)	Friction Angle (degrees)	Tension (MPa)	Permeability Coefficient (m²/Pa s)	Porosity
Sandstone	2650	6.3	4.2	3.8	35	1.7	1.0e−12	0.23
Coal	1500	3.6	2.3	2.0	28	0.8	1.0e−10	0.32
Mudstone	2620	4.5	2.8	2.4	30	1.0	1.0e−11	0.28
Limestone	2630	5.8	3.7	3.0	33	1.4	1.0e−09	0.35

TABLE 4.4 A Variety of Simulation Analysis Programs

No.	Fixed Factor	Changed Factor	Notes
1	$H = 2$ m, $\theta = 30$ degrees, $h = 5$ m	P: 2, 3, 4 MPa	H is impermeable rock thickness; θ is fault dip angle; h is fault displacement; and P is water pressure.
2	$P = 3$ MPa, $\theta = 30$ degrees, $H = 2$ m	H: 1, 3, 5 m	
3	$P = 3$ MPa, $\theta = 30$ degrees, $h = 5$ m	H: 2, 3, 4 m	
4	$P = 3$ MPa, $H = 2$ m, $h = 5$ m	θ: 30, 45, 60 degrees	

floor was 3.0 MPa, the plastic failure of the surrounding rock of the roadway was shown in Fig. 4.4. When the step-by-step roadway excavation approaches to the fault along the seam floor, the roadway water inrush occurred.

As shown in Fig. 4.17A, when the horizontal distance from the roadway working face to the fault was about 4.0 m, there were increasing plastic failure in the roadway floor and the top of the impermeable rock near the fault under the mining disturbance stress and the confined aquifer water pressure. Nevertheless, had the fault been activated, the water inrush did not appear since the water inrush failure channel had not been formed.

As shown in Fig. 4.17B and C, as the roadway excavation advanced gradually and the horizontal distance to the fault was 3.0 m, the plastic failure areas of the surrounding rock were mainly in the floor and the sidewall of the roadway. When the equivalent intensity of the roadway floor is lower than that of the front seam, the failure elements connected and the water inrush failure channel formed. Therefore, the water inrush of the roadway floor occurred.

As shown in Fig. 4.18A, when the roadway floor water inrush occurred, the maximum displacement emerged in the floor near the fault like a beam-bending deformation. Meanwhile, there emerged the convex displacement field to the free face in the roadway. Overall, it can be seen from Fig. 4.18B, the displacement in the floor was obviously larger than that of other positions.

As shown in Fig. 4.19, with the excavation progress, both displacements of the roof and the floor in the roadway showed an increasing tendency, and near the water inrush position the displacement curves became flat. With the increasing pressure of the confined water, the displacement increasing rate in the floor was more significant than that in the roadway roof. The reasons for this phenomenon were mainly that the normal distance from the roadway floor to the fault was getting smaller and smaller, and the fault was easy to slip and rebound, which resulted in significant displacements in the roadway floor. On the other hand, since the roof was less affected by the confined water, the roof sinking displacement was mainly due to the strain energy releasing of the surrounding rock during the roadway excavation.

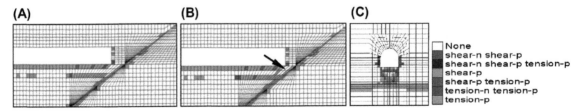

FIGURE 4.17 Plastic failure elements in the surrounding rocks. (A) Before water inrush, (B) after water inrush, and (C) cross section after water inrush.

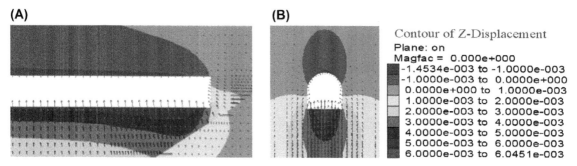

FIGURE 4.18 Vertical displacement field in the surrounding rocks. (A) Longitudinal section and (B) cross section.

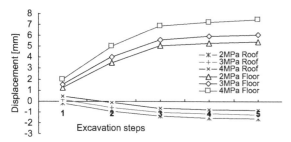

FIGURE 4.19 Vertical displacement curves during the roadway excavation.

As shown in Fig. 4.20, there emerged a wide range of tensile stress in the floor and much larger compressive stress concentration areas in the front of the working face. All these mechanical effects were consistent with the displacement field shown in Fig. 4.18.

2.5.2 Different Water Pressure Conditions

Simulation program No. 1: As shown in Fig. 4.21, with the confined water pressure gradually increasing, the plastic failure elements and the damage extent increased in the impermeable rock. At the same time, the plastic failure elements and the damage extent in the plane also showed an increasing tendency. These results indicate the further activation of the fault hydraulic conductivity. Moreover, the water inrush mode had a gradual transition, ie, inrush from the working face (or bottom) to that from the floor, which could cause the water inrush to come earlier and with more force.

2.5.3 Different Fault Displacements

Simulation program No. 2: As shown in Fig. 4.22, with the increase of the fault displacement, the plastic failure areas gradually reduced in the impermeable layer, but the plastic failure along the fault plane increased a bit, which indicated that the roadway had some lag of water inrush, and the water inrush became more difficult.

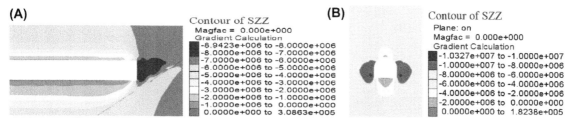

FIGURE 4.20 Vertical principal stress field of roadway excavation. (A) Longitudinal section and (B) cross section.

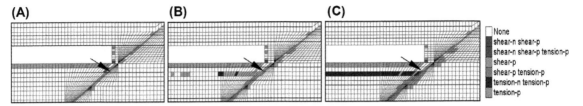

FIGURE 4.21 Plastic failure elements under different water pressures. (A) 2 MPa, (B) 3 MPa, and (C) 4 MPa.

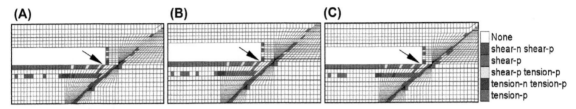

FIGURE 4.22 Plastic failure elements under different fault displacements. (A) 1 m, (B) 3 m, and (C) 5 m.

2.5.4 Different Impermeable Rock Thickness

Simulation program No. 3: As shown in Fig. 4.23, with the increase of the thickness of the impermeable rock, the plastic failure areas obviously reduced in the impermeable layer, but the plastic failure elements along the fault plane almost remained the same. The water inrush mode had a gradual transition, but the inrush from the floor occurred mainly from the working face (or bottom). When the impermeable layer thickness exceeded 3.0 m, it was of little influence to water inrush. So, there was a critical value of the impermeable layer thickness to induce water inrush during the roadway excavation.

2.5.5 Different Fault Dip Angles

Simulation program No. 4: As shown in Fig. 4.24, with the enlargement of the fault dip angles, the plastic failure areas gradually reduced in the impermeable layer and the fault plane, the fault activation became more difficult. the water inrush mode had a gradual transition, the inrush from the floor to that mainly from the working face (or bottom). Therefore, the large angle fault could postpone the water inrush time relatively.

In the process of the roadway excavation, the smaller dip angle fault was prone to water inrush because the smaller dip angle fault reduced the impermeable rock thickness to a large extent; on the other hand, the smaller dip

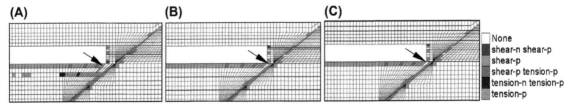

FIGURE 4.23 Plastic failure elements under different impermeable rock thickness. (A) 2 m, (B) 3 m, and (C) 4 m.

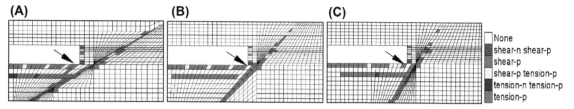

FIGURE 4.24 Plastic failure elements under different fault dip angles. (A) 30 degrees, (B) 45 degrees, and (C) 60 degrees.

angle fault could be easily activated. Therefore, a wider waterproof coal pillar should be set when encountering a smaller dip angle fault during the roadway excavation.

2.6 Conclusion

Under the mining disturbance stress and the confined aquifer water pressure, when the shear force of the fault plane is greater than the maximum shear capacity of the fault material, the fault can be activated. Roadway water inrush must meet two conditions: (1) the fault is activated; (2) the fractured channel of water inrush is formed.

According to the spatial location and the chronological order, there are three modes of water inrush: inrush from the floor, inrush from the working face, and inrush from both the floor and the working face.

Under the conditions of different water pressure, impermeable rock thickness, fault displacement, and fault dip angles, the roadway water inrush modes and their evolution characteristics are obviously different. The results are of reference and guidance to design safety coal pillars and roadway safe excavation.

3. LINING RELIABILITY ANALYSIS FOR HYDRAULIC TUNNEL

3.1 Introduction

Since underground projects have the characteristics of invisibility, complexity, uncertainty, and so forth, the underground project design and risk assessments have always been challenging technical problems in engineering practice.

In the late 1940s, probability methods were used to research various uncertain problems in underground engineering. In the 1970s, reliability methods had been extended to more application fields. Some problems have not been solved yet due to geotechnical engineering complexity. Thus, probability methods in geotechnical engineering are still in the development stages; a lot of concepts are not clear and the computational methods are not simple enough. Reliability methods are facing questions and opposition from some geotechnical technicians. There are many technical problems which have to be further studied.

At present, reliability analysis application in underground engineering is undergoing a process that is from rough to fine, from simple to complex and then back to simple, and is getting into practicality. For example, Most and Knabe (2010) conducted the reliability analysis for shallow foundation engineering using classical reliability methods. Based on the first-order reliability method (FORM), the higher order analysis methods and other optimum methods for more complex problems were tried to improve the accuracy of structure reliability by many experts and scholars. Amirat et al. (2006) put forward and used the Monte Carlo method to carry out reliability evaluation for underground pipelines in complicated conditions. Shi and Peng (2008) researched the multiarch tunnel lining reliability using the Monte Carlo stochastic finite element method. Chen et al. (2011b) obtained probability values of landslide instability under different conditions using the Monte Carlo method. To find the methods that have less calculating amounts than the Monte Carlo analysis method and which can be applied to various linear and nonlinear problems, Hou et al. (2009) put forward that the response surface method could be used to calculate tunnel first-lining function, and the first-lining reliability index was calculated by the FORM method. Allaix and Carbone (2011) did reliability analysis for rock slope using the stochastic response surface method. As the finite element method was widely applicable, uncertain factors were brought into the finite element calculation, then various stochastic finite element methods which were more practical and reasonable were formed (Farah et al., 2011). Moreover, as cross subjects and various new theories came out, various artificial intelligence methods were brought into reliability analysis, such as Zheng et al. (2007), who tried to solve geotechnical engineering reliability problems by using the artificial immune algorithm. Qi et al. (2008) did reliability analysis for tunnel excavation deformation using the genetic algorithm. Ji et al. (2012) tried to solve geotechnical engineering reliability problems.

Though the specialists and scholars in China and abroad have made some achievements on reliability analysis theory, because of the difference and particularity in different engineering practices, how to make reliability analysis methods more practical is still a difficult problem. Taking the shallow hydraulic tunnel under the Qinhuangdao Qin-Fu expressway as the background, considering influence factors like vehicle load and water pressure, combining the coupling characteristics of the tunnel and its surrounding rock, reliability analysis will be conducted for the tunnel structure stability, which is of important theoretical significance and practical value to disaster estimation and security risk assessment of the shallow hydraulic tunnel and similar engineering.

3.2 General Engineering Situation

The Qin-Fu expressway located at Haigang District in Qinhuangdao City is an important part of the road network. The Yin-Qing-Ji-Qin No. 5 hydraulic tunnel underlies the overpass project where the Qin-Fu expressway connects with Huanghe street. The expressway length that surface vehicle load influences is 308 m.

The strata where the Yin-Qing-Ji-Qin No. 5 hydraulic tunnel is located from top to bottom are 1.50-m thick quaternary topsoil and 8.0—10.0-m thick weathering gneiss. The surrounding rock of the hydraulic tunnel is weak, which is in a weathering fragmentation situation and poor stability, joints, and fracture development. The Protodyakonov's coefficient of hardness is 1.5—2.5.

The physical mechanics parameters of rock and soil are shown in Table 4.5. The groundwater influence is not considered in the project because of a low water table.

The No. 5 hydraulic tunnel is a shallow-buried project, which section is a direct-wall round arch, its net size is width × height = 2.7 m × 2.05 m, and its support is the cast-in-place concrete lining with thickness of 0.35 m. The reinforcing steel bars are double-deck. The main reinforcement bars are class-II with spacing of 0.15 m and the distribution reinforcing bars are class-I with spacing of 0.20 m; specifications are $\Phi 18$ and $\Phi 12$, respectively. Concrete design

TABLE 4.5 Physical Mechanics Parameters of Rock and Soil

Name	Density (kg/m^3)	Elastic Modulus (MPa)	Poisson Ratio	Cohesion (kPa)	Friction Angle (degrees)
Topsoil	1900	30.0	0.30	18.0	23
Weathering rock	2450	80.0	0.28	35.0	27

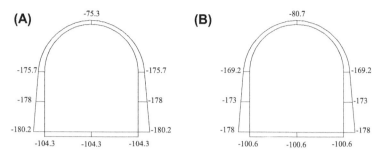

FIGURE 4.25 Axial forces distribution of the hydraulic tunnel structure (unit: kN). (A) With water pressure and (B) without water pressure.

grade is C25 with the lining protective covering thickness of 30 mm. The vehicle load is assumed an even load of 4.0×10^4 N/m.

3.3 Key Parts and Reliability Analysis Equation

3.3.1 Key Parts of the Tunnel Structure

To determine the key bearing parts of the hydraulic tunnel, considering two operating conditions of water pressure, axial forces, and bending moments of the tunnel lining structure were calculated and analyzed by the Lizheng software; the calculation results were shown in Figs. 4.25 and 4.26.

3.3.2 Reliability Analysis Equation of the Tunnel Structure

Based on reliability theory, the stability limit equilibrium equation of the hydraulic tunnel structure can be shown as:

$$Z = g(x_1, \ldots, x_n) = 0 \qquad (4.11)$$

where $g(x_1, \ldots, x_n)$ is the performance function for the hydraulic tunnel, Z is the state variables for the hydraulic tunnel, and $x_n (i = 1, 2, \ldots, n)$ is the variable factor influencing the reliability of the hydraulic tunnel.

Under the carrying capacity limit state of the tunnel structure, considering two random state variables, the reliability limit equilibrium

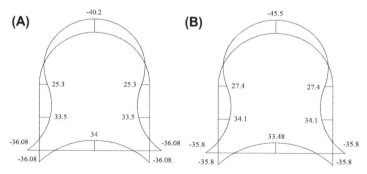

FIGURE 4.26 Bending moments distribution of the hydraulic tunnel structure (unit: kN m). (A) With water pressure and (B) Without water pressure.

equation of the hydraulic tunnel structure can be expressed as follows:

$$Z = R - S \quad (4.12)$$

where R is the supporting resistance, it is a state function of the supporting parameters of the tunnel; S is the load effect, it is a state function of the surrounding rock parameters of the tunnel.

Solving the supporting resistance of the tunnel, the lining can be as the flexural members, and the supporting resistance is given:

$$R = \sigma_{max} \quad (4.13)$$

where σ_{max} is the maximum bending stress of the tunnel lining cross section, the unit is MPa.

According to material mechanics, Eq. (4.14) is obtained:

$$\sigma_{max} = \frac{M_{max}}{W} \quad (4.14)$$

where M_{max} is the maximum bending moment of the tunnel lining cross section, the unit is kN m. W is flexural modulus of the tunnel lining cross section, the unit is m^3.

Taking the lining cross section as the rectangle with height h and width b, then the bending modulus can be written $W = bh^2/6$.

According to the Concrete Structures Design (GB50010-2002), the calculation formulas are:

$$M_{max} = M_{u,max} = \alpha_{s,max} \cdot \alpha_1 f_c b h_0^2 \quad (4.15)$$

$$\alpha_{s,max} = \xi_b(1 - 0.5\xi_b) \quad (4.16)$$

where $M_{u,max}$ is the maximum bending moment of the rectangular cross section of reinforced concrete, the unit is kN m. $\alpha_{s,max}$ is the maximum resistance coefficient of cross section moment. α_1 is the equivalent rectangular stress coefficient. f_c is the design value of concrete compressive strength, and the unit is MPa. b is the calculated length of the tunnel lining (cross-section width), and the unit is m. h_0 is the vertical distance of longitudinal tensile reinforcement force points to the edge of the lining cross-section compression zone, called the effective height of the lining

cross section; the unit is m. ξ_b is relatively height of compression zone boundaries, which is related to the steel and concrete strength grade.

Combining Eqs. (4.13)–(4.16), it yields:

$$R = 6\alpha_1 \xi_b (1 - 0.5\xi_b) f_c \frac{h_0^2}{h^2} \quad (4.17)$$

And the tunnel load effect S can be described by the following formula:

$$S = \gamma R_0 \left[\left(\frac{(p_0 + c_b \operatorname{ctg}\varphi)(1 - \sin\varphi)}{c_b \operatorname{ctg}\varphi} \right)^{\frac{1-\sin\varphi}{2\sin\varphi}} - 1 \right] - p_a \quad (4.18)$$

where γ is the density of the surrounding rock of the tunnel, the unit is kN/m^3. R_0 is the equivalent radius of tunnel excavation, and the unit is m. p_0 is in situ stress, and the unit is MPa. c_b is the surrounding rock cohesion, and the unit is MPa. φ is the internal frictional angle of the surrounding rock of the tunnel, and the unit is degrees. p_a is the water pressure acting on the tunnel lining, the unit is MPa.

Combining Eqs. (4.11), (4.12), (4.17), and (4.18), then the reliability limit equilibrium equation of the tunnel structure can be obtained:

$$g(x_1, \ldots, x_n) = 6\alpha_1 \xi_b (1 - 0.5\xi_b) f_c \frac{h_0^2}{h^2} - \gamma R_0$$

$$\left[\left(\frac{(p_0 + c_b \operatorname{ctg}\varphi)(1 - \sin\varphi)}{c_b \operatorname{ctg}\varphi} \right)^{\frac{1-\sin\varphi}{2\sin\varphi}} - 1 \right]$$

$$+ p_a = 0$$

(4.19)

3.4 Reliability Analysis of Tunnel Structure and Application

3.4.1 Reliability Analysis of Hydraulic Tunnel Structure

Taking the smaller variability of parameters in Eq. (4.19) as definite quantities, such as $\alpha_1 = 1.0$, $\xi_b = 0.55$, $\varphi = 27$ degrees, and $p_a = 0.43$, then these definite quantities were substituted in Eq. (4.19), thus the reliability model of the hydraulic tunnel structure can be obtained as follows:

$$g(f_c, h_0, h, \gamma, R_0, p_0, c_b) = 2.39 f_c \frac{h_0^2}{h^2} - \gamma R_0$$
$$\left[\left(\frac{p_0 + 1.96 c_b}{3.59 c_b}\right)^{0.6} - 1\right]$$
$$+ 0.43 = 0$$

(4.20)

To simplify Eq. (4.20), let $x_1 = f_c$, $x_2 = \frac{h_0^2}{h^2}$, $x_3 = \gamma R_0$, $x_4 = \left(\frac{p_0 + 1.96 c_b}{3.59 c_b}\right)^{0.6}$, then by Eq. (4.20) yields

$$g(x_1, x_2, x_3, x_4) = 2.39 x_1 x_2 - x_3(x_4 - 1) + 0.43$$

(4.21)

Let the means of the various random variables f_c, h_0, h, γ, R_0, p_0, and c_b are $\bar{f}_c$, $\bar{h}_0$, $\bar{h}$, $\bar{\gamma}$, $\bar{R}_0$, $\bar{p}_0$, and $\bar{c}_b$, respectively. Accordingly, their standard deviations are $\sigma(f_c)$, $\sigma(h_0)$, $\sigma(h)$, $\sigma(\gamma)$, $\sigma(R_0)$, $\sigma(p_0)$, and $\sigma(c_b)$, respectively.

According to the mean first-order second moment method (MFOSM), the mean $\bar{x}_i$ and the standard deviation $\sigma(x_i)$ of x_i can be obtained as follows:

$$\bar{x}_1 = \bar{f}_c$$

(4.22)

$$\sigma(x_1) = \left[\left.\frac{\partial x_1}{\partial f_c}\right|_{\bar{f}_c} \sigma(f_c)\right]^{2 \times \frac{1}{2}} = \sigma(f_c)$$

(4.23)

$$\bar{x}_2 = \frac{\bar{h}_0^2}{\bar{h}^2}$$

(4.24)

$$\sigma(x_2) = \left(\left[\left.\frac{\partial x_2}{\partial h_0}\right|_{\bar{h}_0} \sigma(h_0)\right]^2 + \left[\left.\frac{\partial x_2}{\partial h}\right|_{\bar{h}} \sigma(h)\right]^2\right)^{\frac{1}{2}}$$
$$= \left(\left[\frac{2\bar{h}_0}{\bar{h}^2}\right]^2 \sigma^2(h_0) + \left[\frac{2\bar{h}_0^2}{\bar{h}^3}\right]^2 \sigma^2(h)\right)^{\frac{1}{2}}$$

(4.25)

$$\bar{x}_3 = \bar{\gamma} \cdot \bar{R}_0$$

(4.26)

$$\sigma(x_3) = \left(\left[\left.\frac{\partial x_3}{\partial \gamma}\right|_{\bar{\gamma}} \sigma(\gamma)\right]^2 + \left[\left.\frac{\partial x_3}{\partial R_0}\right|_{\bar{R}_0} \sigma(R_0)\right]^2\right)^{\frac{1}{2}}$$
$$= \left([\bar{R}_0 \sigma(\gamma)]^2 + [\bar{\gamma} \sigma(R_0)]^2\right)^{\frac{1}{2}}$$

(4.27)

$$\bar{x}_4 = \left(\frac{\bar{p}_0 + 1.96 \bar{c}_b}{3.59 \bar{c}_b}\right)^{0.6}$$

(4.28)

$$\sigma(x_4) = \left(\left[\left.\frac{\partial x_4}{\partial p_0}\right|_{\bar{p}_0} \sigma(p_0)\right]^2 + \left[\left.\frac{\partial x_4}{\partial c_b}\right|_{\bar{c}_b} \sigma(c_b)\right]^2\right)^{\frac{1}{2}}$$
$$= \left(\left[\frac{1}{5.98 \bar{c}_b}\left(\frac{\bar{p}_0 + 1.96 \bar{c}_b}{3.59 \bar{c}_b}\right)^{-0.4}\right]^2 \sigma^2(p_0) \right.$$
$$\left. + \left[\frac{\bar{p}_0}{5.98 \bar{c}_b^2}\left(\frac{\bar{p}_0 + 1.96 \bar{c}_b}{3.59 \bar{c}_b}\right)^{-0.4}\right]^2 \sigma^2(c_b)\right)^{\frac{1}{2}}$$

(4.29)

The mean $\bar{g}$ of the function $g(x_i)$ is calculated as:

$$\bar{g} = 2.39 \bar{x}_1 \bar{x}_2 - \bar{x}_3(\bar{x}_4 - 1) + 0.43$$

(4.30)

The standard deviation $\sigma(g)$ of the function $g(x_i)$ is calculated as:

$$\sigma(g) = \left(\left[\left.\frac{\partial g}{\partial x_1}\right|_{\bar{x}_1} \sigma(x_1)\right]^2 + \left[\left.\frac{\partial g}{\partial x_2}\right|_{\bar{x}_2} \sigma(x_2)\right]^2 \right.$$
$$\left. + \left[\left.\frac{\partial g}{\partial x_3}\right|_{\bar{x}_3} \sigma(x_3)\right]^2 + \left[\left.\frac{\partial g}{\partial x_4}\right|_{\bar{x}_4} \sigma(x_4)\right]^2\right)^{\frac{1}{2}}$$

(4.31)

where $\left.\frac{\partial g}{\partial x_1}\right|_{\bar{x}_1} = 2.39 \bar{x}_2$, $\left.\frac{\partial g}{\partial x_2}\right|_{\bar{x}_2} = 2.39 \bar{x}_1$, $\left.\frac{\partial g}{\partial x_3}\right|_{\bar{x}_3} = -(\bar{x}_4 - 1)$ and $\left.\frac{\partial g}{\partial x_4}\right|_{\bar{x}_4} = -\bar{x}_3$.

Using MFOSM method, the reliability index β of the hydraulic tunnel underlying Qin-Fu expressway can be calculated as:

$$\beta = \frac{\bar{g}}{\sigma(g)} = \frac{2.39\overline{x_1 x_2} - \overline{x_3}\left(\overline{x_4} - 1\right) + 0.43}{\left(\left[2.39\overline{x_2}\sigma(x_1)\right]^2 + \left[2.39\overline{x_1}\sigma(x_2)\right]^2 + \left[\left(\overline{x_4} - 1\right)\sigma(x_3)\right]^2 + \left[\overline{x_3}\sigma(x_4)\right]^2\right)^{\frac{1}{2}}} \quad (4.32)$$

3.4.2 Engineering Application

According to field investigation, exploration data, and indoor and outdoor test results, seven groups parameter sample values of the dangerous lining sections were obtained, as listed in Table 4.6. Accordingly, the variation of the calculated parameters, such as mean and standard deviation as listed in Table 4.7.

According to the results in Table 4.7 and Eq. (4.32), the reliability index of the hydraulic tunnel structure was obtained: $\beta = 1.197$.

Assuming random state variables that affected the structural stability of the tunnel follow a normal distribution, so the tunnel structure failure probability is described as follows:

$$P_f = \int_{-\infty}^{-\beta} \frac{1}{\sqrt{2\pi}} \exp\left(-\frac{t^2}{2}\right) dt \quad (4.33)$$

where P_f is the tunnel structure failure probability, t is the variable of the reliability index, and β is the reliability index.

According to Eq. (4.33), the failure probability P_f of the tunnel structure is less than 11.5%, that is to say the reliability of the hydraulic tunnel structure is greater than 88.5%. According to geotechnical engineering specifications, the reliability of the permanent structure to meet the engineering requirements must be greater than 90%. Although the reliability of the hydraulic tunnel underlying Qin-Fu expressway is close to 90%, but less than 90%. To ensure the hydraulic tunnel long-term stability, effective engineering treatment countermeasures must be taken for reinforcement of the tunnel structure and its surrounding rocks, thus which can reduce and eliminate the harmful effects on the tunnel structure by the Qin-Fu expressway vehicle loads and other influence factors. After being taken effective reinforcement measures for the tunnel structure and its surrounding rock by grouting in advance, the shallow hydraulic tunnel project is operating well in the fine conditions now.

TABLE 4.6 The Various Parameters of the Sample Values

No.	f_c (MPa)	h_0 (m)	h (m)	γ (kN/m³)	R_0 (m)	p_0 (MPa)	c_b (MPa)
1	10.3	0.27	0.30	25.82	1.63	0.22	0.0346
2	11.9	0.32	0.35	24.01	1.70	0.251	0.035
3	11.2	0.38	0.41	23.43	1.94	0.285	0.0347
4	12.36	0.36	0.38	25.22	1.73	0.26	0.038
5	12.8	0.33	0.36	24.38	1.61	0.243	0.0364
6	11.85	0.30	0.33	23.50	1.72	0.255	0.0296
7	11.0	0.28	0.31	24.72	1.58	0.236	0.0353

TABLE 4.7 The Mean and Standard Deviation Calculations of the Variation Parameters

Item	f_c (MPa)	h_0 (m)	h (m)	γ (kN/m³)	R_0 (m)	p_0 (MPa)	c_b (kPa)
$\bar{g}$	11.63	0.319	0.349	24.44	1.70	0.250	34.8
$\sigma(g)$	0.72	0.032	0.032	0.816	0.13	0.022	2.2

3.5 Conclusions

Based on the interaction analysis of the tunnel structure and its surrounding rock, the reliability limit equilibrium equation and the stability reliability index formula for the shallow hydraulic tunnel underlying the Qin-Fu expressway had been derived. Through calculation analysis, the reliability of the tunnel structure was greater than 88.5%, which was less than 90% of permanent structure required. Therefore, to ensure the hydraulic tunnel long-term stability, some effective engineering treatment measures must be taken.

4. DISTURBANCE DEFORMATION OF AN EXISTING TUNNEL

4.1 Introduction

To solve the growing road traffic congestion, new tunnels adjacent to existing tunnels have been constructed. As limited by topography, geological conditions, and other restrictions, the new tunnel was often arranged adjacent to the old tunnel with small spacing. The so-called small-spacing tunnels are double tunnels, and the net gap between the two tunnels is less than a specified value. The rock and soil between the new tunnel and the existing tunnel could be disturbed many times due to the construction of the new tunnel adjacent to the existing tunnel, which usually results in a complicated superimposed stress and the rock and soil moving in different directions. This may induce a harmful deformation and additional structural internal forces to the existing tunnel, and thus the safe construction of the new tunnel and the safe operation of the existing tunnel will be affected (Wang and Xie, 2008; Wang et al., 2012).

As early as the 1940s, foreign scholar Terzaghi (1942) had analyzed subway tunnels with small spacing in Chicago. In the 1970s, Ghaboussi and Ranken (1977) systematically analyzed the force of rock pillar between the two tunnels, the surface subsidence and internal forces of tunnel lining and other variation characteristics of the parallel tunnels with small spacing were analyzed also by using the finite element method. Considering the construction sequence and drainage factors, Amir-Faryar et al. (2012) studied the adverse effects due to the neighbor tunnel construction by the numerical simulation method. Then Giardina et al. (2012) researched the harmful subsidence to the surface buildings close to the tunnels using similar model test, and so forth.

Although the study on these issues started relatively late in China, the development is very fast. Jiang et al. (2000) studied the construction process of the large-span flat tunnels by physical model tests and numerical simulation methods. Combined with practical engineering, Wang et al. (2010) carried out the model test and numerical simulation to study the influence of the perpendicular undercross shield tunnel construction on the existing tunnel. Dong et al. (2010) developed a geomechanics model testbed, and analyzed the stress of partition wall and the mechanical characteristics of the supporting structure in multiarch tunnel during the tunnels construction. Du and Huang (2009) conducted the numerical analysis of the displacement field of the eccentrically compressed tunnel with small spacing.

In summary, although domestic and overseas scholars have done a great deal of research on the small-spacing tunnels and obtained a lot of achievements, this research is still in the experience exploratory stage due to the geological conditions and influencing factors being complicated. Theoretical research still lags behind the engineering practice. The deformation of the surrounding rocks produced by tunnel excavation not only reflects rock pressure, rock structure, excavation methods, construction techniques, and many other factors, but also reflects the combined effects of these factors. Therefore, the disturbance deformation effect on the existing tunnel of the asymmetric tunnels

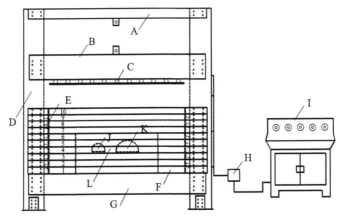

FIGURE 4.27 Sketch of geomechanical model test-bed and the tunnels with small spacing. (A) Upper beam; (B) moving beam; (C) vertical servo loader; (D) vertical beam; (E) lateral servo loader; (F) channel steel; (G) bottom beam; (H) oil separator; (I) hydraulic loading control test-bed; (J) the existing tunnel; (K) the new excavating tunnel; and (L) the rock pillar between two tunnels.

with small spacing has been studied based on a practical engineering.

4.2 Physical Model Test and Numerical Simulation

4.2.1 Engineering Background

The section design dimensions of a two-lane road tunnel are 12-m width and 8-m height, and the surrounding rock is residual soil and full-weathered tuff, with joint fissure developing well and bad overall stability, prone to collapse, off the block. Therefore, the classification of the surrounding rock is grade V. The groundwater in the surrounding rock is loose soil pore water and bedrock fissure water, and the hydrogeological condition is simple.

In the right of the two-lane road tunnel, the three-lane road tunnel (18-m width and 11-m height) was constructed by using the heading-and-bench excavation method with 9 m away from the existing tunnel. The positional relationship between two tunnels is shown in Fig. 4.27.

According to the Code for Design of Road Tunnel in China (JTG D70-2004), the physical and mechanical parameters for grade V of the surrounding rock of the tunnel are shown in Table 4.8.

4.2.2 Similar Model Test

This test was carried out in the self-developed geomechanics model test-bed. As shown in Fig. 4.27, the steel frame structure consists of the upper beams, the moving beam, the bottom beam, and two lateral vertical beams. The internal net size of the steel frame is 4.0-m long, 0.4-m wide, and 2.0-m high. There are 11 vertical servo-loading cylinders arranged on the moving beam, and six horizontal servo-loading cylinders arranged on two lateral vertical beams, respectively. The total load of vertical servo-loading

TABLE 4.8 Physics and Mechanics Parameters of the Surrounding Rock of the Tunnel

Grade	Density (kN/m³)	Elasticity Modulus (GPa)	Poisson Ratio	Cohesion (MPa)	Friction Angle (degrees)
IV	22–24	13–6	0.3–0.35	0.2–0.7	27–39
V	19–22	1–2	0.35–0.45	0.05–0.2	20–27
VI	16–19	<1	0.4–0.5	<0.05	<20

cylinders is 600 kN and that of the horizontal servo-loading cylinders is 400 kN.

The front and back of the test-bed are constrained with No. 20 channel steel. To observe easily, the tempered glass was arranged on the front and back of the model, respectively. and the excavation holes, which sizes were determined according to the tunnels cross section, were reserved in the bottom of glass (Fig. 4.27).

The geometry similarity ratio of the model is 1:33 and the density similitude ratio is 1:1. According to the similarity criteria, the stress similarity ratio of 1:33 was calculated. By repeatedly proportioning tests, an ordinary sand and barite powder with the ratio of 1:1 as aggregate were used, and choosing an engine oil which weight accounted for 6% of the aggregate as a cementing agent, thus the similar materials were made.

The test steps were as follows:

Step 1: To form a model tank, the numbered channel steels from 1 to 10 were assembled to test-bed by the serial number step by step (Fig. 4.27).

Step 2: According to material proportioning, the similar materials were weighed and added to the model tank (the thickness of each layer to 2.5 cm, mass of 88 kg, tamping density to 2.2 g/cm^3), with a spatula to flatten the material surface and making the surface remain flat. Then the materials were tamped with a compaction device, and the similar materials were filled again and tamped again to each marked line.

Step 3: When installing No. 4 channel steel after No. 1–3 channel steels were installed, the two tunnel models were placed. Then, the similar material was filled layer by layer until it reached the position of No. 6 channel steel.

Step 4: After removing No.1 to 6 channel steels on the front, 20-mm organic glass and plastic plates were laid, and then No.–6 channel steels were installed again in its former order.

Step 5: The filling and tamping layer by layer was continuously repeated, until the model tank was filled.

Step 6: To install displacement meter and wiring on the front of the model, the No. 4 and 5 channel steels were removed from the model tank in the test.

Step 7: When the loading up to 100 kN in the vertical direction remained stable, the right tunnel was excavated by four steps, and the vertical displacement was recorded in the process of the excavation (Fig. 4.28).

4.2.3 Numerical Simulation Verification

Based on the physical model test mentioned earlier, the numerical model of the asymmetric tunnels with small spacing was built using the Fast Lagrangian Analysis of Continua (FLAC) technique. The computational model was 80-m length and 64-m height, and divided into 26,250 elements.

The lateral horizontal movements of the computational model were restricted, and its bottom was fixed. The top border was applied uniform loads to simulate the overlying strata weight. It was assumed that material failure of the surrounding rock conforms to Mohr–Coulomb strength criterion. The physical and mechanical parameters of the surrounding rocks of two tunnels were given in Table 4.8.

The process of the numerical simulation was as follows: First, the balance of the initial stress field was calculated, and the displacement and velocity fields were cleared. Then, the excavation process of the large section tunnel was simulated step by step. The maximum vertical displacements (roof and bottom) of the existing tunnel were monitored during the whole calculation.

As shown in Fig. 4.29, on the one hand, with the excavation of the large section tunnel step by step, due to the influence of the rock and soil between two tunnels, the vertical displacement fields of the surrounding rock of the existing tunnel (roof and bottom) gradually deflected

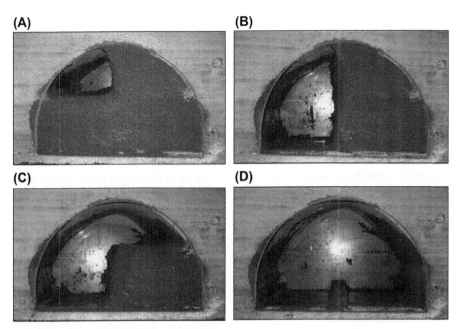

FIGURE 4.28 Photo of the excavation process of the new tunnel step by step. (A) Excavation 1, (B) excavation 2, (C) excavation 3, and (D) excavation 4.

to the new tunnel, and especially after the fourth step excavation of the large section tunnel, the roof displacement fields between the two tunnels appeared a connected domain. On the other hand, the vertical displacements (roof and bottom) on the existing tunnel responded from small to large during the excavation process.

As shown in Fig. 4.30, with the excavation of the large section tunnel step by step, the monitoring curves of the maximum vertical displacement on the existing tunnel (roof and bottom) in the physical model test and the numerical simulation results were basically the same in trend, which verified the results of the physical model test.

4.3 Disturbance Deformation of the Existing Tunnel

To reveal the disturbance deformation effect of the asymmetric tunnels with small spacing, supposing the surrounding rocks as homogeneous material (the material failure conforms to Mohr–Coulomb strength criterion) and heterogeneous material (the material failure conforms to ubiquitous-joint strength criterion), and only considering the temporary support on the existing tunnel, a series of numerical experiments were carried out under different conditions respectively, including the different lithologies, different depths, different spacing, and different dip angle of the dominant joints variation in the surrounding rock of the two tunnels. The computational analysis schemes under different conditions are shown in Table 4.9.

To reveal the disturbance deformation effect on the existing tunnel of the asymmetric tunnels with small spacing, the disturbance effect index α was defined, and the index α was the ratio of the maximum horizontal displacement of the left and right of the rock pillar between two tunnels. In general, higher value of α indicates a

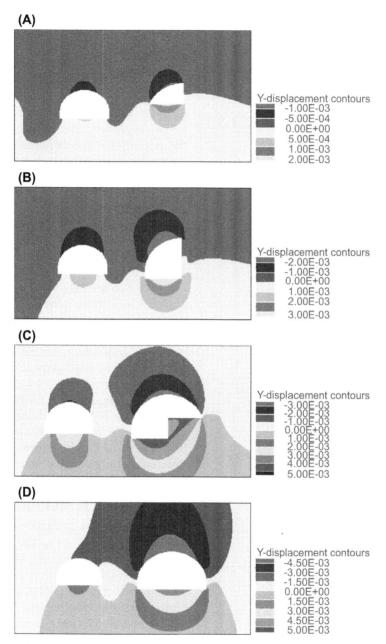

FIGURE 4.29 Vertical displacement fields of the tunnels. (A) Excavation 1, (B) excavation 2, (C) excavation 3, and (D) excavation 4.

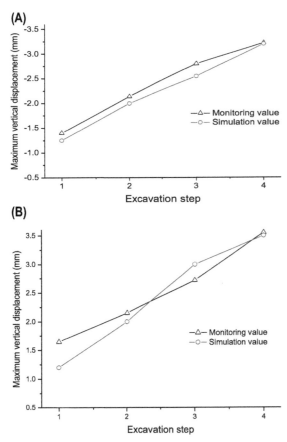

FIGURE 4.30 Vertical displacement fields of the existing tunnel. (A) Roof and (B) bottom.

TABLE 4.9 Scheme for Computational Analysis Under Different Conditions

No.	Analysis Schemes	Changing Factors	Notes
1	Different spacing	B = 4.5, 6.0, 9.0 m	B is the width of the rock pillar;
2	Different lithologies	$ = IV,V,VI	$ is the grade of surrounding rock;
3	Different depths	H = 40, 60, 80, 100 m	H is the depth of tunnels;
4	Different dip angles	Θ = 0, 25, 50, 75 degrees	Θ is the dip angle of the dominant joints in the surrounding rock.

more severe disturbance deformation effect on the existing tunnel due to the new tunnel excavation step by step.

4.3.1 Different Spacing Variations

As shown in Figs. 4.31 and 4.32, with the net distance between tunnels from 9.0 m, 6.0 to 4.5 m decreasing, the vertical displacement of the existing tunnel roof linearly increased significantly due to the large section tunnel excavation, and the displacement of the existing tunnel bottom changed relatively small. As to the disturbance effect index α, it varied from 0.90, 1.05 up to 1.08, indicating that the disturbance deformation effect was in a gradual intense trend.

4.3.2 Different Lithologies Variations

As shown in Figs. 4.33 and 4.34, with the grade of rock mass from IV, V to VI weakening, the vertical displacement of the existing tunnel roof dramatically increased nonlinearly with the large section tunnel excavation step by step, and the vertical displacement of the existing tunnel bottom also increased significantly, but the increasing magnitude of the bottom is far below that of the roof. As to the disturbance effect index α, it varied from 0.54, 0.90 up to 1.20, indicating that the disturbance deformation effect under different grade of the surrounding rock is significant.

4.3.3 Different Depths Variations

As shown in Figs. 4.35 and 4.36, with the depth from 40, 60, 80 to 100 m increasing, the vertical displacement of the existing tunnel roof was greater than that of the bottom with the large section tunnel excavation, and the vertical displacements of roof and bottom approximately linearly increased slowly, but the variation of the value-added magnitude was not significant.

As to the disturbance effect index α, it varied from 0.88, 0.95, 0.97 to 0.98, and the disturbance deformation effect was gradually increasing, but the variation amplitude was small.

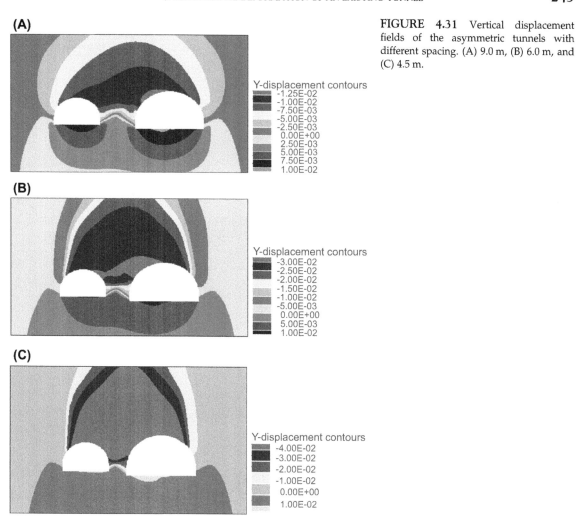

FIGURE 4.31 Vertical displacement fields of the asymmetric tunnels with different spacing. (A) 9.0 m, (B) 6.0 m, and (C) 4.5 m.

4.3.4 Different Dip Angles Variations

The weak intercalation in stability analysis of the tunnel is crucial because the shear strength of the broken rock in the surrounding rock is mostly determined by the weak structural plane or weak intercalation.

Considering the dip angle of the dominant joints, and supposing the surrounding rock conform to the ubiquitous-joint strength criterion, the numerical experiments on disturbance deformation effect of the asymmetric tunnels with small spacing were carried out under different dip angles of the dominant joints.

As shown in Figs. 4.37 and 4.38, with the dip angle of the dominant joints from 0, 25, 50 to 75 degrees increasing, the vertical displacement of roof of the existing tunnel linearly increased gradually with the large section tunnel excavation. Due to the influence of the structural plane, the displacement characteristics of the tunnel roof appeared steep, and the vertical displacement of the tunnel bottom produced small

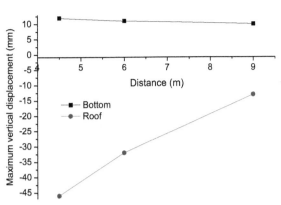

FIGURE 4.32 Vertical displacement curves of the existing tunnel with different spacing.

FIGURE 4.34 Vertical displacement curves of the existing tunnel under different rocks.

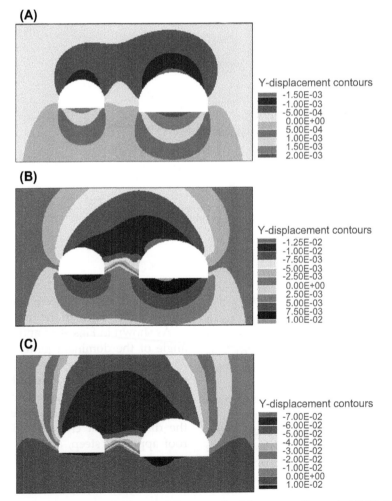

FIGURE 4.33 Vertical displacement fields with different grade of the surrounding rock. (A) IV, (B) V, and (C) VI.

4. DISTURBANCE DEFORMATION OF AN EXISTING TUNNEL

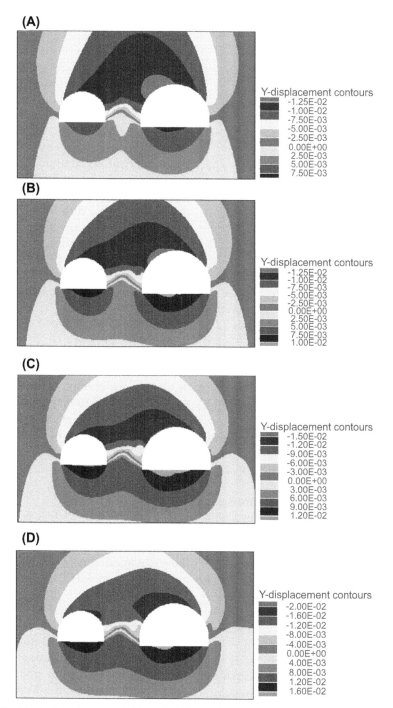

FIGURE 4.35 Vertical displacement fields with different depths. (A) 40 m, (B) 60 m, (C) 80 m, and (D) 100 m.

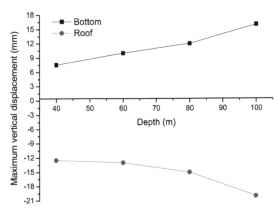

FIGURE 4.36 Vertical displacement curves of the existing tunnel under different depths.

variations. As to the disturbance effect index, it varied from 0.96, 1.00, 1.07 to 1.13, and the disturbance deformation effect was gradually increasing.

4.4 Discussion

It is of importance to conduct a reasonable evaluation of the deformation effect of the intermediate rock mass between the two tunnels with small spacing under the disturbance loads, which directly influences the construction and operation safety of the two tunnels.

The disturbance loads on the intermediate rock mass between the two tunnels with small spacing have three characteristics. The first is rate effect, with the disturbance loads being applied to the rock mass causing the mechanical response related to the loading rate. The second is recycling effect, with the repeated disturbance loads being applied to the rock mass causing the mechanical response due to the tunnels excavation step by step. The third is an interaction between the intermediate rock mass and the supports of the two tunnels. Further research to reveal the deformation effect of the intermediate rock mass between the two tunnels with small spacing should be done from three aspects as mentioned earlier.

In this paper, the physical model test and the numerical method are combined to conduct the research. First, the similar model test for the two tunnels with small spacing was carried out by the self-developed geomechanics model test-bed. Second, the excavation process of the new tunnel step by step was verified by the numerical simulation method. Third, a series of numerical experiments were accomplished according to the different lithologies, different depths, different spacing of the two tunnels, and the dip angle of the dominant joints in the surrounding rock. The deformation effect and the deformation evolution characteristics were revealed for the two tunnels with small spacing, but there is still further research to be done.

The interaction between the intermediate rock mass and the supports of the two tunnels, looked at in essence, is an energy transfer effect under the disturbance loads. Therefore, from the point of view of the energy analysis, it may be a new research method to conduct the research of the deformation effect of the intermediate rock mass between two tunnels with small spacing under the disturbance loads.

4.5 Conclusions

The similar model test for the two tunnels with small spacing was carried out by the self-developed geomechanics model test-bed, and the excavation process of the new tunnel step by step was verified by the numerical simulation method.

The numerical experiments results showed that the disturbance deformation effect changed from weak to strong, and the disturbance effect index presented the gradual increment trend with the lithology from good to poor, tunnel depth from shallow to deep, and the tunnel spacing from large to small.

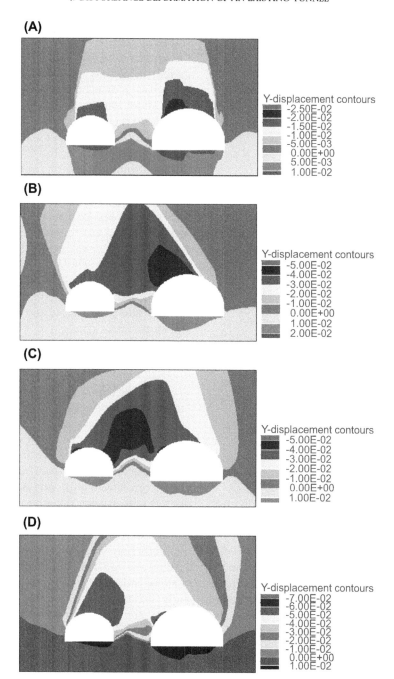

FIGURE 4.37 Vertical displacement fields with different dip angles of the dominant joints. (A) 0 degrees, (B) 25 degrees, (C) 50 degrees, and (D) 75 degrees.

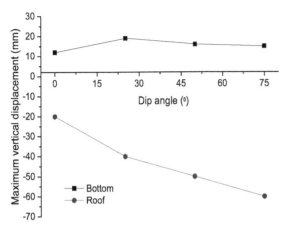

FIGURE 4.38 Vertical displacement curves of the existing tunnel under different dip angles.

Based on the ubiquitous-joint strength criterion, the deformation effect and the deformation evolution characteristics were revealed with the dip angle of the dominant joints from small to big for the two tunnels with small spacing.

5. ENERGY DISSIPATION CHARACTERISTICS OF A CIRCULAR TUNNEL

5.1 Introduction

Since the first record of rockburst at a tin mine in Britain in 1738, rockburst has always been a big puzzle in mining development engineering. Because it occurs suddenly with characteristics of dynamic failure, rockburst often threatens the safety of construction workers and equipment, causes destruction of engineering structures, and affects the project progress; it is still one of the problems in underground engineering on a worldwide basis (Shabarov, 2001; Singh, 1988).

Rockburst is often encountered during excavation of deep-buried underground engineering under high-stress conditions. It is considered as a dynamic instability phenomenon caused by the sudden release of elastic strain energy stored in the surrounding rock mass as a result of the excavation unloading induced stress dissipation in the stress redistribution process of the tunnel wall. There have been many laboratory experiments and theory analyses to research rockburst. In China, the analysis of the energy release in rockburst was made by Wang et al. (2008) through finite element numerical modeling including the influence of the excavation rate. By laboratory experiments it was confirmed that energy dissipation leads to reduction of rock strength, and energy release is the true cause of rock catastrophic failure (Zhao and Xie, 2008). Some scholars have executed a large number of experiments by means of the unloading method, acoustic emission monitoring method, energy criterion, and catastrophe theory (Xu et al., 2002; Zhang and Xu, 2002; Pan et al., 1997, 2006). Exploration and research on prediction and prevention of rockburst were also conducted (Xu, 2005; Yang and Luo, 2007; Xu and Wang, 2003; Du et al., 2007). In other countries, Jaeger and Cook (1979) conducted an early experimental study of rockburst; Ryder and Ozbay (1990), Spottiswoode (1988), and others found that the quantity of mine earthquakes, energy release rate, and rockburst events have a good correlation with excess shear stress, but not yet with energy release rate (ERR).

Although there have been some progress in studies of rockburst mechanism and research on prediction and prevention of rockburst, the results are far from mature. Along with the shallow mineral resources becoming increasingly exhausted, deep mining will inevitably become the main direction of mining development in the future in China. In deep mining, researching effective prevention methods to mining dynamic geological disasters, such as rockburst and others, have both important theoretical value to ensure the safety of mining and also important strategic significance to develop and use deep mineral resources. As unloading and continuous loading follow different stress paths and have different failure effects in the dynamic instability of underground engineering, the energy accumulation, migration, and dissipation process in the surrounding rock of the tunnel under high-stress conditions correspondingly have different characteristics. Thus it is always difficult to reach a result corresponding to practical engineering by following loading mechanics in researching on the mechanical characteristics of excavation unloading failure and its stability. It shows that the process of excavation unloading of a tunnel under high-stress conditions is much closer to practical engineering. With application of the energy calculation in Universal Distinct Element Code (UDEC) in this chapter, the principal stress distribution and energy dissipation characteristic of the surroundings rock during the circular tunnel excavation were analyzed along with variation of the excavated radius, stress level, and stress uneven degree, which has some referential value to revealing the rockburst mechanism and its evolutionary behavior.

5.2 Energy Balance Equation and Energy Release Rate Calculation

5.2.1 Energy Balance Equation

Salamon (1984) held that after tunnel excavation, the energy balance equation due to excavation unloading is:

$$W + U_m = U_c + W_r \quad (4.34)$$

where W is the work done by the shifting of external and gravitational forces working on the convergence and deformation of the rock mass due to tunnel excavation, U_m is the strain energy stored in the mined rock, U_c is the strain energy reaccumulated in the surroundings rock after tunnel excavation, and W_r is the dissipated energy of various forms owing to tunnel excavation.

5.2.2 Energy Release Rate Calculation

The ERR calculation can be done by using the energy computation in UDEC to get the numerical solution which agrees well with the analytical solution. The calculation error is less than 3% and can meet the requirement of engineering calculation.

To effectively reflect the variation of energy release quantity caused by excavation of arbitrary shape and volume, Hodgson and Joughin (1967) proposed the concept of energy release rate (ERR) that stands for the energy release quantity per volume excavation, which is often used as a quantitative index for the rockburst evaluation of surrounding rock mass.

$$k = dw/dv \quad (4.35)$$

where k is ERR, dw is the energy release caused by rock excavation, and dv is the rock volume which is digged out.

5.3 Energy Analysis of the Principal Stress Difference

5.3.1 Analysis of the Energy Release Quantity

5.3.1.1 COMPUTATION MODEL

Considering the excavation of a circular tunnel, with 1000-m deep in a mine, of which the surroundings rock is sandstone of good integrity which density is 2500 kg/m^3, elastic modulus is 90 GPa, Poisson's ratio is 0.3, cohesion is 10 MPa, friction angle is 43 degrees, and tensile strength is 2.0 MPa. A 10 m radius circular model is established, in a self-weight field of hydrostatic pressure state, with a fully clamped boundary. The computational model and its meshes are shown in Fig. 4.39.

5.3.1.2 SIMULATION ANALYSIS SCHEMES

First, along with the excavation radius changes of the circular tunnel from 0.5, 1.0, 1.5, and to 2.0 m, the energy release quantity w of

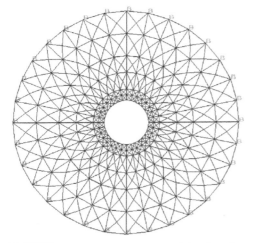

FIGURE 4.39 Computational model and meshes.

the tunnel surroundings rock is calculated, respectively, and the variation of ERR k is analyzed.

Second, to the surrounding rock stress σ increasing from 25, 50, 75 to 100 MPa, in which vertical stress and horizontal stress increase synchronously, the energy release quantity w is calculated respectively and the variation of ERR k is analyzed.

Finally, as the unevenness of the surrounding rock stress is changing by keeping the vertical stress σ_v constant and by increasing the horizontal stress σ_h from 25, 35, 45 to 55 MPa, the energy release quantity w is calculated correspondingly and the variation of ERR k is analyzed; after that, analysis of the maximum principal stress difference and its distribution characteristics in the tunnel surroundings rock are carried out.

5.3.1.3 CHARACTERISTICS OF THE ENERGY RELEASE QUANTITY

Under three conditions, along with the increase of the excavation radius, of the surrounding rock stress σ, and of the horizontal stress σ_h, it can be seen from Fig. 4.40 that for the circular tunnel excavation in deep mining, the energy dissipation quantity increases rapidly in the

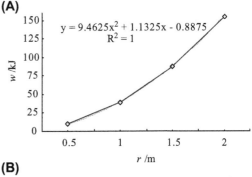
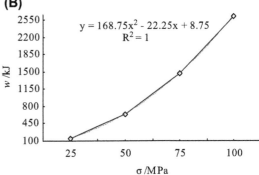
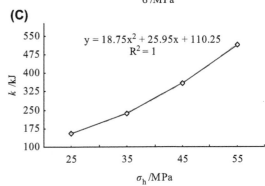

FIGURE 4.40 The curves of the energy dissipation quantity in the tunnel surrounding rock. (A) The excavated radius changing, (B) the hydrostatic state of stress changing, and (C) only horizontal stress changing.

form of concave quadratic curves, and the curve increases significantly in the wake of the stress level growth. The results indicate that with the continuous stress level improvements in deep mining, the energy dissipation quantity during tunnel excavation will show a remarkable growth of nonlinear instability curve, and the potential harmfulness of rockburst will increase.

5.3.2 Characteristics of Energy Release Rate in the Surrounding Rock

Fig. 4.41 shows that for the excavated circular tunnel in deep mining, the energy dissipation rate will increase in the form of concave quadratic curves along with the enhancement of stress level and of stress field unevenness; on the other hand, the excavation radius variation from small to big does not lead to a great change in the energy dissipation rate, which only fluctuates around a constant. Namely, ERR is insensitive to the excavated radius changing, and this result confirms that there is not a direct connection between the rockburst event and ERR.

In the quadratic polynomial fitting curve method, the quadratic coefficient determines the curvature variance of the concave curve. From the curves of energy dissipation quantity and of energy dissipation rate (Figs. 4.40 and 4.41) it can be seen that, in contrast to other factors, stress level is the determining key factor to the concave curve growth. This also reflects the main cause, for rockburst phenomenon in underground works in deep mining being more than that in shallow mining, is the increment of stress level.

5.4 The Principal Stress Distribution Characteristics

5.4.1 The Maximum and Minimum Principal Stress Distribution

As shown in Fig. 4.42, keeping the vertical stress of the surrounding rock invariance at 25 MPa, as the horizontal stress of the surrounding rock rises from 25, 35, 45, to 55 MPa, the principal stress σ_p generated from the surroundings rock will be varying along with the change of its distance to the inner wall, l of the circular tunnel. The maximum principal stress gradually decreases in the shallow rock (distance from the

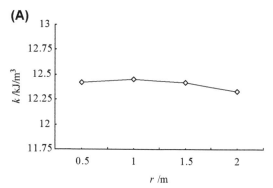

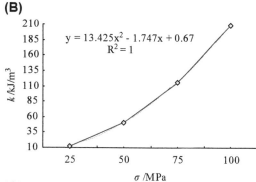

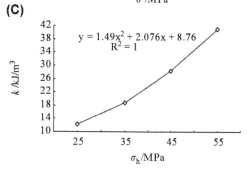

FIGURE 4.41 The curves of the energy dissipation ratio in the tunnel surrounding rock. (A) The excavated radius changing, (B) the hydrostatic state of stress changing, and (C) only horizontal stress changing.

tunnel rim less than about 5.0 m), and increases in the deep rock (distance from the tunnel wall greater than about 5.0 m). Meanwhile, the minimum principal stress has little change. Thus the results come to: (1) Transference of the maximum principal stress difference, generated

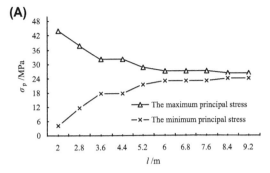
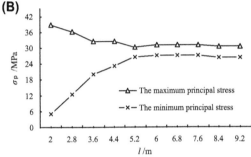
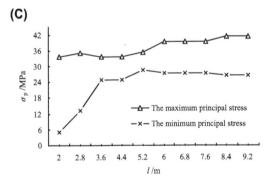
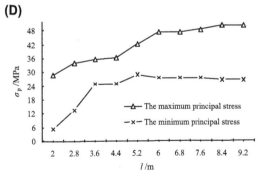

FIGURE 4.42 The principal stresses distribution curves in the tunnel surrounding rock. (A) $\sigma_v = \sigma_h$, (B) $\sigma_v = 1.4\sigma_h$, (C) $\sigma_v = 1.8\sigma_h$, and (D) $\sigma_v = 2.2\sigma_h$.

in the surrounding rock of the excavated circular tunnel from the near to the distant of its wall; (2) positions of the maximum principal stress difference emerged from discontinuous transference, which is identical to the alternating of tension and compression stress and cracking phenomenon in the surrounding rock discovered by scholars.

As the magnitude of maximum principal stress difference (to represent shear stress) and its location are closely related to the phenomenon of rockburst, consequently, the gradual increase and further concentration of the principal stress difference from the near to the distant of its wall in the surrounding rock of the excavated tunnel will be the potential direct factors to induce rockburst.

5.4.2 Concentrating Zones of the Principal Stress Difference

As shown in Fig. 4.43, keeping the vertical stress of the surrounding rock invariance at 25 MPa, as the horizontal stress rising from 25, 35, 45, to 55 MPa, the concentrating zones of the principal stress difference come about in a horizontal direction of the surrounding rock on both sides, and the scope and magnitude of the concentrating zones show growth trends along with the horizontal stress increment. Meanwhile, the concentrating zones also can be seen in the roof and have greater scope and magnitude than those in horizontal direction of the surrounding rock on both sides. Thus, we can say that with the gradual increase of the horizontal stress in surrounding rock, rockburst should be monitored at the tunnel roof.

The numerical simulation results show that, when quasistatic instability occurs after the circular tunnel is excavated, the accumulation, migration, and dissipation of the energy in the surroundings is a process of gradual gestation, then development, and finally outburst, along with the time increasing. Engineering practices demonstrate that rockburst generally lags behind the excavation several hours to dozens

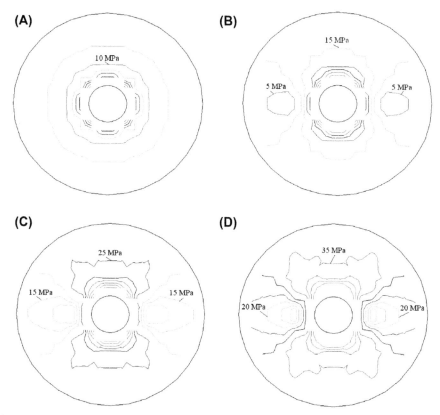

FIGURE 4.43 Zones of the principal stresses difference in the surrounding rock. (A) $\sigma_v = \sigma_h$, (B) $\sigma_v = 1.4\sigma_h$, (C) $\sigma_v = 1.8\sigma_h$, and (D) $\sigma_v = 2.2\sigma_h$.

of hours (Xu and Wang, 2003). Besides, after the tunnel is excavated, the generation of the principal stress concentration phenomenon and the redistribution process of the principal stress are both related to the stress conditions and geological environments of the excavated tunnel.

5.5 Conclusions

The distribution of the principal stress difference and the characteristic of energy dissipation in a tunnel's surrounding rock are analyzed according to the cases of the excavated radius, the stress level, and the nonuniform stress field of the circular tunnel with UDEC technique, some main conclusions can be obtained.

In the process of the circular tunnel being excavated in deep mining, the growth curves of the energy dissipation quantity and the energy dissipation ratio were concave with stress level increasing and the nonuniform stress field aggravating. The growth curve of the energy dissipation quantity was also concave, but the energy dissipation ratio with the excavated radius increasing was almost holding constant.

After excavation of the circular tunnel, the concentrating zones of the principal stress difference are extended and gradually migrated from the near to the distant from the tunnel wall in the surrounding rock with the nonuniform stress field aggravating. The intensity increment of the principal stress difference provides conditions for gestation and accumulation of rockburst energy.

When the horizontal stresses are greater than vertical stresses, the concentrating zones of the principal stress difference in the roof are usually having serious consequences than those on both sides of the excavated tunnel, and for this reason, strengthened attention and monitoring of rockburst should be given to the tunnel roof.

6. PRESSURE-ARCH EVOLUTION AND CONTROL TECHNIQUE

6.1 Introduction

In the construction process of all kinds of tunnels, the safe and efficient construction and lower support costs of the tunnel project are very important. Therefore, it is of the practical value to confirm and reveal the distribution morphology and the evolution characteristics of the pressure-arch of the tunnel and make full use of the self-bearing capacity of the tunnel surrounding rock.

The formation processes and evolutionary characteristics of the pressure-arch of the tunnel have received wide attention from scholars and professionals both nationally and internationally. Polillo et al. (1994) conducted sand arching experiments at the Colorado School of Mines on a similar well configuration and observed the effect of these variations on the stability of the sand surrounding the wellbore. Goel et al. (1996) found the tunnel size effect on support pressure is significant in flat-roofed underground openings and suggested two equations for the associated support pressure estimation. Prokhorov (1999) found that the arch effect coming from the cylindrical design could cause a loosely packed interior region inside the pressings by the large alumina pressing experiment. Hashash and Whittle (2002) discussed the soil arching effect and presented a detailed interpretation of the evolutionary stresses around a braced excavation in a deep soft layer. Jin and Shin (2003) studied the anisotropic reinforcing mechanism of the umbrella arch reinforcement method in a tunnel. Shahin et al. (2004) carried out the three-dimensional model tests of tunnel excavation and the corresponding numerical analyses to investigate the influence of tunnel excavation on surface settlement and earth pressure surrounding a tunnel. Hernández-Montes et al. (2005) addressed the influence of equilibrium, creep, and shrinkage as they affect the design of the buried arches from a theoretical perspective. Li (2006) proposed that the unstable section of the stope be reinforced with bolt-shotcrete ribs based on the pressure-arch in the failed rock. Poulsen (2010) conducted coal pillar load calculation by pressure-arch theory and near-field extraction ratio. Kim et al. (2013) found that the lateral earth pressure decreases within active displacement of a wall at deep excavation, and the arching effect is more significant for deep excavation than for shallow excavation by using experimental tests and a theoretical analysis.

Chinese scholars Liu et al. (2007) measured the distribution of stress in the wall rock mass with earth-pressure cells during the tunnel excavation and suggested an innovated rock mass pressure release rate idea. Zhu et al. (2008b) analyzed the construction sequence optimization and supporting structure of shallow multi-arch tunnels under unsymmetrical pressure conditions. Chen et al. (2011a,b) developed a three-dimensional numerical simulation model to gain an understanding of adjustment of tunneling stress. Jiang and Yin (2012) analyzed the earth pressure acting on the shield tunnel lining and the soil arching and unloading effects due to tunneling. Wang et al. (2014a) and Yang et al. (2015) not only revealed the geometry of the pressure-arch and its mechanical evolution characteristics, but also analyzed the mechanical stability and the instability failure modes of the pressure-arch through a highway tunnel. Furthermore, they conducted a mechanics evolution characteristics analysis of the pressure-arch in the fully mechanized mining field. Li et al. (2014) analyzed the supporting structure of pressure to a small-spacing shallow bias tunnel in soft rock by testing a typical section in the field. Sun et al. (2014) discussed the factors affecting soil arching and stress-sharing by the theoretical analysis and the experimental method. Xing et al. (2014) carried out the development and changes law of the soil arch under different support action through the measurement of radial and circumferential pressure data of soil around the tunnel under a different support force. Xie et al. (2014) put forward the anchor-spray injection to strengthened bearing arch support technology integrated with a high-strength anchor-bearing arch, thick steel mesh spray-up arch, and lag-grouting reinforcement arch, and clarified its arching and strengthening mechanism.

In summary, although some achievements on the pressure-arch of the surrounding rock of the tunnel have been obtained, some studies, such as the arching conditions, the mechanics evolution process, and the stress states that influence the pressure-arch and how to control the pressure-arch artificially have not formed a scientific and mature theoretical system, and more research is need to deepen this study. Therefore, combined with practical engineering, it is necessary to study the pressure transformation law, the arching form, and the three-dimensional evolution characteristics of the pressure-arch and the artificial control technology to the pressure-arch.

6.2 Pressure-Arch of Tunnel and Research Method

6.2.1 Arching Mechanism of Pressure-arch

After the highway tunnel excavation, the stress state of the surrounding rock in the excavation face changed from three-dimensional to two-dimensional. As shown in Fig. 4.33, with the secondary stress state of the surrounding rock of the tunnel adjusting by itself, the radial stress gradually increased from zero on the free face to the original stress in the deep of the surrounding rock, and after the tangential stress experienced a peak value, it gradually changed to the original stress of the surrounding rock. The stress state was constantly adjusted by itself to resist the uneven deformation of the surrounding rock and finally the stress concentration area formed in the surrounding rock, which was known as the pressure-arch (Wang et al., 2015).

6.2.2 Boundary Parameters of Pressure-arch

To facilitate the study, the element stress variable e was defined:

$$e = \frac{\sigma_1 - \sigma_3}{\sigma_3} \times 100\% \qquad (4.36)$$

where σ_1 and σ_3 were the maximum and the minimum principal stress of the unit, respectively, after the tunnel was excavated.

As shown in Fig. 4.44, the boundary corresponding to the peak of the maximum principal stress was defined as the inner boundary of the pressure-arch, and the boundary corresponding to the stress variable e was equal to 10% and was defined as the outer boundary of the

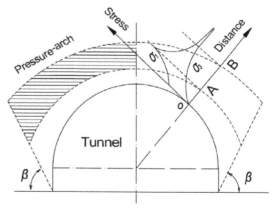

FIGURE 4.44 Schematic pressure-arch and its boundaries.

pressure-arch. The lateral boundaries of the pressure-arch was found on both lateral walls of the tunnel, which extended toward the deep of the surrounding rock along the rock rupture angle β and met with the inner and outer boundaries of the pressure-arch. According to the Protodyakonov's theory and Rankine theory, the rock rupture angle was $\beta = 45 \text{ degrees} + \frac{\varphi}{2}$, where φ was the inner friction angle of the surrounding rock of the tunnel.

6.2.3 Research Method

The evolutionary characteristics and control technique analysis of the pressure-arch of a highway tunnel were as follows.

First, the computational model was built using FLAC^{3D} based on a practical highway tunnel. Then the forming process, distributional pattern, and evolution characteristics of the pressure-arch of the highway tunnel under different stress states were analyzed according to the lateral pressure coefficients of the highway tunnel from small to large.

Second, the numerical simulation of the step-by-step excavation tunnel was carried out along the tunnel axis, the distribution morphology and variation characteristics of the pressure-arch of different tunnel parts were revealed, including of the front of the working face, the roof, and floor of the tunnel. Then the stability of the tunnel surrounding rock was evaluated.

Finally, by changing the bolt support parameters and the bolt-anchor combined support design, the control techniques for artificial reinforcement pressure-arch were analyzed.

6.3 Pressure-Arch Evolution Characteristics

6.3.1 The Computational Model

The dimensions of the cross section of a highway tunnel were 24-m wide and 13-m high, the surrounding rock was mainly composed of weathered sandstone, and the stability of the surrounding rock was poor. The hydrogeological condition of the tunnel surrounding rock was simple.

As shown in Fig. 4.45, the computational model was built using FLAC^{3D} at the depth of 80 m under the ground, and the model was in the hydrostatic stress state. The dimensions of the model were 80-m long, 2-m wide, and 64-m high in the x-, y-, and z-axes, respectively. The model was divided into 10,048 elements. The horizontal displacements of four lateral boundaries of the model were restricted, and its bottom was fixed. The vertical load was applied to the top of the model, which was equal to the weight converted from the overburden thickness. The material of the model was supposed to meet the Mohr–Coulomb strength criterion, and the physics and mechanics parameters were selected as listed in Table 4.10.

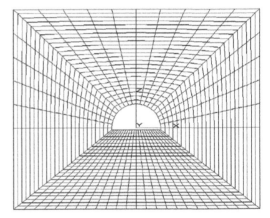

FIGURE 4.45 The computational model and its meshing.

TABLE 4.10 Physical and Mechanical Parameters of the Surrounding Rock of the Tunnel

Unit Weight (kN/m³)	Elasticity Modulus (GPa)	Poisson Ratio	Cohesion (MPa)	Friction Angle (degrees)	Tension (MPa)
23.0	5.0	0.32	1.2	33	0.5

6.3.2 Pressure-Arch Evolution Along the x–z Plane

Based on the definition of inner and outer boundary of the pressure-arch, the pressure-arch axis was determined (λ represented the lateral pressure coefficient of the surrounding rock of the tunnel). As shown in Fig. 4.46, when $\lambda = 0.25$, the tensile stress appeared on the top of the tunnel, and the pressure-arch only formed in the localized region, showing a discontinuous distribution pattern. As shown in Fig. 4.47, when $\lambda = 0.34$, the pressure-arch was thin, and its shape was slightly convex along the top of the tunnel, and obviously convex on both sides, showing butterfly-shaped distribution patterns. As shown in Fig. 4.48, when $\lambda = 0.72$, the pressure-arch showed the vaulted arch, and its axis was a part of parabola. As shown in Fig. 4.49, when $\lambda = 1.0$, the tunnel was in the hydrostatic stress state and the pressure-arch showed even distribution characteristics in range of the rupture angle of the surrounding rock of the tunnel.

As shown in Figs. 4.50–4.52, with the gradual increase of λ, the pressure-arch was compressed in the horizontal direction, the pressure-arch overall showed upward convex. When $\lambda = 1.7$, the vault of the pressure-arch was slightly sunken, showing a bimodal distribution pattern.

As shown in Fig. 4.53A, when $0.34 < \lambda < 1$, with the value of λ decreasing, the shape of pressure-arch gradually changed from the flat arch to the spire arch, and the axis of pressure-arch gradually lowered, ie, the position of the pressure-arch moved down overall. As shown in Fig. 4.53B, when $\lambda > 1$, with the value of λ

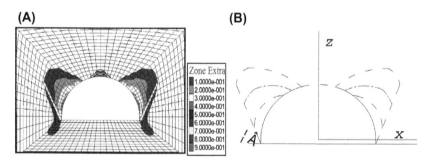

FIGURE 4.46   Pressure-arch boundary when $\lambda = 0.25$. (A) The stress contour of pressure-arch and (B) the simplified figure of pressure-arch.

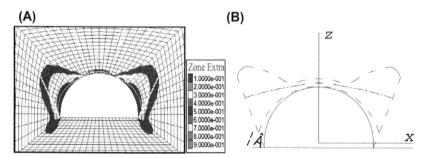

FIGURE 4.47   Pressure-arch boundary when $\lambda = 0.34$. (A) The stress contour of pressure-arch and (B) the simplified figure of pressure-arch.

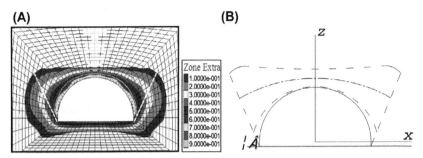

FIGURE 4.48 Pressure-arch boundary when $\lambda = 0.72$. (A) The stress contour of pressure-arch and (B) the simplified figure of pressure-arch.

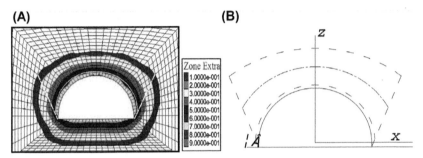

FIGURE 4.49 Pressure-arch boundary when $\lambda = 1.00$. (A) The stress contour of pressure-arch and (B) the simplified figure of pressure-arch.

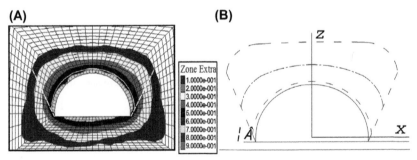

FIGURE 4.50 Pressure-arch boundary when $\lambda = 1.30$. (A) The stress contour of pressure-arch and (B) the simplified figure of pressure-arch.

increasing, the shape of pressure-arch changed from upward convex to sunken, and the pressure-arch axis gradually rose, ie, the position of the pressure-arch overall moved upward then got away from the excavation face of the tunnel, and the arch foot of the pressure-arch was closer to the edge of the tunnel. When $\lambda = 1.7$, the axis of the pressure-arch was sunken at the top of the arch, and the overall shape of the arch showed a bimodal distribution.

In summary, when $\lambda < 0.34$, since the tensile stress appeared, the pressure-arch could not be

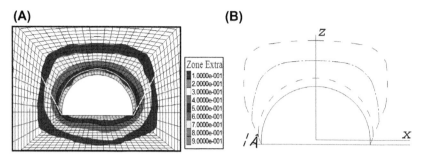

FIGURE 4.51 Pressure-arch boundary when $\lambda = 1.50$. (A) The stress contour of pressure-arch and (B) the simplified figure of pressure-arch.

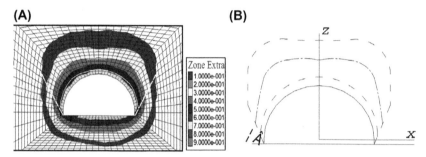

FIGURE 4.52 Pressure-arch boundary when $\lambda = 1.70$. (A) The stress contour of pressure-arch and (B) the simplified figure of pressure-arch.

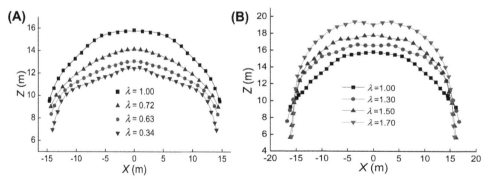

FIGURE 4.53 The change trend of pressure-arch axis with λ. (A) $\lambda < 1.00$ and (b) $\lambda > 1.00$.

formed at the top of the tunnel, and two local pressure-arch zones formed on both sides of the tunnel; when $0.34 < \lambda < 1$, the pressure zone of the surrounding rock at the top of the tunnel gradually expanded and showed a flat arch shape; when $\lambda > 1$, the pressure-arch of the tunnel showed upward convexity, and when λ reached a certain level, the arch vault appeared sunken, and the pressure-arch axis showed a bimodal distribution pattern.

6.4 Pressure-Arch Distribution Along the y-Axis

6.4.1 Three-Dimensional Computational Model

The three-dimensional numerical model was built using FLAC^{3D} as shown in Fig. 4.54. The model was 120-m long, 100-m wide, and 103-m high, and the model was divided into 42,000 elements. The horizontal displacements of four lateral boundaries of the model were restricted, and its bottom was fixed. The upper surface of the model was the load boundary, on which the vertical load was applied to simulate the weight of the overburden 80-m uniform load. The material of the model was supposed to meet the Mohr–Coulomb strength criterion, and moreover, it was supposed that the model was in the hydrostatic stress state, namely the lateral pressure coefficient λ was 1.0.

6.4.2 Pressure-Arch Distribution in Hydrostatic Stress State

As shown in Fig. 4.55, when $\lambda = 1.0$, with the tunnel excavation length variation from 6, 12 to 18 m, the pressure-arch of the tunnel extended outwards gradually, the pressure-arch in front of the working face gradually thickened, and the pressure-arch at the top of the working surface developed obliquely upward with a wedge-shaped bulge, thickened, and then the peak appeared. The sag morphology of the pressure-arch at the bottom of the working surface was more and more obvious. The distribution morphology of the pressure-arch near the working face was nonuniform outward expansion (the white line was the inner boundary of the pressure-arch).

As shown in Fig. 4.56, with the variation of excavation length from 6, 12 to 18 m, the difference values between the maximum and the minimum principal stress of the unit near the working face of the tunnel gradually increased,

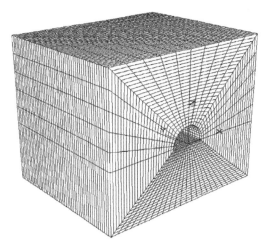

FIGURE 4.54 The three-dimension numerical model and its meshes.

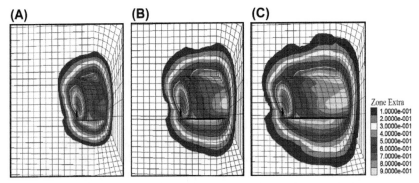

FIGURE 4.55 Pressure-arch distribution with different excavation length. (A) 6 m, (B) 12 m, and (C) 18 m.

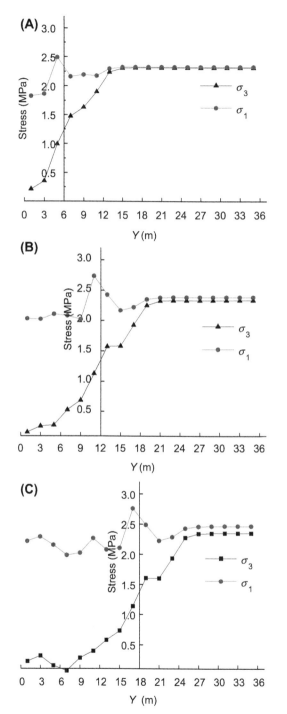

FIGURE 4.56 Principal stress curves with different excavation length. (A) 6 m, (B) 12 m, and (C) 18 m.

and the shear stress τ_{max} characterized the difference of the principal stress of the unit was

$$\tau_{max} = \frac{\sigma_1 - \sigma_3}{2} \quad (4.37)$$

Thus, with the increase of the excavation length of the tunnel, the shear stress concentration degree near the working face of the tunnel increased gradually; in other words, the instability of the surrounding rock near the working tunnel face tended to be enhanced.

6.4.3 Pressure-Arch Distribution Under Different Stress States

As shown in Fig. 4.57, when $\lambda < 1.0$, above the working face of the tunnel the pressure-arch uplifted, showing a wedge-shaped distribution, the pressure-arch of the horizontal direction thickened, the sunken shape of the pressure-arch on the bottom of the tunnel became obvious, and behind the working face the top of pressure-arch became thin. When $\lambda > 1.0$, the pressure-arch on the roof of the tunnel working face uplifted, showing an asymmetrical shape of bimodal distribution, and the pressure-arch in the horizontal direction became thin. The pressure-arch of the tunnel floor appeared sunken, and the pressure-arch of the roof behind the working face thickened. When $\lambda = 1.0$, the distribution morphology of the pressure-arch was between these two cases.

The pressure-arch on the tunnel cross section was shown in Fig. 4.58. When $\lambda < 1.0$, the pressure-arch displayed a pointed-arch. When $\lambda = 1.0$, the pressure-arch displayed a domed arch. When $\lambda > 1.0$, the pressure-arch displayed a flat arch. When $\lambda = 1.0$, the shape of the pressure-arch was the most regular, and its range was minimum.

6.5 Active Control Technology of Pressure-Arch

As shown in Fig. 4.59, for a full-length anchored bolt the rock mass formed a vesicular

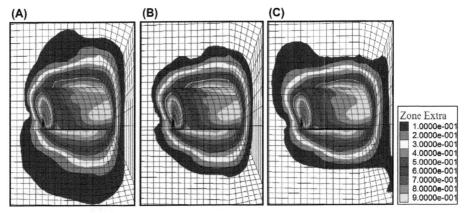

FIGURE 4.57 Pressure-arch distribution under different stress states. (A) $\lambda = 1.1$, (B) $\lambda = 1.0$, and (C) $\lambda = 0.9$.

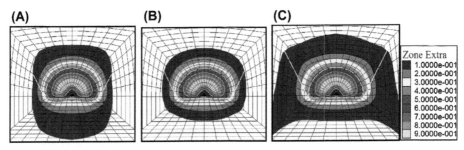

FIGURE 4.58 Pressure-arch distribution along $X-Z$ plane under different stress states. (A) $\lambda = 1.1$, (B) $\lambda = 1.0$, and (C) $\lambda = 0.9$.

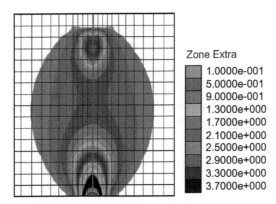

FIGURE 4.59 The concentrated stress areas for the full-length anchored bolt.

stress concentration area in the range of bolt support. If the spacing and the row spacing of the bolts were reasonably arranged, the loose rock mass with large deformation could be linked together to form an active reinforcement compression area.

6.5.1 Pressure-Arch Variation Under the Combined Bolts Support

6.5.1.1 SIZE EFFECT OF THE COMBINED BOLTS SUPPORT

As shown in Fig. 4.60, when the bolt length changed from 1.6, 2.4 to 3.2 m, the stress state of the surrounding rock near the tunnel was obviously adjusted by the combined bolts support. The physical and mechanical parameters of the bolts were listed in Table 4.11.

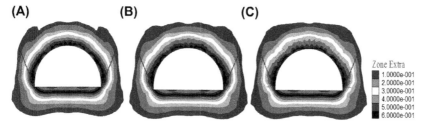

FIGURE 4.60 The reinforced pressure-arch with different bolt lengths. (A) 1.6 m, (B) 2.4 m, and (C) 3.2 m.

TABLE 4.11 The Physical and Mechanical Parameters of the Bolt

Young's Modulus (GPa)	Cross-Sectional Area (cm^2)	Grout Exposed Perimeter (cm)	Grout Cohesion Strength (kN/m)	Tension Yield Strength (kN)	Grout Friction Angle (degrees)
196	4.90	8.80	100	320	28

As shown in Fig. 4.61, when the bolt length changed from 1.6, 2.4 to 3.2 m, the central axis line of the reinforcement pressure-arch moved slightly downward, namely the artificial-reinforced arch gradually closed to the tunnel excavation face, which was of positive significance to maintain the stability of the surrounding rock of the tunnel.

6.5.1.2 SUPPORT EFFECT OF THE COMBINED BOLTS

After the tunnel excavation, for the combined bolts support, the morphology of the reinforced arch was influenced by the lateral pressure coefficient. As shown in Fig. 4.62, the support length of the combined bolt was 3.2 m; when $\lambda > 1.0$ and $\lambda < 1.0$, the morphology distribution of the central axis line of the pressure-arch showed a sunken dome. With the λ changed from $\lambda = 1.0$ to $\lambda < 1.0$, the pressure-arch span at the arch foot changed from narrow to wide, and the top of the arch body showed a descending trend.

6.5.2 Pressure-Arch Variation Under Bolt-Cable Coupling Support

The distribution morphology of the pressure-arch of the tunnel was shown in Fig. 4.63 under

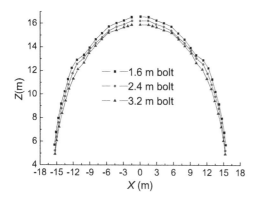

FIGURE 4.61 Pressure-arch distribution with different bolt lengths.

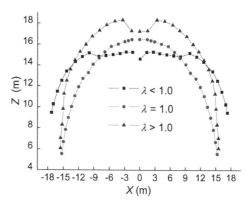

FIGURE 4.62 Pressure-arch distribution with different λ.

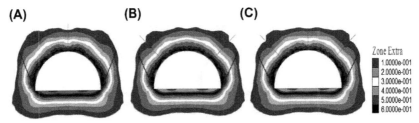

FIGURE 4.63 The artificially reinforced arch with different cable numbers. (A) Single cable, (B) two cables, and (C) three cables.

the bolt-cable coupling support. (The cable was 9 m long and the bolt was 2.4-m long; the main parameters of the cable are listed in Table 4.12.) When a single cable, two cables, and three cables, respectively, combined with the bolts, the distribution differences of the artificial-reinforced arch in the surrounding rock of the tunnel were obvious. With the increase of the cable numbers, the distribution of the artificial-reinforced arch was more uniform, which could fully mobilize the deep rock mass strength and maintain the stability of the surrounding rock of the tunnel.

6.6 Conclusions

With the change of the lateral pressure coefficient from small to large, the pressure-arch of the tunnel mainly was in three forms: the tensile stress appeared at the top of the tunnel, forming the partial pressure-arch ($\lambda < 0.34$); the pressure-arch approached to the tunnel working face, and the vault showed flat ($0.34 < \lambda < 1$); the top pressure-arch of the tunnel uplifted upward ($\lambda > 1$), and when λ increased a certain value the arch vault showed a bimodal distribution pattern.

With the principal stress difference near the working face of the tunnel increasing, the stability of the surrounding rock near the working face of the tunnel became worse for the tunnel being excavated by step-by-step along the tunnel axis. Relative to $\lambda > 1$ or $\lambda < 1$, when $\lambda = 1$, the surrounding rock near the working face of the tunnel was the most likely to remain stable.

It was beneficial to regulate the rock stress state near the tunnel with the increase of the combined bolts parameters. Under the bolt-cable coupling support, the cable was used to mobilize the strength of the deep rock mass, and the artificial bearing arch composed of the bolt-cable coupling support is very conducive to the stability of the tunnel.

7. SKEWED EFFECT OF THE PRESSURE-ARCH IN A DOUBLE-ARCH TUNNEL

7.1 Introduction

The double-arch tunnel is a special tunnel and its lining structures between two adjacent tunnels are supported by the middle wall. With a small area that is easy to connect with the line, the double-arch tunnel has a large advantage in the mountainous highway construction. Owing to the large span and large rock pressure, the two

TABLE 4.12 The Physical and Mechanical Parameters of the Cable

Young's Modulus (GPa)	Cross-Sectional Area (cm^2)	Grout Exposed Perimeter (cm)	Grout Cohesion Strength (kN/m)	Tension Yield Strength (kN)	Grout Friction Angle (degrees)
200	6.16	10.05	175	400	30

single tunnels of the double-arch tunnel excavated interfere mutually, which complicates the supporting system, let alone in the weak surrounding rock with frequent collapse or roof caving accident (Wang and Xie, 2008). Thus, the evolution process and the skewed effect of the pressure-arch under the stepped excavation are essential problems and also very difficult to deal with, which have been the concerns of scholars and technical personnel in the engineering field.

In the past years, the skewed pressure problems under the tunnel construction have been conducted by several scholars. For example, Kovari (1994) first found there was a pressure-arch effect in the loose rock with the tunnel excavation. M.M. Protodyakonov proposed the collapsing arch theory of the loose rock (Bandis et al., 1983). Terzaghi (1955) proposed the existence conditions of the pressure-arch of the excavated sand chamber by the experiment. Rabcewicz proposed that a self-bearing structure could form in the surrounding rock of the chamber in the NATM (Muller and Fecker, 1980). Huang et al. (2002) proposed the identification method of the upper and lower boundaries of the pressure-arch. In 2011, based on the analytical method and the numerical simulation method, Fraldi and Guarracino (2011) provided an idea to analyze instability problems of the tunnel. The domestic scholars, such as Li et al. (2007) monitored and controlled the construction deformation of the urban shallow-buried large-span multiarch tunnel under the complex geological conditions, and proposed the engineering measures to control the large deformation. By establishing the three-dimensional numerical calculation model, Zhu et al. (2008b) analyzed the construction sequence and the supporting force characteristics of the shallow-buried tunnel. Combining the physical model experiment and numerical simulation, Jin and Cui (2012) analyzed the temporal-spatial effect of the force and deformation of the middle wall and the supporting structure of the six-lane multiarch tunnel. Based on the monitoring data and numerical model of MIDAS/GTS, Ji et al. (2011) analyzed the dynamic characteristics of the rock pressure and deformation of the supporting structures. With the similar model test, Yang et al. (2013) analyzed the disturbance deformation effect of the existing tunnel when an adjacent large cross-section highway tunnel was excavated step by step. Chen et al. (2011a,b) conducted the three-dimensional stress redistribution and ground arch development during tunneling. Wang et al. (2009, 2014a,b) analyzed the deformation and the stress evolution characteristics of the surrounding rock of the tunnel and mining field.

In conclusion, there have been many research achievements on the surrounding rock pressure of highway tunnels both in China and abroad, but mature theories and systems for the mechanical evolution process and the instability modes of the surrounding rock pressure have not yet formed and need to be further developed. Regarding one highway double-arch tunnel as an example, the morphological evolution mechanism of the pressure-arch in a tunnel during step-by-step excavation is analyzed; by defining the skewed coefficient and introducing the concept of strain energy entropy, the skewed effect and the energy dissipation characteristics of the surrounding rock of the double-arch tunnel are revealed; under different working conditions, the sensitive factors which influence the stress distribution of a double-arch tunnel are also analyzed. These studies provide a basis for the supporting design and safety construction of a double-arch tunnel.

7.2 Pressure-Arch Morphological Characterization

To facilitate the study, the element stress variable e was defined (Huang et al., 2002):

$$e = \frac{\sigma_{\max} - \sigma_{\min}}{\sigma_{\max}} \times 100\% \quad (4.38)$$

where $\sigma_{\max}$ and $\sigma_{\min}$ were the maximum and minimum principal stress of the surrounding rock unit, respectively, after the double-arch tunnel had been excavated.

As shown in Fig. 4.64, corresponding to the peak point A of the maximum principal stress of the surrounding rock, the boundary was defined as the inner boundary of the pressure-arch. When the stress variable e was equal to 5%, the boundary corresponding to point B was defined as the outer boundary of the pressure-arch. On both lateral walls of the double-arch tunnel, extending toward the deep of the surrounding rock and meeting with the inner and outer boundaries of the pressure-arch along the rock rupture angle β, the closed region was the pressure-arch area. According to Protodyakonov's theory and the Rankine theory, the rock rupture angle was $\beta = 45$ degrees $+ \frac{\varphi}{2}$, where φ was the inner friction angle of the surrounding rock of the double-arch tunnel.

For ease of analysis, the pressure-arch of the double-arch tunnel was divided into three zones as shown in Fig. 4.64. The characteristic parameters of the pressure-arch were defined as the vault thickness S_1, the waist thickness S_2, and the skewback thickness S_3 of the pressure-arch, separately.

The mechanics and engineering significance of the pressure-arch of the double-arch tunnel were as follows: during step-by-step excavation of the double-arch tunnel, the surrounding rock experienced stress redistribution repeatedly, and finally, a skewed complex self-bearing pressure-arch formed near the excavation area of the double-arch tunnel. The morphological characteristics of the pressure-arch and the arch thickness in separate zones could demonstrate the disturbance effect of the surrounding rock. Much more, the studies on the mechanical evolution process and the skewed characteristic of the pressure-arch provided a basis for supporting design and safety construction of the double-arch tunnel.

7.3 Computational Model and Analysis Schemes

7.3.1 The Computational Model

The shape and dimension of the designed cross section of the double-arch tunnel were shown in Fig. 4.65A. The tunnel with 80 m burial depth was excavated by bench method, and its surrounding rock was mainly composed of medium weathered sandstone, while the hydrogeological conditions were simple.

Since there is some difficulty in modeling and meshing for complicated 3D engineering in FLAC^{3D}, first, we built the engineering-geological model using ANSYS, and then the complicated engineering-geological model was imported into FLAC^{3D} as shown in Fig. 4.65B. The dimensions of the model were 70-m long, 4-m wide, and 40-m high in the x-, y-, and z-axis, separately. The model was divided into 9536 elements. Since the infinite long tunnel can be treated as a plane problem through the

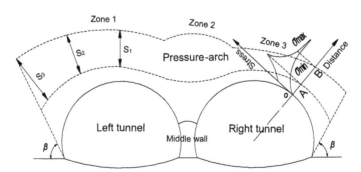

FIGURE 4.64 Pressure-arch zones and morphological parameters of the double-arch tunnel.

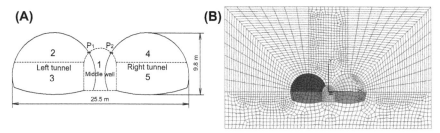

FIGURE 4.65 Excavation steps of the tunnel and the computational model. (A) Excavation steps of the tunnel and (B) computational model and its meshing.

simplified method, therefore, the axial length of the model was thin. The horizontal displacements of four lateral boundaries of the model were restricted, and its bottom was fixed. The upper surface of the model was the load boundary, and on it the vertical load converted from the weight of the overlying rock mass was applied. The material of the model was supposed to meet the Mohr–Coulomb strength criterion, and moreover, it was supposed that the model was in the hydrostatic stress state; namely, the lateral pressure coefficient λ was 1.0. The physics and mechanics parameters were selected as listed in Table 4.13.

7.3.2 Simulation Analysis Schemes

The double-arch tunnel was excavated by bench method to analyze the evolution characteristics of the pressure-arch. The bench method construction was divided into five steps, and the construction sequence is shown in Fig. 4.65A:

Step 1 : After the initial stress field of the calculation model was balanced, the displacement and the velocity field were cleared and then the pilot tunnel one was excavated;

Step 2 : After constructing the middle wall, the upper cross-section 2 of the left tunnel was excavated;

Step 3 : The lower cross-section 3 of the left tunnel was excavated;

Step 4 : The upper cross-section 4 of the right tunnel was excavated;

Step 5 : The lower cross-section 5 of the right tunnel was excavated.

Based on the numerical simulation mentioned earlier, the pressure-arch shape and evolution process of the pressure-arch could be revealed. Then, by defining the skewed coefficient and introducing the concept of strain energy entropy, the skewed effect and the strain energy dissipation characteristics of surrounding rock of the double-arch tunnel were analyzed.

7.3.3 Pressure-Arch Analysis Under Different Conditions

1. Geometric size effect: To reveal the size effect of the pressure-arch, the evolution characteristics of the pressure-arch of two-lane and three-lane tunnels during step-by-step excavation were compared and analyzed.
2. Excavation sequence effect: The excavation sequence of bench method and expanding method were 1-2-3-4-5 and 1-2-4-3-5, respectively. The comparative analysis of

TABLE 4.13 Physical and Mechanical Parameters of the Surrounding Rock

Density (kg/m³)	Elasticity Modulus (GPa)	Poisson Ratio	Cohesion (MPa)	Friction Angle (degrees)	Tension (MPa)
2500	20	0.3	1.5	48	0.5

evolution characteristics of the pressure-arch under two different methods could reveal the nonlinear response characteristics with the different excavation sequence.
3. Stress state effect: Considering the impact of the lateral pressure, three kinds of stress state were designed. The comparative analysis of evolution characteristics of the pressure-arch in double-arch tunnel under different stress states could reveal the skewed effects of the pressure-arch.

7.4 Evolution Characteristic of Pressure-Arch

7.4.1 Evolution Process of the Pressure-Arch

The evolution process analysis of the pressure-arch in double-arch tunnel under bench excavation method was conducted as following.

Step 1: As shown in Fig. 4.66A, after the middle pilot tunnel was excavated, due to the stress self-adjusting effect of the surrounding rock, a symmetrical pressure-arch zone 2 gradually formed at the top of the middle pilot (The white dotted lines on both sides of the middle pilot tunnel were the side boundaries of the pressure-arch, and the white-dotted curve boundary close to the excavation face was its inner boundary, and its outer boundary was the graphic outer contour.).

Step 2: As shown in Fig. 4.66B, after the middle wall was constructed, the upper cross-section 2 of the left tunnel was excavated, zone 1 of the pressure-arch on the top of the left tunnel was formed, zone 2 of the pressure-arch was offset to the right, and the pressure-arch thickness increased. There was an obvious boundary between zone 1 and zone 2 of the pressure-arch.

Step 3: As shown in Fig. 4.66C, after the lower cross-section 3 of the left tunnel was excavated, the larger area of the surrounding rock was disturbed, the vault of zone 1 of the pressure-arch was lifted up, and the thickness of zone 1 increased; zone 2 of the pressure-arch started to develop to the left, the vault was lifted up and the thickness of the pressure-arch increased obviously, and the stress concentration of the

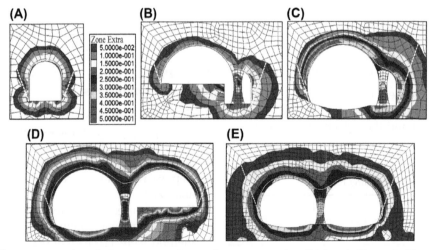

FIGURE 4.66 Pressure-arch evolution process under step-by-step excavation. (A) Step 1, (B) step 2, (C) step 3, (D) step 4, and (E) step 5.

surrounding rock on the top of the middle pilot tunnel was significant, demonstrating the stress of the middle wall greatly increased.

Step 4: As shown in Fig. 4.66D, after the upper cross-section 4 of the right tunnel was excavated, the disturbance area of the surrounding rock was further enlarged, the vault of zone 1 of the pressure-arch continued to rise, and the thickness of the pressure-arch further increased; zone 2 of the pressure-arch continued to develop to the left and the top, but its center was offset to zone 3 of the pressure-arch. At this time, zones 1, 2, and 3 of the pressure-arch were connected to form a skewed asymmetric combination pressure-arch.

Step 5: As shown in Fig. 4.66E, after the lower cross-section 5 of the right tunnel was excavated, with the vault lifting and the thickness of zone 3 increasing, the area of the pressure-arch continued to increase and adjust, and finally there was stress concentration in zone 2, and the area of zone 1 was significantly greater than that of zone 3 of the pressure-arch, which suggested the stress on the vault of the left tunnel of double-arch tunnel was higher than that of the right tunnel, namely the combination pressure-arch of the double-arch tunnel showed a skewed distribution.

As shown in Fig. 4.67, during step-by-step excavation of the bench method, the height and thickness of the combination pressure-arch in double-arch tunnel showed a gradual growth trend. Overall, the pressure-arch height of zone 1 was largest and its increasing rate was fastest; the pressure-arch height and increasing rate of zone 3 were minimal; and the pressure-arch height and increasing rate of zone 2 were medium. The changes of these curves demonstrated that the combination pressure-arch of the double-arch tunnel was a skewed distribution.

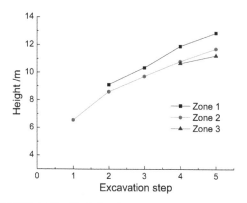

FIGURE 4.67 Vault height variation curves in different zones of the pressure-arch.

7.4.2 Skewed Effect Analysis of the Pressure-arch

The deformation of the tunnel can intuitively reflect the stress state of the surrounding rock, so during step-by-step excavation of the bench method, two displacement monitoring points P_1 and P_2 were set up on the top of the middle wall of the double-arch tunnel as shown in Fig. 4.65A.

Seen from Fig. 4.68, the vertical displacements of points P_1 and P_2 were similar after the double-arch tunnel was excavated step-by-step, but there was an obvious difference between the horizontal displacements of points P_1 and P_2. After the excavation of the middle pilot tunnel, the points P_1 and P_2 were offset to the right and the left with a small horizontal displacement value; with the second and the third step excavation in the left tunnel, the horizontal displacements of points P_1 and P_2 offset to the left and both of the two points' displacement values showed a growth trend; after the fourth and fifth step excavation in the right tunnel, the horizontal displacement of P_2 began to offset the right, but the horizontal displacement value was small, and the horizontal displacement of P_1 had first a slight increase and then leveled off. Overall, the horizontal displacement of P_1 was greater than that of P_2.

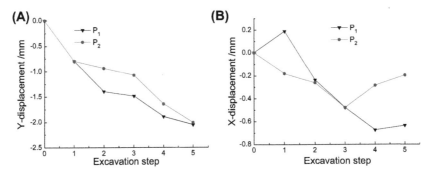

FIGURE 4.68 Displacement curves of the monitoring points. (A) Vertical displacement curves and (B) horizontal displacement curves.

To reflect the skewed effect of the surrounding rock stress of the double-arch tunnel, the skewed coefficient was defined:

$$k = \left|\frac{d_{P_1}}{d_{P_2}}\right| \quad (4.39)$$

where d_{P_1} and d_{P_2} are the horizontal displacement of points P_1 and P_2, respectively. If $k > 1$, the surrounding rock deformation of the double-arch tunnel offsets to the left tunnel, which shows that the rock pressure of the double-arch tunnel is a skewed distribution to the left; if $k = 1$, the surrounding rock deformation of the left and the right sides is symmetrical, which suggests the rock pressure of the double-arch tunnel is a uniform distribution characteristic; if $k < 1$, the surrounding rock deformation offsets to the right tunnel, which suggests the rock pressure is a skewed distribution to the right. As shown in Fig. 4.69, during step-by-step excavation of the bench method, the skewed coefficient of the double-arch tunnel changed gradually from 1.0 to 3.5 and showed that the rock pressure skewed effect of the double-arch tunnel was significant, and the horizontal displacement of the middle wall of the double-arch tunnel was mainly offset to the left tunnel.

7.5 Evolution Characteristic Analysis of the Pressure-Arch

7.5.1 Geometric Size Effect

As shown in Figs. 4.70 and 4.71, under the same conditions, when the double-arch tunnels of two-lane and three-lane were excavated step-by-step of the bench method, it was found that the pressure-arches of two-lane and three-lane highways were skewed distributions. For the double-arch tunnel of a two-lane highway, the pressure-arch height in zones 1, 2, and 3

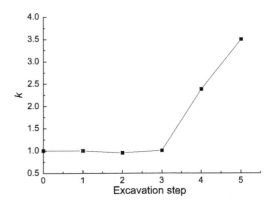

FIGURE 4.69 Skewed coefficient variations of the double-arch tunnel.

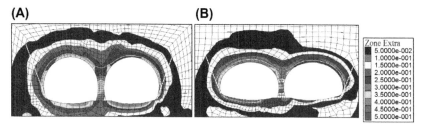

FIGURE 4.70 Pressure-arch shape of different span double-arch tunnel. (A) Two-lane tunnel and (B) three-lane tunnel.

showed a gradual decreasing trend; and for the double-arch tunnel of a three-lane highway, the pressure-arch height in zones 1 and 2 were very close, and the trend for a pressure-arch height of zone 3 was drastically reduced.

Seen from Fig. 4.72, during the step-by-step excavation, the skewed coefficient of the pressure-arch of the two-lane double-arch tunnel changed gradually from 1.0 to 3.5, while that of the three-lane double-arch tunnel gradually increased from 1.0 to 8.0. Therefore, it could clearly be seen with the increase of the tunnel span, the energy accumulation and release of the pressure-arch at the top of the middle wall and the left tunnel were sensitive, and the skewed effect of the pressure-arch of a three-lane double-arch tunnel was much more significant than that of a two-lane one, which should be taken into consideration seriously in the supporting design and construction safety of the double-arch tunnel.

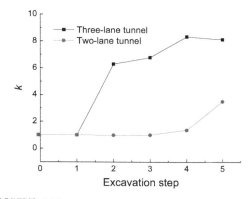

FIGURE 4.72 Skewed coefficient variations of different tunnels.

7.5.2 Construction Sequence Effect

As shown in Figs. 4.73 and 4.74, it was found that the pressure-arch height in zone 1 and zone 2 slightly increased after the double-arch tunnel was excavated by using the expand method, and the increasing rate of the pressure-arch height in zone 3 was bigger, which made the pressure-arch distribution more uniform than that of the bench method; namely, the skewed effect caused by step excavation using the expand method was relatively smaller than that of the bench method (Fig. 4.75).

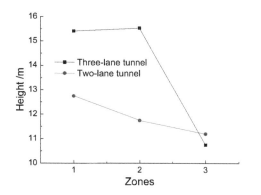

FIGURE 4.71 Vault height variations of different tunnels.

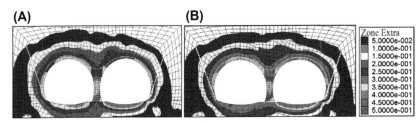

FIGURE 4.73 Pressure-arch shape of different excavation methods. (A) Bench excavation method and (B) expand excavation method.

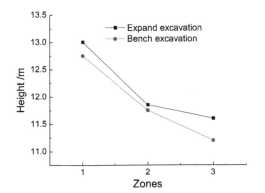

FIGURE 4.74 Vault height variations of different excavation.

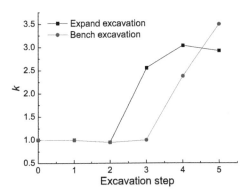

FIGURE 4.75 Skewed coefficient variations of different excavation methods.

7.5.3 Stress State Effect

As shown in Figs. 4.76 and 4.77, with the change of the lateral pressure coefficient from high to low—the arch height of zones 1, 2, and 3 of the pressure-arch almost showed ladder-like changes from high to low—the shape change of the pressure-arch was obvious. When λ was greater than 1, the arch heights of zones 1 and 3 of the pressure-arch showed a substantial upward bulge, and were significantly higher than that of zone 2; When λ was less than 1, relative to arch height of zone 1 of the pressure-arch, the arch heights of zones 2 and 3 of the pressure-arch drastically reduce; When λ equaled to 1, relative to the two cases mentioned earlier, the arch height variation of the pressure-arch was between these two cases, but in this stress state the arch height variation of the pressure-arch of the double-arch tunnel was smaller. Therefore, when λ was greater than 1 and when λ was less than 1, the influences on zones 1 and 3 of the pressure-arch of the double-arch tunnel were greater and the skewed distribution characteristics of the pressure-arch of the double-arch tunnel were obvious.

As shown in Fig. 4.78, with the change of the lateral pressure coefficient of the tunnel surrounding rock from high to low, the skewed effect coefficient of the pressure-arch of the double-arch tunnel changed from 3.7, 3.5–8.9, therefore, when λ equaled to 1, the skewed effect of the pressure-arch was relatively small; when λ was less than 1, the skewed effect of the pressure-arch was remarkable.

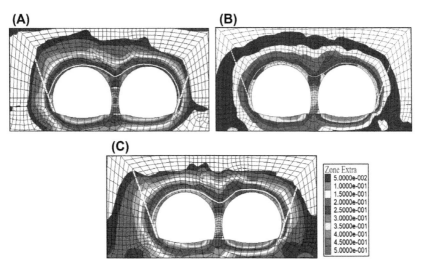

FIGURE 4.76 Pressure-arch shape under different stress states. (A) $\lambda = 1.1$, (B) $\lambda = 1.0$, and (C) $\lambda = 0.9$.

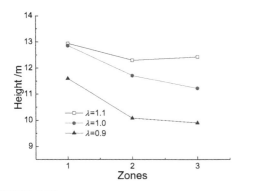

FIGURE 4.77 Arch height variations of different stress states.

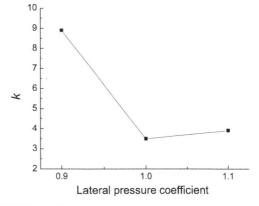

FIGURE 4.78 Skewed coefficient variations of different stress states.

7.6 Conclusion

Along with the step-by-step excavation, the surrounding rock stress of the double-arch tunnel interacted with and then formed the complex pressure-arch. The arch height of the pressure-arch of the double-arch tunnel manifested the skewed distribution characteristics which were almost diminishing, ladder-like, from the left tunnel to the right tunnel.

Based on the concepts of the skewed coefficient, we can see from the skewed effect of the pressure-arch that the double-arch tunnel mainly offsets to the left tunnel. When the double-arch tunnel was excavated step by step, there was the obvious size effect of the pressure-arch. The nonlinear response characteristics in different excavation sequences and the skewed effect with the change of the stress states were sensitive.

References

Allaix, D.L., Carbone, V.I., 2011. An improvement of the response surface method. Structural Safety 33 (2), 165–172.

Amirat, A., Mohamed-Chateauneuf, A., Chaoui, K., 2006. Reliability assessment of underground pipelines under the combined effect of active corrosion and residual stress. International Journal of Pressure Vessels and Piping 83 (2), 107–117.

Amir-Faryar, B., Sherif, A.M., O'Connell, D.P., Suter, K.E., Finnen Jr., R.E., Miller, K.C., 2012. Effect of construction sequence on adjacent underground structures. Electronic Journal of Geotechnical Engineering 17, 3739–3752.

Bandis, S.C., Lumsden, A.C., Barton, N.R., 1983. Fundaments of rock joint deformation. International Journal of Rock Mechanics and Mining Sciences & Geomechanics Abstracts 20, 249–268.

Chen, C.N., Huang, W.Y., Tseng, C.T., 2011a. Stress redistribution and ground arch development during tunneling. Tunnelling and Underground Space Technology 26 (1), 228–235.

Chen, W., Xu, Q., Wang, Z.Y., 2011b. Application of monte–carlo method in reliability analysis of slope stability. Subgrade Construction 4, 135–1374.

Dong, F.T., Song, H.W., Guo, Z.H., Lu, S.M., Liang, S.J., 1994. Roadway support theory based on broken rock zone. Journal of China Coal Society 19 (1), 21–22.

Dong, C.Z., Qu, C., Yang, J.H., Shao, W.P., 2010. Development and application of a geo-mechanics model testbed. Research and Exploration in Laboratory 29 (3), 14–16.

Du, J.H., Huang, H.W., 2009. Dynamic simulation analysis of displacement field of bias closely spaced tunnel with altitude difference. Rock and Soil Mechanics 30 (4), 1102–1108.

Du, Z.J., Xu, M.G., Liu, Z.P., 2007. Analysis on energy source of rock burst and its prevention principle. Mining Research and Development 27 (3), 8–9.

Farah, K., Ltifi, M., Hassis, H., 2011. Reliability analysis of slope stability using stochastic finite element method. Procedia Engineering 10, 1402–1407.

Fraldi, M., Guarracino, F., 2011. Evaluation of impending collapse in circular tunnels by analytical and numerical approaches. Tunneling and Underground Space Technology 26 (4), 507–516.

Fu, H.X., 2003. How to calculate radius of plastic region by applying and reforming kastner formula through practice of Jiazhuqing tunnel. Journal of Railway Engineering Society 1, 94–95.

Ghaboussi, J., Ranken, R.E., 1977. Interaction between two parallel tunnels. International Journal for Numerical and Analytical Methods in Geomechanics 1 (1), 75–103.

Giardina, G., Marini, A., Hendriks, M.A.N., Rots, J.G., Rizzardini, G.E., 2012. Experimental analysis of a masonry façade subject to tunnelling-induced settlement. Engineering Structures 45, 421–434.

Goel, R.K., Jethwa, J.L., Dhar, B.B., 1996. Effect of tunnel size on support pressure. International Journal of Rock Mechanics and Mining Sciences & Geomechanics Abstracts 33 (7), 749–755.

Han, J., Shi, L.Q., Yu, X.G., Wei, J.C., 2009. Mechanism of mine water-inrush through a fault from the floor. Mining Science and Technology (China) 19 (3), 276–281.

Hashash, M.A., Whittle, A.J., 2002. Mechanisms of load transfer and arching for braced excavations in clay. Journal of Geotechnical and Geoenvironmental Engineering 128 (3), 187–197.

Hernández-Montes, E., Aschheim, M., Gil-Martín, L.M., 2005. The buried arch structural system for underground structures. Structural Engineering and Mechanics 20 (1), 69–83.

Hodgson, K., Joughin, N.C., 1967. The relationship between energy release rate, damage and seismicity in deep mines. In: Fairhurst, C. (Ed.), Proceedings of 8th Symposium on Rock Mechanics, Minnesota, 1966, in Failure and Breakage of Rock, pp. 194–203.

Hou, G.Y., Han, R., Huang, X.Z., 2009. Reliability analysis of the tunnel and determination of its index based on response surface method. Chinese Journal of Underground Space and Engineering 5 (5), 965–971.

Hu, Y.Q., Zhao, Y.S., Yang, D., 2007. 3D solid–liquid coupling experiment study into deformation destruction of coal stope. Journal of Liaoning Technical University 26 (4), 520–523.

Huang, Z.P., Broch, E., Lu, M., 2002. Cavern roof stability-mechanism of arching and stabilization by rock bolting. Tunneling and Underground Space Technology 17 (3), 249–261.

Jaeger, J.C., Cook, N.G.W., 1979. Fundamental of rock mechanics. Science Paperbacks 22 (4), 1102–1111.

Ji, M.W., Wu, S.C., Gao, Y.T., Ge, L.L., Li, X.J., 2011. Construction monitoring and numerical simulation of multi-arch tunnel. Rock and Soil Mechanics 32 (12), 3787–3795.

Ji, J., Liao, H.J., Low, B.K., 2012. Modeling 2-D spatial variation in slope reliability analysis using interpolated autocorrelation. Computers and Geotechnics 40, 135–146.

Jiang, M.J., Yin, Z.Y., 2012. Analysis of stress redistribution in soil and earth pressure on tunnel lining using the discrete element method. Tunnelling and Underground Space Technology 32, 251–259.

Jiang, S.P., Liu, H.Z., Xian, X.F., 2000. Physical simulation and numerical analysis on dynamic construction behavior of large-span flat tunnel. Chinese Journal of Rock Mechanics and Engineering 19 (5), 567–572.

Jin, X.G., Cui, B.W., 2012. Simulating and testing study on construction temporal-spatial effect of six lanes multi-arch tunnel. Advanced Materials Research 446–449, 2149–2155.

Jin, B.G., Shin, H.S., 2003. A study on anisotropic reinforcing mechanism of umbrella arch reinforcement method in tunnelling. Journal of the Korean Geotechnical Society 19 (6), 245–259.

Kang, H.M., Li, X.H., Li, T.L., Jin, X.G., Zhang, D.M., 2001. Construction and safety monitoring in section across coal layer of Huayingshan tunnel. Chinese Journal of Rock Mechanics and Engineering 20 (S1), 936–939.

Kim, K.Y., Lee, D.S., Cho, J., Jeong, S.S., Lee, S., 2013. The effect of arching pressure on a vertical circular shaft. Tunnelling and Underground Space Technology 37, 10–21.

Kong, H.L., Miao, X.X., Wang, L.Z., Zhang, Y., Chen, Z.Q., Hang, Y., Chen, Z.Q., 2007. Analysis of the harmfulness of water-inrush from coal seam floor based on seepage instability theory. Journal of China University of Mining and Technology 17 (4), 453–458.

Kovari, K., 1994. Erroneous concepts behind the New Austrian Tunneling Method. Tunnels & Tunneling 11, 38–41.

Li, Y.Y., Li, S.Q., 2008. Analysis of surrounding rock bearing mechanism in deep soft rock roadways with associate support of anchoring and grouting. Journal of Hunan University of Science & Technology (Natural Science Edition) 23 (1), 10–14.

Li, E.B., Wang, D., Wang, Y., 2007. Monitoring and control of construction deformation of urban shallow-buried large-span double-arch tunnel under complex condition. Chinese Journal of Rock Mechanics and Engineering 26 (4), 833–839.

Li, K., Mao, X.B., Chen, L., Zhang, L.Y., 2011. Research on fault activation and risk analysis of water inrush in mining floor above confined aquifer. Chinese Quarterly of Mechanics 32 (2), 261–268.

Li, Y., Zhe, L.R., Wei, Z., Lin, L., 2014. Experimental analysis of supporting structure of pressure to small spacing shallow bias tunnel in soft rock. Electronic Journal of Geotechnical Engineering 19, 4729–4741.

Li, C.C., 2006. Rock support design based on the concept of pressure arch. International Journal of Rock Mechanics and Mining Sciences 43 (7), 1083–1090.

Liu, Z.C., Li, W.J., Sun, M.L., Zhu, Y.Q., 2006. Monitoring and comprehensive analysis in F_4 section of Wuqiaoling tunnel. Chinese Journal of Rock Mechanics and Engineering 25 (7), 1502–1511.

Liu, T., Shen, M.R., Gao, W.J., Tan, D.Y., 2007. Rock mass pressure release rate analysis for double-arch tunnel. Chinese Journal of Underground Space and Engineering 3 (1), 50–54.

Liu, Z.C., Zhu, Y.Q., Li, W.J., Liu, B.X., 2008. Mechanism and classification criterion for large deformation of squeezing ground tunnels. Chinese Journal of Geotechnical Engineering 30 (5), 690–697.

Liu, H.L., Yang, T.H., Yu, Q.L., Chen, S.K., Wei, C.H., 2010. Numerical analysis on the process of water inrush from the floor of seam 12 in Fangezhuang coal mine. Coal Geology & Exploration 38 (3), 27–31.

Most, T., Knabe, T., 2010. Reliability analysis of the bearing failure problem considering uncertain stochastic parameters. Computers and Geotechnics 37 (3), 299–310.

Muller, L., Fecker, F., 1980. The elementary thought and primary principle of New Austrian Tunneling Method. Underground Space 6, 26–32.

Pan, Y.S., Zhang, M.T., Wang, L.G., Li, G.Z., 1997. Study on rockburst by equivalent material simulation tests. Chinese Journal of Geotechnical Engineering 19 (4), 49–56 (in Chinese).

Pan, Y., Zhang, Y., Yu, G.M., 2006. Mechanism and catastrophe theory analysis of circular tunnel rock burst. Applied Mathematics and Mechanics 27 (6), 115–123.

Polillo, A., Vassilellis, G.D., Graves, R.M., Crafton, J.W., 1994. Simulation of sand arching mechanics using an elasto-plastic finite element morphologyulation. SPE Advanced Technology Series 2 (1), 76–85.

Poulsen, B.A., 2010. Coal pillar load calculation by pressure arch theory and near field extraction ratio. International Journal of Rock Mechanics and Mining Sciences 47 (7), 1158–1165.

Prokhorov, I.Y., 1999. Arch effect in high isostatic pressure compacts. Journal of the European Ceramic Society 19 (15), 2619–2623.

Qi, C.Q., Xu, R.P., Wu, J.M., Yu, J., 2008. Reliability analysis of rock mass deformation in tunnel excavation based on genetic algorithm. Journal of Engineering Geology 16 (2), 258–262.

Ryder, J.A., Ozbay, M.U., 1990. Methodology for designing pillar layouts for shallow mining. In: Proceedings of the ISRM International Symposium on Static and Dynamic Considerations in Rock Engineering, p. 273.

Salamon, M.D.G., 1984. Energy considerations in rock mechanics: fundamental results. Journal of The South African Institute of Mining and Metallurgy 84 (8), 233–246.

Shabarov, A.N., 2001. On formation of geodynamic zones prone to rock bursts and tectonic shocks. Journal of Mining Science 37 (2), 129–139.

Shahin, H.M., Nakai, T., Hinokio, M., Yamaguchi, D., 2004. 3D effects on earth pressure and displacements during tunnel excavations. Soils and Foundations 44 (5), 37–49.

Shi, C.H., Peng, L.M., 2008. Study on system reliability of double span tunnel during different construction stages. Rock and Soil Mechanics 29 (5), 1299–1304.

Singh, S.P., 1988. Burst energy release index. Rock Mechanics and Rock Engineering 21 (2), 149–155.

Spottiswoode, S.M., 1988. Total seismicity and the application of ESS analysis to mine layouts. Journal of The South African Institute of Mining and Metallurgy 88 (4), 109−116.

Sun, M.J., Hu, X.L., Tan, F.L., Zhang, Y.M., 2014. Status and progress of soil arching effect research. Electronic Journal of Geotechnical Engineering 19, 4293−4300.

Tang, J.H., Bai, H.B., Yao, B.H., Wu, Y., 2011. Theoretical analysis on water-inrush mechanism of concealed collapse pillars in floor. Mining Science and Technology (China) 21 (1), 57−60.

Tao, B., Wu, F.Q., Guo, Q.L., Guo, G.M., Yang, X.Y., 2006. Research on rheology rule of deep-buried long Wuqiaoling tunnel under high crustal stress by monitoring and numerical analysis. Chinese Journal of Rock Mechanics and Engineering 25 (9), 1828−1834.

Terzaghi, K., 1942. Liner-plate tunnels on the Chicago (Ill.) subway. Proceedings of American Society of Civil Engineers 8 (6), 862−899.

Terzaghi, K., 1955. Evaluation of coefficient of subgrade reaction. Geotechnique 4, 297−326.

Wang, Y.Q., Xie, Y.L., 2008. Research and development of multi-arch tunnel in China. Highway 6, 216−219.

Wang, Y.H., Chen, L.W., Shen, F., 2008. Numerical modeling of energy release in rockburst. Rock and Soil Mechanics 29 (3), 790−794.

Wang, S.R., Zhang, H.Q., Shen, N.Q., Cao, H.Y., 2009. Analysis of deformation and stress characteristics of highway tunnels above mined-out regions. Chinese Journal of Rock Mechanics and Engineering 28 (6), 1144−1151.

Wang, Y., He, C., Zeng, D.Y., Su, Z.X., 2010. Model test and numerical simulation of influence of perpendicular undercross shield tunnel construction on existing tunnel. Journal of the China Railway Society 32 (2), 79−85.

Wang, Y.L., Tan, Z.S., Chen, Y., 2012. Field measurement and analysis of asymmetric shallow-buried twin tunnels with ultra-small spacing. China Engineering Science 14 (11), 24−28.

Wang, S.R., Li, N., Li, C.L., Hagan, P., 2014a. Mechanics evolution characteristics analysis of pressure-arch in fully-mechanized mining field. Journal of Engineering Science and Technology Review 7 (4), 40−45.

Wang, S.R., Li, C.L., Liu, Z.W., Fang, J.B., 2014b. Optimization of construction scheme and supporting technology for HJS soft rock tunnel. International Journal of Mining Science and Technology 24 (6), 847−852.

Wildemeersch, S., Brouyère, S., Orban, P.H., Couturier, J., Dingelstadt, C., Veschkens, M., Dassargues, A., 2010. Application of the hybrid finite element mixing cell method to an abandoned coalfield in Belgium. Journal of Hydrology 392 (3−4), 188−200.

Wu, Q., Zhang, Z.L., Ma, J.F., 2007. A new practical methodology of the coal floor water bursting evaluatingl-The master controlling index system construction. Journal of China Coal Society 32 (1), 42−47.

Wu, Q., Zhu, B., Shou, S.Q., 2011. Flow-solid coupling simulation method analysis and time independent of lagging water-inrush near mine fault belt. Chinese Journal of Rock Mechanics and Engineering 30 (1), 93−104.

Xie, S.R., Xie, G.Q., He, S.S., Zhang, G.C., Yang, J.H., Li, E.P., Sun, Y.J., 2014. Anchor-spray-injection strengthened bearing arch supporting mechanism of deep soft rock roadway and its application. Journal of China Coal Society 39 (3), 404−409.

Xing, X.K., Zhang, J.W., Teng, D.T., Liu, J., Sun, G.X., 2014. Experimental research on distribution regularity of the soil tunnel's pressure arch under different supporting stress. Chinese Journal of Underground Space and Engineering 10 (4), 789−793.

Xu, L.S., Wang, L.S., 2003. Research on rockburst character and prevention measure of Erlang mountain highway tunnel. China Journal of Highway and Transport 16 (1), 74−76.

Xu, S.L., Wu, W., Zhang, H., 2002. Experimental study on dynamic unloading of the confining pressures for a marble under triaxial compression and simulation analyses of rock burst. Journal of Liaoning Technical University 21 (5), 612−615.

Xu, C.G., 2005. Present situa tion of rockburst foreca sting and its countermea sures. Modern Tunnelling Technology 42 (6), 80−85.

Yang, C.X., Luo, Z.Q., 2007. Analyses and control of unstable mode of laneway with rock-burst possibility in deep mine. Mining and Metallurgical Engineering 27 (2), 1−4.

Yang, J.H., Wang, S.R., Li, C.L., Li, Y., 2013. Disturbance deformation effect on the existing tunnel of asymmetric tunnels with small spacing. Disaster Advances 6 (13), 269−277.

Yang, J.H., Wang, S.R., Wang, Y.G., Li, C.L., 2015. Analysis of arching mechanism and evolution characteristics of tunnel pressure-arch. Jordan Journal of Civil Engineering 9 (1), 125−132.

Yin, S.X., 2009. Modes and mechan ism for water inrushes from coal seam floor. Journal of Xi'an University of Science and Technology 29 (6), 661−665.

Yu, X.F., 1982. On the theory of axial variation and basic rules of deformation and fracture of rocks surrounding underground excavations. Uranium Mining and Metallurgy 1, 3−7.

Zhang, Y.B., Xu, D.Q., 2002. The analysis of the experiment of rock burst on different rocks. Journal of Hebei Institute of Technology 24 (4), 8−11.

Zhang, C., Lou, Y.M., Sun, S.H., Qi, Z.X., 2003. Application and discussion on floor arch board technology in project with water inrush and swelling soft rock. Coal Science and Technology 31 (6), 53−55.

Zhao, Z.H., Xie, H.P., 2008. Energy transfer and energy dissipation in rock deformation and fracture. Journal of Sichuan University (Engineering Science Edition) 40 (2), 26–31.

Zhao, D.A., Li, G.L., Chen, Z.M., Li, S.Y., Xia, W.C., 2009. Three-dimensional FE regression analysis of multivariate geostress field of Wushaoling tunnel. Chinese Journal of Rock Mechanics and Engineering 28 (S1), 2687–2694.

Zheng, J.J., Guo, J., Li, F.H., 2007. Immune algorithm for reliability analysis of geotechnical engineering. Chinese Journal of Geotechnical Engineering 29 (5), 785–788.

Zhu, Q.H., Feng, M.M., Mao, X.B., 2008a. Numerical analysis of water inrush from working-face floor during mining. Journal of China University of Mining and Technology 18 (2), 159–163.

Zhu, Z.G., Qiao, C.S., Gao, B.B., 2008b. Analysis of construction optimization and supporting structure under load of shallow multi-arch tunnel under unsymmetrical pressure. Rock and Soil Mechanics 29 (10), 2747–2752.

Further Reading

Gao, M.S., Zhang, N., Dou, L.M., Wang, K., Kan, J.G., 2007. Study of roadway support parameters subjected to rock burst based on energy balance theory. Journal of China University of Mining & Technology 36 (4), 426–430.

Jaeger, J.C., Cook, G.W., 1979. Fundamentals of Rock Mechanics. Chapman and Hall Press, London, pp. 466–470.

Liu, Y., 2008. Numerical analysis of breaking depth of coal floor caused by mining pressure. Journal of Xi'an University of Science and Technology 28 (1), 11–14.

Wang, S.R., Chang, M.S., 2012. Reliability analysis of lining stability for hydraulic tunnel under internal water pressure. Disaster Advances 5 (4), 166–170.

Wang, S.R., Wang, H., 2012. Water inrush mode and its evolution characteristics with roadway excavation approaching to the fault. TELKOMNIKA 10 (3), 505–513.

CHAPTER 5

Slope Engineering

1. THREE-DIMENSIONAL DEFORMATION EFFECT AND OPTIMAL EXCAVATED DESIGN

1.1 Introduction

In the present work, we investigate the stability of rock slope at Antaibao open-pit coal mine that is situated 145 km southwest of Datong in China as a typical engineering example using Fast Lagrangian Analysis of Continua three-dimensional (FLAC3D) technique (Wang, 2005). Attention has been particularly focused on its three-dimensional (3D) deformation effect due to excavation and backfill process. Much of this work has been motivated by optimal design requirements that rely on the knowledge of 3D nonlinear continuum mechanics as well as the underlying geological data including fault, joint, and bedding structure in the slope.

Prior to the 3D analysis, the excavated scheme (so-called old scheme) of the slope that was with a relatively gentle slope angle of 30 degrees was originally designed on the geological section 73200 by Chen and Associates in terms of two-dimensional (2D) conventional limit equilibrium method (Griffiths and Lane, 1999). Unfortunately, a significant amount of coal could not be mined from the underground near the section, and a great quantity of surficial loess-alluvium on the top of the slope would be moved out if this old scheme were used. In fact, conventional limit equilibrium methods cannot be used to describe the process effect induced by excavating or filling in geotechnical engineering (Griffiths and Fenton, 2004). It is, therefore, of very great interest that the slope stability enhanced by a relatively steep angle of excavated slope can be justified as a result of important dealing with the 3D effect of slope under mining and backfilling. Generally, a design scheme with the steeper angle of excavated slope can be obviously evaluated by means of 3D deformation analysis, ie, maintained an appropriate length of excavated axis with time as excavation and backfill proceeded.

Currently, there is a well-developed framework for the estimation of slope stability. Developments in the characterization of complex rock slope deformation and failure using numerical modeling techniques were reviewed recently by Stead et al. (2006). Two key ingredients are (1) the so-called conventional limit equilibrium methods that assume to be of rigid plasticity for geomaterial and only present a factor of safety (FOS) of stability for 2D-plane slope or 3D slope with preassumed shape failure mechanisms (Cheng and Zhu, 2004); and (2) the so-called elastoplastic methods that include strength-reduced technique, simulating the deformation process of slope by numerical techniques, such as finite element and finite difference, which are much more available to complicated slopes, in practice, with complexity of geometry shape and material properties. The former has been broadly accepted in geotechnical engineering practice, particularly in stability evaluation of 2D slope. Fundamental questions aside, 3D problems in practical geotechnical engineering make the evaluating task of the limit equilibrium methods cumbersome, while rendering practically difficult the derivation of the FOS. In sharp contrast with this state of affairs, the latter is being popularized, for it can be used not only to estimate slope stability, but also to evaluate progressive failure mechanism of slope even under consideration of mining or backfilling. Furthermore, the process effect induced by excavation or backfill that is closely associated with construction sequence and step strongly needs to be simulated by a kind of numerical software with advanced constitutive models of geomaterials.

On the other hand, it is also noted that from the viewpoint of geological mechanics a nonlinear dynamical response of excavated slope is strongly dependent on mechanical properties of rock masses, excavation sequence of rock mass, rock joint pattern and fractured geometry, in situ stress field, underground temperature, and underground water activity as well as their coupling effect (Li et al., 2006). In recognition of these synthetic respects on the character of complex rock slope deformation, numerical discontinuum techniques can be used in practice. As Stead et al. pointed out, it must be recognized however that conventional discontinuum models also have inherent limitations, for example, failure frequently followed or preceded by creep, progressive deformation, and extensive internal disruption of the slope mass. Bearing these in mind, the combined use of limit equilibrium and numerical modeling technique, such as finite difference technique is advocated herein to maximize the advantage of both and develop an optimal design scheme of excavation slope for the section 73200 of the mine.

An outline of the remainder of this chapter is as follows. In Section 1 and 2, the geology settings and the geohydrology on the Antaibao area is described, respectively. The physicomechanical properties of the slope and the backfilling material used in modeling are discussed in Section 3. After introducing the basic principle of 3D deformation effect, numerical simulations that deal with excavation process, 3D effect, and slope optimization design is analyzed by FLAC3D in Section 4. Finally, conclusions are presented.

1.2 Engineering Background

Antaibao surface mine is in the Pingshuo coal district of northern Shanxi Province and was operated jointly by the People's Republic of China (PRC) and Island Creek Coal Company (ICC) of the United States of America. Antaibao project is one of the first group joint projects started by China and America in the 1980s. It is a supersurface mine and its annual output reaches 15 million tons.

Antaibao surface mine is representative loessial hilly ground, and its absolute elevation is 1190—1470 m. The stratum there contains three surficial deposits and five bedrock formations. The principle coal reserves are the No. 4, No. 9, and No. 11 seams in the Taiyuan Formation. The Taiyuan Formation is overlain by the Shihezi and Shanxi Formations, however, in the development pit, most of the Shihezi and Shanxi have been removed by erosion. The Taiyuan Formation is underlain by Benxi Formation and the Shuo Xian Formation (Fig. 5.1).

Antaibao surface mine is located on the southeast limb of the Erpu anticline. There are folds, faults, and ring structures. The sedimentary bedrock was cut by several joint sets.

1.3 Computation Model and Mechanical Parameters

The 3D computational work presented is carried out by using the 3D finite difference code FLAC3D developed by Itasca Consulting Group. The software that is characterized with nonlinear

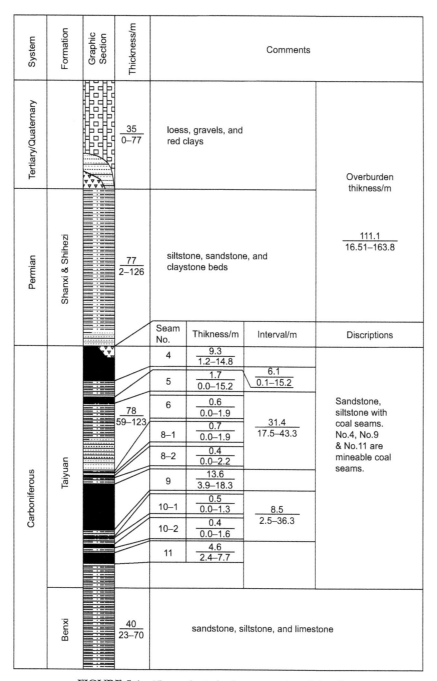

FIGURE 5.1 The geological columnar section of Antaibao.

large deformation and a number of geomaterial models can be readily used to modeling of stress and deformation around either open mining excavations or backfill processes.

Based on the geological section 73200 of the west edge slope of Antaibao surface mine, the mesh of the 3D computation model is built. Under consideration of excavated process step-by-step and steep slope mining, the model that is divided by 52,250 elements and 57,120 nodes consists of 554 m in length (x-direction), 500 m in width (y-direction), and 300 m in depth (z-direction), shown in Fig. 5.2. The horizontal movement of the four sides of the model is restrained, and its bottom side is fastened.

The physicomechanical properties of the soils, altered rock in the ring structure and sedimentary rock, and backfilling materials used in the numerical simulations were based on evaluation of the exploratory drilling and the laboratory test results. The properties used in the simulations are summarized in Table 5.1. A nonlinear Mohr–Coulomb material model with a tension cutoff was used in the present simulations.

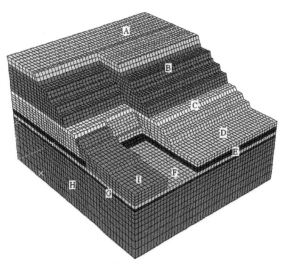

FIGURE 5.2 Distribution of rock materials in a typical computational model. A: Loess; B: weathering sandstone No. 2; C: coal No. 4; D: shale No. 1; E: coal No. 9; F: shale No. 2; G: sandstone, H: fine sandstone; I: the backfilled material.

The coal mine is located at an elevation of 1190–1470 m and has a projected depth of 300 m. The stress field acted on the rock masses is basically determined by gravity and is weakly affected by the local tectonic stresses which were almost completely released, particularly to the relatively shallow mining studied here. This point has further been confirmed by observing the variations of the rock mass stresses after excavating. Although the tectonic stress field has been neglected in the present study, the results of modeling will still be presented with excellent precision.

1.4 Three-Dimensional Deformation Effect of the Slope

A motivation for the present study is to demonstrate the feasibility of excavation scheme with the increase of the excavated slope angle from 30 to 47 degrees on the geological section 73200 in the coal mine. The engineering rock masses that were originally in an equilibrium state under gravity would become an open slope during the mining activity, in which their stress states would be progressively adjusted. As the axial length of slope increases with excavating, the 3D constraint effect is progressively becoming weak. Three-dimensional model can be thus approximately simplified as a plane problem to be modeled while a critical length of excavated slope axis is approached. Obviously, the critical axial length can be definitely referred to as a standard in determination of utilizing a 2D or 3D model for evaluating the deformation characteristics of the excavated slope.

Thus, 10 calculation cases that identify with the properties of material and the boundary condition but correspond to different axial length of excavated slope, ie, 60, 100, 140, 180, 220, 260, 300, 340, 380, and 420 m, were exclusively carried out to gain fresh insight into the 3D effect of slope to be rather different from the conventional limit equilibrium method. Among these simulations, every excavation was implemented

TABLE 5.1 Physical and Mechanical Properties of Rock Mass

Rock Definition	Density (kg/m^3)	Bulk Modulus (MPa)	Shear Modulus (MPa)	Tensile Strength (kPa)	Cohesion (kPa)	Frictional Angle (degrees)
Loess	1960	150	78	1.0	50	23
Weathering sandstone No. 2	2380	800	500	120	800	30
Coal No. 4	1440	500	300	70	300	26
Shale No. 1	2550	1300	800	160	1000	32
Coal No. 9	1330	600	350	90	300	27
Shale No. 2	2560	800	600	140	600	30
Sandstone	1320	600	400	120	300	29
Fine sandstone	2600	2500	2100	1200	3500	39
The backfilled material	2000	350	160	0	200	25

from top to bottom by five steps. Both the largest displacement and horizontal components all increase with the axial length increasing the excavated slope, as shown in Fig. 5.3. It is observed from Fig. 5.3 that increasing ratio of horizontal displacement is of significance before reaching the axial length, 160 m, of excavation; afterward, this ratio slowly varies and approaches zero. It is found that the approximate critical length, ie, 160 m, is just consistent with the height of the excavated slope. Consequently, it can be predicted that the 3D effect of the slope is of prominence under an evident two-side constraint if the axial length of excavated slope is smaller than its corresponding height, whereas such 3D effect is barely noticeable.

1.5 Optimization Design of the Excavated Slope

1.5.1 The Excavated Designs

As the conventional limit equilibrium methods that do not take care of the material deformation can only be used to describe the strength state of slope material. To excavate slope, nonlinear dynamic process closely relies upon the deformation response of slope under excavating. It is noted that, from the discussions in Section 3, the key point about an optimal excavated scheme is how to make the best of the 3D effect of excavated slope. Keeping this in mind, we now optimize the design scheme of excavated slope on the geological section 73200 of the mine by FLAC3D.

In the first case (Case I), the axial length of excavated slope advance is 300 m with two slope angles, 30 and 47 degrees, simulating separately. In the second case (Case II), the axial length is 100 m, the others the same as in the first case. In the final case (Case III), another part of 100 m is again excavated after backfilling the part of 100 m along with the axial direction of slope initially excavated, the others also the same as in the first case. Computational mesh and its optimal construct schemes are given in Fig. 5.4. The locations of monitoring points are represented by Arabic numbers.

1.5.2 Optimization Analysis of the Designs

Different maximum displacements and their corresponding horizontal components (x-direction) are closely related with the

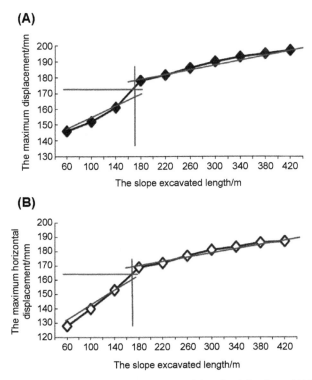

FIGURE 5.3 The displacement curves with the different excavated length of the slope. (A) The maximum displacement curves; (B) the maximum horizontal displacement curves.

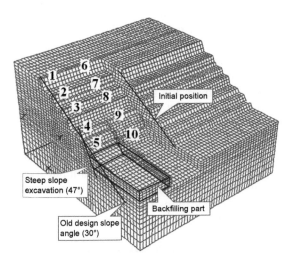

FIGURE 5.4 Computation mesh of excavated slope on the northwestern edge of Antaibao surface mine.

excavated slope angles, shown in Fig. 5.5. Since the backfilling materials in the final case again play a role of active constraint on the excavated surface, the deformations of the excavated slope with either slope angle 30 or 47 degrees are efficiently suppressed; refer to their corresponding results in Fig. 5.6. From the viewpoint of inelasticity mechanics, the plastic zone in the final case is also obviously smaller than that of the other cases and there are few elements which are in shear or tensile failure state, illustrated in Fig. 5.7, respectively, in which the term none means elastic state of rock material at present on that area, shear-n shear failure state at present, shear-p shear failure state in the past, and tension-n, tension-p holding the tensile failure state at present or in the past.

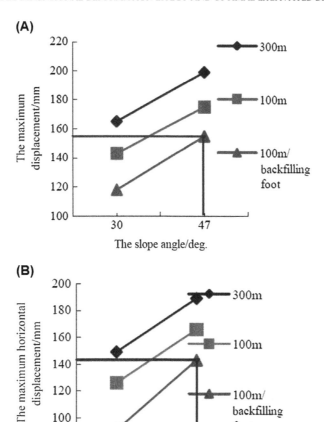

FIGURE 5.5 The displacements with the different excavated condition of the slope. (A) The maximum displacement results; (B) the maximum horizontal displacement results.

To observe in detail the surface deformation of the steep slope in the final case, the representative displacement curves for 10 monitoring points on the slope surface (these points also marked in Fig. 5.4) are shown in Figs. 5.8 and 5.9, which vary with the process of excavating and backfilling during iterating computation. After excavating, larger displacements occurred on these points without considering the creep effect of the rock mass. However, the backfilling instantaneously compensated the imbalance due to excavation which could result in the deformation of the slope, so that after backfilling on the partial initial excavated surface the total axial length, 100 m, of the excavated slope was decreased and the deformation was also reduced by comparing with the former two cases. If, admittedly, the instantaneous backfilling began up to the excavated part restored, the excavated scheme was supposed to be approximately optimal for it not only made the best of waste rock masses or soils, but also saved a

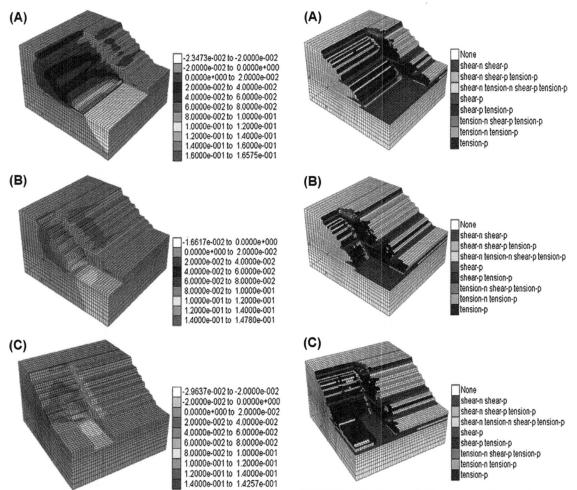

FIGURE 5.6 Horizontal displacement fields corresponding to the cases (*x*-direction). (A) Case I; (B) Case II; (C) Case III.

FIGURE 5.7 Failure field corresponding to the cases. (A) Case I, (B) the Case II, (C) Case III.

great amount of money for the investors. While the axial length of the slope approximately equals to its corresponding height, the 3D effect is of inevitable prominence. It is noted that the running range of the excavator in the in situ mine is at least 69.6 m along with the axis direction of slope for a normal operation of mining.

1.6 Conclusions

Timely proper backfilling for a mine that has been excavated can enhance the stability of this mine. Alternately putting excavation and backfill operation, it is possible to implement an excavated scheme of relatively steep slope angle through considering 3D deformation effect

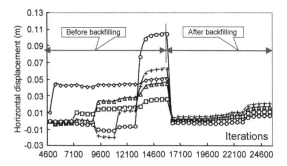

FIGURE 5.8 The variation displacement curves of the backfilled foot slope.

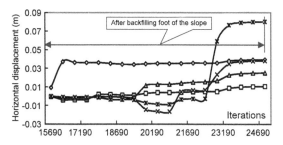

FIGURE 5.9 The variation displacement curves of the excavated slope.

under the conditions of complex geometries and geomaterials at some rock slopes, by using advanced numerical techniques. Accordingly, in rock engineering practice, such valuable 3D effect should be sufficiently utilized to reduce the harmful deformation of slope.

We have applied FLAC3D to evaluate the stability of Antaibao open-pit coal slope and presented a relatively steep scheme of excavation which has successfully been put in practice on the geological section 73200 of the coal mine, thereby allowing the mine to accrue over RMB 216 million in additional income. In practice, a strategy of fast excavation and instantaneous backfilling at Antaibao open-pit coal mine has been made under mining operation. Since the irreversible process of energy dissipation and energy release under mining is significantly influenced by excavation sequence, the excavated schemes of rock slope were optimized by means of large deformation analysis tools, ie, FLAC3D, to give an optimal choice for mine engineers and investigators.

The combined use of limit equilibrium and numerical modeling technique, such as finite difference technique or finite element method is of importance to maximize the advantage of both and develop an optimal design scheme of excavation slope in rock engineering. Mechanical deformation patterns and plastic states as well as the coupling effect of slope due to fluid flow action are globally considered to help better understand the rationality of support design scheme of excavation construction. It is noted that the steep slope excavated scheme proposed for the geological section 73200 is in fact feasible under the condition of no rainfall considered. Consequently, the construction period at the mine was limited from October to June of the coming year.

2. STABILITY ANALYSIS OF THREE-DIMENSIONAL SLOPE ENGINEERING

2.1 Introduction

Compound mining is the mining engineering which is combined by surface and underground mining techniques. According to the sequential operation of the two mining methods in our and other countries, there are four classifications as follows (Niu, 1990):

1. From deep underground mining to shallow surface mining.
2. From shallow surface mining to deep underground mining.
3. Deep underground mining and shallow surface mining at the same time.
4. Different from the earlier-mentioned patterns.

The northwest edge slope engineering (NWESE) of Antaibao surface mine (Fig. 5.10) is the last nonsynchronous compound pattern

FIGURE 5.10 The northwest edge slope engineering of An Tai Bao surface mine.

which process is carrying through underground mining inside the NWESE after surface mining.

It is difficult to do underground mining on the NWESE because there is large faulted zone through the edge slope, and large discharged earth field on the edge slope, and the worked-out section forming large hollowed-out areas due to mining the No. 4 coal seam in the slope.

Furthermore, the structure of surface layer of the slope is incompact, and its intensity is lower for the long-term exposure. Part of the slope may collapse at any time. It is a hazardous site and spot to the workers and equipment in the mining process on the bottom of the slope.

After surface mining, the NWESE was shaped like a large funnel, and plenty of rainwater, groundwater seepage, and weathering rock on the base of the edge slope made the NWESE more unstable. Presumably, after the No. 4 and the No. 9 coal seam in the slope are excavated, the rock stratum above the coal bed will break down, delaminate, and expand upward with the scale of the mining engineering development. So, it is the key technique to appropriately evaluate the stability of the NWESE under such conditions, as the subsidence due to underground mining can lead to the discharged earth field collapse. Another is the NWESE stability in the process of underground mining inside the NWESE after the surface mining.

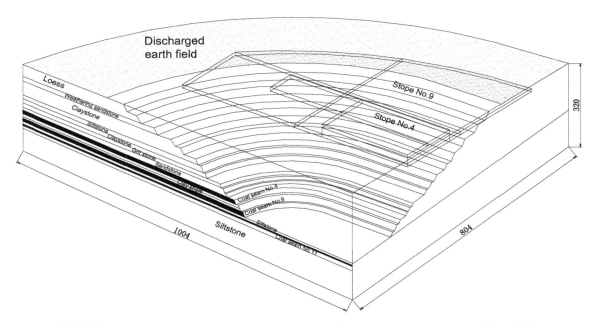

FIGURE 5.11 The engineering geological model of the northwest edge slope engineering (Wang, 2005).

Therefore, adopting a comprehensive analysis method and considering various factors, it is of important theoretical meaning and application value to evaluate the stability of the NWESE under such conditions as before, during, and after mining.

2.2 Engineering Background

Antaibao surface mine is in the Pingshuo coal district of northern Shanxi Province and was operated jointly by the PRC and ICC of the United States of America. Antaibao is one of the first group joint projects started by China and America in the 1980s. It is a supersurface mine and its annual output reaches 15 million tons (Fig. 5.10).

2.2.1 Geological Model

Based on geotechnical investigations, soil samples, and rock core obtained in the exploratory holes drilled by the PRC and ICC, and considering the pattern of the NWESE and its engineering excavation characteristics, the 3D engineering geological model from surface mining to underground mining of the NWESE is generalized and established as shown on Fig. 5.11.

2.2.2 Computation Model

Based on the ring type NWESE, the mesh of the 3D computation model is built and shown on Fig. 5.12.

The ranged space of the computation model is 1004-m long, 804-m wide, and 320-m high. It is meshed to 65,780 elements and contains 70,272 nodes.

The horizontal movement of the model is restrained, and its bottom side is fastened. The upside of the model is load boundary to simulate the heaping effect caused by discharged earth field. The soil and rock material strength criterion is the Mohr–Coulomb rule.

2.2.3 Parameters of the Model

It is always a challenge going with geotechnical engineering to transfer the rock mechanics

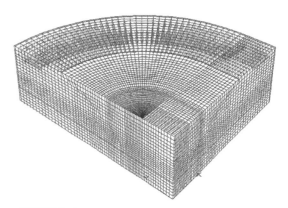

FIGURE 5.12 The mesh of the computation model.

parameters to the engineering rockmass mechanics parameters. On basis of the in situ investigation, field engineering coring, and laboratory experiment, the author researched the rockmass mechanics parameters by combining rockmass rating and basic quality systems (Liu and Tang, 1999); the comprehensive evaluation results were provided the basis data for the design and construction of the northwest slope engineering. Mechanical parameters of the calculating model are shown in Table 5.2.

2.2.4 Analysis Schemes

According to the NWESE in the transition from surface mining to underground mining, the stability of the edge slope was analyzed under different conditions. The computation procedures were as follows:

Case I: The response analysis of the step-by-step excavation and backfilling foot of the slope in the self-weight field.
Case II: Research on the characteristics of the rock displacement after mining No. 4 coal seam and rain filtering in the edge slope.
Case III: Optimizing analysis on the mining sequences to No. 9 coal seam in 3D edge slope.
Case IV: Predicting and analysis on the deformation tendency of the edge slope after mining No. 9 coal seam.

TABLE 5.2 Mechanical Parameters of the Model

Name	Density (kg/m³)	Elastic Modulus (GPa)	Poisson's Ratio	Tensile Strength (kPa)	Cohesive Strength (MPa)	Friction Angle (degrees)
Loess	1960	0.15	0.42	1.0	0.05	20
Weathering-stone	2300	2.0	0.36	50	1.0	33
Sandstone	2380	4.2	0.32	100	1.2	35
Claystone	2490	2.8	0.34	10	0.1	31
Siltstone	2320	4.6	0.32	150	1.0	36
Sandstone	2380	5.5	0.30	250	1.5	39
Seam No.4	1440	1.0	0.38	100	0.5	26
Shale	2450	2.4	0.33	100	1.2	34
Siltstone	2600	4.8	0.32	300	1.6	40
Shale	2580	3.0	0.35	100	1.2	35
Seam No.9	1330	1.2	0.36	100	0.6	27
Sandstone	2380	6.9	0.28	300	1.5	38
Seam No.11	1400	1.3	0.35	110	0.63	25
Shale	2460	3.5	0.27	150	1.3	35
Siltstone	2600	12	0.25	1000	3.7	43

To analyze and predict the deformation tendency of the edge slope under different engineering conditions, three sections are intercepted as monitoring sections from the line-slope (section 1, $y = -100$ m), the joint of the line and the ring slope (section 2, $y = 0$ m), and the middle of the ring slope (section 3, $y = 141$ m) of the computation model. These sections are shown on Fig. 5.13.

2.3 Results Analysis

2.3.1 Step-by-Step Excavation and Backfilling Foot

Simulating analysis process was made according to the actual slope excavation of Antaibao surface mine. First, the slope was

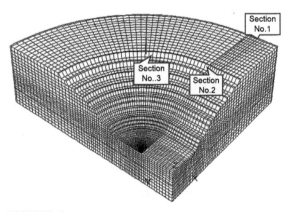

FIGURE 5.13 The monitoring sections of the computation model.

2. STABILITY ANALYSIS OF THREE-DIMENSIONAL SLOPE ENGINEERING

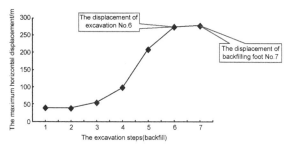

FIGURE 5.14 Maximum horizontal displacement curves of the slope excavating and backfilling foot.

excavated from top to bottom by six steps, forming the ring type NWESE of surface mine. Second, the slope was backfilled to the No. 4 coal seam at the seventh step.

It is known from Fig. 5.14 that the maximum horizontal displacement initially increases slowly and then quickly from the fourth step as the edge

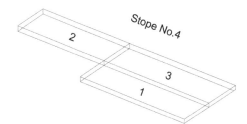

FIGURE 5.15 Mining sequences of No. 4 coal.

slope is excavated deeper and deeper. But the significant displacement tendency was controlled effectively after backfilling foot to the No. 4 coal seam at the seventh step of the slope.

2.3.2 Displacements After Mining No. 4 Coal and Rain Filtering

After mining No. 4 coal seam (the mining sequences are shown on Fig. 5.15), there are two obviously different regions on line-slope. The displacement tendency of the edge slope is funnel form subsidence to the goaf hollow at the slope elevation from 100 m to the top, and the maximum displacement of the subsidence is 380 mm (Fig. 5.16).

And from the bottom to 100 m, the local displacement of the slope in the pit side presents large on the bottom and small at the 100 m top. On the condition of rain filtering, the deformation of the edge slope after mining No. 4 coal seam show the subsidence mainly. The maximum subsidence is about 116 mm.

2.3.3 Mining Sequences Optimization of No. 9 Coal

To decrease the subsidence caused by mining and the harmful deformation to the slope

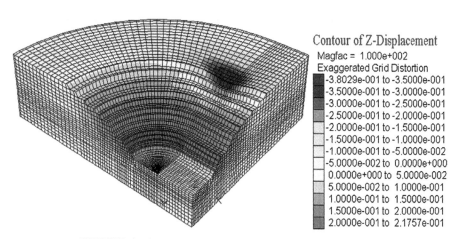

FIGURE 5.16 Subsidence of the slope after mining No. 4 coal.

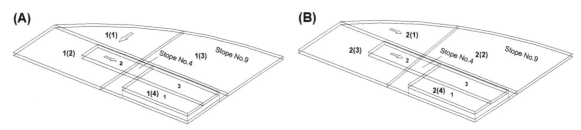

FIGURE 5.17 The mining sequences to the No. 9 coal seam. (A) The first mining sequence; (B) the second mining sequence.

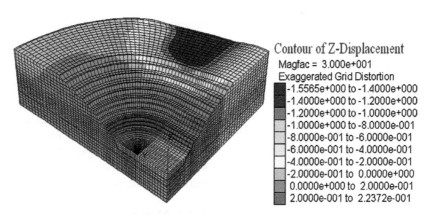

FIGURE 5.18 Subsidence of the slope after mining No. 9 coal.

stability, the mining sequences to No. 9 coal seam are analyzed and shown on Fig. 5.17.

The maximum subsidence is 1.56 m when mining No. 9 coal seam in the first mining sequence and 1.64 in the second. The latter is not only larger than the former in value, but also causes more harmful deformation; the first mining sequence should be chosen. The subsidence of the slope after mining No. 9 coal seam in the first mining sequence is shown on Fig. 5.18.

2.3.4 Deformation Prediction Analysis of the Slope

The monitoring point maximum horizontal displacement with the slope profile in elevation after mining No. 9 coal seam is shown in Fig. 5.19. The two monitoring points at 170 and 130 m elevation of the slope section No. 1 show displacement to the goaf hollow, respectively, and monitoring points above 98 m on sections No. 2 and No. 3 show displacement to the hollow area of the slope. The displacements to the hollow area of the other points on the three monitoring sections are smaller. Though there is larger displacement to the hollow area near the top of the slope, it is helpful to the slope stability. So, the deformation situation is stable to the slope in general.

Fig. 5.20 shows that after mining No. 4 coal seam, displacement of the hollow area appears at the monitoring point at 130 m of the slope, which leads to a little displacement of the monitoring point in pit side at 170 m of the local slope, and mining the No. 4 coal seam has little effect on the horizontal displacements of monitoring points below 98 m. After mining the No. 9 coal seam, obvious displacement of the hollow area

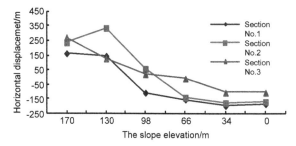

FIGURE 5.19 Maximum horizontal displacement with the slope profile after mining No. 9 coal.

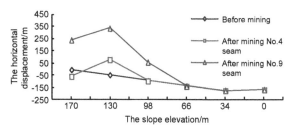

FIGURE 5.20 Maximum horizontal displacement with the slope profile before and after mining (section $y = 0$).

of the monitoring points above 66 m appear, the maximum horizontal displacement of points is at 130 m, and mining the No. 9 coal seam has little effect on the horizontal displacements of the monitoring points below 66 m.

Because mining the No. 9 coal seam will lead to subsidence to the edge slope, it is necessary to choose a certain section ($y = -25$ m) of the computation model of the edge slope and analyze the variable characteristics of the displacement field during mining of the No. 9 coal seam in the first mining sequence.

After mining the No. 9 coal seam, there are two obviously different regions in the edge slope (Fig. 5.21). Slip deformation from large to small appears in the pit side of the edge slope from the bottom to 100 m; the edge slope appears with gradient subsidence to inside of the slope from 100 m. Vertical subsidence deformation appears in the other part of the edge slope.

2.4 Conclusions

Based on the ring type NWESE of Antaibao surface mine, the author established the engineering geological model and analyzed the stability of the 3D model under different engineering conditions by using FLAC3D, the following conclusions can be made.

1. Through the response analysis of the step-by-step excavation and backfilling of the slope in the self-weight field, the results show that the edge slope is stable.
2. The complicated displacement and deformation characteristics of the edge slope are illuminated after mining No. 4 seam and in the condition of rain filtering.
3. The mining sequences to the No. 9 seam in the 3D edge slope were optimized, and by analyzing the displacement of monitoring points of the slope after mining the No. 9 seam in the first mining sequence, the results show that the edge slope is stable too.

In summary, the NWESE under the condition of surface mining and underground mining, the affection on the deformation of the edge slope caused by underground mining will become larger as the underground mining with time. The displacement of the edge slope showed subsidence mainly to the hollow area, and the displacement is helpful to the stability of the edge slope as a whole. Engineering practice proved that the NWESE would get 1.5 million tons of coal by underground mining and increase its profit by RMB 226 million. It provides practical experience for other similar engineering.

3. FRACTURE PROCESS ANALYSIS OF KEY STRATA IN THE SLOPE

3.1 Introduction

The law of overburden failure under mining has great importance to researchers (He et al., 2007). For strata control, the key strata theory

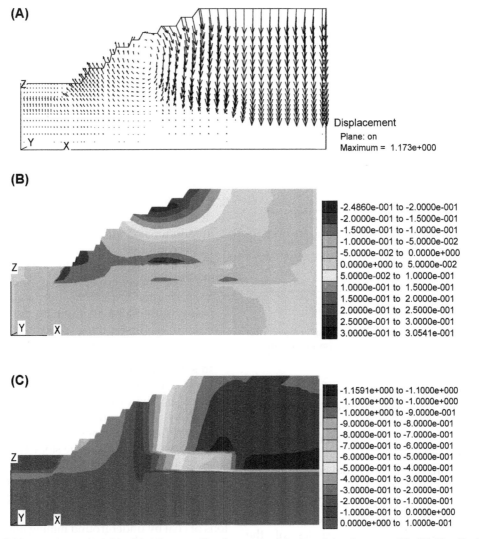

FIGURE 5.21 Displacement field of the slope profile after mining No. 9 coal (section $y = -25$). (A) The displacement vectors field; (B) the x-displacement field; (C) the z-displacement field.

about overlying strata movement is put forward by Qian Minggao and others (Qian et al., 1996; Xu and Qian, 2001). The Pingshuo open-pit mine is a major 100-million-ton coal production base in China, where surface–underground combined mining was put to use to exploit shallow horizontal thick coal, and it has made considerable economic gains as a result of its safe and effective mining model (Ma et al., 2009; Wang and Feng, 2009).

Based on the Winkler assumption and key strata theory, the key stratum is regarded as thick plate on an elastic foundation under the surface−underground combined mining, which traditionally was regarded as an elastic beam. And then the engineering model and mechanical model are built, formulas for calculating the stress and deformation are derived by using the reciprocal theorem method, and the analysis of a practical engineering example has been done.

3.2 Key Strata of Surface−Underground Combined Mining

The key stratum in overburden of surface-underground combined mining can be defined as a hard stratum according to key strata theory, which is thicker and stronger than others and controls the deformation and destruction of overlying strata formed by mining. The force and deformation of the key stratum will affect the stability of roadways and coal pillars in the mining field as well as the stability of the slope engineering.

In accordance with the definition and the deformation characteristics of the key stratum, identifying the position of a key stratum is required to meet the conditions of stiffness and strength. The specific identification method can be described as: identify the positions of hard strata first, then compare the intervals of roof breaking of hard strata to determine the position of the key stratum.

Hard strata deflections are smaller than the lower strata and which don't coordinate with the lower strata in deformation. Hard strata positions can be identified by the next upgrade. Assume that the first stratum is a hard stratum and the $m +$ first stratum is the second hard stratum. As the $m +$ first stratum is a hard stratum, its deflection is smaller than the lower strata; the lower strata need not bear loads of strata above the $m +$ first stratum, and there must be the relation:

$$q_{m+1} < q_m \quad (5.1)$$

where $q_m = \frac{E_1 h_1^3 \sum_{i=1}^{m} h_i \gamma_i}{\sum_{i=1}^{m} E_i h_i^3}$ is the load the first stratum bears when loads are calculated to the mth stratum; $q_{m+1} = \frac{E_1 h_1^3 \sum_{i=1}^{m+1} h_i \gamma_i}{\sum_{i=1}^{m+1} E_i h_i^3}$ is the load the first stratum bears when loads are calculated to the $m +$ first stratum. In the formulas of q_m and q_{m+1}, h_i, γ_i, and E_i represent the thickness, unit weight, and elastic modulus of the ith stratum, respectively.

Assume that q_k is the load which the kth stratum bears:

$$q_k = \frac{E_{k,0} h_{k,0}^3 \sum_{j=0}^{m_k} h_{k,j} \gamma_{k,j}}{\sum_{j=0}^{m_k} E_{k,j} h_{k,j}^3} \quad (k = 1, 2, ..., n) \quad (5.2)$$

where subscript k represents the kth hard stratum; subscript j represents the number of the strata in the soft strata the kth hard stratum controls; m_k is the total of soft strata the kth hard stratum controls; $E_{k,j}$, $h_{k,j}$, and $\gamma_{k,j}$ represents the elastic modulus, unit weight, and thickness of the jth stratum, respectively, in the soft strata controlled by the kth hard stratum; h_k is the thickness of the kth hard stratum.

The interval of roof breaking of the kth hard stratum can be calculated by the formula as following:

$$l_k = h_k \sqrt{\frac{2\sigma_k}{q_k}} \quad (k = 1, 2, ..., n) \quad (5.3)$$

where σ_k is the tensile strength of the kth hard stratum.

If the kth hard stratum is a key stratum, and its interval of roof breaking should be smaller than all hard strata above it. This relationship can be described as:

$$l_{i+1} > l_i \quad (k = 1, 2, ..., n) \quad (5.4)$$

Eq. (5.1) seems like a comparison of loads in the form but it is indeed the stiffness criterion of the key stratum that the deflection of the upper stratum is smaller than the lower stratum.

Eq. (5.4) is the strength criterion of the key stratum that the interval of roofing breaking of the lower stratum is smaller than the upper stratum. So, the strata will be key strata only when they meet the stiffness and strength criteria (Xu and Qian, 2001).

By theoretical analysis, there are similar key strata in the overlying strata in the mines where surface–underground combined mining of shallow horizontal thick coal is applied.

3.3 Engineering Mechanical Model

There is a surface–underground combined mining engineering in the area that between Antaibao open-pit mine and Anjialing No. 2 mine in Pingshuo mining area. The thickness of No. 9 coal is 11.75–14.04 m—13.2 m on average. Fully mechanized long-wall top-coal mining is implemented, and the fully caving mining method is adopted to control the roof. The simplified engineering model is shown in Fig. 5.22. The calculation should conform to medium-thick plate theory when the thickness of the key stratum is not smaller than eighth of the longest edge in the plate.

The edge that exposes partly near the slope can be generalized to a simple supported edge, and the edge above the mined-out area can be generalized to a free edge. Since the formation of large mined-out area, the loads that the key stratum bears have been increasing, the key stratum gradually entered the stage of deformation and failure, and plastic hinges appear on the other two edges, the boundary conditions change gradually from fixed support to a simple support (Wang et al., 2008a). Then the key stratum can be generalized into a mechanical model of a plate with three simple supported edges and a fixed supported edge. Since the slope above the key stratum is step-shaped, its weight can be considered as a triangle load. The key stratum is regarded as thick plate on an elastic foundation, then a mechanical model is obtained as shown in Fig. 5.23.

3.4 Formula Derivation

When a thick bending plate on elastic foundation whose foundation modulus is k bears distributed load q, its control equation is

$$D\nabla^4 w = q - kw - \frac{h^2}{10}\frac{2-v}{1-v}\nabla^2(q - kw) \quad (5.5)$$

where w is the deflection of the plate; q is the load the plate bears; v is Poisson's ratio; h is the thickness of the plate; D is the flexural rigidity of the plate; and $D = \frac{Eh^3}{12(1-v^2)}$.

The deflection and twist angle of the free edge, respectively, can be supposed as:

$$(w)_{x=a} = \sum_{n=1,2}^{\infty} b_n \sin\frac{n\pi y}{b} \quad (5.6)$$

$$(\omega_y)_{x=a} = \sum_{n=1,2}^{\infty} f_n \sin\frac{n\pi y}{b} \quad (5.7)$$

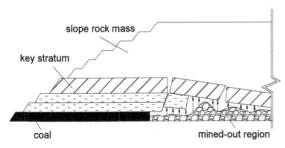

FIGURE 5.22 Engineering model.

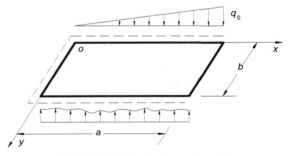

FIGURE 5.23 Mechanical model.

Deflection surface equation of the plate in the model is derived by using the reciprocal theorem method (Li et al., 2000):

$$w(\xi, \eta) = \int_0^a \int_0^b \left(q - \frac{h^2}{10}\frac{2-v}{1-v}\nabla^2 q\right) w_1(\xi, \eta) dx dy$$
$$- \int_0^b (Q_{1x})_{x=a}(w)_{x=a} dy$$
$$- \int_0^b (M_{1xy})_{x=a}(\omega_y)_{x=a} dy$$

(5.8)

where $w_1(\xi,\eta)$ is the deflection on the point (ξ,η) of the virtual basic system; $(Q_{1x})_{x=a}$ and $(M_{1xy})_{x=a}$ are the shear force and bending moment on the boundary of the virtual basic system.

The triangle distributed load can be expanded into a dual-trigonometric series:

$$q = \frac{q_0}{a}x = \frac{-8q_0}{\pi^2}\sum_{m=1,2}^{\infty}\sum_{n=1,2}^{\infty}\frac{\cos m\pi}{mn}\sin\frac{m\pi x}{a}\sin\frac{n\pi y}{b}$$

(5.9)

Then,

$$\nabla^2 q = \frac{8q_0}{\pi^2}\sum_{m=1,2}^{\infty}\sum_{n=1,2}^{\infty}\frac{\cos m\pi}{mn}$$
$$\times\left[\left(\frac{m\pi}{a}\right)^2 + \left(\frac{n\pi}{b}\right)^2\right]\cdot\sin\frac{m\pi x}{a}\sin\frac{n\pi y}{b}$$

(5.10)

where $\rho_n = \sqrt{\left(\frac{n\pi}{b}\right)^4 + \frac{k}{D} + \frac{k}{D}\frac{h^2}{10}\frac{2-v}{1-v}\left(\frac{n\pi}{b}\right)^2}$; $\eta_n = \left(\frac{n\pi}{b}\right)^2 + \frac{k}{D}\frac{h^2}{20}\frac{2-v}{1-v}$.

When $\rho_n > \eta_n$, there will be $\alpha_n = \sqrt{\frac{1}{2}(\rho_n + \eta_n)}$, $\beta_n = \sqrt{\frac{1}{2}(\rho_n - \eta_n)}$, and the virtual basic solution can be transformed into:

$$w_1(\xi,\eta) = \frac{4a^3}{Db\pi^4}\sum_{m=1,2}^{\infty}\sum_{n=1,2}^{\infty}$$
$$\times\frac{\sin\frac{m\pi\xi}{a}\sin\frac{n\pi\eta}{b}}{m^4 + 2m^2\eta_n\left(\frac{a}{\pi}\right)^2 + \rho_n^2\left(\frac{a}{\pi}\right)^4}$$
$$\cdot\sin\frac{m\pi x}{a}\sin\frac{n\pi y}{b}$$

(5.11)

Eq. (5.11) can also be written as:

$$w_1(\xi,\eta) = \frac{4b^3}{Da\pi^4}\sum_{m=1,2}^{\infty}\sum_{n=1,2}^{\infty}$$
$$\times\frac{\sin\frac{m\pi\xi}{a}\sin\frac{n\pi\eta}{b}}{n^4 + 2n^2\eta_m\left(\frac{b}{\pi}\right)^2 + \rho_m^2\left(\frac{b}{\pi}\right)^4}$$
$$\cdot\sin\frac{m\pi x}{a}\sin\frac{n\pi y}{b}$$

(5.12)

After the expressions of virtual basic solutions, the deflection and twist angle of the free edge in actual system being substituted into Eq. (5.8), and $\sin\frac{n\pi\eta}{b}$ being transformed into dual-trigonometric series, the expression of deflection surface equation of the thick bending plate on elastic foundation can be obtained as

$$w(\xi,\eta) = \sum_{m=1,2}^{\infty} \frac{-2q_0}{m\pi D} \frac{\cos m\pi}{(\alpha_m^2+\beta_n^2)^2} \left\{ \left[1 + \frac{h^2}{10}\frac{2-v}{1-v}\left(\frac{n\pi}{b}\right)^2\right] - \frac{2\alpha_m\beta_m \text{ch}\frac{\alpha_m b}{2}\cos\frac{\beta_m b}{2} + (\alpha_m^2-\beta_m^2)\text{sh}\frac{\alpha_m b}{2}\sin\frac{\beta_m b}{2}}{2\alpha_m\beta_m\left(\text{sh}^2\frac{\alpha_m b}{2}+\cos^2\frac{\beta_m b}{2}\right)} \cdot \right.$$

$$\left\{ \left[1 - \frac{h^2}{10}\frac{2-v}{1-v}\left(\alpha_m^2-\beta_m^2-\frac{m^2\pi^2}{a^2}\right)\right]\text{sh}\alpha_m\left(\eta-\frac{b}{2}\right)\cdot\cos\beta_m\left(\eta-\frac{b}{2}\right) + \frac{h^2}{10}\frac{2-v}{1-v}2\alpha_m\beta_m\text{sh}\alpha_m\left(\eta-\frac{b}{2}\right)\cdot\sin\beta_m\left(\eta-\frac{b}{2}\right)\right\} - $$

$$\frac{2\alpha_m\beta_m\text{sh}\frac{\alpha_m b}{2}\sin\frac{\beta_m b}{2} - (\alpha_m^2-\beta_m^2)\text{ch}\frac{\alpha_m b}{2}\cos\frac{\beta_m b}{2}}{2\alpha_m\beta_m\left(\text{sh}^2\frac{\alpha_m b}{2}+\cos^2\frac{\beta_m b}{2}\right)} \cdot \left\{\left[1-\frac{h^2}{10}\frac{2-v}{1-v}\left(\alpha_m^2-\beta_m^2-\frac{m^2\pi^2}{a^2}\right)\right]\text{sh}\alpha_m\left(\eta-\frac{b}{2}\right)\cdot\right.$$

$$\left.\sin\beta_m\left(\eta-\frac{b}{2}\right) - \frac{h^2}{10}\frac{2-v}{1-v}2\alpha_m\beta_m\text{ch}\alpha_m\left(\eta-\frac{b}{2}\right)\cos\beta_m\left(\eta-\frac{b}{2}\right)\right\}\right\}\sin\frac{m\alpha_m\xi}{a} + \sum_{n=1,2}^{\infty}\frac{1}{2\alpha_n\beta_n(\alpha_n^2+\beta_n^2)(\text{ch}^2\alpha_n a - \cos^2\beta_n a)}\cdot$$

$$\left\{[\text{sh}\alpha_n\xi\cos\beta_n\xi(\alpha_n\text{ch}\alpha_n a\sin\beta_n a + \beta_n\text{sh}\alpha_n a\cos\beta_n a) - \text{ch}\alpha_n\xi\sin\beta_n\xi(\alpha_n\text{sh}\alpha_n a\cos\beta_n a - \beta_n\text{ch}\alpha_n a\sin\beta_n a)]\cdot\right.$$

$$\left[-\alpha_n^3 + 3\alpha_n\beta_n^2 + \alpha_n\left(\frac{n\pi}{b}\right)^2\right] - [\text{sh}\alpha_n\xi\cos\beta_n\xi\cdot(\alpha_n\text{ch}\alpha_n a\sin\beta_n a - \beta_n\text{sh}\alpha_n a\cos\beta_n a) + \text{ch}\alpha_n\xi\sin\beta_n\xi\cdot$$

$$\left.(\alpha_n\text{sh}\alpha_n a\cos\beta_n a + \beta_n\text{ch}\alpha_n a\sin\beta_n a)]\cdot\left[-3\alpha_n^2\beta_n + \beta_n^3 + \beta_n\left(\frac{n\pi}{b}\right)^2\right]\right\}\sin\frac{n\pi\eta}{b}(b_n) + $$

$$\frac{(1-v)}{2}\sum_{n=1,2}^{\infty}\frac{1}{\alpha_n\beta_n(\alpha_n^2+\beta_n^2)(\text{ch}^2\alpha_n a - \cos^2\beta_n a)} \cdot \left\{[\text{sh}\alpha_n\xi\cos\beta_n\xi(\alpha_n\text{ch}\alpha_n a\sin\beta_n a + \beta_n\text{sh}\alpha_n a\cdot\right.$$

$$\cos\beta_n a) - \text{ch}\alpha_n\xi\sin\beta_n\xi(\alpha_n\text{sh}\alpha_n a\cos\beta_n a - \beta_n\text{ch}\alpha_n a\sin\beta_n a)]\alpha_n\left(-\frac{n\pi}{b}\right) - $$

$$[\text{sh}\alpha_n\xi\cos\beta_n\xi(\alpha_n\text{ch}\alpha_n a\sin\beta_n a + \beta_n\text{sh}\alpha_n a\cdot\cos\beta_n a) + \text{ch}\alpha_n\xi\sin\beta_n\xi(\alpha_n\text{sh}\alpha_n a\cos\beta_n a + $$

$$\left.\beta_n\text{ch}\alpha_n a\sin\beta_n a)]\beta_n\left(\frac{n\pi}{b}\right)\right\}\sin\frac{n\pi\eta}{b}(f_n)$$

(5.13)

The stress function is

$$\varphi(\xi,\eta) = \sum_{n=1,2}^{\infty}\left\{-D(1-v)\left[\frac{5}{h^2}\frac{n\pi}{b}(b_n) + \left(\frac{5}{h^2}+\frac{n^2\pi^2}{b^2}\right)\cdot(f_n)\right]\right\}\frac{\text{ch}\delta_n\xi}{\delta_n\text{sh}\delta_n a}\cos\frac{n\pi\eta}{b}$$

(5.14)

where $\delta_n = \sqrt{\left(\frac{n\pi}{b}\right)^2 + \frac{10}{h^2}}$.

The shear force, the bending moment, and the torsion at the edge are all zero. The condition that the bending moment is zero was used when the reciprocal theorem was applied, so all the functions still need to meet the conditions as following:

$$(Q_x)_{x=a} = 0, \quad (M_{xy})_{x=a} = 0.$$

Here are the calculation formulas of the thick bending plate on elastic foundation:

$$Q_x = -D\frac{\partial}{\partial x}\nabla^2 w - \frac{h^2}{10}\frac{2-v}{1-v}\frac{\partial(q-kw)}{\partial x} + \frac{\partial\varphi}{\partial y}$$

(5.15)

$$Q_y = -D\frac{\partial}{\partial y}\nabla^2 w - \frac{h^2}{10}\frac{2-v}{1-v}\frac{\partial(q-kw)}{\partial y} - \frac{\partial\varphi}{\partial x}$$

(5.16)

$$M_x = -D\left(\frac{\partial^2 w}{\partial x^2} + v\frac{\partial^2 w}{\partial y^2}\right) - D\frac{h^2}{5}\frac{\partial^2}{\partial x^2}\nabla^2 w$$
$$-\frac{h^2}{10}\frac{2-v}{1-v}(q-kw) - \frac{h^4}{50}\frac{2-v}{1-v}\frac{\partial^2(q-kw)}{\partial x^2}$$
$$+\frac{h^2}{5}\frac{\partial^2 \varphi}{\partial x \partial y}$$

(5.17)

$$M_y = -D\left(\frac{\partial^2 w}{\partial y^2} + v\frac{\partial^2 w}{\partial x^2}\right) - D\frac{h^2}{5}\frac{\partial^2}{\partial y^2}\nabla^2 w$$
$$-\frac{h^2}{10}\frac{2-v}{1-v}(q-kw) - \frac{h^4}{50}\frac{2-v}{1-v}\frac{\partial^2(q-kw)}{\partial y^2}$$
$$-\frac{h^2}{5}\frac{\partial^2 \varphi}{\partial x \partial y}$$

(5.18)

$$M_{xy} = -D(1-v)\frac{\partial w}{\partial x \partial y} + \frac{h^2}{10}\left(\frac{\partial Q_x}{\partial y} + \frac{\partial Q_y}{\partial x}\right)$$

(5.19)

Two equations will be obtained after the deflection surface equation and the stress function is substituted into the boundary conditions; then, b_n and f_n can be calculated in MATLAB.

According to Eq. (5.20), the deformation and force anywhere of the thick bending plate on elastic foundation can be calculated.

$$\left.\begin{aligned}\sigma_x &= \frac{12M_x}{h^3}z \\ \sigma_y &= \frac{12M_y}{h^3}z \\ \sigma_z &= -\frac{3(q-kw)}{4}\left[\frac{2}{3} - \frac{2z}{h} + \frac{1}{3}\left(\frac{2z}{h}\right)^3\right] \\ \tau_{xy} &= \frac{12M_{xy}}{h^3}z \\ \tau_{xz} &= \frac{3Q_x}{2h}\left[1 - \left(\frac{2z}{h}\right)^2\right] \\ \tau_{yz} &= \frac{3Q_y}{2h}\left[1 - \left(\frac{2z}{h}\right)^2\right]\end{aligned}\right\}$$

(5.20)

3.5 Example Analysis

In the engineering model, the key stratum above the mined-out region is sandstone. As a calculation example, the measurement of the stratum has a length of 160 m, width of 60 m, and height of 20 m. The unit weight γ is 2.38×10^3 kN/m³; the elastic modulus E is 5.5 GPa; Poisson's ratio v is 0.3; the compression strength σ_c is 51 MPa; the tensile strength σ_t is 0.3 MPa; the cohesion c is 1.5 MPa; the internal friction angle φ is 38 degrees; the foundation modulus k is 120 MPa.

The points in the following figures which are used to show the calculation results are all points on the centerline of the plate surface.

3.5.1 Stress Analysis of the Key Stratum

The maximum principal stress distribution is shown in Fig. 5.24; in the graph, all the curves are S-shaped curve, and there are a trough and a wave in each curve. By the horizontal line through the point of zero stress, the figure can be divided into two areas: (1) tensile region, which is next to the mined-out region, and (2) compressive region, which is near the slope.

The points of key strata with different lengths, which have the same distance from the free edge,

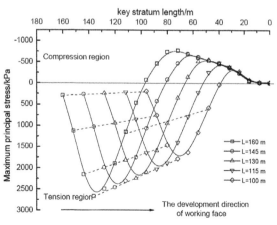

FIGURE 5.24 Maximum principal stress distribution.

are connected by dashed lines in the figure. The slopes of the dashed lines are all positive, and the feature illustrates that the value of the maximum principal stress at the points which have the same distance from the free edge decreases with the working face advancing. The greater the value of the slope, the faster the maximum principle stress decreases. From the moving tendency of the trough and the wave, it is clear that the maximum tensile stress and the maximum pressure stress decrease while the working face increases. Before reaching the wave, the value of the maximum principal stress increases gradually at the same point.

Suppose the length of the key stratum is l, the length of the tensile region is l_t, and the length of the compressive region is l_c. The distribution of tension stress and compression stress in detail is given in Table 5.3.

As we can see from the table, with the length of the key stratum decreasing, the proportion of tension region increases and the proportion of the compression region decreases.

All the shear stress curves are U-shaped, and there is a trough on each curve (Fig. 5.25). The points of key strata with different lengths, which have the same distance from the free edge, are connected by dashed lines in the figure. The slopes of the dashed lines are all positive, and the feature illustrates that the value of the shear stress at the points which have a same distance from the free edge decreases while the working face advances. The greater the value of the slope, the faster the shear stress decreases. From the moving tendency of the trough, it is clear that the maximum shear stress decreases while the

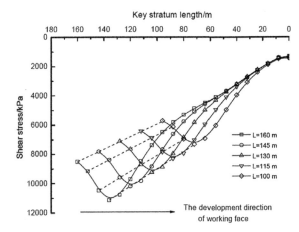

FIGURE 5.25 Shear stress distribution.

working face advances. After the wave, the value of the shear stress increases gradually at the same point.

As shown by comparing Figs. 5.24 and 5.25, the trough's location of the maximum principal stress curves and the shearing stress curves is the same; it means that the locations of the maximum tensile stress and the maximum shearing stress are the same, and they both appear near the mined-out region. The part of the key stratum where the tensile stress surpasses tensile strength or the shear stress surpasses shearing strength will enter the plastic state, which may break down, and the plastic zone will move to the slope. This situation fits in with the fracture of the roof stratum in engineering.

3.5.2 Settlement Analysis of Key Stratum

As shown from the curves in Fig. 5.26, the settlement decreases from the free edge to the simple supported edge, the maximum settlement appears at the free edge, and zero settlement appears at the simple supported edge. All curves can be divided into two parts: the sharp reduction region and the gentle reduction region. In the sharp reduction region, the load is maximal and the constraints are the least.

TABLE 5.3 Proportion of Tension Stress and Compression Stress

Key Stratum Length (m)	160	145	130	115	100
l_t/l	0.394	0.441	0.485	0.593	0.61
l_c/l	0.606	0.559	0.515	0.407	0.39

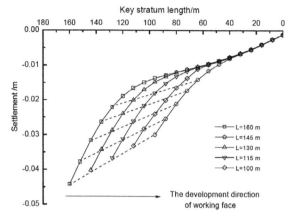

FIGURE 5.26   Settlement curves of key stratum.

The points of key strata with different lengths, which have the same distance from the free edge, are connected by dashed lines in the figure. The slopes of the dashed lines are all positive, and the feature illustrates that the value of the settlement at the points which have the same distance from the free edge decreases while the working face advances. The smaller the value of the slope, the slower the settlement decreases. With working face advancing, the value of the settlement increases gradually at the same point.

The key stratum keeps coordination with the foundation in deformation, so both of them have the same deformation tendency. In engineering, with the working face advancing, the roof's settlement and the mined-out region both increase, and the features of the settlement curves are in agreement with this situation.

3.6 Conclusions

By theoretical analysis, there are key strata in the overlying strata in the mining progress where surface—underground combined mining of shallow horizontal thick coal is applied.

Based on actual projects, considering the spatial effect of the key stratum, the key stratum is regarded as thick plate on an elastic foundation with three simple supported edges and a fixed supported edge; the deflection surface equation is derived by using the reciprocal theorem method, and the method to calculate the deformation and force is given.

By analysis in the example, the characteristics of stress and deformation of the key stratum in the process of the periodic failure are revealed. The research results have an important guiding significance in theory and practical value to evaluate the slope stability and arrange the roadways reasonably.

4. PARAMETERS OPTIMIZATION OF THE SLOPE ENGINEERING

4.1 Introduction

There are many ways to improve labor productivity and economic benefits for coal mining enterprises, but all ways must rely on the technological progress to accelerate the construction of high-yield and high-efficiency mines and to promote the modernization of the coal industry. As China's major mining base with hundreds of million tons of coal production, the Pingshuo mining district has first tried to use the open-underground combined mining in shallow horizontal thick coal, which is a safe and new model and has achieved considerable economic benefits (Ren et al., 2007).

In the process of exploitation of mineral resources in the open-underground combined mining model, two aspects must be considered, one is as much as possible to prevent waste, the other is the stability of the mining slope engineering (Wang and Feng, 2009). At present in China the research on the open-underground combined mining mainly focuses on the metal mines, and the open-underground combined mining model was mainly used in the inclined coal-seam conditions; the study on the open-underground combined mining of the shallow horizontal thick coal is very immature. Since there is neither practical experience nor a mature theory of similar conditions in China and abroad as a reference, a series of new technical challenges appeared in the open-underground combined mining in Pingshuo mining district (Wang et al., 2008a,b; Feng et al., 2009). In the face of the reality that the theory lags behind the practice, the systematic studies for these technical problems are of important theoretical meaning and practical value.

4.2 Engineering Background

The shallow coal seams in Antaibao, Shanxi Province, China, were mined by the open-underground combined mining. The underground working faces B400—404 and B900—904 of the No. 2 coal mine were located in the southeast slope of the Antaibao Opencast Mine. The 9# coal seam working face was mined by using the long-wall fully mechanized top-coal caving method along the dip, which was used to control the roof. The thickness of 9# coal seam working face was from 11.75 to 14.04 m, an average of 13.20 m. The design mining height was 3.20 m, an average of caving height of 10 m, and the ratio of mining height and caving height was 1:3.13. The design working face was 240-m long and the dip of coal seam was from 2 to 7 degrees, with an average dip of 5 degrees, and the surface is the dumping site of Antaibao Opencast Mine.

When the original design parameters were conducted by the open-underground combined mining, the landslides and collapse appeared on the bench of the local slope, which caused the transport platform to work improperly; these problems not only affected the mining production but also threatened the safety of the staff. The cracks and deformation failure of the transport platform are shown in Fig. 5.27.

4.3 Physical Model Test and Numerical Simulation

According to the length and width of the working face, seam thickness, and boundary conditions in the actual project, the test on the engineering model was designed. The geometric similarity constant α_l of the model was 200, the bulk density similarity constant α_γ was 1.6, and the stress similarity constant was $\alpha_\sigma = \alpha_l \alpha_\gamma = 320$.

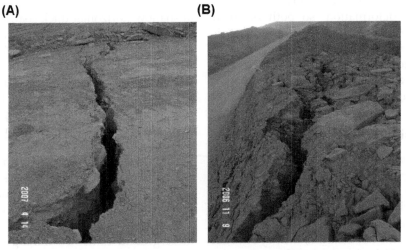

FIGURE 5.27 The cracks and deformation failure of the slope transport platform. (A) Road cracks; (B) roadside cracks.

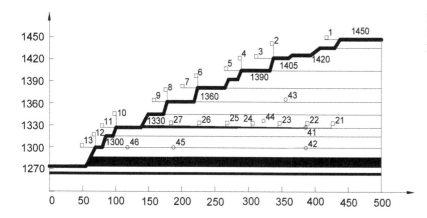

FIGURE 5.28 Arrangement of the displacement sensors and the pressure cells (m).

The sand was used as an aggregate, and the lime and gypsum were used as cement material. And the different strata were separated by using mica. The cross-section of the model, the arrangement of displacement sensors, and the pressure cells are shown in Fig. 5.28. The displacement meters shown in Fig. 5.28 were placed on the monitoring points from 1 to 13 and from 21 to 27, which were represented by a box. The pressure cells were placed on the monitoring points from 41 to 46, which were represented by the circle.

In the test, the overall height of coal seam was mined. The distance from the mining starting position to the toe of the slope was 240 m, and the mining directions were from the starting position to the left and right at the same time.

This test was conducted in State Key Laboratory of Coal Resources and Safe Mining in China. The settlement curves of monitoring points of the slope on elevation 1330 were shown in Fig. 5.29. With the underground mining advancing, the range of the overlying strata collapsing changed from small to large, the subsidence also gradually increased, and the law of the subsidence curves of the overlying strata collapsing was consistent with the conventional mining rock collapse.

Based on the geological section of the physical model test, the 3D numerical model was built by FLAC3D. Through constantly adjusting the tests parameters of rock and soil, the coal seam mining was conducted and the slope failure was simulated to reproduce in the physical model test.

To observe the subsidence of the overlying strata in the mining process, corresponding to the monitoring points at 4# coal roof in the physical model test, the subsidence of the overlying strata was monitored in the numerical model. The overlying strata subsidence curves after 9# coal being mined were shown in Fig. 5.30.

From the contrast of Figs. 5.29 and 5.30, it can be seen that the numerical simulation results and the physical model test results had little difference and the settlement trend in the numerical

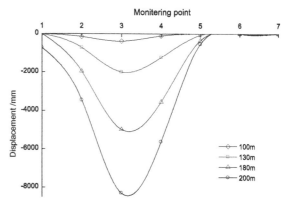

FIGURE 5.29 Settlement curves of monitoring points on elevation 1330 of the slope in the physical model.

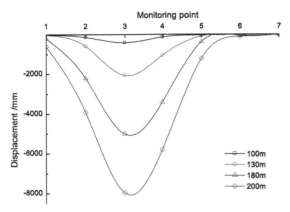

FIGURE 5.30 Settlement curves of the same monitoring points in the numerical model.

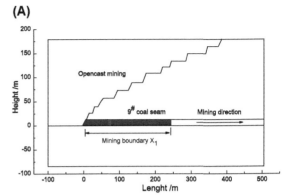

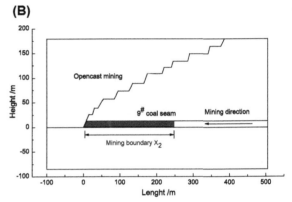

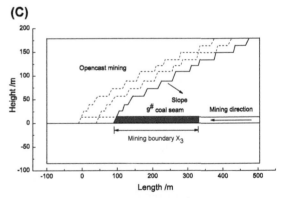

FIGURE 5.31 Sketch of simulation schemes (m). (A) Scheme 1; (B) Scheme 2; (C) Scheme 3.

simulation was consistent with that of the physical model test, indicating that the adjusted parameters of rock and soil, and the numerical simulation results, could reflect the actual project (Wu et al., 2008a).

4.4 Optimization Parameters of Mining Boundary

4.4.1 Simulation Schemes

According to the different sequence and different direction of the opencast mining and the underground mining, the following three kinds of schemes were respectively analyzed and optimized.

Scheme 1: It adopted, first, opencast mining, then underground mining, and the underground mining was along the uphill.

Scheme 2: It adopted, first, opencast mining, then underground mining, and the underground mining was along the downhill.

Scheme 3: It adopted, first, underground mining, then opencast mining, and the underground mining was along the downhill.

4.4.2 The Computational Model

As shown in Fig. 5.31, corresponding to the three simulation schemes, the three computational models were built using FLAC3D, respectively.

The geometric parameters of three computational models are the same with the physical models. The dimensions of the model were 605-m long, 300-m wide, and 264-m high. The model was divided into 39,900 elements. The horizontal displacements of four lateral boundaries of the model were restricted, its bottom was fixed, and the top of the model was free.

The physics and mechanics parameters selected of rock and soil in these computational models were shown as listed in Table 5.4. The material of the model was supposed to meet the Mohr–Coulomb strength criterion (Wang et al., 2007; Griffiths and Fenton, 2004; Stead et al., 2006).

4.4.3 Criteria for the Assessment

In the numerical simulation, the calculation and analysis were carried out according to the following criteria.

The slope of elevation 1390 was the main transport platform. To ensure the vehicles with loads normally pass, it needed to be guaranteed that the settlement difference between the end points of the transport platform was less than 200 mm.

The settlement curves of the monitoring points in the main transport platform were convergent.

The failure scope and the connectivity of the plastic zone of the slope engineering under the open-underground combined mining were known.

TABLE 5.4 Physical and Mechanical Parameters of Rock and Soil

Name	Density (kg/m³)	Elastic Modulus (GPa)	Poisson's Ratio	Tension Strength (MPa)	Cohesion (MPa)	Friction Angle (degrees)
Loess	1960	0.15	0.42	0.001	0.05	20
Sandstone	2300	2.00	0.36	0.05	1.0	33
Sandstone	2380	4.20	0.32	0.10	1.2	35
Mudstone	2490	2.80	0.34	0.01	0.1	31
Siltstone	2320	4.60	0.32	0.15	1.0	36
Sandstone	2380	5.50	0.3	0.25	1.5	39
4# coal	1440	1.00	0.38	0.10	0.5	26
Shale	2450	2.40	0.33	0.10	1.2	34
Siltstone	2600	4.80	0.32	0.30	1.6	40
Shale	2580	3.00	0.35	0.10	1.2	35
9# coal	1330	1.20	0.36	0.10	0.6	27
Sandstone	2380	6.90	0.28	0.30	1.5	38
11# coal	1400	1.30	0.35	0.11	6.3	25
Shale	2460	3.50	0.27	0.15	1.3	35
Siltstone	2600	12.0	0.25	1.00	3.7	43

4.4.4 Calculation Results and Analysis

4.4.4.1 THE CHANGE OF SETTLEMENT DIFFERENCE RATE

The monitoring points were set at the top and foot on the elevation 1390 transport platforms. In the process of achieving the mining safe distance, the settlement difference curves of the top and the foot in three schemes were shown in Fig. 5.32.

Seen from Fig. 5.32, the settlement difference changes of the slope engineering mainly were controlled by the underground mining. In Scheme 1, with the working face advancing toward the uphill, there was a clear steep descent of the settlement difference. In Scheme 2, the growth of settlement difference curves was relatively moderate. The settlement difference curve of Scheme 3 fell rapidly first and then slowed down.

4.4.4.2 CHARACTERISTICS OF THE DISPLACEMENT VECTOR FIELD

Fig. 5.33 shows that the displacement vector on the top of the slope after open-underground combined mining was large, its orientation pointed toward the mined-out areas, and the horizontal displacement appeared near the toe of the slope, but the value was very small (Fig. 5.34). The vertical displacement of the slope mainly occurred in the overlying strata of the mined-out areas and showed the zonal distribution, and the closer it was to the center of the mined-out areas, the greater the settlement value was. Outside of the mined-out areas the upward small displacement occurred near the toe of the slope.

When three schemes achieved the mining safe distance as shown in Fig. 5.33, the displacement vector maximum from small to large, followed by Scheme 3 (0.88 m), Scheme 2 (0.96 m), and Scheme 1 (0.99 m).

4.4.4.3 CHARACTERISTICS OF THE PLASTIC ZONE FIELD

After the open-underground combined mining of three schemes, the differences of plastic zones distribution obviously were seen, indicating that the order of the underground mining had a tremendous impact on the distribution of plastic zones. As seen from Fig. 5.35, when each scheme achieved the mining safe distance, the plastic zones near the transport platform went from small to large, followed by Scheme 3, Scheme 2, and Scheme 1.

4.5 Parameter Optimization of the Mining Boundary

By using three criteria which included the maximum settlement difference of the slope transport platform, the convergence of the displacement curve of monitoring point, and the range of the plastic zones, the slope stability was assessed and the mining boundary stability was judged (Liu et al., 2005; Lin et al., 2007).

The results showed that the mining boundaries of the open-underground combined mining of three schemes were different in the case of the slope stability and the vehicle's normal transportation. The mining boundary distance

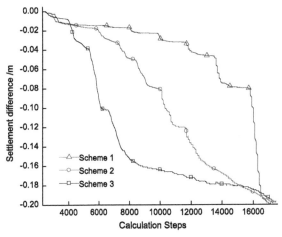

FIGURE 5.32 Settlement curves of monitoring points at the transport platform.

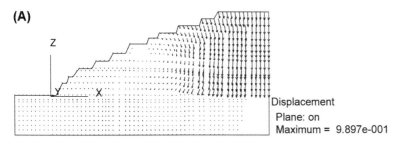

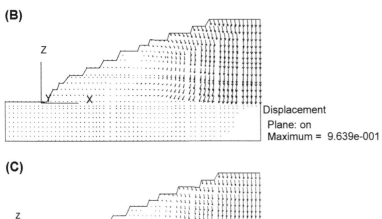

FIGURE 5.33 Displacement vector fields of the open-underground combined mining (m). (A) Scheme 1; (B) Scheme 2; (C) Scheme 3.

of Scheme 1, Scheme 2, and Scheme 3, respectively, was 285, 275, and 265 m.

Scheme 2 had been selected in the northwest slope engineering of the Antaibao Opencast Mine since 2005, the design mining boundary was 276 m, and the mining safety situation had been in good condition till now.

Scheme 1 had been adopted by No. 2 Coal Mine, the design mining boundary was 260 m, due to less than the optimization mining boundary of 285 m, the mining of 4# and 9# coal seams directly caused the three main roadways to be squeezed, the deformation damage was severe, and the subsidence deformation of the slope engineering was so large that the transport platform worked abnormally.

4.6 Conclusions

Combined with the practical engineering, the 3D numerical models were built by FLAC3D based on the physical model test and the mining boundary parameters of slope engineering were optimized under the open-underground combined mining condition.

The results showed that the mining scheme—which was first opencast mining, then underground mining, and the underground mining

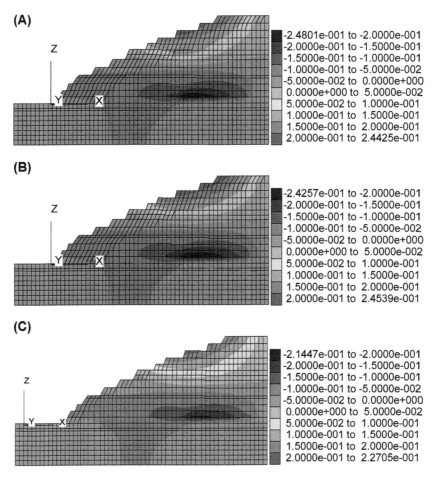

FIGURE 5.34 Horizontal displacement fields of the open-underground combined mining (m). (A) Scheme 1; (B) Scheme 2; (C) Scheme 3.

of shallow thick coal seam was along the downhill—could increase the resources recovery rate and improve the economic benefit under safe conditions by comparing with other mining schemes.

The mining Scheme 1 and mining Scheme 2 had been verified in Antaibao slope engineering, the mining boundary of the former was so small that the operating conditions were poor; while that of the latter was appropriate and the application effect was good.

5. KEY TECHNOLOGIES IN CUT-AND-COVER TUNNELS IN SLOPE ENGINEERING

5.1 Introduction

With the rapid development of China's economy the shallow tunnels on the slope terrain in the mountains gradually increased. Since the cover layer of the shallow tunnel is thinner, the open-cut method is usually used to build

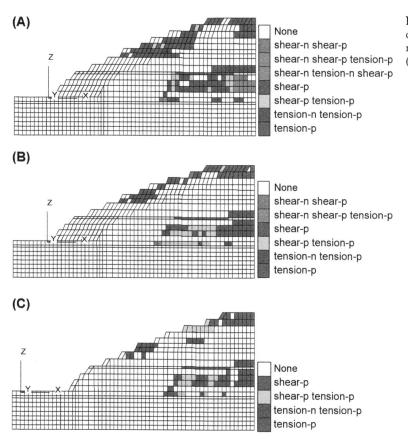

FIGURE 5.35 Plastic zone fields of the open-underground combined mining. (A) Scheme 1; (B) Scheme 2; (C) Scheme 3.

the shallow tunnel. There are potential risks, such as landslide induced by cutting the slope and the noncompacted backfill material during the construction of the cut-and-cover tunnel. It is very important for safety construction to research the key technologies of the shallow tunnel (Yang et al., 2013).

Due to the influence of multiple disturbances of the surrounding rock during shallow tunnel construction, the mechanical behaviors of the cut-and-cover tunnel differ greatly in different construction stages, which often bring great difficulties and potential risks to construction. Therefore, many scholars have conducted studies about these problems. For example, Wu et al. (2008a,b) researched the distribution and the developing law of displacement and stress of the half-buried tunnel under different conditions by the numerical simulation method. Shu et al. (2010) discussed the stability of the tunnel portal section by the method of half-buried tunnel construction based on design and construction of Qiaowu tunnel. Zhang et al. (2010) analyzed the bolt mechanical behavior during the half-dark arch tunnel construction by the numerical simulation method. Li (2011) studied the causes and hazards of unsymmetrical pressure tunnel through case collection and statistical analysis, and the influence factors of critical depth of unsymmetrical pressure tunnel were analyzed as well as the principle of control measures. Dai and Zhang (2013) identified the construction sequence and support scheme for the portal of the shallow tunnel under unsymmetrical pressure combining with the engineering case of Beileyuan tunnel. In addition, Karakus et al. (2007) made finite element analysis for the twin metro tunnel constructed in Ankara clay. Hage and Shahrour (2008) conducted numerical analysis of the interaction between twin-tunnels to reveal the influence of the relative position and construction procedure. Ng et al. (2013) had completed 3D centrifuge modeling of the effects of twin tunneling on an existing pile. To ensure the stability of the cut-and-cover tunnel, the grouting technique is commonly used to reinforce the surrounding rock of the tunnel. At present, it is still in the trial and exploratory stage for the slurry diffusion process within the soil medium, the effective diffusion range, and grouting pressure dissipation (Duan and Li, 2012; Yang and Deng, 2012; Ding et al., 2013). Zhu and Du (2012) analyzed the mechanism of the high pressure and jet grouting for tunnel surrounding soil reinforcement. Lv and Ji (2012) analyzed the deformation and plastic zone of the shallow tunnel based on the mechanism of the grouting reinforcement. Sun et al. (2011), based on the assumption of Bingham fluid and narrow plate model of grouting diffusion, developed a formula for calculating the diffusion radius in soil considering the time-varying behaviors of grout, etc.

Overall, there are still many problems to be studied for the shallow tunnels being built on the slope terrain in the mountains. Therefore, based on a practical engineering, this chapter focused on risk assessment on the construction of the cut-and-cover tunnel and researched the grouting reinforcement mechanism for the tunnel surrounding soil by using particle flow code (PFC) technique.

5.2 Engineering Background

The highway tunnel on the gravel slope containing clay can be used as an example (Fig. 5.36), of which the design clear width is

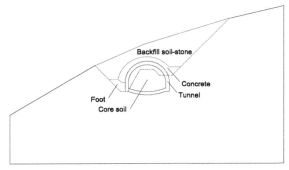

FIGURE 5.36 Cross-sectional schematic of the tunnel.

11.0 m and clear height is 5.0 m, and the design driving speed is 80 km/h.

The surrounding rock of the highway tunnel, belonging to class V, is mainly residual soil and strongly weathered tuff. Its stability is poor with the developed rock fracture, where collapse and landslide are to occur during open-cut construction. The hydrogeological conditions are simple.

The tunnel would be constructed in the following steps: (1) cutting the slope by open-cut method; (2) constructing arch support and the left wall of the tunnel after slope protection; (3) constructing the antislide piles at a distance of the right wall of the tunnel; (4) backfilling the cover of the tunnel layer by layer; (5) excavating the reserved core soil and completing the inverted arch and the right wall construction.

5.3 Mechanics Effect Analysis of the Antislide Pile

To evaluate the mechanics effect on the antislide pile, the PFC was adopted to simulate both conditions of installing antislide pile and no antislide pile during the construction of the cut-and-cover tunnel.

5.3.1 The Computational Model

In 1971, P.A. Cundall proposed the discrete element method, then P.A. Cundall and O.D.L.

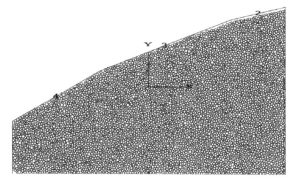

FIGURE 5.37 Particle distribution of the computational model.

Strack developed and launched the commercial particle element programs of PFC2D and PFC3D suitable for rock and soil mechanics. PFC is a micro-/mesomechanics program developed by using the explicit difference algorithm and the discrete element theory. Based on the internal structure of the medium (particles and contacts), PFC can be used to study the macromechanical characteristics and mechanical response of the meso level of mechanical behavior.

According to the engineering geological conditions and the tunnel construction scheme, the computational model was built based on the typical section as shown in Fig. 5.36 (Fig. 5.37). The bottom and both side boundaries of the model were fixed, and the top walls were removed after the particles were completely generated and the initial stress obtained. Interactions within particles representing the rock soil mass were described by a contact-bond model, while the connections of particles composing the foundation and arch were simulated with a parallel-bond model to describe the mechanical behavior of tensile, shear, and bending moment.

The physical and mechanical parameters adopted in the calculation were listed in Table 5.5.

5.3.2 Simulation Analysis Scheme

The simulation analysis scheme of the tunnel construction was as follows:

Step 1: Particle generation and initial self-weight stress field calculation. First, the closed walls were created as the model boundaries. Second, a certain amount of particles within a small radius were generated inside the walls with many voids among them, and then the voids were diminished by a particle radius expansion. Finally, the initial stress equilibrium of the particle assemblies was obtained under the self-weight field.

Step 2: Tunnel excavation and protection slope with core soil reserved. Particles representing the excavation area were deleted, followed by 20% increase of friction coefficient and contact-bond strength of the particles that were within the range of right side slope surface to 3.5 m below, to approximately simulate the bolt reinforcement effect on the slope. Calculation was continued until stress equilibrium to observe the slope deformation due to excavations.

Step 3: Before the construction of left foundation and arch, two cases were considered: with antislide pile or not. The foundation and arch were modeled by

TABLE 5.5 Physical and Mechanical Parameters of the Computational Model

Density (kg/m³)	Maximum Particle (mm)	Minimum Particle (mm)	Particle Stiffness (MN)	Friction Coefficient	Wall Stiffness (MN/m)	Particle Amplification Factor
2200	250	100	100	0.3	10,000	1.67

corresponding shaped particle assemblies with the PFC built-in FISH and Generate commands, in which the particle–particle contact was simulated with parallel-bond model to play a supporting role.

Step 4: Backfilling rubble concrete at the arch foot, the soil-stone at the vault under the condition of antislide pile installation, and that without the pile. Walls were temporarily introduced for generating specific shaped assemblies of backfill particles with the built-in PFC FISH and generate commands, after which the temporary walls were deleted and calculated to the stress equilibrium state.

Step 5: Core soil excavation under the condition of antislide pile installation and that without the pile. The core soil was excavated after the installation of the arch and backfill. As there always was a certain time difference between the excavation and support, the situation that core soil excavated but not yet supported should be considered by at this moment calculating to equilibrium state, the excavation effect on the right side of the slope could be observed.

Step 6: The inverted arch excavation and backfill with the antislide pile installation. The soil particles in tunnel bottom area were deleted, and the particles representing the inverted arch were generated by using the parallel-bond model.

5.3.3 Mechanics Effect of the Antislide Pile

Fig. 5.38 shows the displacement vector field of the slope after the open-cut, from which it can be seen that two slopes were formed at both sides of the tunnel location after 1:1 step-slope excavation, with a low left slope and the right slope at 17 m high. As the rock and soil of the interested area was mainly loose gravel with clay, large displacement toward the free side appeared at the right slope after open-cut and part of which showed a sliding trend. The left low slope shaped to the free side, too, but the displacement was lower.

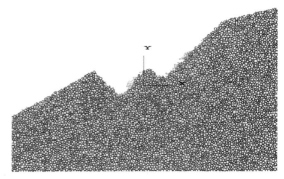

FIGURE 5.38 Displacement vector field after open-cut excavation.

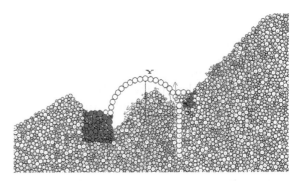

FIGURE 5.39 Tunnel arch construction with antislide pile.

Figs. 5.39 and 5.40 show the construction of tunnel foundation and supporting arch with antislide pile and without, respectively.

Figs. 5.41 and 5.42 show the backfill construction at tunnel vault with antislide pile and without, respectively.

From Figs. 5.43 and 5.44 it can be seen that the maximum displacement of the right slope due to core soil excavation was 381 mm under the installation of the antislide pile, which was smaller than that without the antislide pile (852 mm). Therefore, the antislide pile was necessary for reducing the slide hazard due to core soil excavation.

Fig. 5.45 shows the completed tunnel project under the installation of the antislide pile.

FIGURE 5.40 Tunnel arch construction without antislide pile.

FIGURE 5.41 Backfill soil-stone with antislide pile.

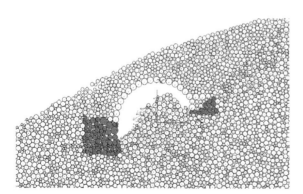

FIGURE 5.42 Backfill soil-stone without antislide pile.

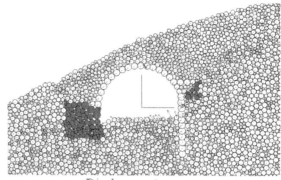

Displacement
Maximum = 3.809e-001

FIGURE 5.43 Displacement vector field with antislide pile.

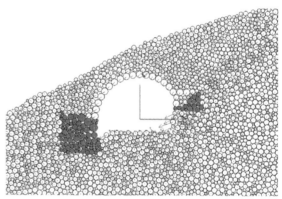

Displacement
Maximum = 8.522e-001

FIGURE 5.44 Displacement vector field without antislide pile.

The scheme with the antislide pile was adopted in field construction as shown in Fig. 5.46.

5.4 Liquid–Solid Coupling Analysis of Fracturing Grouting

5.4.1 Liquid–Solid Coupling in Particle Flow Code

The void geometry in an assembly of circular particles in PFC is regarded as a network

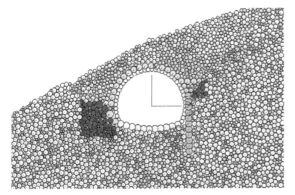

FIGURE 5.45 Displacement vector field of the completed tunnel.

FIGURE 5.46 Setting an antislide pile on site.

structure of identical dots linked by pipe between two adjacent domains.

As far as the fluid is concerned, the pipe is equivalent to a parallel-plate channel, with length L, aperture a, and unit depth (in the out-of-plane dimension). The flow rate (volume per unit time) in a pipe is given by:

$$q = ka^3 \frac{P_2 - P_1}{L} \quad (5.21)$$

where k is a conductivity factor and $P_2 - P_1$ is the pressure difference between the two adjacent domains. The sign convention is such that a positive pressure difference produces a positive flow from domain 2 to domain 1.

1. When the compressive force is taken as positive, the aperture a satisfied the following equation:

$$a = \frac{a_0 F_0}{F + F_0} \quad (5.22)$$

where a_0 is the residual aperture for zero normal force; F_0 is the value of normal force; and F is the value of normal force at which the aperture decreases to $a_0/2$.

2. When the normal force is tensile (intact bond) or it is zero, then the aperture a is simply equal to:

$$a = a_0 + mg \quad (5.23)$$

where m is a dimensionless multiplier and g is the normal distance between the surfaces of the two particles.

Each domain receives flows from the surrounding pipes: $\sum q$. In one time-step Δt, the increase in fluid pressure is given by the following equation, assuming that inflow is taken as positive.

$$\Delta P = \frac{K_f}{V_d}\left(\sum q\Delta t - \Delta V_d\right) \quad (5.24)$$

where ΔP is the increment fluid pressure; K_f is the fluid bulk modulus; and V_d is the apparent volume of the domain.

There are three forms of coupling between the fluid and the solid particles in PFC as follows.

1. Changes in aperture are caused by contact opening and closing, or changes in contact force.
2. Mechanical changes in domain volumes cause changes in domain pressures.
3. Domain pressure exerts tractions on the enclosing particles.

5.4.2 Building Computational Model

The model considered a range of 10×10 m (width × height) in which the center installed a grouting hole with a drainage area of

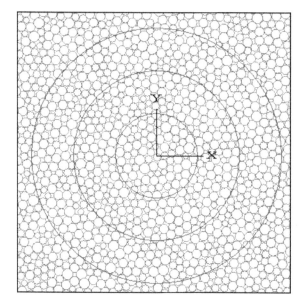

FIGURE 5.47 Particle distribution of the model.

radius 0.1 m and a grouting pressure 1.5 MPa (Fig. 5.47).

To research on the grouting pressure to the surrounding soil and observe the crack generation, development, and extension, three measurement circles were set up with radii of 1.0, 2.0, and 3.0 m, respectively. The computational model built by PFC is shown in Fig. 5.47. The physics and mechanics parameter used in the model are listed in Tables 5.6 and 5.7.

5.4.3 Simulation Analysis Scheme

Step 1: The computational model was built and calculated to stress equilibrium state under soil self-weight, then the displacements were reset to zero.

TABLE 5.7 Parameters Used in Fracturing Grouting Test

Grouting Pressure (MPa)	Permeability Coefficient (m/s)	Aperture for Zero Normal Force (mm)	Normal Force When Aperture Decreases to Half Aperture (Pa)
1.5	0.65	1.0	5000

Step 2: The initial equilibrium of fluid−solid coupling was executed and after which the displacements were reset to zero.

Step 3: Under the given grouting pressure, the fracturing grouting test was simulated with fluid−solid coupling calculation.

5.5 Grouting Effect Analysis in the Surrounding Rock

It can be seen from Fig. 5.48 that the grouting pressure (blue) spread from the center grouting hole to the surrounding soil and became smaller under the given grouting pressure of 1.5 MPa. The compression (red) and tension (black) cracks appeared in the surrounding soil from the near to the distant point. Most cracks occurred within a radius of 1.0 m and very few cracks could be found outside a radius of 2.0 m, which meant that the main controlling scope of grouting fracture was within a radius of 2.0 m.

Figs. 5.49 and 5.50 show the curves of porosity and compressive stress in the surrounding soil of the grouting hole, respectively, under different measurement ranges. The porosity variation in the surrounding soil was on the whole spreading inside-out from large to small (Fig. 5.49) during the grouting test. In the initial stage, the soil

TABLE 5.6 Physics and Mechanics Parameters of the Particle Model

Density (kg/m³)	Maximum Particle (mm)	Minimum Particle (mm)	Particle Stiffness (MN)	Friction Coefficient	Wall Stiffness (MN/m)	Porosity
2000	120	80	0.15	0.2	100	0.15

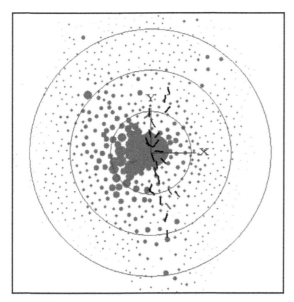

FIGURE 5.48 Grouting pressure and fracture propagation map.

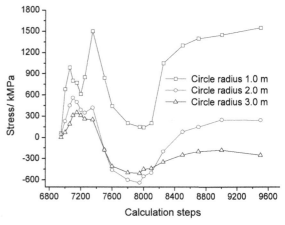

FIGURE 5.50 Particle stress variation around the grouting hole.

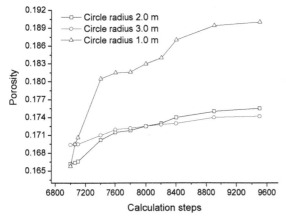

FIGURE 5.49 Porosity variation curves around the grouting hole.

porosity within the 2.0-m radius circle started lower than that within the 3.0-m radius circle and ended higher, which meant the grouting process experienced a gradual development of near fracturing and distant compacting. The particle stress in the soil around the grouting hole was also spreading inside-out from large to small (Fig. 5.50) and experienced the complicated variation of tension and compress during the grouting test.

5.6 Conclusions

For the shallow tunnel construction on mountain slopes, the nonlinear mechanical behavior under loading and unloading conditions (eg, excavation and backfill construction) can be modeled by PFC technique and the particle displacement trends can be displayed graphically. The results showed the construction scheme with antislide pile could effectively control the landslide and reduce risks during the core soil excavation, which is of certain reference value to similar engineering.

The process of crack generation, development, and extension during fracturing grouting process can be simulated by using PFC technique. The grout diffusion, soil compaction, and fracturing effect during the grouting process were analyzed based on a practical engineering, which is of great theoretical and practical meaning when discussing and building the connections between the microvariation of physical and mechanical parameters and the macromechanical response.

6. POTENTIAL RISK ANALYSIS OF A TAILINGS DAM

6.1 Introduction

The tailings pond is a necessary facility for maintaining normal production of a mine as a place for stockpiling tailings. On the other hand, the tailings dam is a major danger for metal and nonmetal mines, because dam failure may occur. A possible tailings accident can cause not only great losses and endanger the lives and properties of downstream residents but also lead to serious environmental pollution.

Many scholars have studied the stability of tailings ponds and dam failure disasters. Shakesby and Richard (1991) have studied the tailings dam accident at the Arcturus gold mine in Zimbabwe and concluded that the dam failure resulted from the poor seepage condition of the dam's foundation, too steep a dam slope and the tailings being in a saturated state under heavy rain. After investigating tailings dam failures in America, Strachan (2001) holds that dam failures resulted from flood overtopping, static or dynamic instability, seepage, internal corrosion, and poor foundation conditions. Chakraborty and Choudhury (2009) investigated the behavior of an earthen tailings dam under static and seismic conditions and found that dam deformation is affected seriously by seismic action and that the underlying input acceleration of the tailings dam had an amplifying effect along the height of the dam. Sjödahl et al. (2005) conducted a safety evaluation of the Enemossen tailings dam in southern Sweden using resistivity measurements. Brett et al. (2010) have completed the geotechnical design of a main creek tailings dam applying the upstream construction method.

Li et al. (2012) summarized two instability models based on the combination of a liquid–solid coupling method with a strength reduction method. One was global instability and the other was local instability, mostly caused by a too-shallow saturation line in the tailings dams (Kang et al., 2013). Chen et al. (2008) provided a comprehensive method to evaluate tailings dam stability through numerical calculation of the seepage stability, static stability, and dynamic stability of a specific project. Yin et al. (2010) studied the regularity of the saturation lines' change under flooding and normal conditions, when the dam was heaped up to about two-thirds of the total height of 120 m. Hu et al. (2004) analyzed the antislide stability of an upstream tailing by changing the dam's height, the saturation line conditions, and the drainage system operating conditions. Li et al. (2005) analyzed the stability of a tailings dam with the Sweden method and the Bishop method and gave the interrelated parameters of the stability and credibility of the dam structure when its level was raised to 510–520 m. Lou et al. (2005) calculated the effect of heightening a dam on the stress-strain isoline through simulation using the finite element method and evaluated the dam's stability with the residual thrust method.

6.2 Engineering Overview

We take the flatland tailings pond of the Sanshan Island gold mine, Shangdong Province, China, shown in Fig. 5.51, as an example. It covers an area of about 0.22 km^2, a catchment area of about 0.21 km^2. The ground level is about 3.1–4.2 m and the reservoir elevation is about 3.5–20.7 m. The starter dam consists of roller compacted sand with a height of 11.0 m, top width of 3.0 m, and outer slope ratio of 1:1.8–1:2.5.

The waste rock has been heaped up from 2010 at the northwest corner of the tailings dam. Its accumulation level is up to 38.5 m, which is 11 m higher than that of the tailings stacking dam (status elevation 27.5 m). The plane shape of the waste rock heap is approximately rectangular, with an east–west length of 260 m and a north–south length of 110 m, covering an area

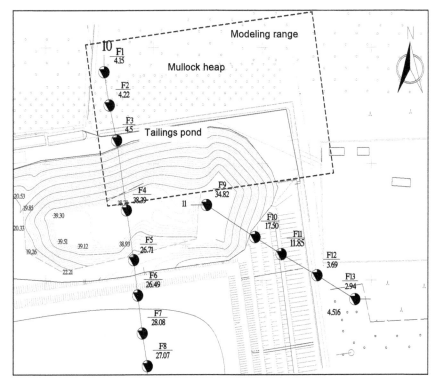

FIGURE 5.51 The plan of the tailings pond and mullock heap.

of about 1.89×10^4 m^2 and occupying a volume of 36.67×10^4 m^3. About 50-m wide at the south side of the mullock heap, it weighs on the north slope of the dam. The mullock heap slide is unaffected until a natural repose slope ratio of 1.0: 1.1—1.0:3.0 and no obvious collapse or crack was found during the site investigation. Fig. 5.51 shows the plan of the tailings pond and mullock heap.

According to the drilling survey result, the stratum is divided into three layers, from top to bottom: mullock material, tailings material, and the natural formation. The mullock heap mainly consists of gravel with silt. The tailings fill dam mainly consists of tailings silt and silty clay. The original ground is composed of medium-coarse sand of alluvial-diluvial and marine deposit genesis and alluvial-diluvial silty clay.

6.3 Three-Dimensional Model and Simulation Analysis Scheme

6.3.1 The Computational Model

In view of the technical difficulties of FLAC3D for complex 3D engineering modeling, the finite element software MIDAS/GTS from South Korea was adopted for geometric modeling of the complex geologic body and mesh generation, followed by model data transformation from MIDAS/GTS to FLAC3D to make up the preprocessing shortcomings of FLAC3D and give full play to its powerful calculating function.

Although the element shape of MIDAS/GTS is basically the same as that of FLAC3D, the node numbering rules and node order are different; therefore, the element and node data exported from MIDAS/GTS have to be rearranged according to a FLAC3D recognizable format and then be imported, thus leading to a data transformation between the two software applications, which can be done through programming. The entire modeling process is shown in Fig. 5.52.

The tailings dam 3D numerical model (Fig. 5.53) was built following the aforementioned procedure. A system of coordinate axes was defined with the origin at the silty clay layer 34 m beneath the natural ground, with the z-axis pointing upward. The model was approximately 280-m long, 250-m wide, and 38.5-m high in the x-, y-, and z-axis, respectively. The present height of the waste rock was 38.5 m and its future height 42.0 m.

The horizontal displacement of the four lateral boundaries of the model was restricted, the bottom was fixed, and the top was free.

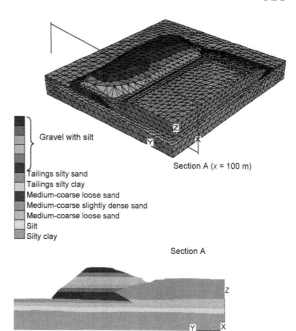

FIGURE 5.53 Three-dimensional model mesh and its material sets.

Only the geomaterial dead weight was taken into account to obtain the initial stress field. The material of the model was supposed to meet the Mohr—Coulomb strength criterion and the physics and mechanics parameters were selected as listed in Table 5.8 according to the field test and the laboratory test.

6.3.2 Liquid—Solid Coupling in FLAC3D

The fluid—solid coupling behavior involves two mechanical effects in FLAC3D. First, changes in pore pressure cause changes in the effective stress of the solid. Second, the fluid in a zone reacts to mechanical volume changes by a change in pore pressure.

The variables of fluid flow through porous media, such as pore pressure, saturation, and the specific discharge are related through the fluid mass—balance equation, Darcy's law for fluid transport, a constitutive equation specifying the fluid response to changes in pore

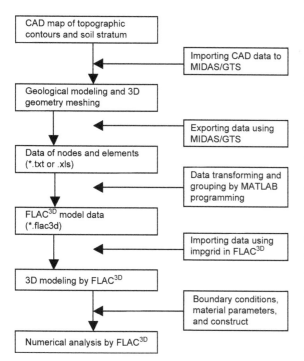

FIGURE 5.52 Flow chart of the 3D modeling process.

TABLE 5.8 Physics and Mechanics Parameter of the Model

Name	Density (kg/m^3)	Elasticity Modulus (GPa)	Poisson Ratio	Cohesion (kPa)	Friction Angle (degrees)	Permeability Coefficient (cm/s)
Gravel with silt	2250	50	0.32	1	35	1.34×10^{-1}
Tailings silty sand	1710	15	0.40	3	25	2.93×10^{-3}
Tailings silty clay	1910	8	0.35	19	8	2.30×10^{-6}
Medium-coarse loose sand	1950	70	0.30	3	30	1.70×10^{-2}
Medium-coarse slightly dense sand	1960	75	0.30	3	32	3.20×10^{-2}
Medium-coarse loose sand	1870	63	0.30	3	29	2.80×10^{-3}
Silt	2190	47	0.32	30	20	1.20×10^{-4}
Silty clay	2030	80	0.30	35	15	1.50×10^{-6}
Water	1000					

pressure, saturation, volumetric strains, and an equation of state relating pore pressure to saturation in the unsaturated range. Assuming the volumetric strains are known, substitution of the fluid mass—balance equation into the fluid constitutive relation, using Darcy's law, yields a differential equation in terms of pore pressure and saturation that may be solved for particular geometries, properties, boundaries, and initial conditions.

In summary, possible causes of tailings dam failure include flood overtopping, slope instability, seepage failure, structural damage, seismic liquefaction, and so forth. In general, the stability of a tailings dam being influenced by the seepage field cannot be ignored. The current study is mainly based on the assumption of the 2D plane strain. There is little literature on 3D stability of the tailings seepage and deformation, so a 3D numerical model was built based on a flatland tailings pond project, to conduct a liquid—solid coupling analysis of the potential risk due to successive preloading at the front of the dam. Some engineering countermeasures will be put forward corresponding to the evaluation results.

6.3.3 Simulation Analysis Schemes

The numerical simulation consisted of four steps as follows:

Step 1: The initial seepage field and initial stress field were calculated under the current operating water level of 27.0 m and then the displacements were reset to zero.
Step 2: The heap process was divided into six steps up to the present level of 38.5 m to analyze the deformation characteristics of the tailings dam under the current conditions.
Step 3: The heap height was increased with an additional accumulation to a height of 42.0 m and the deformation characteristics analysis of the tailings dam was repeated.
Step 4: A stability evaluation and potential risk analysis of the tailings dam were carried out by using the liquid—solid coupling method considering different preloading conditions, and the corresponding safety factors were formulated.

6.4 Liquid–Solid Coupling Analysis of the Tailings Dam

6.4.1 Pore Pressure Distribution Under Current Operating Level

To facilitate the analysis, a vertical cross-section at $x = 100$ m (section A) was defined, as shown in Fig. 5.52. All of the results occurred in section A under the current operating water level of 27.0 m.

It can be seen from the pore pressure distribution (Fig. 5.54) under the condition of the current operating water level that the stable seepage line in the tailings dam extends from the embankment to the outer toe of the slope of the mullock heap, where failure occurs more easily due to the seepage of groundwater.

6.4.2 Deformation Characteristics of the Tailings Dam

The calculation results of the displacement fields are displayed in this section, where Figs. 5.55 and 5.56 correspond to the current elevation of the mullock heap (38.5 m), while

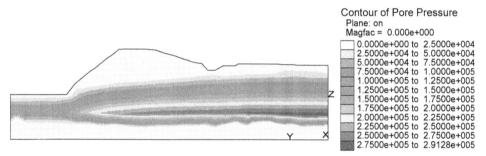

FIGURE 5.54 Pore pressure distribution under the current water level.

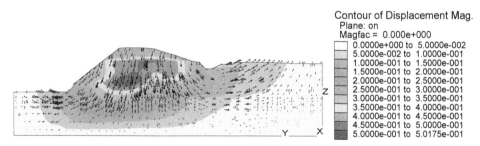

FIGURE 5.55 Displacement field and *arrow* under the current height.

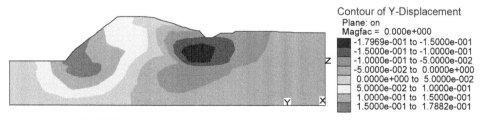

FIGURE 5.56 Horizontal displacement field under the current height.

Figs. 5.57 and 5.58 correspond to the additional heaped elevation (42.0 m). By comparative analysis it can be seen that:

Apart from its consolidation deformation owing to self-gravity under gradual accumulation of the mullock heap, the outside ground surface, starter dam, and outer slope of the tailings fill dam are loaded and crushed with different deformation characteristics as a result of their different stiffness levels. The outer slope toe of the mullock heap consists mainly of surface settlement and lateral uplift. The tailings dam deforms inward under pressure with a certain lateral deformation that results in a little uplift of the tailings silty sand. From the viewpoint of magnitude, the horizontal displacement of 180 mm toward the inner tailings dam is close to the 179 mm in the opposite direction in Figs. 5.55 and 5.56, but the horizontal displacement of 199 mm toward the inner tailings dam is smaller than the 227 mm in the opposite direction in Figs. 5.57 and 5.58. The aforementioned deformations developed significantly after the height of the mullock heap was increased.

The results suggest three potential failure modes of the tailings dam under preloading: (1) the compressive shear zone in the outside ground is likely to induce a sliding failure through the outer slope toe of the tailings dam in case additional loading is continued; (2) local compression and shear failure could appear during the deformation process of the tailings dam under gradual preloading; (3) an uplift failure may occur in the tailings embankment as the load increases when the mullock heap is heightened.

6.4.3 Safety Risk Analysis of the Tailings Dam

According to China technical codes, the safety factor of the tailings dam in this example project should be not less than 1.25 under normal operating conditions. The internal shear strength reduction method of FLAC3D was adopted to

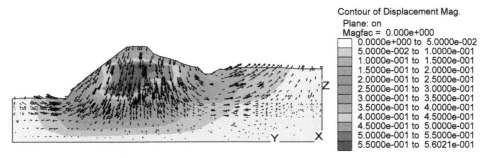

FIGURE 5.57 Displacement field and *arrow* under the future height.

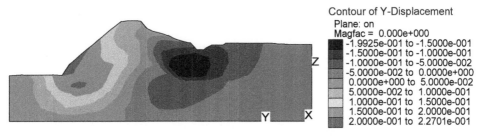

FIGURE 5.58 Horizontal displacement field under the future height.

calculate the safety factor of the tailings dam for different heap heights and to determine the potential slip surface position. The maximum shear strain increment of the tailings dam is 15.2 under the present heap height 38.5 m (Fig. 5.59), which would increase to 24.7 when the mullock heap is heightened to 42.0 m (Fig. 5.60), and the potential slip surface would change from 1 (Fig. 5.59) to 2 (Fig. 5.60). The FOS of the tailings dam is 1.28 under the present heap height 38.5 m (Fig. 5.59), which would reduce to 1.23 when the mullock heap is heightened to 42.0 m (Fig. 5.60). This means that there is a lack of safety reserve and the heap height should not be increased.

6.5 Conclusions

It is concluded from the liquid–solid coupling analysis that there are three potential failure modes of the tailings dam under preloading: (1) the compressive shear zone in the outside ground is likely to induce a sliding failure through the outer slope toe of the tailings dam; (2) local compression and shear failure could appear during the deformation process of the tailings dam under gradual preloading; (3) an uplift failure may occur in the tailings embankment when the mullock heap is heightened. Under the present conditions, the tailings dam meets the safety requirements, whereas it does not in the event of additional heaping.

According to these results, the following engineering countermeasures are put forward: the height of the present heap must be cut to satisfy the stability requirements under the condition of rain infiltration, and a sound monitoring and regular inspection should be established to ensure the safety of the tailings pond operation. Engineering practice has shown that these measures achieve good results.

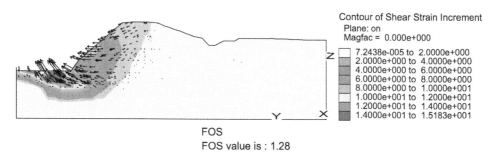

FIGURE 5.59 Potential slip surface and safety factor under the current height.

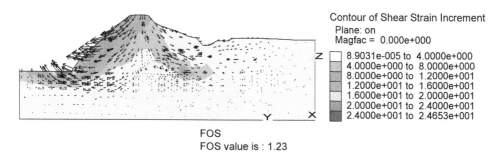

FIGURE 5.60 Potential slip surfaces and safety factor under future height.

7. A NEW LANDSLIDE FORECAST METHOD

7.1 Introduction

Microdata generally has two forms of expression. First, cross-sectional data (horizontal data), which manifests as a set of different factual situations at the same time point. Second, longitudinal data, which manifests as a set of the various situations of the same observational unit at different time points. With the rapid development of science and technology, statistical method research of longitudinal data has made considerable progress. Different from cross-sectional data, longitudinal data is a particular form of repeated measures data which is obtained from repeatedly taking the same type of measurement over time and space on the same group of experimental individuals. One possible objective in analyzing longitudinal data sets is to distinguish certain factors which affect the outcomes, and control them. In addition, recognitions can be made of the interactions between factors influencing the outcomes in the analysis, and therefore the analysis of longitudinal data is very important. Longitudinal studies have very extensive application in many fields, such as medicine, epidemiology, economics, and social sciences (Lin and Carrol, 2001; Tian and Xue, 2008).

The loss of the equilibrium state of an open-pit slope is generally always starting with tension fracture on the slope surface, followed by a gradual increase of the tension fracture, up to the equilibrium state limit in which a small external force perturbation may lead to a bad geological hazard, such as a landslide. So it is of important theoretical significance and practical engineering value to research further the hazards of slope landslide and its forecast, fully utilizing the deformation data of landslide tension fractures. This chapter intends to propose a new method of disaster source recognition and forecast to an open-pit slope landslide by combination of partial linear model of longitudinal data from statistics and nonlinear catastrophe theory on the basis of observational data of field slope displacement, by which landslide prediction and forewarning can be carried out to the slope of an open-pit coal mine.

7.2 Longitudinal Data Analysis Method

7.2.1 Estimation of Regression Function

To establish notation useful in the next section as well, we let horizontal displacement of slope surface, x_{ij}, and vertical displacement, y_{ij}, represent repeated measurements over time and space on the same group of experimental individuals, ie, longitudinal data. The partial linear model of longitudinal data adopted in this paper has the form as follows, ie,

$$y_{ij} = x_{ij}\beta + g(t_{ij}) + e_{ij} \qquad (5.25)$$

where $i, 1, 2,\ldots, n; j, 1, 2,\ldots, m; (x_{ij}, t_{ij}) \in R^p \times R$, are the random or fixed design point ranges; β is a p-dimensional vector of unknown parameters; $g(\cdot)$ is an unknown regression function defined at closed interval $[0,1]$; and e_{ij} is random error.

We let $e_i = (e_{i1}, e_{i2},\ldots, e_{im})'$ be an error vector of ith individual with mth element e_{im}; $\{e_i\}$ are independent of each other; mean value $E(e_i) = 0$; covariance $\text{Var}(e_i) = \Sigma$(unknown). Based on $\{(t_{ij}, y_{ij}, x_{ij})', 1 \leq i \leq n, 1 \leq j \leq m\}$, the unknown regression parameter β and regression function $g(\cdot)$ in model (1) are estimated by application of feasible generalized least-square method, and after a series of derivation, partial linear model of longitudinal data can be obtained with the form as $y_{ij} = x_{ij}\beta + g(t_{ij})$.

7.2.2 Regression Analysis of the Slope Horizontal Displacement

Horizontal displacement of slope surface, x, and vertical displacement, y, contain abundant influence information of various environmental factors on slope (eg, geologic and geomorphic conditions and inducing factors, such as rainfall, groundwater, and blast working). As time goes on, the displacement x and y will show variation of the slope displacement; in other words, there should be a certain relationship between time and displacement: $x = x(t)$ and $y = y(t)$.

In view of the favorable properties, such as local support, nonnegativity, three recursion of B-spline function, and of the B-spline quasiinterpolation operator, which can reflect general variation law of data and can be sufficiently smooth, cubic B-spline quasiinterpolation operator Q is introduced with the form as $Qf := \sum_{i=0}^{N-2} f(x_{i+2}) B_{i,3}(x)$, on the basis of field measurement data, where the B-spline definition is given by recursion formula as follows (Cheng and Cai, 1997).

Let $X = \{x_0, \ldots, x_m\}$ be a nondecreasing sequence, and the sequence can be used as a B-spline node, then the ith p-order B-spline basis function $B_{i,p}(x)$ can be defined as

$$B_{i,0}(x) = \begin{cases} 1 & (x_i \leq x \leq x_{i+1}) \\ 0 & \text{(others)} \end{cases},$$

$$B_{i,p}(x) = \frac{x - x_i}{x_{i+p} - x_i} B_{i,p-1}(x) \quad (5.26)$$
$$+ \frac{x_{i+p+1} - x}{x_{i+p+1} - x_{i+1}} B_{i+1,p-1}(x)$$

For the appearance of 0/0, the quotient is considered to be equal to zero. Then the approximate differential expression of function $y = y(x)$ has the form

$$Q'f := \sum_{i=0}^{N-2} f(x_{i+2}) B'_{i,3}(x) \quad (5.27)$$

The local support of cubic B-spline $B_{i,3}(x)$ is $[x_i, x_{i+4}]$, and according to Eq. (5.26) we obtain the expression of $y = y(x)$ on the support minimal interval $[x_i, x_{i+4}]$, which is represented by linear combination of $B_{m-4,3}(x)$, $B_{m-3,3}(x)$, $B_{m-2,3}(x)$, and y_{m-2}, y_{m-1}, y_m. After rearranging terms, we obtain a cubic polynomial with the form as $y = b_3 x^3 + b_2 x^2 + b_1 x + b_0$, where x denotes time t, y denotes horizontal displacement x, so this cubic polynomial can be written as $X = b_3 t^3 + b_2 t^2 + b_1 t + b_0$. Therefore, the partial linear model of longitudinal data, $y_{ij} = x_{ij}\beta + g(t_{ij})$, can be represented as a relationship between vertical displacement and time, t, ie, $y = (b_3 t^3 + b_2 t^2 + b_1 t + b_0)\beta + g(t)$; in other words, $y = f(t)$.

7.3 Catastrophe Mechanism of Landslide and Forecast Method

7.3.1 Catastrophe Mechanism Analysis of Landslide

Slope landslide catastrophe is a system which transforms from gradual changing to catastrophe, from quantitative change to qualitative change, affected by interaction of internal and external causes, and it's also a nonlinear evolution and uncontinuous change phenomenon, in which catastrophe is the ultimate failure precursor. Catastrophic theory is a new branch of mathematics to research discontinuous phenomenon, concluding various phenomenon into different categories of topologic structure and discussing the discontinuous characteristics of all kinds of critical points, by which the catastrophe problem that possibly appears in continuum system can be solved effectively. It is mainly used to solve nonlinear problems of prominent discontinuous phenomenon, such as landslide, earthquake, stope collapse, rock burst, and so forth.

In the mechanics range, the cusp catastrophic model of catastrophic theory is the most widely applied that has a 3D phase space because of possessing two control parameters and a state variable, whose regular function form of its potential function is $v(x) = x^4 + px^2 + qx$, in which x is a state variable (x represents a time variable here), and p and q represent control parameters. Correspondingly, all critical points of $V(x)$ constitute set M, called the equilibrium surface

(Fig. 5.61), which satisfies the following relationship:

$$\operatorname{grad}_x V = 4x^3 + 2px + q = 0 \quad (5.28)$$

The figure of M in (x, y, z) space is a smooth surface containing fold or pleat, and thus it has a different number of equilibrium positions in different regions. Obviously, points on the equilibrium surface that have vertical tangents satisfy the equation:

$$\operatorname{grad}_x(\operatorname{grad}_x V) = 12x^2 + 2p = 0 \quad (5.29)$$

The singularity set of state is composed of all the points with vertical tangents on the equilibrium surface, and the bifurcation set, defined by the projection of the singularity points in control surface $p - q$, is composed of all the points who make sudden changes in state variable x.

By combining Eq. (5.28) with Eq. (5.29) and eliminating x, the bifurcation set can be determined as $F = 8p^3 + 8q^2 = 0$, by which the control parameter surface is divided into three parts as $F > 0$, $F < 0$, and $F = 0$, as illustrated in Fig. 5.62.

If phase point P is just on the edge of the fold (where the top and bottom sheets fold over to form the middle sheet), a sudden jump of state x will take place as the points inevitably jump into another sheet. Thus the discriminant for slope rock mass failure can be obtained:

$$F = 8p^3 + 8q^2 \quad (5.30)$$

Where the slope rock mass is stable when $F > 0$; it is under a critical state when $F = 0$; and landslide takes place when $F < 0$.

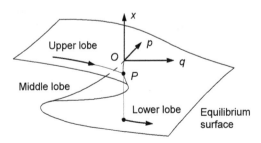

FIGURE 5.61 Profile of equilibrium.

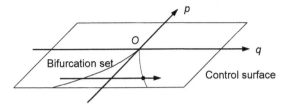

FIGURE 5.62 Sketch of bifurcation set.

7.3.2 Forecast Method of Landslide Catastrophe

The function $y = f(t)$, which can be obtained by using the partial linear model of longitudinal data mentioned before, is an expression of slope state information including various information of the slope rock slide. In other words, it is another expression of cusp catastrophic function $v(x) = x^4 + px^2 + qx$.

By using the Taylor series expansion method, and by truncating the front five terms, the univariate function $y = f(t)$ can be reduced to

$$y = \sum_{i=0}^{4} a_i t^i \quad (5.31)$$

with $a_i = \frac{1}{i!} \frac{\partial^i y(t)}{\partial t^i}\bigg|_{t=0}$.

Let $t = a_4^{1/4} x - n$; Eq. (5.31) can be transformed into a standard form of cusp catastrophe, $v(x) = x^4 + px^2 + qx$, in which $n = -\frac{a_3}{4a_4^{3/4}}$, $p = 6n^2 + \frac{3na_3}{a_4^{3/4}} + \frac{a_2}{a_4^{1/2}}$, $q = 4n^3 + \frac{3n^2 a_3}{a_4^{3/4}} + \frac{2na_2}{a_4^{1/2}} + \frac{a_1}{a_4^{1/4}}$, with the help of variable substitution. And then, F value can be derived from Eq. (5.30) on the basis of the obtained p and q, to predict stability of slope rock at time t.

7.4 Engineering Application of the New Forecast Method

Generally, the evolutionary process of slope catastrophe is relatively complicated, and it needs to set the observation network monitoring the slope displacement to master the deformation of the slope. With the nonworking slope of an opencast coal mine in Inner Mongolia as a study area, observation network of surface deformation is established, as shown in Fig. 5.63.

The monitoring line is usually arranged in the central part of the sliding mass, along the direction of the maximum sliding velocity that predicted, in most cases roughly being perpendicular to the strike direction of the opencast slope. On every monitoring line we arrange two to four points, and we also set a relatively fixed point distal to the slope every two to three monitoring lines. Interval distance between the monitoring line and the observation stake is 15—30 m. For slopes that have different deformation failure mechanisms, monitoring points of slope displacement are usually set according to the following principles: (1) for landslide of creepage sliding-tension type and sliding-tension type, monitoring point displacement and crack depth data on posterior border of the slope top are always the forecast object selection and the monitoring focus on the tension crack; (2) for a sliding-bending landslide, displacement data of the monitoring point on the frontal-bending uplift part are the forecast objects, and the uplift part is the monitoring emphasis; (3) for the bedding displacement—bending failure slope, uplifts in front of the basal slope are the key parts restricting slope stability, and time prediction of failure based on displacement monitoring data of these parts is close to the actual failure time of the model.

Research presented in this chapter focuses on landslide forecast based mainly on the monitoring point displacement that is near the tension crack of the posterior border of a slope top and the parts of the frontal-bending uplift. An observation station was established in the nonworking wall of the opencast coal mine slope from October 2005, where the monitoring data of the slope surface displacement of stake No. 3 from January 2006 to February 2007 is shown in Table 5.9.

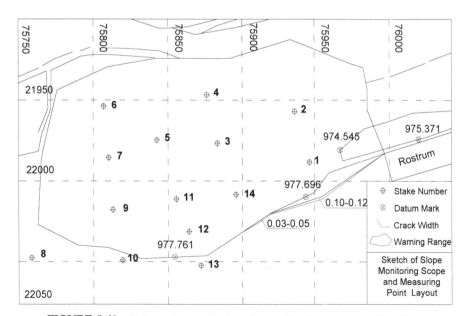

FIGURE 5.63 Deformation monitoring of ground surface in nonworking slope.

TABLE 5.9 Observation Values of Stake No. 3

Observation time/date	2006.1.18	2.02	2.14	2.28	3.21	4.03	4.19	5.03	5.10	5.31
Horizontal displacement/mm	1.017	1.356	1.592	1.644	1.7	1.75	1.98	2.298	6.778	7.328
Vertical displacement/mm	0.862	0.998	1.05	1.064	1.06	1.098	1.27	1.602	5.141	5.164
Observation time/date	2006.6.13	7.05	8.07	8.29	9.11	10.12	11.09	07.2.12	2.26	
Horizontal displacement/mm	7.556	8.309	8.598	9.02	9.088	9.222	9.289	9.356	9.424	
Vertical displacement/mm	5.338	6.024	6.058	6.218	7.123	8.061	7.156	7.318	7.48	

The process of forecast is:

1. Regression coefficient β and regression function $g(\cdot)$ are obtained by processing the monitoring value with the help of the partial linear model of longitudinal data.
2. The partial linear model of longitudinal data, $y = f(t)$, which can be used to forecast the value of x and y on March 12, 2007, is derived after getting the relation between x and t with B-spline method.
3. The partial linear model of longitudinal data from March 12 can be obtained by listing the predictive value of x and y into the frontal observation series and repeating steps 1 and 2.
4. With the help of Taylor series expansion of the forecast model obtained from step 3 and the help of expression $a_i = \frac{1}{i!}\frac{\partial^i y(t)}{\partial^i t}\bigg|_{t=0}$, it can be obtained that $a_0 = 4.134$, $a_1 = -0.00413$, $a_2 = 0.00476$, $a_3 = 0.00652$, $a_4 = 0.000213$.
5. From a_i it can be calculated that $p = 0.32615$, $q = -0.03419$, which lead to $F = 8p^3 + 27q^2 = 1.6861 > 0$. by substitution of p and q into Eq. (5.30). As $F > 0$, it goes to the conclusion that stake No. 3 would be stable on March 12 and no landslide would take place.
6. By repeating steps 1–5, it is calculated to be $p = -0.19753$, $q = -0.05819$, $F = 8p^3 + 27q^2 = -3.1514 < 0$, on March 26. Then, a judgment of instability of stake No. 3 and landslide on the day can be made because

TABLE 5.10 Stability Decision of all Stakes in Nonworking Slope

Time	March 12						March 26					
Stake Number	F Value	Stability State	Stake Number	F Value	Stability State	Stake Number	F Value	Stability State	Stake Number	F Value	Stability State	
3	1.686	Stable	12	5.762	Stable	3	−3.151	Unstable	12	3.161	Stable
4	2.631	Stable	14	2.895	Stable	4	1.414	Stable	14	1.534	Stable
5	3.142	Stable	16	4.761	Stable	5	1.798	Stable	16	2.913	Stable
6	3.842	Stable	17	2.581	Stable	6	2.103	Stable	17	1.313	Stable
7	2.943	Stable	18	5.330	Stable	7	1.642	Stable	18	2.702	Stable
8	4.867	Stable	19	5.798	Stable	8	2.562	Stable	19	3.362	Stable
9	4.314	Stable	20	3.661	Stable	9	2.342	Stable	20	1.943	Stable

of $F < 0$. And the corresponding actual situation of the opencast slope on March 29 was that stake No. 3 slid and the monitoring point was lost. In the same way, F values of other observation stakes on March 12 and March 26 can be obtained as listed in Table 5.10.

It is seen from Table 5.10 that on March 12 for all stakes $F > 0$, the slope would be stable and no landslide would take place; on March 26, for stake No. 3 $F < 0$, for other stakes $F > 0$, there would not be whole slope sliding but only rock movement in a small range near the stake No. 3, which is corresponds to the actual situation. For large and middle-scale landslide, one monitoring point is not enough to control its dynamic changes, but several or more. Due to various reasons, dynamic displacements of monitoring points cannot be synchronous, not even identical, so the forecast results obtained from monitoring data of different points may be different. In such a case, there is no other choice but to select concentrated periods of most points as the forecast results.

7.5 Conclusions

Based on monitoring data of slope displacement, a new method for landslide catastrophe forecast is proposed by combining longitudinal data analysis methods and statistics with nonlinear catastrophic theory. The feasibility of this new method is verified through monitoring data of an opencast coal mine slope. Field practice shows that the new method for landslide catastrophe forecast enriches the theory of landslide forecast.

References

Brett, D., Ahmedzeki, A., Longey, R., 2010. Geotechnical design of main creek tailings dam. Australian Geomechanics Journal 45 (3), 31–40.
Chakraborty, D., Choudhury, D., 2009. Investigation of the behavior of tailings earthen dam under seismic conditions. American Journal of Engineering and Applied Sciences 2 (3), 559–564.
Chen, D.Q., Wang, L.G., Li, G., 2008. Analysis of tailing dam's stability. Journal of Liaoning Technical University (Natural Science) 27 (3), 349–361.
Cheng, X.J., Cai, M.F., 1997. A grey-catastrophe model and its application in acoustic emission monitoring. China Mining Magazine 6 (2), 37–39.
Cheng, Y.M., Zhu, L.J., 2004. Unified formulation for two dimensional slope stability analysis and limitations in factor of safety determination. Soils and Foundations 44 (6), 121–127.
Dai, G., Zhang, R.G., 2013. On portal construction technique of Beileyuan shallow tunnel under unsymmetrical pressure. Shanxi Architecture 39 (15), 174–176.
Ding, X.M., Wang, A.H., Liu, H.L., Wu, Y.D., Kong, G.Q., 2013. Disaster controlling technology for soft ground deformation of existing highway. Disaster Advances 6 (S2), 170–175.
Duan, B.F., Li, L., 2012. Grouting reinforcement technique for subsurface excavation of group metro tunnels construction in complex environment. Disaster Advances 5 (4), 105–109.
Feng, J.Y., Zhu, J.M., Liu, X.Q., Zhang, H.T., 2009. Determination of reasonable horizontal distance between open-off-cut and slope under the combined mining mode of open-pit with underground mining. Journal of Mining & Safety Engineering 26 (1), 66–69.
Griffiths, D.V., Fenton, G.A., 2004. Probabilistic slope stability analysis by finite elements. Journal of Geotechnical and Geoenvironmental Engineering 130 (5), 507–518.
Griffiths, D.V., Lane, P.A., 1999. Slope stability analysis by finite elements. Geotechnique 49 (3), 387–403.
Hage, C.F., Shahrour, I., 2008. Numerical analysis of the interaction between twin-tunnels: influence of the relative position and construction procedure. Tunnelling and Underground Space Technology 23 (2), 210–214.
He, G.L., Hong, F., Wang, Y.P., 2007. Mechanical analysis of coal pillar and roof system in goaf. Journal of Architecture and Civil Engineering 24 (1), 31–35.
Hu, M.J., Guo, A.G., Chen, S.Y., 2004. Reflections on anti-slide stability analysis of a gangue dam. Rock and Soil Mechanics 25 (5), 769–773.
Kang, H.Z., Jia, K.W., Ma, W.H., 2013. Experimental study on compressive strength and elastic modulus of ferrous mill tailing concrete. Journal of Engineering Science and Technology Review 6 (5), 123–128.
Karakus, M., Ozsan, A., Basarir, H., 2007. Finite element analysis for the twin metro tunnel constructed in Ankara Clay, Turkey. Bulletin of Engineering Geology and the Environment 66 (1), 71–79.

Li, B., 2011. Study on disease characteristics of unsymmetrical pressure tunnel and its control measures. Journal of Railway Science and Engineering 8 (6), 59–63.

Li, G.Z., Li, P.L., Xu, H.D., 2005. Tailing dam stability analysis based on structural confidence level. Gold 26 (6), 48–50.

Li, M., Shi, J.X., Wang, C.X., Tang, H.M., Ye, S.Q., Zhu, H., 2006. Study on coupling effect and coupling parameter of the landslide. Chinese Journal of Rock Mechanics and Engineering 25 (S1), 2650–2655.

Li, Q., Zhang, L.T., Qi, Q.L., Zhou, Z.L., 2012. Instability characteristics and stability analysis of a tailings dam based on fluid-solid coupling theory. Rock and Soil Mechanics 33 (S2), 243–250.

Lin, X.H., Carrol, R.J., 2001. Semiparametric regression for clustered data using generalized estimating equations. Journal of the American Statistic Association 96, 1045–1056.

Lou, J.D., Li, Q.Y., Chen, B., 2005. Tailing-dam math simulation and stability analysis. Journal of Hunan University of Science & Technology (Natural Science Edition) 20 (2), 58–61.

Li, W.L., Fu, B.L., Yang, Z.A., 2000. Solution of a bengding thick rectangular plate with four free edges on an elastic foundation. Journal of Tianjin University 33 (1), 72–76.

Lin, H., Cao, P., Gong, F.Q., 2007. Analysis of locations and displacement modes of monitoring points in displacement mutation criteria. Chinese Journal of Geotechnical Engineering 29 (9), 1433–1437.

Liu, J.L., Luan, M.T., Zhao, S.F., Yuan, F.F., Wang, J.L., 2005. Discussion on criteria for evaluating stability of slope in elastoplastic FEM based on shear strength reduction technique. Rock and Soil Mechanics 26 (8), 1345–1348.

Liu, Y.R., Tang, H.M., 1999. Rock Mechanics. China University of Geosciences Press, Wuhan.

Lv, W., Ji, X.M., 2012. Numerical simulation of double liquid grouting reinforcement for shallow-buried tunnel. Journal of Shenyang Jianzhu University (Natural Science) 35 (2), 225–229.

Ma, J.Y., Liu, P.P., Wang, S.R., Zhu, J.M., 2009. Optimized analysis on boundary distance of mine with surface and underground combined mining. Coal Science and Technology 7 (10), 27–30.

Ng, C.W., Lu, H., Peng, S.Y., 2013. Three-dimensional centrifuge modelling of the effects of twin tunnelling on an existing pile. Tunnelling and Underground Space Technology 35, 189–199.

Niu, C.J., 1990. Modern Theory and Practice of Surface Mining. Science Press, Beijing.

Qian, M.G., Miao, X.X., Xu, J.L., 1996. Theoretical study of key stratum in ground control. Journal of China Coal Society 21 (3), 225–230.

Ren, L.M., Liu, X.Q., Yang, Y.Q., 2007. Simulation research on the slope failure of horizontal thick coal seam under the condition of combined mining. China Mining Magazine 16 (12), 72–73.

Stead, D., Eberhardt, E., Coggan, J.S., 2006. Developments in the characterization of complex rock slope deformation and failure using numerical modelling techniques. Engineering Geology 83 (2), 217–235.

Shakesby, R.A., Richard, W.J., 1991. Failure of a mine waste dump in Zimbabwe: causes and consequences. Environmental Geology and Water Sciences 8 (2), 143–153.

Shu, Z.L., Xu, D.X., Xu, K., Liu, X.R., Li, S., 2010. Study on construction of portal section of half-buried tunnel. Chinese Journal of Underground Space and Engineering 6 (5), 1060–1064.

Sjödahl, P., Dahlin, T., Johansson, S., 2005. Using resistivity measurements for dam safety evaluation at enemossen tailings dam in southern Sweden. Environmental Geology 49 (2), 267–273.

Strachan, C., 2001. Tailings dam performance from USCOD incident-survey data. Mining Engineering 53 (3), 49–53.

Sun, F., Zhang, D.L., Chen, T.L., 2011. Fracture grouting mechanism in tunnels based on time-dependent behaviors of grout. Chinese Journal of Geotechnical Engineering 33 (1), 88–93.

Tian, P., Xue, L.G., 2008. The rth mean consistency of estimators in semiparametric regression model for longitudinal data. Mathematica Applicata 21 (3), 535–541.

Wang, J.A., Shang, X.C., Liu, H., 2008a. Study on fracture mechanism and catastrophic collapse of strong roof strata above the mined area. Journal of China Coal Society 33 (8), 851–855.

Wang, S.R., Wei, X., He, M.C., 2008b. New computational method to the stability factor of the 3D slope and its application to mining engineering. Journal of Mining & Safety Engineering 25 (3), 277–280.

Wang, S.R., 2005. Research on Stability and Countermeasure of Slope Engineering in Antaibao Mine. University of Science and Technology Beijing, Beijing.

Wang, S.R., Feng, J.L., 2009. 3–D deformation effect and optimal excavated design of surface mine under mining engineering. Journal of coal science and engineering 15 (4), 361–366.

Wang, S.R., Zhou, H.B., Wu, C.F., Liu, C.Y., 2007. Research on rock mechanics parameters by using comprehensive evaluation method of rock quality grade-oriented. Rock and Soil Mechanics 28 (S1), 202–206.

Wu, J.P., Zhu, J.M., Chen, X.Y., 2008a. Simulation research on optimization of boundary parameter under the condition of open-underground mining. China Mining Magazine 17 (9), 79–82.

Wu, Y.M., Wang, F., Zou, Z.M., Lu, K.C., 2008b. Mechanical character analysis and in-situ monitoring for portal section of half-buried tunnel. Chinese Journal of Rock Mechanics and Engineering 27 (S1), 2873–2882.

Xu, J.L., Qian, M.G., 2001. Study and application of dominant stratum theory for control of strata movement. China Mining Magazine 10 (6), 54–56.

Yang, H.C., Deng, J.W., 2012. The selection of supplemental grouting methods for disaster prevention in rapid transit underground tunnel excavation. Disaster Advances 5 (4), 271–277.

Yang, J.H., Wang, S.R., Li, C.L., Li, Y., 2013. Disturbance deformation effect on the existing tunnel of asymmetric tunnels with small spacing. Disaster Advances 6 (13), 269–277.

Yin, G.Z., Li, Y., Wei, Z.A., Jing, X.F., Zhang, Q.G., 2010. Regularity of the saturation lines' change and stability analysis of tailings dam in the condition of flood. Journal of Chongqing University (Natural Science Edition) 33 (3), 72–86.

Zhang, Z., Mo, H.J., Tan, Y., 2010. Analysis of the bolt mechanical behavior during the half-dark arch tunnel construction. Sichuan Architecture 30 (3), 178–181.

Zhu, H., Du, J.H., 2012. Theory analysis and engineering application of tunnel surrounding rock grouting. Journal of Shenyang Jianzhu University (Natural Science) 28 (3), 498–500.

Further Reading

Wang, S.R., Zhang, H.Q., Zou, Z.S., Wang, P., Yu, T., 2015. Potential risk analysis of tailings dam under preloading condition and its countermeasures. Journal of Engineering and Technological Sciences 47 (1), 46–56.

Wang, S.R., Zhang, Y.B., 2009. New method for landslide forecast and its application based on longitudinal data and catastrophic theory. Journal of the China Coal Society 34 (5), 640–644.

Wang, S.R., Chang, M.S., 2012. Reliability analysis of lining stability for hydraulic tunnel under internal water pressure. Disaster Advances 5 (4), 166–170.

Yang, J.H., Wang, S.R., Zhang, H.Q., Cao, C., 2014. Particle-scale analysis of key technologies of the cut-and-cover tunnel on the slope. Journal of Engineering Science and Technology Review 7 (4), 46–52.

CHAPTER 6

Mining Geomechanics

1. ANALYTICAL ANALYSIS OF ROOF-BENDING DEFLECTION

1.1 Introduction

The deformation properties and instability mechanism of the roof in the shallow mined-out areas are the current challenging problems which should be worked out urgently in engineering practice (Wang et al., 2009).

Some scholars treat the roof as elastic rock beam for analytical analysis and research of the roof deformation characteristics of the mined-out areas. For example, Zhang (2009) simplified the roof as elastic beam and analyzed the creep process based on the creeping damage theory. Jiang et al. (2009) improved the beam model and discussed the influence to the roof thickness caused by horizontal stress and rock fractures based on structure stability theory and damage theory. Qin and Wang (2005) regarded the roof as elastic beam and analyzed the instability process of the mechanical system of stiff roof and coal pillar using the catastrophe theory. Swift and Reddish (2002) considered the roof as elastic beam and analyzed the stability factors to the roof in the mined-out areas. The differences between the elastic beam hypothesis and the practical roof made it difficult to reflect the actual conditions of the roof stress and its deformation.

Other scholars treated the roof as an elastic thin plate to do such analysis and research. For example, Lin et al. (2008) analyzed the overlying strata and obtained the strength condition of the key layer breaking instantly through the elastic thin slab theory and plastic limit analysis method. Wang et al. (2008) regarded the roof as thin plate and then analyzed the roof fracture process and its effect on the collapse of the mined-out areas. Li et al. (2010) simplified the overlying strata of the steeply inclined seam as thin plate, obtained the prediction deformation formula based on the thin plate bending theory with elasticity mechanics. Li et al. (2008) treated the stiff roof as elastic thin plate to study the stress distribution rules and the fracture mechanism of the roof while considering the initial boundary conditions and the periodical ground pressure. Liu et al. (2009) assumed the roof as thin plate and explored the stability of the stiff roof and pillar system. In practical engineering, the research was limited to using thin plate theory because the roof of the mined-out areas was usually in the state of the thick plate.

Although some scholars regarded the roof as thick plate and achieved some conclusions (Zhu et al., 2006; He, 2009), the numerous problems need to be solved urgently through engineering practice. The authors assumed the roof as thick plate and analyzed the characteristic of roof bending deflection with mixed boundary condition under uniform load by the reciprocal theorem method based on Reissner's thick plate theory to provide the technological supports for practical projects.

1.2 The Computational Model

In the northeast edge of the Antaibao Surface Mine, there were the mined-out areas of Jingyang mine exploited by the room-and-pillar stoping method. Based on the field investigation and the collected information, the engineering model of the mined-out areas was built as in Fig. 6.1.

The roof was usually regarded as the thick plate with four fixed edges or with a simple support, which were different depending on boundary conditions. Thus, it was more appropriate to simplify the roof as the thick plate with mixed boundary conditions (Fig. 6.1B); the engineering mechanics model is shown in Fig. 6.2.

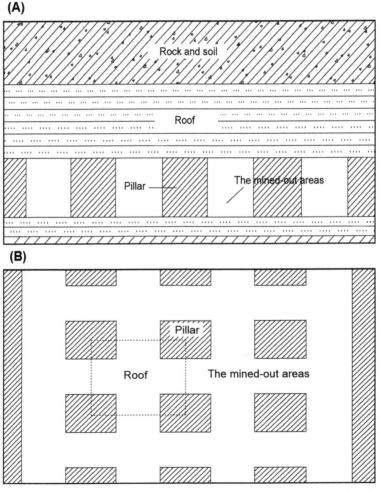

FIGURE 6.1 The sketch of engineering model. (A) Section map of the mined-out areas; (B) overhead view of the mined-out areas.

1. ANALYTICAL ANALYSIS OF ROOF-BENDING DEFLECTION

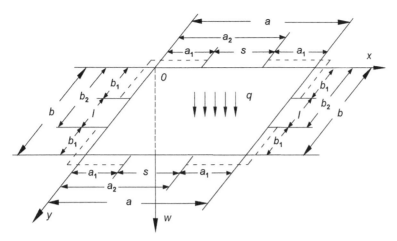

FIGURE 6.2 The sketch of engineering mechanics model (real system of thick plate with mixed boundary).

In the engineering mechanics model, it was assumed that the long side of the thick rectangular plate is a; the short side is $b (b \leq a)$; the supporting length along the long side direction of the plate is a_1; the free length is s; the supporting length along the short side direction is b_1; and the free length is l. The thickness of the roof is h; its elastic modulus is E; and the Poisson's ratio is v. The dead load of the overlying strata is regarded as uniform load q which is distributed on the upper surface of the roof. Thus, the engineering mechanics model was generalized as in Fig. 6.2.

1.3 Deriving Formulas for the Roof Bending Deflection

1.3.1 Basic Equation

On the base of Reissner's thick plate theory, the basic static equations are shown as follows:

$$D\nabla^4 W = q(x,y) - \frac{2-v}{1-v}\frac{h^2}{10}\nabla^2 q(x,y) \quad (6.1)$$

$$\nabla^2 \varphi - \left(\frac{10}{h^2}\right)\varphi = 0 \quad (6.2)$$

where D is the flexural rigidity of the roof; ∇ is Laplace operator; W is the deflection of the roof; $q(x,y)$ is the uniform load distributed on the roof; v is the Poisson's ratio; h is the thickness of the roof; and φ is the stress function.

$$D = \frac{Eh^3}{12(1-v^2)} \quad (6.3)$$

where E is the elastic modulus of the roof.

1.3.2 Basic System of the Thick Plate

Fig. 6.3 shows the basic system of the thick rectangular plate supported under concentrated loads, that is, the transverse two-dimensional

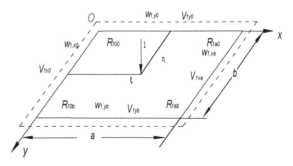

FIGURE 6.3 Basic system for the thick rectangular plate.

(2D) Dirack-Delta function $\delta(x - \xi, y - \eta)$ acts in the convective coordinate (ξ, η) of the thick plate with four edges simply supported. And w_{1x0}, w_{1xa}, w_{1y0}, and w_{1yb} are the four-sided rotation angles of the basic system, respectively. V_{1x0}, V_{1xa}, V_{1y0}, and V_{1yb} are the equivalent shear forces, respectively. R_{100}, R_{1a0}, R_{1ab}, and R_{10b} are the corner loads, respectively (the subscript 1 of w_{1x0} means basic system and the subscript $x0$ means the edge with $x = 0$; the rest have the nearly same meanings).

1.3.3 Real System of the Thick Plate

As seen from Fig. 6.3, the thick plate with the center free and the rest with four edges supported under uniform load, that is, the real system of the thick rectangular plate with a mixed boundary. Deflections and twist angles are w_{x0}, w_{xa}, w_{y0}, w_{yb}, ω_{x0}, ω_{xa}, ω_{y0}, and ω_{yb}, respectively.

For the real symmetric system, the deflection w_{y0} and twist angle ω_{y0} on free boundary $y = 0$ are equal to w_{yb} and ω_{yb} on free boundary $y = b$, and the deflection w_{x0} and twist angle ω_{x0} on free boundary $x = 0$ are equal to w_{xa} and ω_{xa} at free boundary $x = a$. Thus, the deflection formula is written as follows:

$$W = W(\xi, \eta) \quad (6.4)$$

The deflection equations are:

$$w_{y0} = w_{yb} = \sum_{i=1,3}^{\infty} A_i \cdot \sin\frac{i\pi(x - a_1)}{s} \quad (6.5)$$

$$w_{x0} = w_{xa} = \sum_{j=1,3}^{\infty} B_j \cdot \sin\frac{j\pi(y - b_1)}{l} \quad (6.6)$$

The twist angles are:

$$\omega_{yx0} = \omega_{yxa} = \sum_{k=1,3}^{\infty} C_k \cdot \sin\frac{k\pi(y - b_1)}{l} \quad (6.7)$$

$$\omega_{xy0} = \omega_{xyb} = \sum_{f=1,3}^{\infty} D_f \cdot \sin\frac{f\pi(x - a_1)}{s} \quad (6.8)$$

where $x \in [a_1, a_2]$, $y \in [b_1, b_2]$, $a_2 = a_1 + s$, $b_2 = b_1 + l$.

The stress function is:

$$\varphi(\xi, \eta) = \sum_{n=1,3}^{\infty} \left[E_n \text{ch}\delta_n \xi + F_n \text{ch}\delta_n(a - \xi) \right] \cos \beta_n \eta$$

$$+ \sum_{m=1,3}^{\infty} \left[G_m \text{ch}\gamma_m \eta \right.$$

$$\left. + H_m \text{ch}\gamma_m(b - \eta) \right] \cos \alpha_m \xi \quad (6.9)$$

Where $\alpha_m = \frac{m\pi}{a}$, $\beta_n = \frac{n\pi}{b}$, $\gamma_m = \sqrt{\alpha_m^2 + \frac{10}{h^2}}$, $\delta_n = \sqrt{\beta_n^2 + \frac{10}{h^2}}$. A_i, B_j, C_k, D_f, E_n, F_n, G_m, and H_m are undetermined coefficients.

1.3.4 Driving the Deflection Equation

The reciprocal theorem method is used between the basic system and the real system, and the result is written as follows:

$$W(\xi, \eta) + \int_{a1}^{a2} Q_{1yb} \cdot w_{yb} dx - \int_{a1}^{a2} Q_{1y0} \cdot w_{y0} dx$$

$$+ \int_{b1}^{b2} Q_{1xa} \cdot w_{xa} dy - \int_{b1}^{b2} Q_{1x0} \cdot w_{x0} dy$$

$$+ \int_{a1}^{a2} M_{1xyb} \cdot \omega_{xyb} dx - \int_{a1}^{a2} M_{1xy0} \cdot \omega_{xy0} dx$$

$$+ \int_{b1}^{b2} M_{1yxa} \cdot \omega_{yxa} dy - \int_{b1}^{b2} M_{1yx0} \cdot \omega_{yx0} dy$$

$$= \int_0^a \int_0^b q \cdot W1(x, y; \xi, \eta) dx dy$$

$$(6.10)$$

Substituting the related boundary values Eq. (6.5) to Eq. (6.8) of the real system, the related boundary values Q_{1y0}, Q_{1yb}, Q_{1x0}, Q_{1xb}, M_{1xy0}, M_{1xyb}, M_{1yx0}, M_{1yxa}, and the basic solution $W_1(x,y,\xi,\eta)$ (Fu and Tan, 1995) of the basic system into Eq. (6.10). Then the deflection equation is reached as follows after calculating:

$$H_{nk(b,l)} = \frac{(-1)^k \sin\frac{b_1+s}{a}n\pi - \sin\frac{b_1}{a}n\pi}{n^2 - \left(\frac{b}{l}k\right)^2} \quad (6.15)$$

The boundary conditions of the real system are as follows:

$\eta = 0, b$ and $\xi \in [a_1, a_2]$: $M_{\xi\eta} = 0$, $Q_\eta = 0$, $M_\eta = 0$; $\xi = 0, a$ and $\eta \in [b_1, b_2]$: $M_{\eta\xi} = 0$, $Q_\xi = 0$, $M_\xi = 0$.

$$W(\xi,\eta) = \frac{4q}{Da}\sum_{m=1,3}^{\infty}\left\{1 + \frac{1}{2\cosh\frac{1}{2}\alpha_m b}\left[\alpha_m\left(\eta-\frac{b}{2}\right)\sinh\alpha_m\left(\eta-\frac{b}{2}\right) - \left(2+\frac{1}{2}\alpha_m\tanh\frac{1}{2}\alpha_m b\right)\cosh\alpha_m\left(\eta-\frac{b}{2}\right)\right]\right\}\frac{1}{\alpha_m^5}\sin\alpha_m\xi$$

$$-\frac{2l}{\pi b}\sum_{j=1,3}^{\infty}j\cdot B_j\sum_{n=1,3}^{\infty}\frac{[\operatorname{sh}\beta_n(a-\xi)+\operatorname{sh}\beta_n\xi]}{\operatorname{sh}\beta_n a}\sin\beta_n\eta\cdot\phi_{jn}(b,l) - \frac{2s}{\pi a}\sum_{i=1,3}^{\infty}i\cdot A_i\sum_{m=1,3}^{\infty}\frac{[\operatorname{sh}\alpha_m(b-\eta)+\operatorname{sh}\alpha_m\eta]}{\operatorname{sh}\alpha_m b}\sin\alpha_m\xi\cdot\phi_{im}(a,s)$$

$$+\frac{(1-\upsilon)}{\pi}\sum_{n=1,3}^{\infty}\left\{\beta_n a\operatorname{cth}\beta_n a[\operatorname{sh}\beta_n(a-\xi)+\operatorname{sh}\beta_n\xi]-\beta_n(a-\xi)\operatorname{ch}\beta_n(a-\xi)-\beta_n\xi\operatorname{ch}\beta_n\xi\right\}\frac{n\cdot\sin\alpha_m\xi}{\alpha_m\operatorname{sh}\alpha_m b}\cdot\sum_{k=1,3}^{\infty}C_k$$

$$\cdot H_{nk}(b,l) + \frac{(1-\upsilon)}{\pi}\sum_{m=1,3}^{\infty}\left\{\alpha_m b\operatorname{cth}\alpha_m b[\operatorname{sh}\alpha_m(b-\eta)+\operatorname{sh}\alpha_m\eta]-\alpha_m(b-\eta)\operatorname{ch}\alpha_m(b-\eta)-\alpha_m\eta\operatorname{ch}\alpha_m\eta\right\}\frac{m\cdot\sin\beta_n\eta}{\beta_n\operatorname{sh}\beta_n a}\cdot\sum_{f=1,3}^{\infty}D_f$$

$$\cdot H_{mf}(a,s)$$

(6.11)

where:

$$\phi_{im(a,s)} = \frac{(-1)^i \sin\frac{a_1+s}{a}m\pi - \sin\frac{a_1}{a}m\pi}{i^2 - \left(\frac{s}{a}m\right)^2} \quad (6.12)$$

$$\phi_{jn(b,l)} = \frac{(-1)^j \sin\frac{b_1+l}{b}n\pi - \sin\frac{b_1}{b}n\pi}{j^2 - \left(\frac{l}{b}n\right)^2} \quad (6.13)$$

$$H_{mf(a,s)} = \frac{(-1)^f \sin\frac{a_1+s}{a}m\pi - \sin\frac{a_1}{a}m\pi}{m^2 - \left(\frac{a}{s}f\right)^2} \quad (6.14)$$

Due to the twist angles assumed in the corners of each edge, that is ω_{x0}, ω_{xa}, ω_{y0}, and ω_{yb}. The related equations on the boundary should be satisfied $\eta = 0, b$: $\omega_\xi = \omega_{\xi\eta 0}$ and $\xi = 0, a$: $\omega_\eta = \omega_{\eta\xi 0}$.

Where $M_\eta = 0$ and $M_\xi = 0$ have been satisfied. For the symmetric conditions $E_n = -F_n$, $G_m = -H_m$. Therefore, we can get six equations according to the six boundary conditions and obtain six undetermined coefficients. The deflection, shear, moment, and rotation angle are showed as follows:

$$\frac{4q}{b} \cdot \frac{\tanh\frac{a \cdot \beta_n}{2}}{\beta_n^2} + \frac{16D(v-1)}{b\pi} \sum_{m=1,3}^{\infty} \frac{\beta_n^2 \cdot m \cdot \alpha_m^2}{\alpha_m^2 + \beta_n^2} \sum_{f=1,3}^{\infty} D_f \frac{\sin\frac{m\pi a_1}{a}}{m^2 - \left(\frac{af}{s}\right)^2} + \frac{4D(v-1)}{\pi} \cdot \frac{(1-\cosh\beta_n a)n\beta_n^2}{\sinh\beta_n a} \sum_{k=1,3}^{\infty} C_k$$

$$\cdot \frac{\sin\frac{n\pi b_1}{b}}{n^2 - \left(\frac{bk}{l}\right)^2} - E_n \beta_n (1-\cosh\delta_n a) + \frac{4}{b} \sum_{m=1,3}^{\infty} \frac{G_m \beta_n \gamma_m \sinh\gamma_m b}{\alpha_m^2 + \beta_n^2} = 0 \qquad n = 1, 3, \cdots$$

(6.16)

$$\frac{4q}{a} \cdot \frac{\tanh\frac{b \cdot \alpha_m}{2}}{\alpha_m^2} + \frac{16D(v-1)}{a\pi} \sum_{n=1,3}^{\infty} \frac{\beta_n^2 \cdot n \cdot \alpha_m^2}{\alpha_m^2 + \beta_n^2} \cdot \sum_{f=1,3}^{\infty} C_k \cdot \frac{\sin\frac{n\pi b_1}{b}}{n^2 - \left(\frac{bk}{l}\right)^2} + \frac{4D(v-1)}{\pi} \cdot \frac{(1-\cosh\alpha_m b)m\alpha_m^2}{\sinh\alpha_m b} \sum_{f=1,3}^{\infty} D_f$$

$$\cdot \frac{\sin\frac{m\pi a_1}{a}}{m^2 - \left(\frac{af}{s}\right)^2} + G_m \alpha_m (1-\cosh\gamma_m b) - \frac{4}{a} \sum_{m=1,3}^{\infty} \frac{E_n \alpha_m \delta_n \sinh\delta_n a}{\alpha_m^2 + \delta_n^2} = 0 \qquad m = 1, 3, \cdots$$

(6.17)

$$\frac{16}{\pi^2} \sum_{m=1,3}^{\infty} m \cdot \alpha_m \sum_{i=1,3}^{\infty} i \cdot A_i \frac{\left(\sin\frac{m\pi a_1}{a}\right)^2}{\left[i^2 - \left(\frac{sm}{a}\right)^2\right]\left[m^2 - \left(\frac{af}{s}\right)^2\right]} + \frac{16h^2 a}{5s\pi^2} \sum_{m=1,3}^{\infty} m^2 \alpha_m^2 D_f \frac{\left(\sin\frac{m\pi a_1}{a}\right)^2}{\left[m^2 - \left(\frac{af}{s}\right)^2\right]^2} - \frac{4ah^2}{5D\pi s(1-v)}$$

$$\cdot \sum_{m=1,3}^{\infty} m \cdot G_m \gamma_m \sinh\gamma_m b \frac{\sin\frac{m\pi a_1}{a}}{m^2 - \left(\frac{af}{s}\right)^2} + D_f = 0 \qquad f = 1, 3, \cdots$$

(6.18)

1. ANALYTICAL ANALYSIS OF ROOF-BENDING DEFLECTION

$$\frac{16}{\pi^2}\sum_{n=1,3}^{\infty} n\cdot\beta_n \sum_{j=1,3}^{\infty} j\cdot B_j \frac{\left(\sin\frac{n\pi b_1}{b}\right)^2}{\left[j^2-\left(\frac{nl}{b}\right)^2\right]\left[n^2-\left(\frac{bk}{l}\right)^2\right]} + \frac{16h^2 b}{5l\pi^2}\sum_{n=1,3}^{\infty} n^2\beta_n^2 C_k \frac{\left(\sin\frac{n\pi b_1}{b}\right)^2}{\left[n^2-\left(\frac{bk}{l}\right)^2\right]^2} - \frac{4bh^2}{5D\pi l(1-\upsilon)}$$

$$\cdot \sum_{n=1,3}^{\infty} n\cdot E_n \delta_n \sinh\delta_n a \frac{\sin\frac{n\pi b_1}{b}}{n^2-\left(\frac{bk}{l}\right)^2} + C_k = 0 \qquad k=1,3,\cdots \tag{6.19}$$

$$\frac{2q}{b}\left\{\frac{(1-\upsilon)}{\beta_n^3}\left(\tanh\frac{\beta_n a}{2} - \frac{\beta_n a}{2\cosh^2\frac{\beta_n a}{2}}\right) - \frac{\upsilon h^2 \tanh\frac{\beta_n a}{2}}{5\beta_n}\right\} = D(1-\upsilon)\frac{16s}{ab\pi}\sum_{m=1,3}^{\infty}\frac{\alpha_m^3}{\alpha_m^2+\beta_n^2}\sum_{i=1,3}^{\infty} i\cdot A_i \frac{\sin\frac{m\pi a_1}{a}}{i^2-\left(\frac{sm}{a}\right)^2}$$

$$+D(1-\upsilon)\frac{4l}{b\pi}\tanh\frac{\beta_n a}{2}\beta_n^2 \sum_{j=1,3}^{\infty} j\cdot B_j \frac{\sin\frac{n\pi b_1}{b}}{j^2-\left(\frac{nl}{b}\right)^2} - D(\upsilon-1)^2 \frac{16}{b\pi}\sum_{m=1,3}^{\infty}\frac{m\cdot\alpha_m^2 \beta_n^2}{\left(\alpha_m^2+\beta_n^2\right)^2}\sum_{f=1,3}^{\infty} D_f \frac{\sin\frac{m\pi a_1}{a}}{m^2-\left(\frac{af}{s}\right)^2}$$

$$+D(1-\upsilon)\frac{16h^2}{5b\pi}\sum_{m=1,3}^{\infty}\frac{m\cdot\alpha_m^4}{\alpha_m^2+\beta_n^2}\sum_{f=1,3}^{\infty} D_f \frac{\sin\frac{m\pi a_1}{a}}{m^2-\left(\frac{af}{s}\right)^2} + D(1-\upsilon)\frac{4h^2}{5\pi} n\beta_n^3 \tanh\frac{\beta_n a}{2}\sum_{k=1,3}^{\infty} C_k \frac{\sin\frac{n\pi b_1}{b}}{n^2-\left(\frac{bk}{l}\right)^2} \tag{6.20}$$

$$-D(\upsilon-1)^2 \frac{2n}{\pi}\left[\beta_n^2 a + \beta_n(1-\beta_n a\coth\beta_n a)\tanh\frac{\beta_n a}{2}\right]\sum_{k=1,3}^{\infty} C_k \frac{\sin\frac{n\pi b_1}{b}}{n^2-\left(\frac{bk}{l}\right)^2} + \frac{h^2}{5} E_n\left(\beta_n^2+\delta_n^2\right)\sinh^2\frac{\delta_n a}{2}$$

$$-\frac{2h^2}{5b}\sum_{m=1,3}^{\infty} G_m \frac{\alpha_m^2+\gamma_m^2}{\gamma_m^2+\beta_n^2}\gamma_m \sinh\gamma_m b \qquad n=1,3,\cdots$$

$$\begin{aligned}
&\frac{2q}{a}\left\{\frac{(1-\upsilon)}{\alpha_m^3}\left(\tanh\frac{\alpha_m b}{2} - \frac{\alpha_m b}{2\cosh^2\frac{\alpha_m b}{2}}\right) - \frac{\upsilon h^2 \tanh\frac{\alpha_m b}{2}}{5\alpha_m}\right\} = D(1-\upsilon)\frac{4s}{a\pi}\tanh\frac{\alpha_m b}{2}\alpha_m^2 \sum_{i=1,3}^{\infty} i\cdot A_i \frac{\sin\frac{m\pi a_1}{a}}{i^2 - \left(\frac{sm}{a}\right)^2} \\
&+ D(1-\upsilon)\frac{16l}{ab\pi}\sum_{n=1,3}^{\infty}\frac{\beta_n^3}{\alpha_m^2 + \beta_n^2}\sum_{j=1,3}^{\infty} j\cdot B_j \frac{\sin\frac{n\pi b_1}{b}}{j^2 - \left(\frac{nl}{b}\right)^2} - D(\upsilon-1)^2\frac{16}{a\pi}\sum_{n=1,3}^{\infty}\frac{n\cdot \alpha_m^2 \beta_n^2}{\left(\alpha_m^2 + \beta_n^2\right)^2}\sum_{k=1,3}^{\infty} C_k \frac{\sin\frac{n\pi b_1}{b}}{n^2 - \left(\frac{bk}{l}\right)^2} \\
&+ D(1-\upsilon)\frac{16h^2}{5a\pi}\sum_{n=1,3}^{\infty}\frac{n\cdot \beta_n^4}{\alpha_m^2 + \beta_n^2}\sum_{k=1,3}^{\infty} C_k \frac{\sin\frac{n\pi b_1}{b}}{n^2 - \left(\frac{bk}{l}\right)^2} + D(1-\upsilon)\frac{4h^2}{5\pi}m\cdot \alpha_m^3 \tanh\frac{\alpha_m b}{2}\sum_{f=1,3}^{\infty} D_f \frac{\sin\frac{m\pi a_1}{a}}{m^2 - \left(\frac{af}{s}\right)^2} \quad (6.21) \\
&- D(\upsilon-1)^2 \frac{2m}{\pi}\left[\alpha_m^2 b + \alpha_m(1-\alpha_m b\coth\alpha_m b)\tanh\frac{\alpha_m b}{2}\right]\sum_{f=1,3}^{\infty} D_f \frac{\sin\frac{m\pi a_1}{a}}{m^2 - \left(\frac{af}{s}\right)^2} - \frac{h^2}{5}G_m\left(\gamma_m^2 + \alpha_m^2\right)\sinh^2\frac{\gamma_m b}{2} \\
&+ \frac{2h^2}{5a}\sum_{n=1,3}^{\infty} E_n \frac{\delta_n^2 + \beta_n^2}{\alpha_m^2 + \beta_n^2}\beta_n \sinh\beta_n a \quad m = 1, 3, \cdots
\end{aligned}$$

Solving the equations from Eq. (6.16) to Eq. (6.21) will get the coefficients A_i, B_j, C_k, D_f, E_n, and G_m, and then the deflection, shear, moment, and rotation angle can be worked out (Liu, 2012).

As a numerical example, the model was assumed that side length of a square thick plate is $a = b = 10$ m; the length of the supported edge is $a_1 = b_1 = 3$ m; and the length of free section is $s = l = 3$ m. The poison's ratio of the plate is $\upsilon = 0.3$, and its elastic modulus is $E = 50$ GPa. The uniform load acing on the plate is $q = 0.3$ MPa. The deflection is supposed to be obtained when the width-thickness ratio is $h/a = 0.1, 0.2$, and 0.3.

According to these equations, it is easy to get the deflection values of any place by using MATLAB software. Under the MATLAB environment, the deflection values along the z direction can be reached when the points $x/a = 0.5$, $y/b = 0.1, 0.2, \ldots$, and 1.0; $x = 0$, $y/l = 0.1, 0.2, \ldots$, and 1.0 are chosen.

In the process of solving the coefficient equations, the approximation solution with can be obtained with sufficient accuracy through picking up the finite terms as shown in Table 6.1. The

TABLE 6.1 Maximum Deflection and the Relative Errors of Thick Plate at $x/a = 0.5$ and $y/b = 0.5$ With Different c

ANSYS (mm)	c = 40		c = 60		c = 80	
	W (mm)	Error (%)	W (mm)	Error (%)	W (mm)	Error (%)
3.45	3.58	3.97	3.54	2.77	3.52	2.12
0.51	0.54	6.69	0.53	5.48	0.53	4.54
0.20	0.21	5.22	0.21	4.06	0.21	3.55

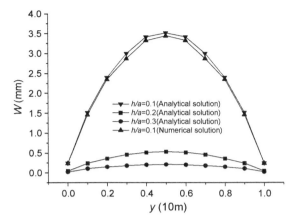

FIGURE 6.4 The deflection curves of thick plate at $x/a = 0.5$.

satisfactory results are obtained to keep relative errors less than 5.0% through calculation when c (c means the coefficients m, n, i, j, k, and f) is equal to 80.

1.4 Results and Analysis

1.4.1 Verification for Analytical Solution

The thick plate with mixed boundary condition under uniform load is simulated by using

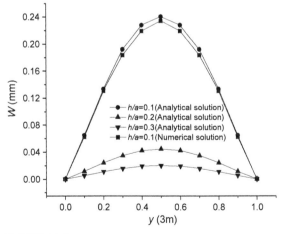

FIGURE 6.5 The deflection curves of thick plate at $x = 0$.

ANSYS. A square plate with side length $a = b = 10$ m is selected, and the thickness–width ratio is $h/a = 0.1$, 0.2, and 0.3, respectively. The numerical model was computed under the same condition under the uniform loads being applied, and the displacement of z direction can be obtained, that is, the deflection values of the roof of the mined-out areas.

As shown in Figs. 6.4 and 6.5, the deflection curves of the thick plate at $x/a = 0.5$ and at $x = 0$ with different thickness–width ratios had been worked out.

The result indicates that whether on the boundary or in the place of the deflection maximum, the error is less than 5.0% between the numerical solution and the analytical solution, which is acceptable in practical engineering.

1.4.2 Comparative of Two Different Boundary Conditions

The square plates of side length $a = b = 10$ m, with four supported edges and with mixed boundary conditions, are selected, and the length of free sections and the supported sections of the thick plate with mixed boundary conditions are $a_1 = b_1 = 3$ m and $s = l = 3$ m, respectively. The uniform load q applied on the upper surface of the plate is equal to 0.3 MPa. The poison's ratio of the plate is $v = 0.3$, and its elastic modulus is $E = 50$ GPa, The thickness–width ratio of the plate is $h/a = 0.1$, 0.2, and 0.3. The results are shown in Figs. 6.6 and 6.7.

1.5 Conclusions

Based on Reissner's thick plate theory, the roof bending deflection with mixed boundary conditions in the mined-out areas under uniform load was analyzed through the reciprocal theorem method. The function of roof bending deflection was achieved and the analytical solution was worked out.

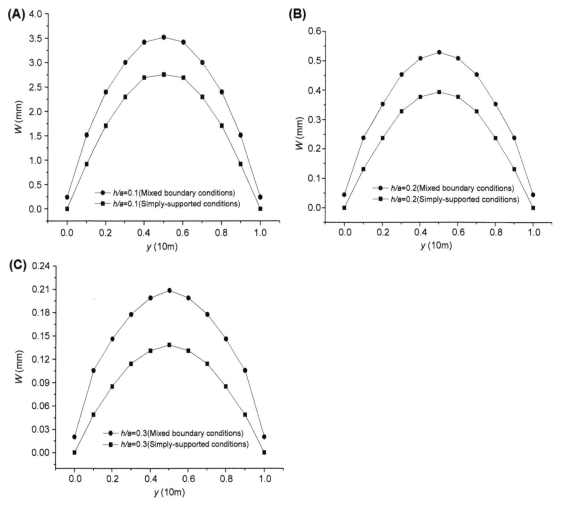

FIGURE 6.6 The deflection curves of thick plate at $x/a = 0.5$. (A) $h/a = 0.1$; (B) $h/a = 0.2$; and (C) $h/a = 0.3$.

Comparing the numerical solution with the analytical solution, the thick plate theory is appropriate for analyzing the characteristics of roof bending deflection under the same condition.

Through analyzing the different characteristics of roof bending deflection with two different boundary conditions, it shows that roof bending deflection with mixed boundary conditions is bigger than that with supported boundary conditions, and the deformation scope of roof-bending deflection with mixed boundary conditions is apparently greater than that of supported boundary conditions with an increase in the thickness—width ratio.

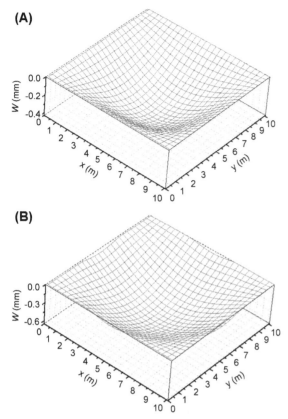

FIGURE 6.7 The deflection surface of thick plate with different boundary conditions. (A) $h/a = 0.2$ thick plate with simply supported boundary conditions; (B) $h/a = 0.2$ thick plate with mixed boundary.

2. ANALYTICAL SOLUTION OF THE ROOF SAFE THICKNESS

2.1 Introduction

Massive shallow mine-out areas are left in the Antaibao Surface Mine. Rock strata has distorted and the roof has degraded with the passage of time, and the roofs of shallow mined-out areas are weathered. During the process of digging downward step-by-step in the surface mine, the overlying strata thickness of the roof of mined-out areas gradually decreases. The mechanical construction equipment formed a loading condition on the surface of the mined-out areas at the same time. Under the mechanical construction load, the roof of the mined-out areas could be activated, leading to the evolution progressive damages and producing new decline and collapse. When the accumulated damage from the roof developed, the miners and the excavation equipment on the surface could unexpectedly fall into the mined-out areas, which was a hidden danger to the workers and greatly influenced safe production.

It is essential to determine the safe thickness of the roof of the mined-out areas reasonably for the surface mine excavation. On the one hand, if the value selected is too large it will lead to increasing drilling, substantial increases in treatment costs to the mined-out areas, and greatly reduced production efficiency; on the other hand, if the value selected is too small, there is not enough security from isolation thickness, the workers and mining equipment are unsafe, and it may lead to heavy casualties and equipment damage, resulting in mine shutdown and other consequences (Tian and Nie, 2009). Therefore, It is of important theoretical meaning and practical value to reasonably determine the safe thickness of the roof and to ensure safe surface mine production.

Currently, for the security evaluation on the roof thickness, there are mainly semiquantitative analyses, numerical analyses, and forecasting methods used in China and abroad (Swift and Reddish, 2002; Wang et al., 2009; Palei and Das, 2009). Because of the rockmass structural parameters, rock engineering parameters, and the complexity and uncertainty of the physical and mechanical parameters of rock, most of the numerical calculation results cannot be directly applied to engineering design. The model prediction method can reflect the safe roof thickness of a variety of factors, but the method requires a lot of testing and monitoring of data for statistical analysis, the different factors vary greatly

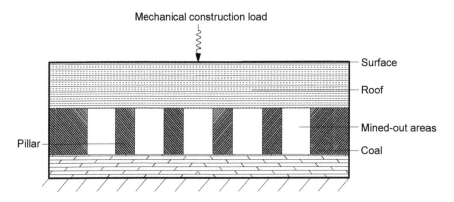

FIGURE 6.8 The sketch of engineering mode.

from one project to another, so the established prediction models have many problems in use and the application of practical engineering. Although the mechanical model has more simplified conditions, the reliability of the traditional semiquantitative analysis is affected to some extent because the method is simple, practical, and workable, so it is still widely applied in practice. In the semiquantitative analysis, many scholars regard the roof of mined-out areas as an elastic beam approximation approach, which greatly simplifies the analysis process, but its clear limitation is that the beam cannot reflect the spatial effect of the roof, and cannot reflect its anisotropy; thus the results should be questioned (Nomikos et al., 2002; Zhang, 2009).

According to the document data, there are many methods for determining safe roof thickness while considering the role of static loading, but the construction dynamic loading conditions are rarely reported. Therefore, based on Reissner's theory, considering the construction machinery loading dynamic and taking rock tensile strength as a control condition, we derived the roof safe thickness formula under concentrated harmonic loading, and the roof safe thickness was verified in the actual engineering practice.

2.2 Generalized Mechanical Model

In the northeast edge of Antaibao Surface Mine lies the mined-out areas of Jingyang mine formed by the room-and-pillar stoping method. Based on the field investigation and the collected information, the engineering model was shown in Fig. 6.8. The overlying strata composed a composite foundation on the mined-out areas by the room-and-pillar mining method with coal pillar rheological weathering; the elastic modulus and the average cross-sectional area of coal pillars would continue to decrease, resulting in a decrease in its effective flexural stiffness. When the coal pillars supporting capacity decreased to a critical value, the midpoint of each side of the roof came into the plastic state, then the formation of plastic hinge expanded. Then we can regard the roof of the mined-out areas as a supported rectangular plate (He et al., 2007).

Among the construction machinery equipment in the Antaibao Surface Mine, the maximum load density of Bucyrus Electrical (BE) shovel is 0.29 MPa. During the construction process, constant shoveling and loading kept the giant shovel arm swinging back and forth. The giant shovel could be considered concentrated harmonic loading, which applied to the roof of the mined-out areas. According to engineering practice, the roof could be regarded as a length, b width of the elastic rectangular plate. For a simple calculation, the dynamic load acted on the center of the roof surface, and the mechanical model is shown in Fig. 6.9.

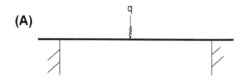

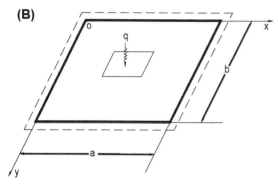

FIGURE 6.9 The sketch of mechanic model (A) Lateral view and (B) Surface view.

2.3 Formula Derivation

2.3.1 Basic Equation

The basic equation on static problems based on Reissner's plate theory is given as follows:

$$D\nabla^4 W = q(x,y) - \frac{2-\nu}{1-\nu}\frac{h^2}{10}\nabla^2 q(x,y) \quad (6.22)$$

where D is the flexural rigidity of plate; ∇ is the Laplace operator; W is the plate deflection; $q(x,y)$ is the load on the plate; ν is the Poisson's ratio; and h is the thickness of the plate.

To solve the shear force, a function ϕ is introduced, which is called the stress function and must satisfy the following equation:

$$\nabla^2 \phi - \frac{10}{h^2}\phi = 0 \quad (6.23)$$

Assuming the concentrated harmonic force P acting on (x_0,y_0), the function is expressed as follows:

$$F(x,y,t) = P\delta(x-x_0, y-y_0)\sin \omega t \quad (6.24)$$

where ω is the vibration frequency and $\delta(x-x_0, y-y_0)$ is the unit concentrated load function acting on (x_0,y_0).

The load function is expanded into the following forms of double trigonometric series:

$$P\delta(x-x_0, y-y_0) = \frac{4p}{ab}\sum_{m=1,2}^{\infty}\sum_{n=1,3}^{\infty} \sin \alpha_m \xi$$

$$\times \sin \beta_n \eta \cdot \sin \alpha_m x_0 \sin \beta_n y_0$$

$$(6.25)$$

where $\alpha_m = \frac{m\pi}{a}, \beta_n = \frac{n\pi}{b}$.

When the plate is forced to vibrate, then the inertial force expression can be written as $q_1 = \rho\frac{\partial^2 W(x,y,t)}{\partial t^2}$, with Eq. (6.24) subtracting q_1, which is substituted into Eq. (6.22). According to D'Alembert theorem, The plate forced vibration control equation is obtained:

$$D\nabla^4 W = F(x,y,t) - \rho\frac{\partial^2 \overline{W}}{\partial t^2}$$

$$-\frac{2-\nu}{1-\nu}\cdot\frac{h^2}{10}\nabla^2\left[F(x,y,t) - \rho\frac{\partial^2 \overline{W}}{\partial t^2}\right]$$

$$(6.26)$$

Excluding $F(x,y,t) = F(x,y,t)\sin \omega t$ damping force interference, the forced vibration steady-state response is assumed:

$$\overline{W}(x,y,t) = W(x,y,t)\sin \omega t \quad (6.27)$$

Substituting Eq. (6.27) into Eq. (6.26), the forced vibration equation under concentrated harmonic loads is:

$$D\nabla^4 W = P\delta(x-x_0, y-y_0) + D\lambda^2 W$$

$$- \frac{kh^2}{10}\nabla^2\left[P\delta(x-x_0, y-y_0) + D\lambda^2 W\right]$$

$$(6.28)$$

where $\lambda^2 = \frac{\rho\omega^2}{D}, k = \frac{2-\nu}{1-\nu}$, ρ is the plate mass density per unit area.

Based on Reissner's plate theory, plate bending moment, shear force, and angle formula can be deduced:

$$M_x = -D\left(\frac{\partial^2 W}{\partial x^2} + \nu\frac{\partial^2 W}{\partial y^2}\right) + \frac{h^2}{5}\frac{\partial Q_x}{\partial x}$$

$$- \frac{h^2}{10}\frac{\nu}{1-\nu}\left(P\delta(x-x_0, y-y_0) + D\lambda^2 W\right)$$

$$M_y = -D\left(\frac{\partial^2 W}{\partial y^2} + \nu \frac{\partial^2 W}{\partial x^2}\right) + \frac{h^2}{5} \frac{\partial Q_y}{\partial y}$$

$$- \frac{h^2}{10} \frac{\nu}{1-\nu} \left(P\delta(x-x_0, y-y_0) + D\lambda^2 W\right)$$

$$Q_x = -D\frac{\partial}{\partial x}\nabla^2 W - \frac{kh^2}{10}\frac{\partial}{\partial x}\left(P\delta(x-x_0, y-y_0)\right.$$

$$\left. + D\lambda^2 W\right) + \frac{\partial \varphi}{\partial y}$$

$$Q_y = -D\frac{\partial}{\partial y}\nabla^2 W - \frac{kh^2}{10}\frac{\partial}{\partial y}\left(P\delta(x-x_0, y-y_0)\right.$$

$$\left. + D\lambda^2 W\right) - \frac{\partial \varphi}{\partial x}$$

$$\omega_x = \frac{\partial W}{\partial x} + \frac{1}{D}\frac{h^2}{5(1-\nu)}Q_x$$

$$\omega_y = \frac{\partial W}{\partial y} + \frac{1}{D}\frac{h^2}{5(1-\nu)}Q_y.$$

2.3.2 Dynamic Fundamental Solutions

As is shown in Fig. 6.10, take a simple supported plate as the basic system. In the current coordinates (ξ, η) acts as the horizontal 2D unit Dirac-δ function $\delta(x-\xi, y-\eta)$ (Li et al., 1998).

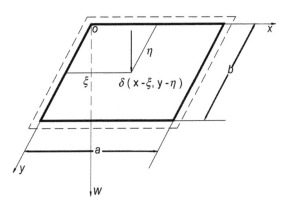

FIGURE 6.10 Basic system of thick plate with four edges supported.

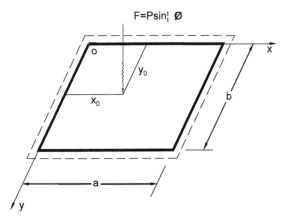

FIGURE 6.11 Real system of thick plate under concentrated load.

Substituting $P\delta(x-\xi, y-\eta)$ into Eq. (6.1), the dynamic fundamental solution of the thick plate can be found:

$$W_1(x, y, \xi, \eta) = \frac{4}{abD} \sum_{m=1,2}^{\infty} \sum_{m=1,2}^{\infty} \frac{\sin \alpha_m \xi \sin \beta_n \eta}{(\alpha_m^2 + \beta_n^2)^2}$$

$$\cdot \sin \alpha_m x \sin \beta_n y$$

(6.29)

Fig. 6.11 shows the real system of thick rectangular plate under the harmonic concentrated load. Applying reciprocal theorem between the real system and basic systems, the deflection expression of the real system can be obtained:

$$W(\xi, \eta) = \int_0^a \int_0^b \left(p\delta(x-x_0, y-y_0)\right.$$

$$\left. + D\lambda^2 W\right) \cdot W_1(x, y, \xi, \eta) - \int_0^a \int_0^b$$

$$\frac{kh^2}{10}\nabla^2 \left[p\delta(x-x_0, y-y_0)\right.$$

$$\left. + D\lambda^2 W\right] \cdot W_1(x, y, \xi, \eta)$$

(6.30)

To make

$$W(\xi, \eta) = \sum_{m=1,3}^{\infty} \sum_{n=1,3}^{\infty} C_{mn} \sin\partial_m \xi \sin \beta_n \eta \quad (6.31)$$

Substituting Eqs. (6.31) and (6.32) into Eq. (6.30), C_{mn} and k_{mn} are obtained.

$$C_{mn} = \frac{40p + 4kh^2 p(\alpha_m^2 + \beta_n^2)}{10abDk_{mn}} \sin\partial_m x_0 \sin \beta_n y_0 \quad (6.32)$$

$$k_{mn} = \left(k_m^2 + k_n^2\right)^2 - \lambda^2 - \frac{kh^2}{10}\lambda^2\left(k_m^2 + k_n^2\right) \quad (6.33)$$

If Eq. (6.12) is substituted in Eq. (6.11), the deflection equation of the roof can be obtained:

$$W(\xi, \eta) = \sum_{m=1,3}^{\infty} \sum_{n=1,3}^{\infty} \frac{40p + 4kh^2 p(\alpha_m^2 + \beta_n^2)}{10abDk_{mn}}$$
$$\cdot \sin \alpha_m \xi \sin \beta_n \eta \sin \alpha_m x_0 \sin \beta_n y_0 \quad (6.34)$$

2.3.3 Stress Function

To make

$$\varphi = \sum_{m=1,2}^{\infty} \left[G_m \mathrm{ch}\gamma_m \eta + H_m \mathrm{ch}\gamma_m(b-\eta)\right]\cos \alpha_m \xi$$
$$+ \sum_{n=1,2}^{\infty} \left[E_n \mathrm{ch}\delta_n \xi + F_n \mathrm{ch}\delta_n(a-\xi)\right]\cos \beta_n \eta$$

$$(6.35)$$

It is easy to verify that Eq. (6.35) can meet the stress control Eq. (6.23), where G_m, H_m, E_n, F_n are undetermined constants.

$$\alpha_m = \frac{m\pi}{a}, \quad \beta_n = \frac{n\pi}{b}, \quad \gamma_m = \alpha_m^2 + \frac{10}{h^2}, \quad \delta_n = \beta_n^2 + \frac{10}{h^2}.$$

Substituting Eqs. (6.34) and (6.35) into the shear force and bending moment formula, under the control of the boundary conditions:

$$\omega_y\big|_{x=0,\,x=a} = 0, \quad M_x\big|_{x=0,\,x=a} = 0, \quad \omega_x\big|_{x=0,\,x=b} = 0, \quad M_y\big|_{x=0,\,x=b} = 0,$$

we can obtain $G_m = H_m = E_n = F_n = 0$ by calculation, and then the stress function $\varphi(\xi, \eta) = 0$ can be obtained also.

2.3.4 Stress Expression

$$\sigma_x = \frac{12M_x}{h^3}z, \quad \sigma_y = \frac{12M_y}{h^3}z,$$
$$\sigma_z = \frac{q}{2}\left[1 - 3\frac{z}{h} + 4\frac{z^3}{h^3}\right] \quad (6.36)$$

where z is the plate height along w-direction.

We can see from the stress expressions, the maximum tensile stress is in $z/2$ of the plate, and the maximum shear stress is in $z = 0$ on the plate surface.

According to the mechanical properties of rock, it has the higher compressive strength but lower tensile strength. Therefore, the minimum safe thickness of the roof of mined-out areas can be obtained by taking the tensile strength criterion as the constrain condition. So, the allowable thickness h of the roof should satisfy the following safe formula:

$$h = \sqrt{\frac{6M_{\max(x,y)}}{\sigma_t}}, \quad H = kh \quad (6.37)$$

TABLE 6.2 Physical and Mechanical Parameters of the Roof

Rock Name	Density (kg/m^3)	Modulus (GPa)	Poisson Ratio	Tensile Strength (MPa)	Cohesive Force (MPa)	Internal Friction Angle (Degrees)
Sandstone	2650	2.20	0.24	0.20	5.80	45

where k is safe amplification factor, $k = 2.0$. H is the safe thickness of roof of mined-out areas.

In practical engineering, although the roof has some tensile damage zones, it is still has a certain capacity, so the calculated solution is safer. Calculated by MATLAB programming, it is easy to obtain the bending moment M. By substituting M into the formula in Eq. (6.37), we can obtain the safe thickness of roof.

2.4 Engineering Application

A surface mine was underlied mined-out areas by room-and-pillar stoping method. The harmonic mechanical load was acting on the center of the roof surface of the mined-out areas; its vibration frequency is 0.1 Hz and load density is 0.29 MPa. The width and the height of the mined-out areas is 10 m × 10 m, respectively, and the width of the pillar is 5.0 m.

The roof rock is regarded as sandstone, and the roof is a supported rectangular plate with four sides. The physical and mechanical parameters of the roof are listed in Table 6.2.

According to Eq. (6.37), m and n in the expression of deflection, shear, and bending moment is 80; the calculated results are listed in Table 6.3, which met the engineering requirements. In

TABLE 6.3 The Calculated Results Under Dynamic Loading

Conditions of Security Control	Allowable Safe Thickness (m)	The Thickness of the Selected Critical Safety (m)
Allowable tensile stress	4.9	9.8

this case, the selected critical safe thickness of the roof was 9.8 m, which was consistent with the construction experience value of 10 m under the same conditions.

2.5 Conclusions

Based on Reissner's theory, considering the construction dynamic loading, the mechanical model of the roof was generalized, and the deflection equation and the internal force expression were derived from the concentration of harmonic dynamic loading on the roof surface of the mined-out areas.

The design formula of the roof's thickness was derived by taking the tensile strength criterion as the constraint condition, and it was verified through practical engineering. The computing method for roof thickness could be the reference and guidance for similar engineering practice.

3. CATASTROPHE CHARACTERISTICS OF THE STRATIFIED ROCK ROOF

3.1 Introduction

There are many shallow mined-out areas left in the Antaibao Surface Mine in Shanxi Province, China. Underground mined-out areas are insidious in nature with irregular spatial distribution, and roof caving and collapse are unpredictable; therefore, the shallow mined-out areas in Antaibao Surface Mine are potential safety hazards during the normal production process (Wang et al., 2012).

In coal-measured sedimentary strata, the stratified composite structure is the most common type of rockmass structures in the roof of the mined-out areas. The stratified rock roof often produces sudden fracture and instability phenomenon due to complexity of overlying rockmass structure, physicomechanical parameters, engineering rockmass constitutive relation, as well as uncertainty of external disturbing loads. Although many studies on the instability mechanism of nonlinear systems, especially for the rock roof of the mined-out areas, have been carried out by scholars, and a number of achievements have been gained, there are still some deficiencies and gaps, including with instability characteristics and the evolution mechanism of the stratified rock roof during the bending deformation process (Zhang et al., 2006; Wang et al., 2010; Tsesarsky, 2012).

With development of nonlinear science, chaos and stochastic resonance (SR) theory appear at the right moment. Chaos is a kind of aperiodic motion mode, which is unique and widespread in the nonlinear system. Since E.N. Lorenze discovered the chaos phenomenon in 1963, the research on chaos has attracted more attention and also became a main focus in the field of nonlinear science (Li et al., 2006; Ayati and Khaloozadeh, 2010; Emadi and Mahzoon, 2012). SR was preliminarily put forward by Benzi and others, which was mainly used to explain the problem of quaternary glaciers, and the study on how stochastic forces work on nonlinear system has become a new hotspot since 1970s. From then on, SR was applied widely to various industry fields (Xiao et al., 2006; Gorji et al., 2011).

Based on the shallow mined-out areas in Antaibao Surface Mine, the theories of SR and chaos were introduced to research the damage from evolution characteristic and instability mechanism of the stratified rock roof during the roof bending deformation process. The research results may have important theoretical significance and practical value to disaster prediction and risk assessment on the rock roof of the shallow mined-out areas.

3.2 Engineering Mechanical Model

In the northeast edge of Antaibao Surface Mine were Jingyang Coal Mine shallow mined-out areas exploited by the room-and-pillar mining method. Since the rock roof of shallow mined-out areas was easily weathered, the stratified rock roof would distort and decline, and roof failure would occur under the gravity and the overlying strata load with the passage of time.

Based on the field investigation and the information collected, the stratified rock roof can be modeled by deformation of the supported beam. The roof gravity and its overlaying strata load are simplified as uniform load, and the value of the uniform load is q. The length of the roof beam is l. The elastic model of the beam is E. The moment inertia is I (all variables are dimensionless). The engineering mechanical mode of the stratified rock roof is built as shown in Fig. 6.12.

The equation of deflection curve for the supported beam of the stratified rock roof is as follows:

$$y = -\frac{qx}{24EI}(l^3 - 2lx^2 + x^3) \quad (6.38)$$

Let X be maximum displacement of centroid of the roof, then Eq. (6.2) can be derived from Eq. (6.38). Eq. (6.39) denotes the relationship between deflection and maximum centroid displacement of the roof beam.

$$y = \frac{16}{5}X\left(\frac{x}{l} - \frac{2x^3}{l^3} + \frac{x^4}{l^4}\right) \quad (6.39)$$

3.3 Deformation Evolution Analysis of the Rock Roof

3.3.1 Deformation Energy Analysis of the Rock Roof

According to elastic theory, the deformation energy induced by bending deformation of the

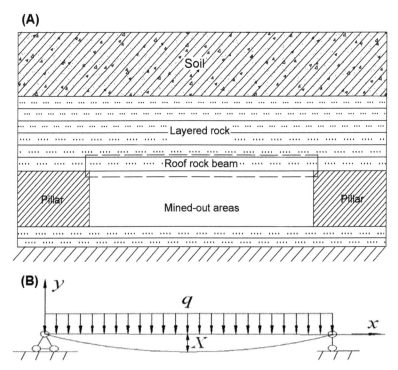

FIGURE 6.12 The mechanical model of simply supported beam for the stratified rock roof. (A) Engineering model; (B) mechanical model.

roof beam included increments of elastic and gravitational potential energy.

The increment of elastic potential energy induced by the bending deformation of the roof beam can be defined by

$$\Delta U_1 = \int M(E, I, \theta) d\theta \qquad (6.40)$$

where $M(E,I,\theta)$ is moment of unit curve radian. The increment of radian is represented as $d\theta$ of which the concrete expression is

$$d\theta = \frac{ds}{r} \approx (y'')^2 \left[1 + (y')^2\right]^{-3/2} dx \qquad (6.41)$$

when the length of roof beam is far greater than X, Eq. (6.40) can be represented as

$$\Delta U_1 \approx \int_0^l M(E, I, \theta) \frac{(y'')^2}{(1+y')^{3/2}} dx \qquad (6.42)$$

The increment of gravitational potential energy induced by bending deformation of the roof beam is

$$\Delta U_2 \approx \int_0^l q(x) y \, dx \qquad (6.43)$$

In summary, the increment of total potential energy is as follows:

$$\Delta U = \Delta U_1 + \Delta U_2$$

$$= \int_0^l M(E,I,\theta)\frac{(y'')^2}{(1+y')^{3/2}}dx + \int_0^l qy\,dx \tag{6.44}$$

Considering the deflection Eq. (6.39) of the roof beam, Eq. (6.44) can be expanded by the Taylor function. The expression is

$$\Delta U = \left[A(E,I,l)\left(\frac{192}{5}\right)^2 X^2 - B(E,I,l)2\left(\frac{192}{5}\right)^2\left(\frac{16}{5}\right)^2 X^4 + \cdots\right] + c_1 X \tag{6.45}$$

where $A(E,I,l) = \int_0^l M(E,I,l)\left(\frac{x^2}{l^6} - \frac{2x^3}{l^7} + \frac{x^4}{l^8}\right)dx$,

$B(E,I,l) = \int_0^l M(E,I,l)\left(\frac{1}{l} - \frac{6x^2}{l^3} + \frac{4x^3}{l^4}\right)\left(-\frac{12x}{l^3} + \frac{12x^2}{l^4}\right)x\,dx$,

$c_1 = \frac{16}{5}\int_0^l q(x)\left(\frac{x}{l} - \frac{2x^2}{l^2} + \frac{x^4}{l^4}\right)dx$

Rounding high order terms for Eq. (6.45) and combining the like terms, then

$$\Delta U = b_1 X^2 - a_1 X^4 + c_1 X \tag{6.46}$$

The coefficient of each item for Eq. (6.46) can be represented as:

$$b_1 = A(E,I,l)\left(\frac{192}{5}\right)^2,$$

$$a_1 = B(E,I,l)2\left(\frac{192}{5}\right)^2\left(\frac{16}{5}\right)^2.$$

3.3.2 Bending Deformation Evolution of the Rock Roof

If the resultant force working on the roof beam of the shallow mined-out areas is considered as that on the centroid of the roof beam, then the resultant force F can be derived from Eq. (6.45).

$$F = -\frac{\partial \Delta U}{\partial X} = 4a_1 X^3 - 2b_1 X - c_1 \tag{6.47}$$

The motion equation of the roof beam is described by $F = M\ddot{X}$. During the evolutionary process of bending deformation of the roof beam, some influence factors are considered, such as damping, damage, and fracture factors within small ranges, and so on, and these factors are assumed to relate to systemic velocity of the roof beam; therefore, the resultant force can be described by the following equation:

$$F = M\ddot{X} + \beta_1 \dot{X} + \beta_2(\dot{X})^2 + \cdots \tag{6.48}$$

Systemic velocity $\ddot{X}$ goes to zero since the evolutionary process of bending and deformation is approximate to the static motion process. Only resultant force F and systemic velocity $\dot{X}$ are taken into consideration and the higher order terms of systemic velocity are neglected, and then combining Eqs. (6.47) and (6.48), which finally get the following equation:

$$\dot{X} = \frac{4a_1}{\beta_1}X^3 - \frac{2b_1}{\beta_1}X - \frac{c_1}{\beta_1} \tag{6.49}$$

Considering the periodic varying factors of annual seasons and each day, such as temperature, humidity, barometric pressure, and so on, all of which are approximately represented as $A\cos \omega t$, and then adding the stochastic factors $-\Gamma(t)$, it can be obtained that the evolutionary motion equation of bending deformation resembles Lagevin equation

$$\dot{X} = aX - bX^3 - c + A\cos \omega t + \Gamma(t) \tag{6.50}$$

where $a = -\frac{2b_1}{\beta_1}$, $b = -\frac{4a_1}{\beta_1}$, $c = -\frac{c_1}{\beta_1}$.

3.4 Stochastic Resonance and Chaos Characteristics Analysis

3.4.1 Stochastic Resonance Phenomenon of the Rock Roof

As for the stratified rock roof of the shallow mined-out areas, if the necessary conditions which produce stochastic resonance (SR) were satisfied, even very small forces $\Gamma(t)$ could have a strong impact on the evolutionary motion of the roof. Based on Fokker–Planck equation, J.H. Lin et al.

had discussed the characteristics of SR, such as Eq. (6.50) (Lin and Lu, 1992; Lu et al., 1993). So the brief discussions on the evolutionary motion equation of bending deformation of roof beam is made in this passage according to the previous studies.

When the roof beam system is in the stable state of evolutionary process ($b > 0$ must be guaranteed in Eq. (6.50)), the SR possibly appear in the system. If $4a^3 - bc^2 > 0$, the system is in the double tendency and also may produce the SR phenomenon during stochastic evolutionary process of the roof beam.

Let $a = 1$, $b = 1$, $c = 0$, then Eq. (6.49) is transferred into Eq. (6.51).

$$\dot{X} = X - X^3 + A \cos \omega t + \Gamma(t) \qquad (6.51)$$

Then the corresponding Fokker–Planck equation is

$$\frac{\partial P(X,t)}{\partial t} = -\frac{\partial}{\partial x}\left[(X - X^3 + A \cos \omega t) P(X,t)\right] + D \frac{\partial^2}{\partial x^2} P(X,t) \qquad (6.52)$$

When the system is in the stable state, the expectation X is as follows:

$$\langle X(t) \rangle = \int X P(X,t) dX \qquad (6.53)$$

The corresponding relation between the peak value X_m and expectation X is

$$\langle X(t) \rangle = X_m \sin \omega t \qquad (6.54)$$

By using numerical computation method, the numerical solutions can be obtained from Eq. (6.52). According to Eqs. (6.53) and (6.54), it can be deduced from $P(X,t)$ that relations between expectation X and signal amplitude A, frequency ω, and noise intensity D (Lu et al., 1993).

If $A = 0.1$, and by varying value of the ω, the corresponding relations between the maximum displacement X_m and noise intensity D can be obtained, as illustrated in Fig. 6.13.

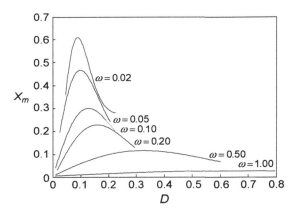

FIGURE 6.13 The resonance curves of the roof beam with different frequency.

As shown in Fig. 6.13, these curve clusters $X_m - D$ present obvious resonant peak and nearly each curve has a maximum value. When drive frequency $\omega = 0.50$, the curve becomes relatively flat; moreover, when $\omega = 1.00$, the curve is close to a straight line. Therefore, the resonant characteristic produced by roof beam is not conspicuous when $0.5 < \omega < 1$, conversely, the resonant peak is more obvious when $\omega < 0.5$. Thus it can be seen that the noise would have little impact on the displacement of the roof beam if the outside noise intensity was not nearby the resonant peak. However, if the noise intensity was near the resonant peak, even very little noise intensity from outside would lead to larger bending deformation. Consequently, the sudden fracture would easily occur for the roof beam if the SRs played a major part in the shallow mined-out areas.

When $4a^3 - bc^2 \leq 0$, the system is in the state of single tendency, there is no SR phenomenon, and the noises from outside will have little impact on the roof system. The maximum centroid displacement of the roof beam will be evolving by means of Eq. (6.54) and the catastrophe fluctuation will never occur.

If $b < 0$, the system will evolve into an unstable situation and the effect of stochastic force $\Gamma(t)$ and periodic force $A \cos \omega t$ can be neglected in an evolving motion of bending deformation of the roof beam, so Eq. (6.50) can be briefly written as follows:

$$\dot{X} = aX - bX^3 - c \quad (6.55)$$

When $4a^3 - bc^2 < 0$, there are one minimum and two maximums in potential energy function of the roof beam. Assuming the minimum is in the stable state and two maximums in the metastable state, if $\dot{X}$ is between two metastable states, the large-scale deformation and breakage will never appear in the roof beam. However, if $\dot{X}$ lies outside of the metastable states, the large-scale deformation and breakage will appear.

When $4a^3 - bc^2 \geq 0$, there is only one peak value in the potential-energy function, and the system will be in the metastable state near the peak. If the centroid displacement of the roof beam is too large, then catastrophe breakage of the roof beam will come into being.

3.4.2 Stochastic Chaos Characteristics of the Rock Roof

Chaos is a kind of aperiodic motion in the nonlinear system, and its later stage results are very sensitive to the initial value variation. There are obvious uncertainties in the chaos response result, and the butterfly effect is the right representation of the chaos.

It was discovered that Eq. (6.50) was very similar to the Duffing equation as research on the chaos characteristic of bending deformation was being carried out. The Duffing equation is described as Eq. (6.56)

$$\ddot{x} - x + \xi\dot{x} + x^3 = \gamma \cos(\omega t) \quad (6.56)$$

The variation form of Eq. (6.56) is

$$\xi\dot{x} = x - x^3 + \gamma \cos(\omega t) - \ddot{x} \quad (6.57)$$

For the quasistatic process of the nonlinear system, $\ddot{x} \approx 0$ in Eq. (6.57). Assuming it is same as stochastic force factors, then $\ddot{x} \approx \Gamma(t)$. Therefore, there are some similar stochastic chaos characteristics between the Duffing equation and Eq. (6.50) to some extent.

As for Eq. (6.51), under such conditions that when drive frequency $\omega = 1.00$ and signal intensity varies in numerical values ($A = 0.10$, $A = 0.15$, and $A = 0.20$), the stochastic chaos scheme is shown in Fig. 6.14, which describe relations between the maximum displacement X

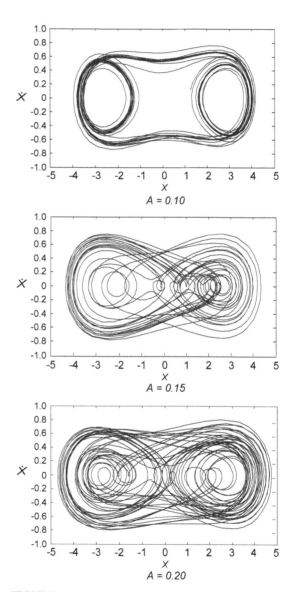

FIGURE 6.14 Random chaotic phase diagrams with different signal intensity under fixed driving frequency.

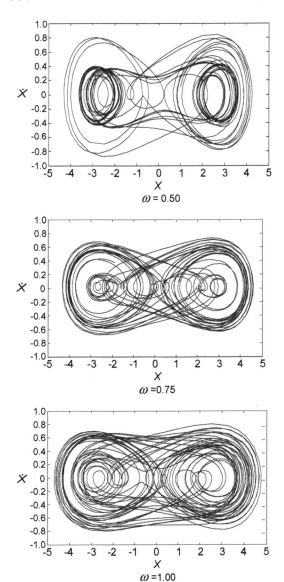

FIGURE 6.15 Random chaotic phase diagrams with different driving frequency under fixed signal intensity.

and the centroid velocity $\dot{X}$. When the signal intensity $A = 0.20$ and drive frequency varies in numerical values ($\omega = 0.50$, $\omega = 0.75$ and $\omega = 1.00$), the stochastic chaos scheme is shown in Fig. 6.15.

The comparisons between Figs. 6.14 and 6.15 show that the roof beam would produce stochastic chaos phenomena under certain conditions. With the increase of environmental signal intensity, the chaos phenomenon becomes more obvious during the evolving process of the roof bending deformation. Similarly, with the increasing of the external signal frequency, the chaos phenomenon becomes more and more obvious as well.

The catastrophe possibility in the rock roof would be increased due to the impact of chaos in mined-out areas. Moreover, most coal-producing areas are located in the north of China where the four seasons climate changes are significant, for which it would also make periodic variation of environment signal significant. Therefore, it is more difficult to predict the catastrophe of the rock roof and assess its risk in the mined-out areas.

3.5 Conclusions

The motion equation that demonstrated the bending deformation of the stratified rock roof had been derived. The results showed that the rock roof of shallow mined-out areas would produce SR and chaos phenomenon during bending deformation process.

With the increasing of environment signal intensity and frequency, the chaos phenomenon produced in the rock roof will become obvious in the mined-out areas during bending deformation process.

The impacts of chaos will result in increasing the catastrophe possibility of the rock roof of shallow mined-out areas; in addition, it would also increase difficulties to predict catastrophe and assess risk in the roof of shallow mined-out areas.

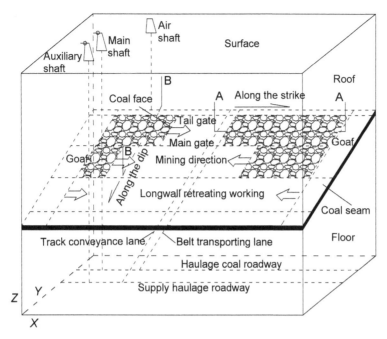

FIGURE 6.16 The schematic drawing of the longwall mining method.

4. PRESSURE-ARCH ANALYSIS IN COAL MINING FIELD

4.1 Introduction

With the development of mining science and technology, the longwall mining method has become popular in mines. As indicated in Fig. 6.16, the longwall mining method along the strike was widely used in Kailuan Coal Mines in China. This mining method is mainly used for roof collapse in gently inclined medium-thickness seam. The length of the coal face along the dip direction is generally 100–200 m, and the advancing length along the strike direction is generally 800–1200 m. After the underground coal was mined-out in the inclined mining field, the pressure-arch with a dip angle was formed because of the self-regulating function of the surrounding rock, and the pressure-arch promoted and ensured the stability of the mining field (Poulsen, 2010; Wang et al., 2014a). With the mining advancing, the cracks in the roof of the coal seam expanded and connected, the immediate roof fell, and the basic roof broke up; then, the stress peak area of the pressure-arch moved forward and the high stress area on the vault of the pressure-arch moved up. The pressure-arch of the mining field constantly moved forward along the direction of the advancing working face, and finally presented a symmetrical shape along the strike of the coal seam, but presented an asymmetric shape along the dip of the coal seam as shown in Fig. 6.17. The stress evolution of the pressure-arch and its stability problem both were the nonlinear problems.

Many scholars have conducted the relevant research. For example, Liu et al. (2004) presented

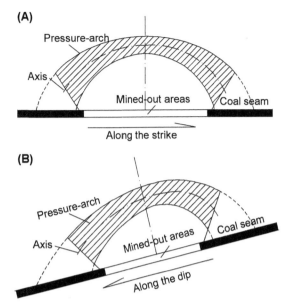

FIGURE 6.17 The pressure-arch along the strike and dip, respectively. (A) Section A—A; (B) Section B—B.

an analytical design method for the truss-bolt reinforcement system and analyzed the arching action by lateral behavior of the inclined roof bolts in reinforcing the fractured roof in coal mines. Shao et al. (2007) indicated that there was an unloaded arch structure existing in the overlying rock and believed that the range of the arch structure was a dynamic expanding process, which was also the reason of the mining pressure appearing on the working face. Xie and Zhao (2009) obtained the parameters about when the arch structure would be formed during the top-coal caving process and discussed a new theory and technology of top-coal caving with vibration to increase the top-coal recovery ratio and to lower the waste content rate. Ren and Qi (2011) studied the instability criterion of the stress arch-shell of the surrounding rock in the mining field, and believed that the stress arch-shell would eventually be instable when the mining height exceeded a certain value. Du and Bai (2012) established the immediate roof's structural mechanics model of the longwall mining under thin bedrock and analyzed the forming mechanism and break laws of the semiarch equilibrium structure of the immediate roof. Wu et al. (2012) believed that the macro stress arch-shell of the pressure-arch showed an asymmetry arch shape in the large dip coal seam after the coal was mined-out, deduced the shape equation of the stress arch-shell of the overlying rocks, and put forward four kinds of instable modes. Xin et al. (2014) found that the asymmetric caving arch was finally formed in top-coal caving in the inclined coal seam by the found that the numerical calculation and field observation of the roof movement. Tikov and Mostafa (2008) conducted the laboratory tests to design and apply the paste backfill technology in the underground mine considering the arching effect of the stress in the mining field. Alehossein (2009) presented a simple model for estimating potential maximum ground surface subsidence caused by underground coal mines based on a triangular zone of major caving, beyond which arching and within which caving was dominant. Sinnott and Cleary (2009) used a three-dimensional (3D) discrete element to investigate the convective motion leading to arching in a vertically vibrated, deep granular bed. Nierobisz (2011) proved that the reduction of costs between 24% and 57% may be achieved in relation to arch support by the use of independent roof bolting in Polish coal mines. Ghosh and Gong (2014) found the arch structure was formed in the caving process and analyzed a new method of vibration technique which was used to break the arch structure to improve the recovery rate in the longwall top-coal mining.

In this chapter, based on the previous results and taking the mining face of No. 9 coal of Fangezhuang of Kailuan Coal Mines as the engineering background and using the structural mechanics model, the instability mechanism of the pressure-arch in the mining field under different dip angles of the coal seam was analyzed, which was of some reference and guiding significance for the safety construction of the coal mining.

4.2 Mechanical Model Analysis

4.2.1 Mechanical Model of Pressure-Arch

The coal production capacity of Fangezhuang Coal Mine is 4.0 Mt/a. The depth of the mining face of No. 9 coal is about 700 m and the average thickness of No. 9 coal is 3.5 m. The seam angle mainly ranges from 5 to 15 degrees (The local dip angle is nearly 45 degrees). The length of the coal face along the dip direction is about 100 m, and the advancing length of the longwall mining along the strike direction is about 650 m. After the coal having been mined-out, the pressure-arch showed a skewed asymmetric shape along the dip of the coal seam with an inclined dip angle, as shown in Fig. 6.17.

Because of the complexity of the underground mining engineering, the inner and outer boundaries of the pressure-arch in the surrounding rock are not smooth but irregular, even having relative dislocation in the multilayer surrounding rock. To understand and treat the complex pressure-arch, the simplified axis of the pressure-arch model was carried out and the engineering model of the pressure-arch in the inclined mining field was shown in Fig. 6.18. An asymmetric shape of the pressure-arch was shown in Fig. 6.19, which was obtained from making a section along the dotted line in Fig. 6.18. According to the Protodyakonov's theory and the Rankine's theory (Yang et al., 2015), the rupture angle θ of the surrounding rock in Fig. 6.18 is defined as $\theta = 45° + \frac{\varphi}{2}$, where φ is the inner friction angle of the surrounding rock in the mining field.

Combined with the structural mechanics model, the pressure-arch at the skewbacks can be seen as the hinge support, and since the strength on the vault may be insufficient the

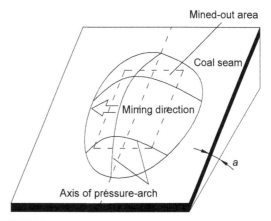

FIGURE 6.18 The simplified engineering model of the pressure-arch.

pressure-arch at the vault also can be seen as a hinged point. Therefore, to study the instability mechanism of the pressure-arch in the mining field, the axis of the pressure-arch can be assumed to be a three-hinged arch with the smooth parabola arch axis.

4.2.2 Mechanical Analysis of Pressure-Arch

As shown in Fig. 6.19, the equation of the arch axis is defined as

$$y = ax^2 + bx + c \qquad (6.58)$$

To simplify the computation, A (0,0), B (L,h), and C (l_c, y_c) being defined, the hypothesis that

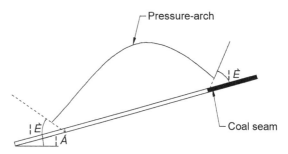

FIGURE 6.19 The simplified pressure-arch along the dip direction.

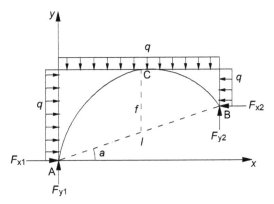

FIGURE 6.20 The simplified mechanical model of the pressure-arch.

point C is the middle point on a line along the x-axis as a special case of the pressure-arch. The angle between AB and the x-axis is α, the arch span is l which represents the length of the coal face along the dip direction and the height of the pressure-arch is defined as f. Then $L = l\cos\alpha$, $h = l\sin\alpha$, $l_c = \frac{l\cos\alpha}{2}$, $y_c = f + \frac{l\sin\alpha}{2}$ can be obtained.

Substituting the coordinate $(0,0)$, (L,h), (l_c,y_c) into Eq. (6.58), it yields the following equation:

$$\begin{cases} c = 0 \\ aL^2 + bL = h \\ al_c^2 + bl_c = y_c \end{cases} \quad (6.59)$$

The axis equation of the pressure-arch can be obtained by solving Eq. (6.59):

$$y = -\frac{4\left[(f + \frac{l\sin\alpha}{2})l\cos\alpha - l\sin\alpha\frac{l\cos\alpha}{2}\right]}{l^3\cos^3\alpha}x^2$$

$$+ \frac{4\left[(f + \frac{l\sin\alpha}{2})l^2\cos^2\alpha - l\sin\alpha\frac{l^2\cos^2\alpha}{4}\right]}{l^3\cos^3\alpha}x$$

(6.60)

As shown in Fig. 6.20, suppose that the pressure-arch is forced by the uniformly distributed load q on the horizontal and vertical directions and the bearing reaction forces are F_{x1}, F_{y1}, F_{x2}, and F_{y2}.

It's easy to get

$$F_{x1} = \frac{ql^2}{8f} - \frac{q(f + l\sin\alpha)}{2} \quad (6.61)$$

$$F_{x2} = \frac{ql^2}{8f} - \frac{q(f - l\sin\alpha)}{2} \quad (6.62)$$

$$F_{y1} = \left(\frac{ql^2}{8f} - \frac{qf}{2}\right)\tan\alpha + \frac{ql\cos\alpha}{2} \quad (6.63)$$

$$F_{y2} = -\left(\frac{ql^2}{8f} - \frac{qf}{2}\right)\tan\alpha + \frac{ql\cos\alpha}{2} \quad (6.64)$$

It is easy to obtain the moment, shear, and axial force at any section K for

$$M_k = \left[\left(\frac{ql^2}{8f} - \frac{qf}{2}\right)\tan\alpha + \frac{ql\cos\alpha}{2}\right]x$$
$$- \left(\frac{ql^2}{8f} - \frac{q(f + l\sin\alpha)}{2}\right)y - \frac{q}{2}x^2 - \frac{q}{2}y^2$$

(6.65)

$$Q_k = \left[\left(\frac{ql^2}{8f} - \frac{qf}{2}\right)\tan\alpha + \frac{ql\cos\alpha}{2} - qx\right]\cos\beta$$
$$- \left[\frac{ql^2}{8f} - \frac{q(f + l\sin\alpha)}{2} + qy\right]\sin\beta$$

(6.66)

$$F_{Nk} = -\left[\left(\frac{ql^2}{8f} - \frac{qf}{2}\right)\tan\alpha + \frac{ql\cos\alpha}{2} - qx\right]\sin\beta$$
$$- \left[\frac{ql^2}{8f} - \frac{q(f + l\sin\alpha)}{2} + qy\right]\cos\beta$$

(6.67)

where β is the angle between the tangent of any section K and the x-axis.

4.3 Failure Models Analysis of Pressure-Arch

Based on the formation mechanism and evolution characteristics of the pressure-arch in the mining field as well as the stability criterion of the three-hinged arch, the instability process of

4. PRESSURE-ARCH ANALYSIS IN COAL MINING FIELD

FIGURE 6.21   The instability process of the pressure-arch.

the pressure-arch is analyzed as shown in Fig. 6.21.

By analyzing the evolution characteristics of the pressure-arch and its mechanical structural features, it is derived that the vault, waists, and skewbacks are the key parts of the pressure-arch in the mining field. And there are three forms of failure models in local damage, namely tension-failure, shear failure, and compression failure. Such cases are the following:

(a) If the following mechanical condition is met, the tensile damage occurs.

$$\frac{F_{Nk}}{A_k} + \frac{M_k}{W_k} > \sigma_t \quad (6.68)$$

where F_{Nk}, A_k, M_k, and W_k are the axial force, the sectional area, the moment, and the sectional flexural modulus at any section K, respectively, and σ_t is the tensile strength of the surrounding rock. The units of F_{Nk}, A_k, M_k, W_k, and σ_t are N, m^2, N m, m^3, and Pa, respectively.

(b) If the following condition is met, the compression failure occurs:

$$\frac{F_{Nk}}{A_k} - \frac{M_k}{W_k} > \sigma_c \quad (6.69)$$

where σ_c is the compression strength of the surrounding rock.

(c) If the following condition is met, the shear failure occurs:

$$\frac{Q_k}{A_k} > \tau \quad (6.70)$$

where Q_k is the shear of any section K, and τ is the shear strength of the surrounding rock.

4.4 Instability Characteristics of Pressure-Arch

To analyze the instability mechanism of the pressure-arch in the mining field under different dip angle of the coal seam, Eqs. (6.55), (6.56), and (6.57) are calculated for visualization using MATLAB, and the resulting data is imported into the graphics software program Origin. The length of the coal face along the dip direction is set to 100 m and the seam angle is set to ranging from 0 to 50 degrees in the simplified computation as an example.

4.4.1 Variation of Characteristics of Pressure-Arch

As shown in Fig. 6.22, when the dip angle of the coal seam was 0 degree, the shapes of the pressure-arch in the mining field were symmetrical to $x = 50$ m. When the dip angle of the seam was not 0 degree, the pressure-arch displayed a

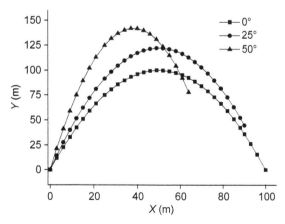

FIGURE 6.22   Variation characteristics of the pressure-arch under different dip angles.

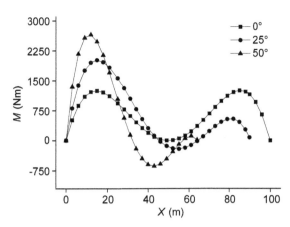

FIGURE 6.23 Moment curves of the pressure-arch under different dip angles.

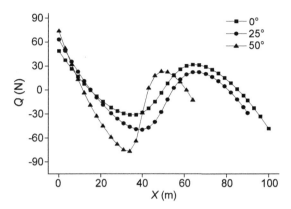

FIGURE 6.24 Shear curves of the pressure-arch under different dip angles.

skewed asymmetrical shape, and the greater the dip angle was, the more apparent the inclination of the pressure-arch was.

4.4.2 Internal Forces Variation of Pressure-arch

As shown in Fig. 6.23, when the dip angle of the coal seam was 0 degree, the bending moment curve of the pressure-arch in the mining field displayed a symmetrical distribution. The bending moment at the waists was greater than that at the vault and skewbacks, thus it is prone to create compression failure at the waists of the pressure-arch. When the dip angle of the seam was not 0 degree, the bending moment curves displayed an asymmetrical distribution. Furthermore, with the dip angle increasing, the bending moment has significantly increased.

As shown in Fig. 6.24, when the dip angle of the seam was 0 degree, the shear force curves of pressure-arch in the mining field displayed the centrosymmetric distribution. The shear at the skewbacks was greater than that at the vault and waists. Thus it is prone to create shear failure at the skewbacks of the pressure-arch. When the dip angle of the seam was not 0 degree, the shear curves displayed an asymmetrical distribution. With the dip angle increasing, the shear force has significantly increased and the peak of shear force gradually shifted from the front skewback of the arch to the vault. Thus it is prone to create shear failure at the vault of the pressure-arch.

As shown in Fig. 6.25, when the dip angle of the seam was 0 degree, the axial force was compression force and the axial force line in the mining field showed a symmetrical distribution. The axial force of the vault was greater than that at the waists and skewbacks. When the dip angle of the seam was not 0 degree, the axial force line showed an asymmetrical distribution. With the dip angle increasing, the tension force

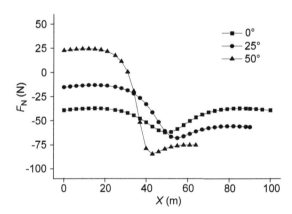

FIGURE 6.25 Axial force curves of the pressure-arch under different dip angles.

appeared at the front skewback of the arch and significantly increased. Therefore, it can be seen that the tension-failure is prone to create at the front skewback.

4.5 Conclusions

The mechanics model of the pressure-arch in the mining filed was built and analyzed, the arch axis equation and the bending moment, shear force, and axial force at any section are obtained. According to the formation mechanism and evolution characteristics of the pressure-arch in the mining field as well as the instability criterion of the pressure-arch, there are three forms of failure models, namely tension-failure, shear failure, and compression failure.

When the dip angle of the seam is 0 degree, the pressure-arch in the mining field is prone to create compression failure at the waists and shear failure at the skewbacks. With the dip angle increasing, the internal force of the pressure-arch displayed an asymmetrical distribution. It is easy to create shear failure at the vault and tension-failure at the front skewback, and at last the local damage causes the dynamic instability in the entire structure.

To obtain the greatest benefit from the roof periodic caving and the support of the mining roadway, it is important to make full use of the capacity of self-bearing and transferring other loads of the pressure-arch, control the key parts of the local damage, and actively regulate the mine pressure according to the instability criterion of the pressure-arch in the longwall mining field under different seam dip angles.

5. ANALYSIS OF ACCUMULATED DAMAGE EFFECTS ON THE ROOF

5.1 Introduction

There have been many shallow mined-out areas left in the Antaibao Surface Mine, Shanxi Province, China. As the surface mine constantly carried out the blasting operation, the explosive stress waves led to a shaking effect, forming the loading environment to the roof of shallow mined-out areas. Under the repeated blasting vibrations, the roof of the mined-out areas could be activated, leading to the evolution of progressive damages and producing new decline and collapse. Thus the miners and the excavation equipment on the surface may unexpectedly fall into the mined-out areas, which was a hidden danger to the workers, and greatly influenced safe production.

According to the documented data, Wang et al. (2007) took the strip mine side slope as background and carried out research on progressive damage to the slope rockmass under the mine-induced seismic load (Wang et al., 2007). Yang and Scovira (2007) studied the dynamic response characteristics of the underground structure and the surrounding rocks under blasting vibration and conducted sonic wave tests on the surrounding tunnel rocks and slope under blasting vibration. Zhao et al. (2008) made the stability analysis of rock and soil slope engineering with chambers under dynamic load. Scholars, such as Yang R., Mogi, and Singh also have done similar studies (Yang and Scovira, 2007; Singh and Roy, 2008).

In summary, although the scholars in China and other countries have conducted various research on the damage effect of rockmass under blasting vibration, and have achieved some beneficial conclusions, the damage evolution process and accumulated damage effects of the roof of mined-out areas was rarely reported. Therefore, taking into consideration the mined-out areas in Antaibao, it is of great importance and value to carry out an analysis of progressive damage effects on the roof of the mined-out areas under blasting vibrations.

5.2 Computational Model and Parameters

5.2.1 The Computational Model

In the northeast edge of Antaibao Surface Mine, underlying the mined-out areas of

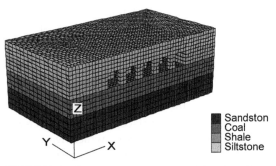

FIGURE 6.26   The computational model and meshes.

Jingyang mine by the room-and-pillar stoping method. Based on the field investigation and the collected information, the 3D computational model was built as shown in Fig. 6.26.

The computational model is 130 m long, 130-m wide, and 52-m high in the x-, y-, and z-axis. The span of mined-out areas is 10 m, and the width of coal pillars is 5 m. The model was restricted to horizontal movement on four sides with its bottom fixed and its top free. The destruction of rock material was assumed consistent with Mohr–Coulomb strength criterion. The model was divided into 110,031 zones and 20,787 nodes.

5.2.2 Parameter Selection

Based on field investigation and laboratory testing of mechanical parameters of the rock specimen, the physical and mechanical parameters of rockmass were selected as listed in Table 6.4.

5.3 Determining Dynamic Load and Boundary Conditions

Among the construction machinery equipment in the Antaibao Surface Mine, the maximum load density of BE shovel is 0.29 MPa. During the construction process, constant shoveling and loading kept the giant shovel arm swinging back and forth. So the giant shovel could be considered as a pseudostatic load, and the dynamic load factor was taken as 2.0; thus the shovel load density was 0.58 MPa, which applied to the roof of the mined-out areas by equivalent uniform load.

The blasting vibration velocity recorded by field monitoring is shown in Fig. 6.27. The blasting vibration waves with different incident angles were applied on the surface of the mined-out areas.

Under blasting vibration, the static boundary was adopted to reduce the reflection influence of blasting shock wave on the model boundaries. Because exerting velocity and acceleration load on the static boundary will make it useless, the velocity load should be converted into stress. For example, to the normal shock wave, the conversion formula is as follows:

$$\sigma_n = 2(\rho C_p) V_n \qquad (6.71)$$

$$C_p = \sqrt{\frac{K + 4G/3}{\rho}} \qquad (6.72)$$

TABLE 6.4   Physical and Mechanical Parameters of Rockmass

Lithology Name	Thickness (m)	Density (kg/m³)	Elastic Modulus (GPa)	Poisson Ratio	Tensile Strength (MPa)	Cohesion (MPa)	Friction Angle (Degrees)
Sandstone	15	2380	5.50	0.30	0.25	1.50	39
Coal	10	1440	1.00	0.38	0.10	0.50	26
Shale	12	2450	2.40	0.33	0.15	1.20	34
Siltstone	15	2600	4.80	0.32	0.30	1.60	40

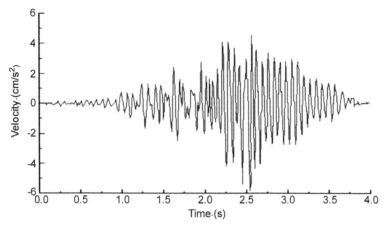

FIGURE 6.27 The blasting vibration velocity recorded by field monitoring.

where σ_n is the normal stress, ρ is medium density, V_n is the particle velocity components in the normal, and C_p is the velocity of P wave.

5.4 Simulation Programs

To reflect the dynamic response characteristics of the roof of mined-out areas under repeated blasting vibrations, the calculation and analysis were conducted as follows:

1. Building a 3D computational model, carrying out the balance calculation of initial stress field before mining, and the displacement and velocity fields were set to zero after the balance. Only initial stress caused by rock and soil weight was considered in this case.
2. Based on the mined-out areas identified by field investigation, drilling, and geological prospecting, the balance calculation of stress induced by the room-and-pillar stoping process was carried out, then the displacement and velocity fields were set to zero.
3. Exerting mechanical load on the surface of mined-out areas, the balance calculation of stress induced by shoveling was carried out, then the displacement and velocity fields were set to zero.
4. Based on the analysis, the equal-time interval blasting vibration waves with different incident angles as 0, 30, 60, and 90 degrees were exerted on the surface of the mined-out areas three times. Then the accumulated damage effect on the roof of mined-out areas was analyzed.

To reveal the dynamic characteristics of displacement of the roof and surrounding rocks of the mined-out areas under blasting vibrations, the monitoring sites were set as in Fig. 6.28.

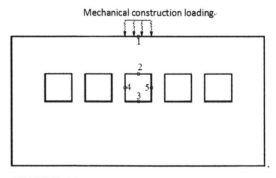

FIGURE 6.28 Schematic diagram of monitor points.

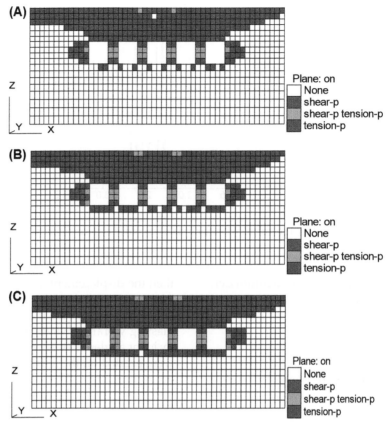

FIGURE 6.29 The plastic zones of the mined-out areas. (A) Under the first blasting vibration; (B) under the second blasting vibration; (C) under the third blasting vibration.

5.5 Computational Results and Analysis

5.5.1 *Plastic Zones Analysis*

Take the blasting vibration wave with 90 degrees incident angle as an example. As is shown in Fig. 6.29, the number of plastic zones in the roof of mined-out areas gradually increased with the growing blasting vibration times, and the plastic zones started to expand bilaterally at the bottom of mined-out areas. Therefore, under the repeated blasting vibrations, the plastic zones would increase cumulatively, and the damage range would expand cumulatively in the roof and surrounding rockmass of the mined-out areas.

As shown in Fig. 6.30, under the blasting vibration waves with different incident angles were exerted on the surface of the mined-out areas three times, the plastic zones produced in the roof and surrounding rockmass were obvious differences. The incident angle is 30, 60, 90, and 0 degrees of the loading conditions by the number of plastic zones. This shows that the different incident angles can cause different cumulative damage effects in the roof and surrounding rockmass.

5. ANALYSIS OF ACCUMULATED DAMAGE EFFECTS ON THE ROOF

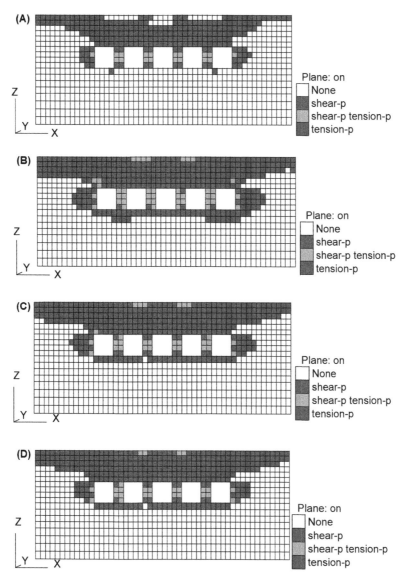

FIGURE 6.30 Plastic failure filed of the mined-out areas under different incident angles. (A) 0 degree incident angle; (B) 30 degrees incident angle; (C) 60 degrees incident angle; (D) 90 degrees incident angle.

5.5.2 Deformation Characteristics Analysis

5.5.2.1 VARIATION OF VERTICAL DISPLACEMENT

Take the blasting vibration wave with 90 degrees incident angle as an example. As is shown in Fig. 6.31, with the increasing times of the blasting vibration, the vertical displacements appeared to step-increase. After the third blasting vibration, the vertical displacement maximum values of monitoring point 1, 2, and 3, respectively, came to 87, 99, and 4.2 mm. It

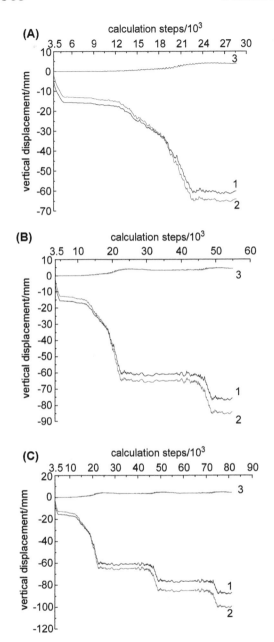

FIGURE 6.31 The vertical displacement curves of monitoring points under blasting vibrations. (A) Under the first blasting vibration; (B) under the second blasting vibration; (C) under the third blasting vibration.

was obvious that the repeated blasting vibration would cause accumulated vertical displacement in the roof and bottom of mined-out areas.

As is shown in Fig. 6.32, under the blasting vibration waves with different incident angles, which were exerted on the surface of the mined-out areas three times, the vertical displacements produced in the roof and bottom showed obvious differences. When the incident angle is 0, 30, 60, and 90 degrees of the loading conditions, then the maximum vertical displacement is 38, 116, 97, and 91 mm, respectively. This shows that the different incident angles can cause different cumulative displacement effects in the roof and bottom. Therefore, during the blasting operation, it is necessary to pay attention on the roof damage due to different incident angles of the blasting vibration waves.

5.5.2.2 VARIATION OF HORIZONTAL DISPLACEMENT

Take the blasting vibration wave with 90 degrees incident angle as an example. As is shown in Fig. 6.33, with the increasing blasting vibration times, the horizontal displacements appeared to step-increase. After the third blasting vibration, the horizontal displacement maximum values of monitoring point 4 and 5 came to 9.0 and 9.2 mm. It was obvious that the repeated blasting vibrations would cause accumulated horizontal displacement in the roof and bottom of mined-out areas.

As is shown in Fig. 6.34, under the blasting vibration waves with different incident angles, which were exerted on the surface of the mined-out areas three times, the horizontal displacements produced in the pillar showed obvious differences. When the incident angle is 0, 30, 60, and 90 degrees of the loading conditions, then the maximum horizontal displacement is 4, 25, 12, and 9.5 mm, respectively. This shows that the different incident angles can cause different cumulative displacement effects

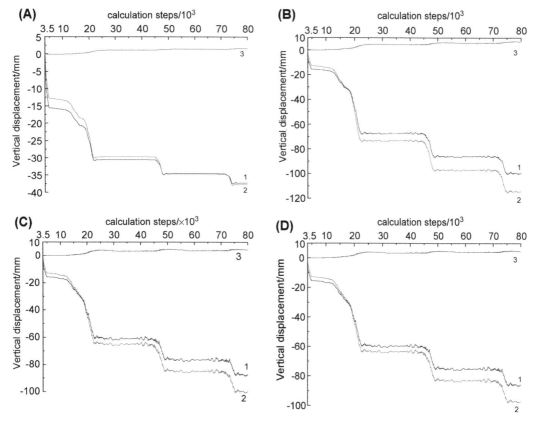

FIGURE 6.32 The vertical displacement curves of monitoring points under different incident angles. (A) 0 degree incident angle; (B) 30 degrees incident angle; (C) 60 degrees incident angle; (D) 90 degrees incident angle.

in the pillar. Therefore, during the blasting operation, it is necessary to pay attention to the pillar damage due to different incident angles of the blasting vibration waves.

5.5.3 Principal Stress Field Analysis

Take the blasting vibration wave with 90 degrees incident angle as an example. As is shown in Fig. 6.35, with the increasing blasting vibration times, the principal stress field in the roof of mined-out areas appeared to redistribute, the maximun principal stress is gradually increased from 2.57, 2.64−2.66 MPa, and the concentration extent of the maximum principal stress gradually upgraded. It was obvious that the dynamic behavior characteristics of the principal stress field were positively correlated with plastic damage in the roof and the surrounding rocks of mined-out areas.

As is shown in Fig. 6.36, under the blasting vibration waves with different incident angles were exerted on the surface of the mined-out areas three times, the maximum principal stress values in the roof and the surrounding rocks showed obvious differences. When the incident angle is 0, 30, 60, and 90 degrees of the loading conditions, then the maximum principal stress is 2.38, 2.74, 2.67, and 2.66 MPa, respectively.

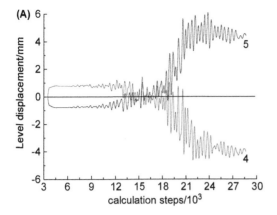
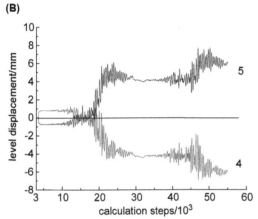
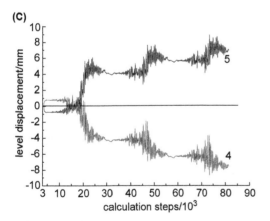

FIGURE 6.33 The horizontal displacement curves of monitoring points under blasting vibrations. (A) Under the first blasting vibration; (B) under the second blasting vibration; (C) under the third blasting vibration.

This shows that the different incident angles can cause different cumulative maximum principal stress effects in the roof and the surrounding rocks.

Based on these results, we can see that the significant cumulative damage effect in the roof and the surrounding rocks were caused by the blasting vibration load. For the rockmass being into the plastic postpeak state, microcracks in the rock could develop and expand under constant loading; thus the macroscopic destruction phenomenon was produced.

In addition, the blasting vibration wave with a small angle of about 30 degrees can cause significant damage effect in the roof and surrounding rock of the mined-out areas than others incident angles. The blasting vibration waves with different incident angles were exerted on the surface of the mined-out areas, the diffraction effects of reflection fields, refraction fields, and scattering fields were coupled with the secondary stress field of engineering rock, which caused the complex dynamic mechanical response characteristics in the roof and the surrounding rock of the mined-out areas.

5.6 Conclusions

Under repeated blasting vibrations, the roof and surrounding rocks of the mined-out areas of the surface mine will produce accumulated displacement, stress, and plastic damage. With the increasing blasting vibration times, the accumulated amount will increase. The special accumulated phenomenon under the repeated dynamic load is obviously different from that under the mechanical load, namely, the dynamic loading accumulated effect.

The results showed the amount of plastic zones accumulated in the roof and the surrounding rocks of mined-out areas, displacement accumulated by step-increase, and the maximum principal stress accumulated were positively

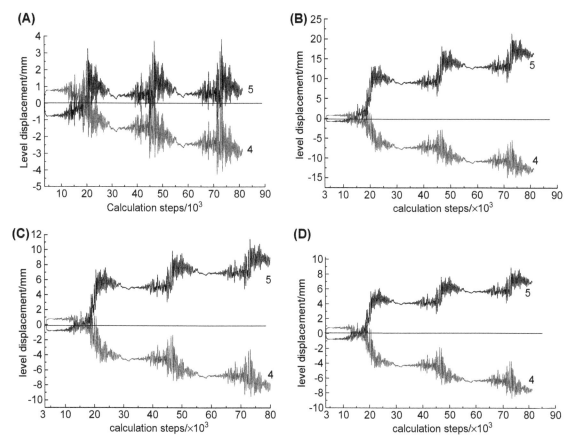

FIGURE 6.34 The horizontal displacement curves under different incident angles. (A) 0 degree incident angle; (B) 30 degrees incident angle; (C) 60 degrees incident angle; (D) 90 degrees incident angle.

correlated with blasting vibration times. It was obvious that the accumulated effect increased by nonlinear relations while blasting vibration times increased.

Under the blasting vibration waves with different incident angles were exerted on the surface of the mined-out areas three times, the accumulated displacement, stress, and plastic damage in the roof and the surrounding rocks showed obvious differences. The blasting vibration wave with a small angle of about 30 degrees can cause more damage in the roof and surrounding rock of the mined-out areas than other incident angles.

During the blasting operation, it is necessary to pay attention to the damage effects due to different incident angles of the blasting vibration waves. We must control the amount of blasting explosive in one-time operation and blasting times, and strengthen warning observation for the roof displacement of mined-out areas to ensure safe production.

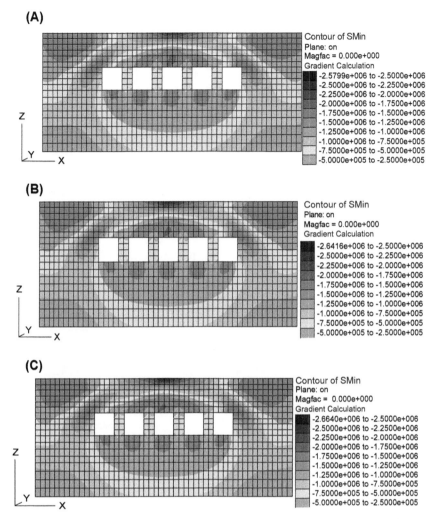

FIGURE 6.35 The principal stress filed of the mined-out areas under blasting vibrations. (A) The first blasting vibration; (B) the second blasting vibration; (C) the third blasting vibration.

6. TUNNEL AND BRIDGE CROSSING THE MINED-OUT REGIONS

6.1 Introduction

The highway network density has been on the increase with the rapid development of expressway construction. A number of highway bridges and tunnels have to cross the mined-out regions under many circumstances. As a result of the uneven settlement deformations of the mined-out regions, destructive phenomena possibly appear, such as fluctuations in highway pavements, deformations, and crackings in lining structures and large deformations in bridge abutments, all of which will go into the serious

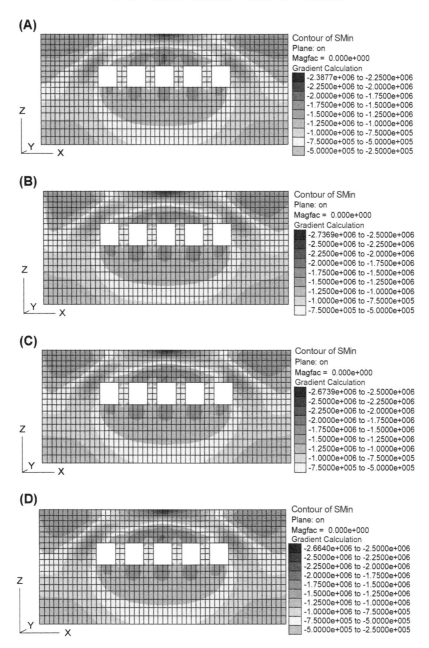

FIGURE 6.36 The principal stress filed of the mined-out areas under different incident angles. (A) 0 degree incident angle; (B) 30 degrees incident angle; (C) 60 degrees incident angle; (D) 90 degrees incident angle.

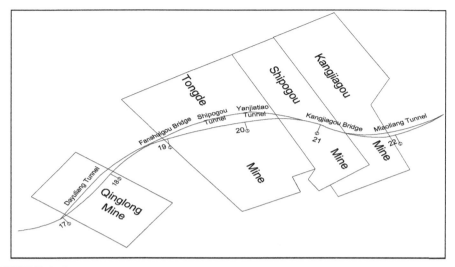

FIGURE 6.37 Schematic diagram of two-bridge and four-tunnel project above mined-out regions.

potential hazards to the operation and maintenance of highway bridges and tunnels (Tong et al., 2004).

It can be found that the existing treatment standards, assessment basis, theoretical calculation, and treating technology on the mined-out regions under highway bridges and tunnels are still at the stage of experience and lacking in systematic and mature theories used for reference according to plentiful document retrievals. Therefore, it is both a challenge and an emergency for harmful effects assessment and treatment technology on the mined-out regions under highway bridges and tunnels (Li et al., 2005; Wen, 2007; Hu et al., 2008).

In this study, the 3D engineering model was built by using the MIDAS/GTS finite element program based on the treatment project for mined-out regions under Kangjiagou bridge and Miaoliang tunnel that in Liulin section of Qingdao-Yinchuan highway. After that, numerical analyses of the ground deformation response, and tunnel lining and bridge structural deformations as well as their structural forces by using 3D Fast Lagrangian Analysis of Continua (FLAC3D) were performed.

It is intended to explore the harmful effects assessment and make effective engineering countermeasures for highway bridges and tunnels crossing mined-out regions.

6.2 Engineering Background

The Lishi-Jundu section of Qingdao-Yinchuan highway is located in Liulin county, Shanxi province, spanning 38.542 km, which consists of two-bridge and four-tunnel projects crossing four mined-out regions. The accumulative total length affected by the mined-out regions is about 5.6 km, and the engineering construction is very difficult both in China and abroad. The schematic diagram of the two-bridge and four-tunnel project above the mined-out regions is generalized in Fig. 6.37.

Kangjiagou Mine is joint-operated and has been mining by vertical shaft development from 1997 to the present, and its annual output achieves 4 Mt. The most prominently mined coal is No. 4 seam, and the seam average thickness is 2.4 m. The lithologic characters of the seam roof and floor are both mudstone and the attitude of the rocks is $249° \angle 4°$. The buried depth of the coal seam is from 170 to 245 m. The mined-out region locates 130–150 m below the designed elevation, and the ratio of mining depth to thickness is 54–63. The mining method is longwall mining, and the recovery rate is about 80%. The surface of the mined-out region is made up of mild clay with colors of pale yellow and brownish red, and collapse craters with diameters about 5 m can be found in the earth surface.

Kangjiagou mined-out region distributes in the Lishi-Jundu section of Qingdao-Yinchuan highway: K_{21+340}-K_{21+910} (Right line); K_{21+340}-LK_{21+900} (Left line). The geomorphic units in the mined-out region belong to the hilly area of Jin-Shan loess plateau, and the microtopography with top surfaces of relative gentleness includes loess ridges, loess mounds, loess tablelands, and loess gullies. The earth surface is covered with loess Q_3 and Q_2, and the footslope bedrock partially outcrops. There are also many loess erosion gullies and steep furrow banks. The topographic relief is of wide scope, and its absolute elevation is 750–1000 m.

6.3 Analysis of Harmful Effects on Highway Tunnel and Bridge

6.3.1 The Computation Model

The finite difference program FLAC3D has become a relatively perfect computational analysis tool for the engineers since it was introduced into China in the early 1990s. However, it has some difficulty in modeling and meshing for complicated 3D engineering. It takes much time to do a large amount of modeling work by using FLAC3D, resulting in more time-consumption and difficulty.

In this research, the process of building FLAC3D computation model can be generalized into three steps:

Step 1: The .DXF files of mined-out region distributions, strata boundaries, and tunnel outlines were read-in through the data interface (put inside MIDAS/GTS) between MIDAS/GTS and CAD, and the corresponding entities were generated based on the imported data. The topographic contour data in .DXF format were read-in by taking advantage of the Terrain Geometry Maker built-in MIDAS/GTS to generate the terrene curved surface by which the geometric entities would have been partitioned. And by combining with embedments and boolean operations, etc., to geometric entities, the 3D geometry model of bridge and tunnel engineering above the mined-out region was generated.

Step 2: The model meshes were divided in MIDAS/GTS, and the grouping work was done at the same time.

Step 3: The 3D computation model of bridge and tunnel engineering above the mined-out region was built in FLAC3D with less preprocessing difficulty by importing the completed model in step two into FLAC3D.

The engineering-geological model imported into FLAC3D is shown in Fig. 6.38, and the corresponding 3D computation model meshes are shown in Fig. 6.39.

The calculation scope includes Kangjiagou bridge and Miaoliang tunnel and the underlying mined-out region, 800-m long, 400-m wide, and 455-m high, consisting of 247,578 elements and 45,909 nodes.

The schematic diagram of the engineering model of Kangjiagou bridge–Miaoliang tunnel engineering above the mined-out region is shown in Fig. 6.40. Four sides of the computation model are horizontal movement limited, the bottom fixed, and the top free. Entity elements are used to simulate all of the rock and soil materials which assumed conforming Mohr–Coulomb strength criteria, and the elastic constitutive model is applied for tunnel preliminary supports and bridge abutments. Bolts, shotcretes, and grid steel frames which make up the preliminary supports are considered as linings of 0.45-m thick.

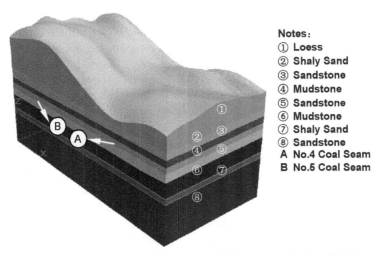

FIGURE 6.38 Three-dimensional engineering-geological model of Kangjiagou bridge—Miaoliang tunnel engineering.

The abutments are built on the earth surface. Physicomechanical parameters of geomaterials for numerical calculation are shown in Table 6.5.

6.3.2 Simulation Analysis Schemes

Numerical simulation analyses of the ground deformation response, and tunnel lining and bridge structural deformations as well as their structural forces were performed using FLAC3D. The processes were first generated in the step-by-step mining region for treatment after mining coal in Kangjiagou. After that, the Miaoliang tunnel was excavated and Kangjiagou bridge was constructed. The analysis schemes are as follows:

1. The initial stress field before mining in Kangjiagou was calculated to balance state according to the field geomorphic characteristics of the mined-out region under

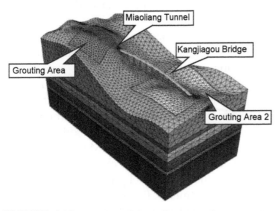

FIGURE 6.39 Meshes of three-dimensional computation model.

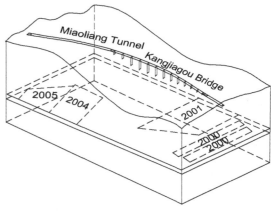

FIGURE 6.40 Schematic diagram of Kangjiagou bridge—Miaoliang tunnel above the mined-out region.

TABLE 6.5 Physicomechanical Parameters of Geomaterials for Numerical Calculation

No.	Name	Density (kg/m³)	Elastic Modulus (GPa)	Poisson Ratio	Tensile Strength (MPa)	Cohesion (MPa)	Internal Friction Angle (Degrees)
1	Loess	1650	0.15	0.32	0.0028	0.075	28
2	Shaly sand	2650	8.50	0.22	1.4000	7.200	40
3	Sandstone	2660	9.20	0.20	1.6000	7.600	41
4	Mudstone	2620	6.50	0.25	1.2000	6.400	38
5	Sandstone	2670	9.50	0.20	1.8000	7.800	42
6	Mudstone	2630	6.70	0.25	1.3000	6.700	40
7	No. 4 coal	1450	5.30	0.28	5.6000	3.800	34
8	Shaly sand	2650	8.50	0.22	1.5000	7.400	40
9	No. 5 coal	1470	5.50	0.28	5.8000	4.000	34
10	Sandstone	2680	12.0	0.18	2.8000	8.600	44
11	Tunne initial supporting	2300	18.5	0.20			
12	Bridge pier and abutment	2500	22.0	0.16			

Kangjiagou bridge—Miaoliang tunnel engineering, and then the displacements were reset to zero where considering only the initial stress induced by geomaterial deadweight.

2. The response character analyses of the surface deformations caused by step-by-step mining were proceeded corresponding to the years based on the mined areas defined by field investigations, drillings, and geophysical prospecting.
3. The grouting treatment technology was carried out after mining coal in Kangjiagou, next to Miaoliang tunnel, was excavated and Kangjiagou bridge was constructed. The response of the ground deformation, tunnel lining, and bridge structural deformation as well as their structural forces were analyzed by using FLAC3D.

6.3.3 Surface Subsidence of the Mined-Out Regions

Simulations were performed and the surface subsidence field was obtained as shown in Fig. 6.41 on the basis of the construction sequence as follows, step-by-step mining, grouting treatment after mining coal in Kangjiagou, the Miaoliang tunnel excavation, and Kangjiagou bridge construction. The center of the surface subsidence basin is located at one side of Miaoliang tunnel and the maximum subsidence is about 123 mm.

It is known from Fig. 6.41 that the two surface subsidence basins show skewed nonclosed oval shape, the central subsidence position has a bias toward a distance to one side of Miaoliang tunnel. Kangjiagou bridge and Miaoliang tunnel both cross the subsidence basin edges bearing harmful tension deformation. Consequently,

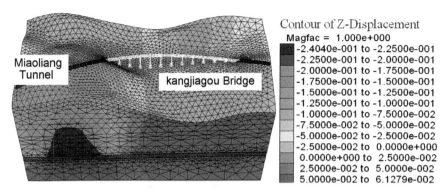

FIGURE 6.41 Map of ground settlement field (unit: m).

security countermeasures should be adopted to ensure the safety of Miaoliang tunnel in the period of construction and later transport operation.

6.3.4 Deformation of the Bridge and Tunnel Engineering

6.3.4.1 VERTICAL DEFORMATION OF TUNNEL LINING STRUCTURE

The key deforming position of Miaoliang tunnel lining structure corresponds to the characteristics of the ground subsidence basin (Fig. 6.42), and the maximum subsidence deformation is 3.7 mm from which it can be seen that the grouting treatment effect on Kangjiagou mined-out region is good, and the influence of subsidence deformation on Miaoliang tunnel brought about by the mined-out region is on the small side.

6.3.4.2 DEFORMATION ANALYSIS OF BRIDGE STRUCTURE

The bridge structure is made up of deck, joist, pier, and abutment (Fig. 6.43A). The settlements of the bridge abutments affected by the underground mined-out region is shown in Fig. 6.43B.

Fig. 6.43 shows that the maximum settlement of abutments reaches 67.5 mm, and that the settlement key parts correspond to the surface deformation characteristics of the underlying mined-out region. Table 6.6 is drawn for the purpose of checking the calculation analysis of the measuring points deformations of Kangjiagou

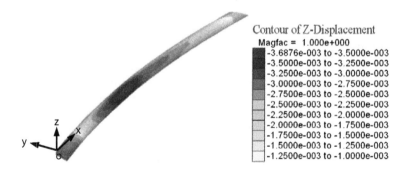

FIGURE 6.42 Vertical deformation field of lining structure of Miaoliang tunnel lining (m).

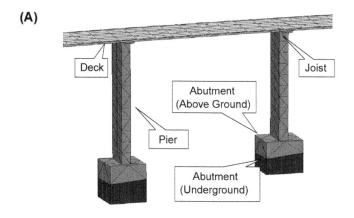

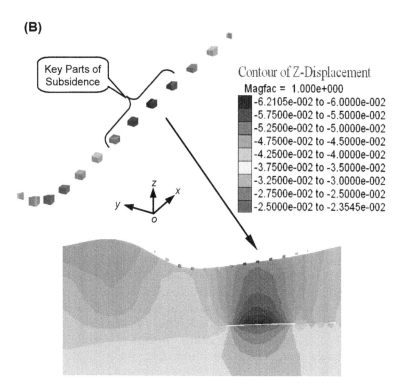

FIGURE 6.43 Bridge structure and settlement of bridge abutment (m); (A) schematic diagram of bridge structure; (B) settlement of bridge abutment.

bridge abutments. The nodal displacement data in the table are extracted from abutment nodes both on bottom and top following x-axis forward.

Based on Code for Design of Ground Base and Foundation of Highway Bridges and Culverts in China (JTGD63-2007), the displacements of simple bridge abutments should satisfy that (1) the

TABLE 6.6 Deforming Statistics of Positive x Direction Deformations of Abutment Foundation

Abutment Nodes	x Direction		y Direction		z Direction	
	Nodal Deformation (mm)	Difference of Adjacent Nodes (mm)	Nodal Deformation (mm)	Difference of Adjacent Nodes (mm)	Nodal Deformation (mm)	Difference of Adjacent Nodes (mm)
1	−17.48		−29.64		−37.32	
2	−17.15	0.33	−24.15	5.49	−32.38	4.94
3	−10.49	6.66	−15.27	8.88	−25.41	6.97
4	0.68	11.17	−5.20	10.07	−24.71	0.69
5	10.95	10.27	4.10	9.30	−29.91	5.20
6	16.92	5.97	12.98	8.89	−38.78	8.87
7	16.02	0.90	20.46	7.48	−48.88	10.10
8	9.45	6.57	26.46	6.00	−56.19	7.31
9	0.37	9.08	29.47	3.01	−58.47	2.27
10	−7.12	7.50	29.50	0.02	−54.75	3.71
11	−11.45	4.33	27.28	2.21	−47.77	6.99
12	−12.30	0.85	24.56	2.72	−39.84	7.92
13	−11.47	−0.83	21.60	2.96	−32.15	7.70

abutment averaging settlements should be no more than $2.0\sqrt{L}$ (unit: cm); (2) the difference of averaging settlements among adjacent abutments should be no more than $1.0\sqrt{L}$ (unit: cm); (3) the horizontal displacements of abutment measuring points should be no more than $0.5\sqrt{L}$ (unit: cm), in which L is on behalf of the least span length among adjacent abutments (unit: m).

The abutments spacing of Kangjiagou bridge is 40 m, and according to these regulations, the abutment uniform settlement should be no more than 126 mm, the difference between uniform settlements of adjacent abutments no more than 63 mm, and the horizontal displacements of abutment measuring points no more than 32 mm.

It can be seen from Table 6.6 that the maximum settlement 58.47 mm is less than 126 mm; the maximum difference between uniform settlements of adjacent abutments 10.10 mm is less than 63 mm, and the maximum horizontal displacement of abutment measuring points 29.64 mm is less than 32 mm, all of which meet the requirements of the code. Therefore, the underlying mined-out region has little deformation influence on Kangjiagou bridge structure after the grouting treatment, and satisfies engineering requirements.

6.3.4.3 STRESS CHARACTERISTIC OF BRIDGE AND TUNNEL STRUCTURE

The maximum principal stress distribution in Miaoliang tunnel lining is shown in Fig. 6.44. Miaoliang tunnel is in the tension part of the mined-out region edges and the maximum tension stress in corresponding positions of the lining structure achieves about 0.50 MPa, so the

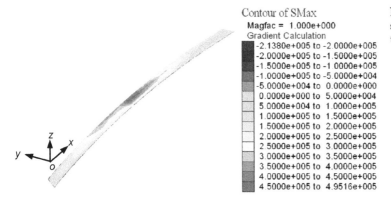

FIGURE 6.44 Maximum principal stresses distribution of the Miaoliang tunnel lining (Pa).

lining support to this part should be reinforced in the tunnel construction.

From Fig. 6.45, it is known that the maximum principal tensile stress reaches 1.67 MPa in the connecting part of abutment 5# and the deck. The connecting part is at an easy rate to subject tensile failure. There are also stress concentrations of different degrees in connecting parts of the deck and abutment 3#, 4#, 6#, and 7#, which is on both sides of abutment 5#, mainly in tension. So, reinforcement measures should be adopted in these key force-bearing parts in the process of design and construction for bridge structures to ensure the safety of the engineering.

6.4 Conclusions

A new procedure is proposed aiming at the difficulty in building complicated 3D model with FLAC3D, to accomplish geometry modeling and grid division by using MIDAS/GTS, and then importing the completed model into FLAC3D through data conversion. This work offers a new solution to building a complicated 3D engineering model with FLAC3D.

The characteristics of the surface subsidence basin and the influence range of harmful deformation induced by mined-out region under the bridge and tunnel structures are analyzed based on the numerical simulation, taking account of the interactions between strata and structures. The studies are important to guide the design of grouting (or supplement) hole for strata strengthening.

It is put forward that effective measures should be taken to control the key parts deformations and forces of the bridge and tunnel structures in the process of design and construction. The conclusions are obtained by analyzing the deformation and stress distribution characteristics of the tunnel lining, and bridge structure as well as abutments above the mined-out region with a view to the time-space-process effect.

7. PRESSURE-ARCH ANALYSIS IN HORIZONTAL STRATIFIED ROCKS

7.1 Introduction

The pressure-arch of the mine opening is the result of the redistribution of stress, which both promotes the stability and reflects the strength of the surrounding rock (Poulsen, 2010; Wu et al., 2012; Wang et al., 2014a,b,c). The surrounding rock of coal excavation is composed of the stratified rocks including primary structural planes, such as the bedding and also the weak interlayers. The physical and mechanical

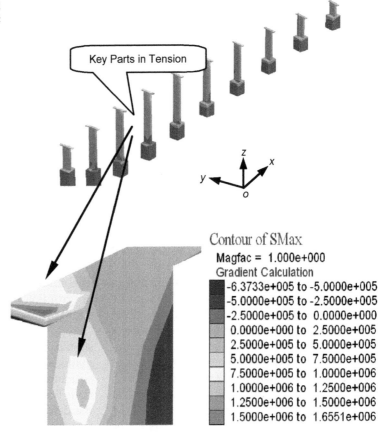

FIGURE 6.45 Maximum principal stresses map of the bridge abutment (Pa).

characteristics of the stratified rocks are transverse isotropic and orthogonal anisotropic materials largely due to the weak interlayers (Li et al., 1997; Wang et al., 2014a,b,c). These structural planes will affect the shape and the extent of the pressure-arch and the stability of the surrounding rock. Therefore, it is of great significance to investigate the distribution characteristics of the pressure-arch in the horizontal stratified rocks.

Many scholars have conducted relevant research on the stratified rocks or on the surrounding rock with weak interlayers using many methodologies, including field investigation, numerical simulations, physical model tests, and analytical analyses. For example, Zhan et al. (2000) revealed the destruction and instability mechanism of the rockmass with weak interlayers through mechanical analysis. Jia et al. (2006) studied the failure mechanism of the stratified rock roof of the tunnel. When a layered system of rock beds is subjected to a sufficiently large extensional strain, joints form in the competent layers. Jain et al. (2007) found that the joint spacing first decreases and then increases with depth for a given applied strain. Zhan et al. (2007) discussed the weak interlayers with different distances that affected the stability of the underground cavern. Yang et al. (2008) studied the deformation and failure of the stratified rocks under different supporting schemes and proposed reasonable supporting parameters. Fortsakis et al. (2012) explored the anisotropic behavior of the stratified rocks and analyzed the effects on the structural planes with different dip angles. Wang et al. (2014a,b,c) found that not only the depth of the tunnel, but also the parameters of the joints had significant effects on the loose zone in the stratified rocks. Li et al. (2014) researched the effects of thin coal seam on the roof of roadways with different thicknesses and positions. They found the deformation and failure mechanism of the surrounding rock to the corresponding supporting measures of the roadway.

In summary, these studies have reached some important conclusions regarding the instability mechanisms of the stratified rocks. However, these studies of failure characteristics of the stratified rocks are not sufficiently detailed, and the research on the pressure-arch in the stratified rocks during coal mining is also lacking. Therefore, it is significant to study the pressure-arch distribution characteristics in the horizontal stratified rocks for coal mining and tunnels.

7.2 Pressure-Arch Forming Mechanism in the Stratified Rocks

As shown in Fig. 6.46, with underground coal being mined-out, three distinct zones gradually form in the overlying rock after the mine excavation: the stress-relief zone, the pressure-arch zone, and the original-rock zone. The horizontal stratified rock containing the weak interlayers can be simplified as a layered structure model as illustrated in Fig. 6.46.

In the layered structure model, the mechanical condition of the adjacent strata, which has produced relative dislocation at point A, can be described as follows:

$$\tau > c + \sigma \cdot \tan \varphi \qquad (6.73)$$

where τ is the shear stress, σ is the normal stress, c is the cohesion, and φ is the internal friction angle of the adjacent layers at point A.

For the adjacent strata to separate at point B, it should meet all three of the following conditions:

1. The lithology of the upper part is harder than that of the lower part.

$$f_{\text{upper}} > f_{\text{lower}} \qquad (6.74)$$

where f_{upper} and f_{lower} are the hardness in the Protodyakonov scale of the upper and lower parts at point B, respectively.

2. The mechanical condition is that the shear stress exceeds the shear strength of the layers at point B, or:

$$\sigma > \sigma_t \qquad (6.75)$$

where σ is the normal stress and σ_t is the tensile strength of the adjacent layers at point B.

3. The deformation condition is given as follows:

$$W_{\text{upper}} > W_{\text{lower}} \qquad (6.76)$$

where W_{upper} and W_{lower} are the vertical deflection of the upper and lower parts at point B, respectively.

After the coal is mined-out, due to lack of rock support, the separation occurs in the overlying rock in the mine opening and the relative slip

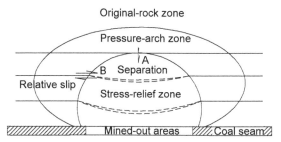

FIGURE 6.46 Pressure-arch in stratified rocks after coal being mined-out.

between layers appears. The deformation and the damage in the stratified rocks result in a stress-relief zone being formed. From the stress-relief zone to the deep rock in the mine opening, the deformation gradually reduces whereas the stress gradually increase until the peak is reached. From the pressure-arch zone to the deeper surrounding rock, the stress gradually returns to the initial state to form a transition zone before reaching the original-rock zone.

7.3 Computational Model and Characteristics of the Pressure-Arch

7.3.1 The Computational Model

As shown in Fig. 6.47, the 3D computational model was built using FLAC3D under the fully mechanized mining conditions. The computational model is 240-m long, 240-m wide, and 200-m high, and the dimensions of the fully mechanized mining panel are 100-m long, 100-m wide and 3.5-m high in the x-, y-, and z-axes, respectively. There is a 600-m thick rockmass above the top of the model with the average density of 2500 kg/m^3, which can be converted into an application of a vertical load of p on the model, ie, $p = 15$ MPa.

The horizontal displacement of the four lateral boundaries of the model was restricted; the vertical load p was applied onto the top of the model; and its bottom was fixed. The material of the model was supposed to meet the Mohr–Coulomb strength criterion, and the physical and mechanical parameters were selected and listed in Table 6.7.

7.3.2 Mechanical Characteristics of the Pressure-Arch

Based on previous research results (Wang et al., 2014a,b,c), the inner boundary of the pressure-arch was fixed at the peak point of maximum principal stress and its outer boundary was fixed at the point where the ratio of maximum principal stress to the difference between the maximum principal stress and the minimum principal stress was equal to 10%.

As shown in Figs. 6.47 and 6.48, along the positive direction on the z-axis along the monitoring line, the greater the vertical distance H between the overlying strata and the coal seam, the lesser

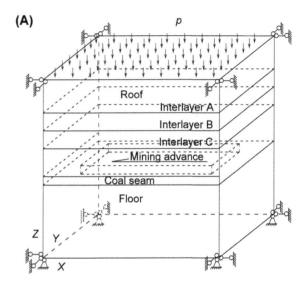

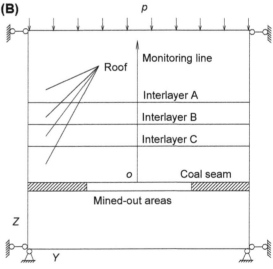

FIGURE 6.47 Schematic of the computational model. (A) Three-dimension model; (B) Y–Z section of the model.

TABLE 6.7 Physical and Mechanical Parameters of the Model

Name	Density (kg/m³)	Elastic Modulus (GPa)	Poisson's Ratio	Tension (MPa)	Cohesion (MPa)	Friction Angle (Degrees)
Layer 1	2600	8.0	0.36	1.2	4.0	36
Layer 2	2600	11.0	0.33	1.5	4.5	38
Layer 3	2600	14.0	0.30	1.7	5.0	40
Layer 4	2600	17.0	0.28	2.0	8.0	43
Roof	2600	23.0	0.24	2.5	9.0	44
Coal seam	1440	3.0	0.38	1.0	3.0	30
Floor	2700	25.0	0.23	3.0	10.0	45
Bedding	—	—	—	0.1	0.3	24
Weak interlayers	2550	2.0	0.40	0.1	0.3	24

the vertical displacement. In addition, as H increases, both the maximum and the minimum principal stress gradually rise. At the position of the structural plane, the maximum principal stress reduces and delamination occurs.

When H reaches 55 m along the monitoring line, the maximum principal stress reaches the peak value and separation no longer occurs at the inner boundary of the pressure-arch. Similarly, when H reaches at 100 m, the ratio of the maximum principal stress to the difference between the maximum and the minimum principal stress was equal to 10%, ie, the stress gradually returns to the initial state, where the outer boundary of the pressure-arch was determined.

As shown in Figs. 6.47 and 6.49, along the positive direction on the y-axis, the displacements

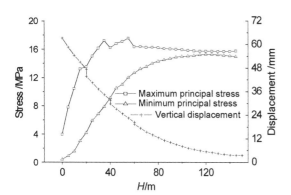

FIGURE 6.48 Stress and vertical displacement on the Y–Z section.

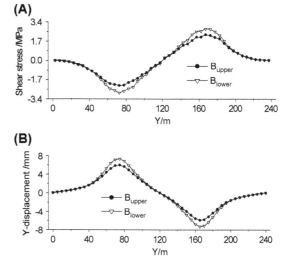

FIGURE 6.49 Shear stress on interlayers B and y-displacement curves. (A) Shear stress on interlayers B; (B) y-displacement curves.

TABLE 6.8 Calculation Projects of the Computational Model

Project	Variable	Value
1	Number of interlayers (N)	0, 1, and 3
2	Distance from the weak interlayers to the coal seam (H)	30, 80, and 120 m
3	Thickness of the weak interlayers (T)	0.01, 0.5, and 1.0 m
4	Lateral pressure coefficient (λ)	0.95, 1.0, and 1.2
5	Spacing of the bedding plane (S)	10, 20, and 30 m

are different. This means that relative slip appear at interlayers B. The relative slip reaches the maximum value at the points of 72.5 and 167.5 m, and the tendencies of shear stress and shear displacement are similar. Thus the inner boundary of the pressure-arch could be determined at the point where the shear stress reached maximum value. Relative slip reaches the minimum value at the points of 42.5 and 197.5 m, and the shear stress gradually return to the initial state; here, the outer boundary of the pressure-arch could be determined.

7.3.3 Numerical Simulation Programs

To reflect the distribution characteristics of the pressure-arch in horizontal stratified rocks under different conditions, five calculation projects were formulated and listed in Table 6.8.

7.4 Result and Discussion

7.4.1 The Number N of Interlayer Variation

As shown in Fig. 6.50A, without stratification of the surrounding rock, the shape of the pressure-arch in the fully mechanized mining field is symmetrical and its boundaries form a

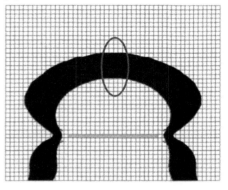

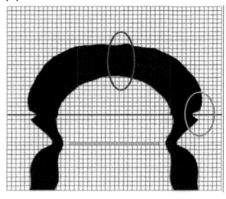

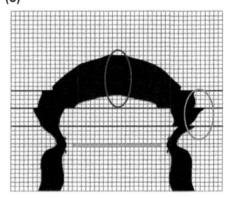

FIGURE 6.50 Shape of pressure-arch variation of N. (A) $N = 0$; (B) $N = 1$; (C) $N = 3$.

smooth curve. Fig. 6.50B shows that when there is one weak interlayer in the surrounding rock, the pressure-arch formed in the upper and lower parts constitute a new pressure-arch. The thickness and height of the new pressure-arch are higher, and it becomes flat at the vault. The outer boundary of the new pressure-arch at the weak interlayers moves inward, forming a dent. In Fig. 6.50C, when there are three interlayers in the surrounding rock with a 20-m spacing, the change in the shape of the pressure-arch is more apparent. The new pressure-arch becomes pointed at the vault. The inner and outer boundaries of the pressure-arch move inward and obvious faulting results in the structural planes. As seen from Fig. 6.50C, the degree of faulting tends to a more significant extent with the interlayers closer to the vault of the pressure-arch. The inner boundary at the vault is limited between the first and the second planes.

As shown from Fig. 6.51A, when there was one weak interlayer in the surrounding rock, both areas A of stress-relief and pressure-arch zones are larger than the other two values and the area of pressure-arch zone is larger that of the stress-relief. In Fig. 6.51B, the thickness R_T of the vault of the pressure-arch tends to increase with the increase of number of interlayers N. It can be seen that the interlayers could change the shape and distribution of the pressure-arch, affecting its formation. The shear failures are easier to produce in the structural planes.

7.4.2 From the Weak Interlayer to Coal Seam H Variation

The distance between the weak interlayer and the coal seam also affects the pressure-arch. As shown in Fig. 6.52A, when H is 30 m, the thickness of the pressure-arch is the greatest and the outer boundary of the pressure-arch spread to the deeper rock. In Fig. 6.52B, when H is 80 m, the weak interlayer has no effect on the inner boundary of the pressure-arch, but makes the outer boundary move inward. Fig. 6.52C shows that when H is 120 m, the

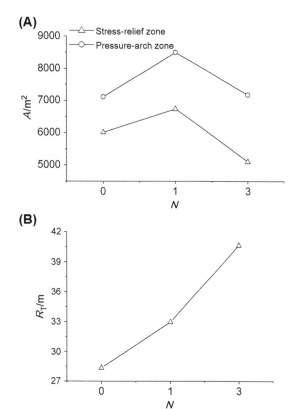

FIGURE 6.51 Zone areas A and vault thickness R_T variation of N. (A) Zone areas A variation of N; (B) vault thickness R_T variation of N.

weak interlayer has no effect on the pressure-arch. Seen from Fig. 6.53, the vault thickness R_T of the pressure-arch and the areas A of the stress-relief and pressure-arch zones nearly both decrease with the increase of the distance H from the weak interlayer to the coal seam. It can be concluded that the weak interlayer almost affect the pressure-arch when the pressure-arch is above the weak interlayer.

7.4.3 The Weak Interlayer Thickness Variation

As shown in Fig. 6.54, as the thickness of the weak interlayer rises, the thickness and the height of the pressure-arch above the weak interlayer

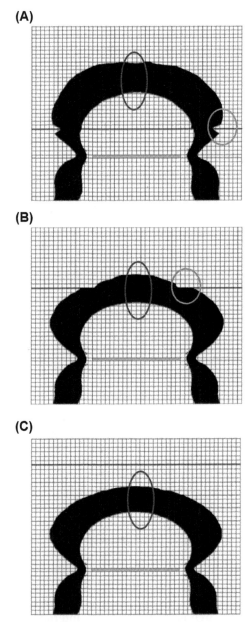

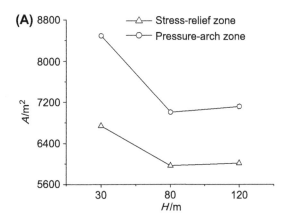

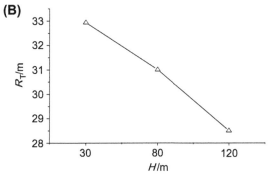

FIGURE 6.53 Zone areas A and vault thickness R_T variation of H. (A) Zone areas A variation of H; (B) vault thickness R_T variation of H.

reduce. This means that a thinner weak interlayer has a negative effect on the carrying capacity of the pressure-arch and more rocks are needed to participate in the transmission of load. As seen from Fig. 6.55, not only the vault thickness R_T of the pressure-arch, but also the areas A of the stress-relief and pressure-arch zones totally decrease with the thickness T increase of the weak interlayer. However, the thickness of the pressure-arch with the weak interlayer increase, which results in the formation of several larger dents on the pressure-arch.

FIGURE 6.52 Shape of pressure-arch variation of H. (A) $H = 30$ m; (B) $H = 80$ m; (C) $H = 120$ m.

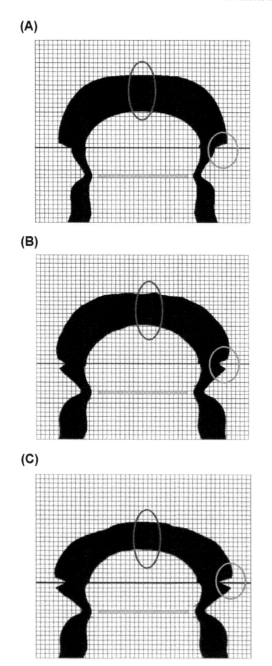

FIGURE 6.54 Shape of pressure-arch variation of T. (A) $T = 0.01$ m; (B) $T = 0.5$ m; (C) $T = 1.0$ m.

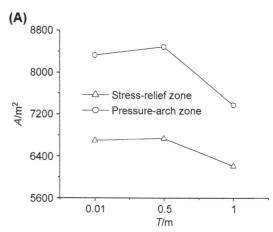

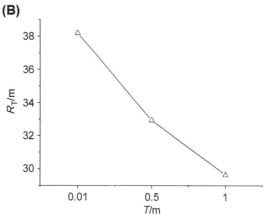

FIGURE 6.55 Zone areas A and vault thickness R_T variation of T. (A) Zone areas A variation of T; (B) vault thickness R_T variation of T.

7.4.4 The Lateral Pressure Coefficient Variation

As shown in Fig. 6.56, as the lateral pressure coefficient λ increases, the height of the pressure-arch increases significantly. The thickness of the vault increases but that of the feet reduce. The shape of the pressure-arch changed gradually from a flat arch to a cuspate arch as the lateral pressure coefficient changes from $\lambda < 1$ to $\lambda > 1$. In Fig. 6.57, with λ changing

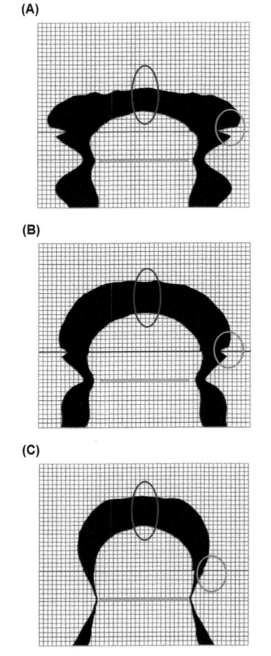

FIGURE 6.56 Shape of pressure-arch variation of λ. (A) $\lambda = 0.95$; (B) $\lambda = 1.0$; (C) $\lambda = 1.2$.

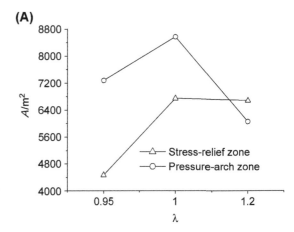

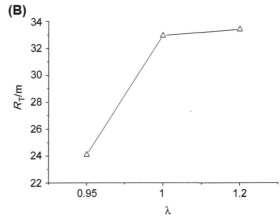

FIGURE 6.57 Zone areas A and vault thickness R_T variation of λ. (A) Zone areas A variation of λ; (B) vault thickness R_T variation of λ.

from 0.95, 1.0 to 1.2, both the areas A of the stress-relief zones and the vault thickness R_T of the pressure-arch increase significantly. However, the area A of the pressure-arch zone first increases then reduces when λ increases. If the lateral pressure coefficient is too small, there is no pressure-arch being formed. Conversely, if the lateral pressure coefficient is too large, no pressure-arch is followed on the sides of the mine excavation. So, a stable pressure-arch could form only when the lateral pressure coefficient is appropriate.

7.4.5 The Bedding Planes Spacing S Variation

As shown in Fig. 6.58, when the spacing is 10 m, tangential dislocations appear in the pressure-arches in every layer. Compared with the spacing of 10, 20, and 30 m, the faulting effects are abated as the spacing becomes larger, and the pressure-arch retract to being thin and flat. In Fig. 6.59, as S changes from 10, 20–30 m, the areas A of pressure-arch zone and the vault thickness of the pressure-arch all reduce. The areas A of stress-relief zone first decrease then increase with increasing S. The results show that structural planes would reduce

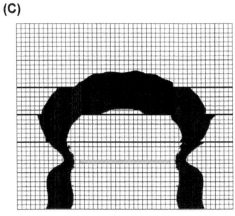

FIGURE 6.58 Shape of pressure-arch variation of S. (A) $S = 10$ m; (B) $S = 20$ m; (C) $S = 30$ m.

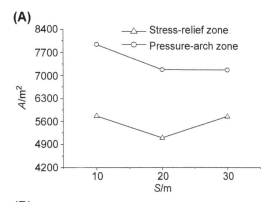

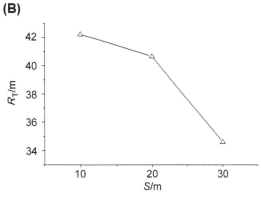

FIGURE 6.59 Zone areas A and vault thickness R_T variation of S. (A) Zone areas A variation of S; (B) vault thickness R_T variation of S.

the bearing capacity of the surrounding rock and more rocks are needed to bear the loads.

7.5 Conclusions

Compared to the surrounding rock without a weak interlayer, the new forming pressure-arch in the stratified rock appears indented at the primary structural planes. Those structural planes limit the development of the pressure-arch and reduce the stability of the surrounding rock.

The weak interlayers influence the pressure-arch when the pressure-arch is above the weak interlayers, and these effects would reduce as the distance between the weak interlayers and the coal seam increase. The thickness of the weak interlayers has an impact on the shape of the pressure-arch.

Furthermore, as the lateral pressure coefficient increases, the height of the pressure-arch increases significantly and the shape changes from a flat arch to a cuspate arch. A stable pressure-arch can form only when the lateral pressure coefficient is appropriate. The spacing of the structural plane affects the distribution of the pressure-arch and the smaller spacing is more disadvantageous in terms of the bearing capacity of the surrounding rock.

8. PRESSURE-ARCH IN A FULLY MECHANIZED MINING FIELD

8.1 Introduction

With the underground coal mining out, the equilibrium of the original 3D stress of the mining field is broken, the stress of the surrounding rock gradually transfers from the mining face to the deep of the rock, and the concentration area of compressive stress will generate within a certain range in the surrounding rock, which forms the pressure-arch in the mining field (Wu et al., 2012). The pressure-arch of the mining field is the result of the stress self-regulating. The pressure-arch usually has mechanics properties of the arch structure, which can sustain its loading and transfer the other loads to the foot of the pressure-arch, and this is of great significance to ensure the stability of the surrounding rock of the mining field (Wang et al., 2013).

Many scholars have done the relevant research on the pressure-arch in the engineering rocks and have made a lot of achievements. For example, Huang et al. (2002) studied the formation mechanism of the natural arch for underground chamber and made a preliminary conclusion. Poulsen (2010) conducted the coal pillar load calculation by pressure-arch theory. Trueman et al. (2008) researched the multiple draw−zone interaction in block caving mines by means of a large physical model. Liang et al. (2012) had defined the pressure-arch of a tunnel, and the inner and outer boundaries of the pressure-arch were determined by the stress analysis, but the determination of outer boundary for the arch body is not clear enough. Xie (2006) had discovered the shell structure of macroscopic stress of the mining field and studied influence factors, such as mining height and other factors on the mechanical characteristics of the surrounding rock of the fully mechanized caving face. Du et al. (2011) discussed the stress variation of the mining field with the working face advancing, but the influence factors on the pressure-arch were not considered. Geng (2009) simulated the variation of the stress arch in the overlying strata after coal mining out by using Universal Distinct Element Code (UDEC). Xu et al. (2006) had derived the deflection formulas of the rock slab based on the pressure-arch theory.

In summary, these studies on the geometry parameters and mechanics evolution characteristics of the pressure-arch are still not deep enough, and the discussions are mainly restricted to 2D problems, so an in-depth study on the mechanics evolution characteristics of the 3D pressure-arch in the fully mechanized mining field is of important theoretical and practical significance.

8.2 Pressure-Arch Geometry Parameters

As shown in Fig. 6.60, after the coal was mined-out, the 3D stress balance of the mining field was broken, then the minimum principal stress at the mining free face became zero, and the stress value gradually increased from the free face to the deep of the surrounding rock until the original stress state. Meanwhile, the maximum principal stress underwent the change from small to the peak value and then decreased to the original stress state. The pressure-arch formed as a result of the 3D stress self-regulating.

Based on the previous research results, the inner boundary of the pressure-arch are fixed at point A, where is the maximum principal stress (Fig. 6.60).

To determine the outer boundary of the pressure-arch, the variable k is defined as

$$k = \frac{\sigma_1 - \sigma_3}{\sigma_1} \quad (6.77)$$

where σ_1 is the maximum principal stress, and σ_3 is the minimum principal stress of the surrounding rock of the mining field. When k is equal to 10%, the corresponding point B is determined, which is at the outer boundary of the pressure-arch (Fig. 6.60).

From the two points in the seam floor which also are at both laterals of the mining field as the starting points, two lines are drawn at the rock rupture angle θ and are extended upward, respectively, until they meet with the inner and outer boundaries of the pressure-arch, thus the lateral boundaries of pressure-arch are determined as shown in Fig. 6.60.

The rupture angle θ of the surrounding rock is defined as

$$\theta = 45° + \frac{\varphi}{2} \quad (6.78)$$

where φ is the inner friction angle of the surrounding rock. The pressure-arch of the surrounding rock is not only in the roof of the mining field but also in both sides and the floor of the mining field. In this chapter, the pressure-arch is determined mainly on the top of the mining field as shown in Fig. 6.60. For the purpose of analysis, the geometry parameters of pressure-arch are shown in Fig. 6.61.

Taking C_1 line and C_2 line as the skewbacks, the arch thickness are h_{f1} and h_{f2} and the peak value of the maximum principal stress are P_{f1} and P_{f2}, respectively. And similarly, taking B_1 line and B_2 line as arch waist, the arch thickness are h_{m1} and h_{m2} and the peak value of the maximum principal stress are P_{m1} and P_{m2}, respectively. The line A represents the vault, where the arch thickness is h_t and the peak value of maximum principal stress is P_t. The dotted line is the central axis of the pressure-arch, and it is called the pressure-arch curve. As shown in Fig. 6.61, the smooth pressure-arch curve formed through the median value of the lines between the inner and the outer boundary along the normal direction of the pressure-arch.

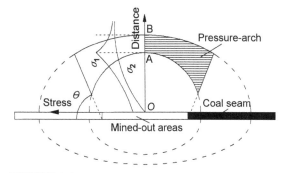

FIGURE 6.60 Schematic boundaries of the pressure-arch.

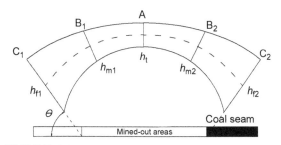

FIGURE 6.61 Geometry parameters of the pressure-arch.

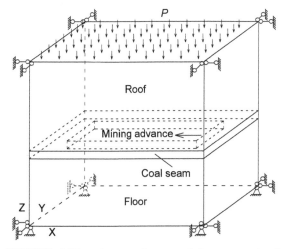

FIGURE 6.62 Schematic diagram of the computational model.

8.3 Three-Dimensional Computational Model

Taking practical engineering in a mine as background, after the rock group on the roof and floor of the coal seam is generalized, the three-dimension computational model was built using FLAC3D under the fully mechanized mining condition.

As shown in Fig. 6.62, the computational model was 240-m long, 240-m wide, and 200-m high, and the dimensions of the fully mechanized mining was 100-m long, 100-m wide, and 3.5-m high in the x-, y-, and z-axis, respectively. The fully mechanized mining was at the depth of 600-m under the ground, and it was in hydrostatic state of stress. There was 600-m thickness rock above the top of the model, and the vertical load p converted from the rock weight is 15 MPa, where the rock average weight is 25 kN/m^3.

The horizontal displacement of four lateral boundaries of the model were restricted, the top of the model was applied the vertical load p, and its bottom was fixed. The material of the model was supposed to meet the Mohr–Coulomb strength criterion, and three groups of the physics and mechanics parameters were selected as listed in Table 6.9.

8.4 Calculation Results and Discussion

8.4.1 Distance Variation of the Mining Advancing

Seen from Fig. 6.63A, for the horizontal seam, along the dip of the coal seam, the geometric shape of the pressure-arch varied from flat arch

TABLE 6.9 Physical and Mechanical Parameters of the Computational Model

Classification	Name	Density (kg/m^3)	Elasticity Modulus (GPa)	Poisson Ratio	Tensile Strength (MPa)	Cohesion (MPa)	Friction Angle (Degrees)
Soft rock	Roof	2200	4.5	0.26	1.20	3.5	40
	Coal	1440	3.0	0.28	1.00	3.0	30
	Floor	2300	5.0	0.25	1.50	4.5	42
Medium rock	Roof	2600	20.0	0.25	2.00	8.0	43
	Coal	1440	3.0	0.28	1.00	3.0	30
	Floor	2700	15.0	0.24	3.00	10.0	45
Hard rock	Roof	2600	60.0	0.24	10.00	35.0	48
	Coal	1440	3.0	0.28	1.00	3.0	30
	Floor	2700	80.0	0.23	12.00	40.0	50

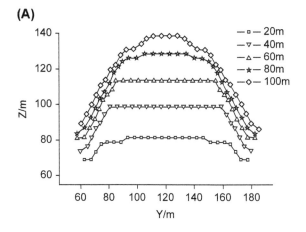

addition, the height of the pressure-arch gradually increased.

For the horizontal coal seam, whether along the strike or the dip of the coal seam, the pressure-arch shapes in the fully mechanized mining field were symmetrical, namely $h_{f1} = h_{f2}$, $P_{f1} = P_{f2}$, $h_{m1} = h_{m2}$, and $P_{m1} = P_{m2}$. Therefore, we can take half of the pressure-arch to study the variation characteristics of arch thickness and the maximum principal stress in the arch.

As shown in Fig. 6.64, with working face advancing the thickness of skewback, arch waist

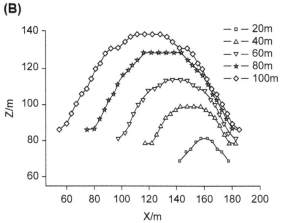

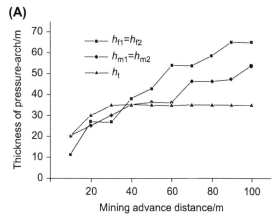

FIGURE 6.63 Shape characteristics of the central axis of the pressure-arch. (A) Along dip of coal seam; (B) along strike of coal seam.

to round arch gradually with the working face advancing. The height of the pressure-arch increased, and the central axis of the pressure-arch gradually moved to the deep of the surrounding rock from the mining face. Seen from Fig. 6.63B, along the strike of the horizontal coal seam, the pressure-arch varied from the pointed arch to well-arched shape, the openings of the pressure-arch increased gradually, and the front foot of the pressure-arch constantly moved forward with working face advancing. In

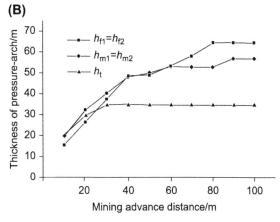

FIGURE 6.64 Thickness variation characteristics of the pressure-arch. (A) Along dip of coal seam; (B) along strike of coal seam.

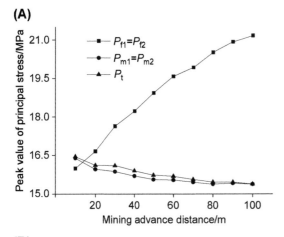

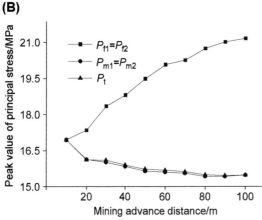

FIGURE 6.65 Peak value of principal stress variation of the pressure-arch. (A) Along dip of coal seam; (B) along strike of coal seam.

the thickness of the skewback constantly increased with the increase of mining advanced distance.

8.4.2 Rock Strength Variation of the Mining Field

Seen from Fig. 6.66, with the strength of the surrounding rock from soft to hard, along the dip of the seam, the pressure-arch shape changed from flat-top arch to round arch, and the height of the arch-pressure was getting away from the mining face. Along the strike of the seam, the pressure-arch shape changed

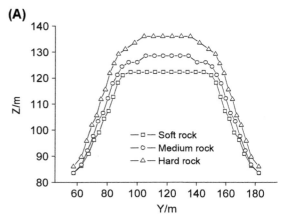

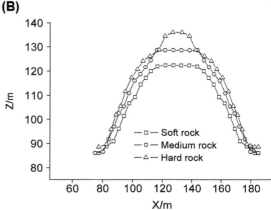

FIGURE 6.66 Shape characteristics of the central axis of the pressure-arch. (A) Along dip of coal seam; (B) along strike of coal seam.

and vault increased, but the vault thickness would no longer continue to increase to a certain value, the overall trend was $h_{f1} = h_{f2} > h_{m1} = h_{m2} > h_t$.

As seen from Fig. 6.65, the maximum principal stress produced in the foot of the pressure-arch. With the working face advancing, the maximum principal stress in the skewback gradually increased, but in other parts the maximum principal stress constantly decreased. The results showed that the pressure of the skewback increased due to loads passing, so

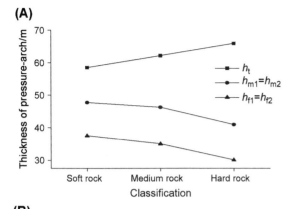
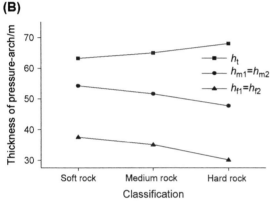

FIGURE 6.67 Thickness variation characteristics of the pressure-arch. (A) Along dip of coal seam; (B) along strike of coal seam.

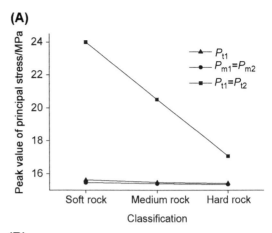
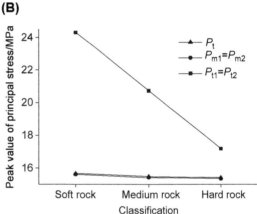

FIGURE 6.68 Peak value of principal stress variation of the pressure-arch. (A) Along dip of coal seam; (B) along strike of coal seam.

from the flat-top arch to the pointed arch gradually.

Overall, with the strength of the surrounding rock from soft to hard, except that the vault thickness had a slight increase, the thickness of the other parts of the pressure-arch decreased (Fig. 6.67). Meanwhile, as shown in Fig. 6.58, the peak value of the maximum principal stress in the arch foot demonstrated a significant decreasing trend. Thus, the higher the rock strength was, the more easily the pressure-arch formed. The arch thickness showed that the arch loading was reducing, and the decrease of peak value of the maximum principal stress in

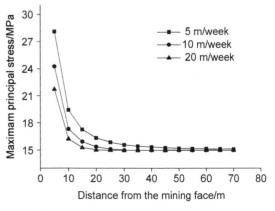

FIGURE 6.69 Maximum stress variation with mining speed.

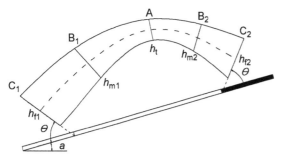

FIGURE 6.70 Asymmetric shape of pressure-arch.

the pressure-arch showed that the mining field was more stable.

8.4.3 Velocity Variation of the Mining Advancing

Assuming three mining speed was 5, 10, and 20 m/week, the calculation results were shown in Fig. 6.69. Seen from Fig. 6.69, the slower the mining speed was, the more obvious stress the concentration phenomenon of the surrounding rock was. Therefore, to improve the mining speed can significantly reduce the stress concentration in the surrounding rock in front of the working face, and also can do some contributions to the stability of the pressure-arch in the mining field.

8.4.4 Dip Angle Variation of the Coal Seam

Fig. 6.70 demonstrates the geometry parameters of the pressure-arch with dip α of the coal seam.

As shown in Figs. 6.71, 6.72, and 6.73, with the dip angle of the coal seam changing from 0, 10, 20–30 degrees along the strike of the seam, the pressure-arch was symmetrical. But along the dip of the seam, the pressure-arch displayed an asymmetric shape, and the vault was tilted and moved to the upward direction. The greater the

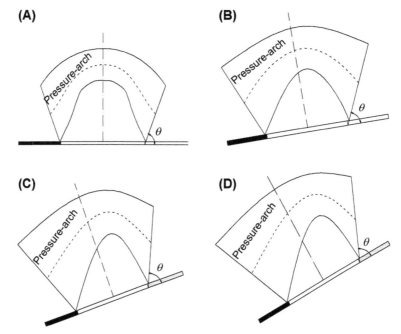

FIGURE 6.71 Pressure-arch shape characteristics with different dips. (A) 0 degree; (B) 10 degrees; (C) 20 degrees; (D) 30 degrees.

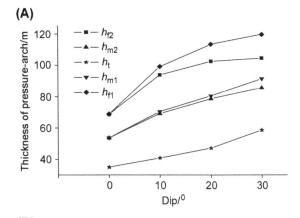

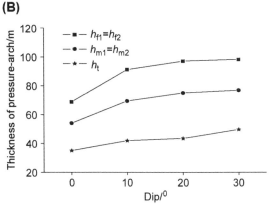

FIGURE 6.72 Thickness variation of the pressure-arch with different dips. (A) Along dip of coal seam; (B) along strike of coal seam.

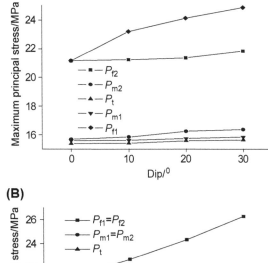

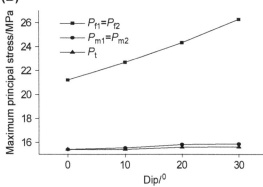

FIGURE 6.73 Peak value of principal stress variation with different dips. (A) Along dip of coal seam; (B) along strike of coal seam.

dip angle was, the more apparent the inclination of the pressure-arch was. The arch thickness and peak stress is not asymmetric, that is $h_{f1} \neq h_{f2}$, $P_{f1} \neq P_{f2}$, $h_{m1} \neq h_{m2}$, $P_{m1} \neq P_{m2}$.

Seen from Figs. 6.71 and 6.72, with the dip angle of the coal seam increased, the arch thickness and the maximum principal stress also increased, and the increasing trend of the thickness was $h_{f1} > h_{f2} > h_{m1} > h_{m2} > h_t$ in general. Meanwhile, the skewback located in the low position of the pressure-arch would bear more loading, and the stress concentration tended to be further intensified (Fig. 6.73).

8.5 Conclusion

For the horizontal seam, with the working face advancing, the geometric shape of the pressure-arch varied from flat arch to round arch gradually in the fully mechanized mining field. The height of the pressure-arch increased, and the thickness of skewback, arch waist, and vault also increased.

With the strength increase of the surrounding rock from soft to hard rock, the thickness and the loading of pressure-arch decreased. To improve the mining speed can significantly reduce the stress concentration in the surrounding rock in front of the working face, and that can have some contribution to the stability of the pressure-arch in the mining field.

With the increase of dip angle of the coal seam, along the dip of the seam, the pressure-arch displayed an asymmetric shape, and the vault was tilted and moved to the upward direction. Meanwhile, the stress concentration tended to be further intensified, and the skewback located in the low position of the pressure-arch would bear more load than the other one.

References

Alehossein, H., 2009. A triangular caving subsidence model. Transactions of the Institutions of Mining and Metallurgy, Section B: Applied Earth Science 118 (1), 1–4.

Ayati, M., Khaloozadeh, H., 2010. Stable chaos synchronisation scheme for non-linear uncertain systems. IET Control Theory and Applications 4 (3), 437–447.

Du, F., Bai, H.B., 2012. Mechanical model for immediate roof structure of thin bedrock in fully-mechanized sublevel caving face. Electronic Journal of Geotechnical Engineering. September 17V, 3075–3088.

Du, X.L., Song, H.W., Chen, J., 2011. Numerical simulation of the evolution of the pressure-arch during coal mining. Journal of China University of Mining & Technology 40 (6), 863–867.

Emadi, H., Mahzoon, M., 2012. Investigating the stabilizing effect of stochastic excitation on a chaotic dynamical system. Nonlinear Dynamics 67 (1), 505–515.

Fortsakis, P., Nikas, K., Marinos, V., Marinos, P., 2012. Anisotropic behaviour of stratified rock masses in tunneling. Engineering Geology 141–142, 74–83.

Fu, B.L., Tan, W.F., 1995. Reciprocal theorem method for solving the problem of thick rectangular plates. Journal of Applied Mathematics and Mechanics 16 (4), 367–379.

Geng, Y.M., 2009. Study on evolution law of stress arch of overlying strata in mines. Journal of Shandong University of Science and Technology (Natural Science) 28 (4), 43–47.

Ghosh, A.K., Gong, Y.X., 2014. Improving coal recovery from longwall top coal caving. Journal of Mines, Metals and Fuels 62 (3), 51–57.

Gorji, M.H., Torrilhon, M., Jenny, P., 2011. Fokker-Planck model for computational studies of monatomic rarefied gas flows. Journal of Fluid Mechanics 680, 574–601.

He, G.L., 2009. Determination of critical thickness of stiff roof in coal mine based on thick plate theory. Chinese Journal of Underground Space and Engineering 5 (4), 659–663.

He, G.L., Li, D.C., Zhai, Z.W., Tang, G.Y., 2007. Analysis of instability of coal pillar and stiff roof system. Journal of China Coal Society 32 (9), 897–901.

Hu, H.F., Hao, B.Y., Kand, L.X., 2008. Effect analysis on treatment of mined-out under the highway. Journal of Xi'an University of Science and Technology 28 (2), 270–273.

Huang, Z.P., Broch, E., Lu, M., 2002. Cavern roof stability-mechanism of arching and stabilization by rock bolting. Tunnelling and Underground Space Technology 17 (3), 249–261.

Jain, A., Guzina, B.B., Voller, V.R., 2007. Effects of overburden on joint spacing in layered rocks. Journal of Structural Geology 29, 288–297.

Jia, P., Tang, C.A., Wang, S.H., 2006. Destroy mechanism of tunnel with stratified roof. Journal of China Coal Society 31 (1), 11–15.

Jiang, X.L., Cao, P., Yang, H., Lin, H., 2009. Effect of horizontal stress and rock crack density on roof safety thickness of underground area. Journal of Central South University (Science and Technology) 40 (1), 211–216.

Li, G.R., She, C.X., Chen, S.H., 1997. Destructive testing and finite element analysis of bending deformation of layered rock slope. Chinese Journal of Rock Mechanics and Engineering 16 (4), 305–311.

Li, H., Liang, B., Li, G., Bai, Y.P., Zhang, C.M., 2010. Prediction on bending deflection of overlying strata caused by steeply-inclined coal seam mining in mountainous area. The Chinese Journal of Geological Hazard and Control 21 (3), 101–104.

Li, H.J., Fu, B.L., Tan, W.F., 1998. Reciprocal theorem method for the forced vibration of elastic thick rectangular. Applied Mathematics and Mechanics 19 (2), 175–188.

Li, S., Xu, W., Li, R.H., 2006. Chaos control of duffing systems by random phase. Acta Physica Sinica 55 (3), 1049–1054.

Li, W.T., Wang, Q., Li, S.C., Wang, D.C., Huang, F.C., Zuo, J.Z., Zhang, S.G., Wang, H.T., 2014. Deformation and failure mechanism analysis and control of deep roadway with intercalated coal seam in roof. Journal of China Coal Society 39 (1), 47–56.

Li, X.H., Jiang, D.Y., Liu, C., Ren, S., 2005. Study on treatment technology of highway tunnel through mined-out area. Rock and Soil Mechanics 26 (6), 910–914.

Li, X.Y., Gao, F., Zhong, W.P., 2008. Analysis of fracturing mechanism of stope roof based on plate model. Journal of Mining & Safety Engineering 2, 180–183.

Liang, X.D., Zhao, J., Song, H.W., 2012. Experimental and numerical analysis on the arching action from stress adjusting in surrounding rocks. Journal of Engineering Geology 20 (1), 96–102.

Lin, H.F., Li, S.G., Cheng, L.H., 2008. Key layer distinguishing method of overlying strata based on the thin slab theory. Journal of China Coal Society 33 (10), 1081–1085.

Lin, J.H., Lu, Z.H., 1992. The local modification scheme of finite difference method used to numerical solution of

Fokker-Planck equation. Journal of Beijing Normal University (Natural Science) 28 (4), 497–501.

Liu, H., Hu, Q.T., Wang, J.A., Li, J.G., 2009. Analysis on stability of pillar and stiff roof system in the gob area. Journal of Coal Science & Engineering 15 (2), 206–209.

Liu, Y., 2012. Fast evaluation of canonical oscillatory integrals. Applied Mathematics & Information Sciences 6 (2), 245–251.

Liu, B., Yue, Z.Q., Tham, L.G., 2004. Analytical design method for a truss-bolt system for reinforcement of fractured coal mine roofs - illustrated with a case study. International Journal of Rock Mechanics and Mining Sciences 42 (2), 195–218.

Lu, Z.H., Lin, J.H., Hu, G., 1993. A numerical study of the Fokker-Planck equation for stochastic resonance problem. Acta Physica Sinica 42 (10), 1556–1566.

Nierobisz, A., 2011. Development of roof bolting use in Polish coal mines. Journal of Mining Science 47 (6), 751–760.

Nomikos, P.P., Sofianos, A.I., Tsoutrelis, C.E., 2002. Structural response of vertically multi-jointed roof rock beams. International Journal of Rock Mechanics & Mining Sciences 39 (1), 79–94.

Palei, S.K., Das, S.K., 2009. Logistic regression model for prediction of roof fall risks in bord and pillar workings in coal mines. Safety Science 47 (1), 88–96.

Poulsen, B.A., 2010. Coal pillar load calculation by pressure arch theory and near field extraction ratio. International Journal of Rock Mechanics and Mining Sciences 47 (7), 1158–1165.

Qin, S.Q., Wang, S.J., 2005. Instability leading to rockbursts and nonlinear evolutionary mechanisms for coal-pillar-and-roof system. Journal of Engineering Geology 13 (4), 437–446.

Ren, Y.F., Qi, Q.X., 2011. Study on characteristic of stress field in surrounding rocks of shallow coalface under long wall mining. Journal of China Coal Society 36 (10), 1612–1618.

Shao, X.P., Shi, P.W., He, G.C., 2007. Analysis on unloaded arch structure of roof in mining steep seams using horizontal section top-coal caving. Journal of University of Science and Technology, Beijing 29 (5), 447–451.

Singh, P.K., Roy, M.P., 2008. Characterisation of blast vibration generated from open-pit blasting at surface and in belowground openings. Transactions of the Institutions of Mining and Metallurgy 117 (3), 122–127.

Sinnott, M.D., Cleary, P.W., 2009. Vibration-induced arching in a deep granular bed. Granular Matter 11 (5), 345–364.

Swift, G.M., Reddish, D.J., 2002. Stability problems associated with an abandoned ironstone mine. Bulletin of Engineering Geology and the Environment 61 (3), 227–239.

Tian, Z.H., Nie, Y.X., 2009. Determination method of the minimum safety thickness of the roof in the complicated mined-out regions. Open Pit Mining Techniques 9 (5), 26–27.

Tikov, B., Mostafa, B., 2008. Design and application of underground mine paste backfill technology. Geotechnical and Geological Engineering 26 (2), 147–174.

Tong, L.Y., Liu, S.Y., Qiu, Y., Fang, L., 2004. Current research state of problems associated with mined-out regions under expressway and future development. Chinese Journal of Rock Mechanics and Engineering 23 (7), 1198–1202.

Trueman, R., Castro, R., Halim, A., 2008. Study of multiple draw-zone interaction in block caving mines by means of a large 3D physical model. International Journal of Rock Mechanics and Mining Sciences 45 (7), 1044–1051.

Tsesarsky, M., 2012. Deformation mechanisms and stability analysis of undermined sedimentary rocks in the shallow subsurface. Engineering Geology 133–134, 16–29.

Wang, J.A., Shang, X.C., Ma, H.T., 2008. Investigation of catastrophic ground collapse in Xingtai gypsum mines in China. International Journal of Rock Mechanics & Mining Sciences 45 (8), 1480–1499.

Wang, L.G., Liu, C., Zhao, N., Wang, J.G., 2007. Research on accumulative destruction of rock mass affected by dynamic. Journal of Bohai University (Natural Science Edition) 28 (1), 1–5.

Wang, S.R., Jia, H.H., Wu, C.F., 2010. Determination method of roof safety thickness in the mined-out regions under dynamic loading and its application. Journal of the China Coal Society 35 (8), 1263–1268.

Wang, S.R., Jia, H.H., Yang, W.B., 2012. Analysis of accumulated damage effects on the roof of mined-out areas under blasting vibration waves. Advanced Science Letters 15 (1), 410–416.

Wang, S.R., Li, N., Li, C.L., Hagan, P., 2014a. Mechanics evolution characteristics analysis of pressure-arch in fully-mechanized mining field. Journal of Engineering Science and Technology Review 7 (4), 40–45.

Wang, S.R., Wang, H., Hu, B.W., 2013. Analysis of catastrophe evolution characteristics of the stratified rock roof in shallow mined-out areas. Disaster Advances 6 (Suppl. 1), 59–64.

Wang, S.R., Xu, D.F., Hagan, P., Li, C.L., 2014b. Fracture characteristics analysis of double-layer rock plates with both ends fixed condition. Journal of Engineering Science and Technology Review 7 (2), 60–65.

Wang, S.R., Zhang, H.Q., Shen, N.Q., Cao, H.Y., 2009. Analysis of deformation and stress characteristics of highway tunnel-bridges above mined-out regions. Chinese Journal of Rock Mechanics and Engineering 28 (6), 1144–1151.

Wang, Z.W., Qiao, C.S., Song, C.Y., Xu, J.F., 2014c. Upper bound limit analysis of support pressures of shallow tunnels in layered jointed rock strata. Tunnelling and Underground Space Technology 43, 171–183.

Wen, Y., 2007. Hazards analysis of goafs underlying highway and the controlling measurements. Soil Engineering and Foundation 21 (1), 35–37.

Wu, Y.P., Wang, H.W., Xie, P.S., 2012. Analysis of surrounding rock macro-stress arch-shell of long-wall face in steeply dipping seam mining. Journal of China Coal Society 37 (4), 559–564.

Xiao, L., Li, F., Wang, S., 2006. Convergence of the Vlasov-Poisson-Fokker-Planck system to the incompressible Euler equations. Science in China (Series A: Mathematics) 49 (2), 255–266.

Xie, G.X., 2006. Influence of mining thickness on mechanical characteristics of working face and surrounding rock stress shell. Journal of China Coal Society 31 (1), 6–10.

Xie, Y.S., Zhao, Y.S., 2009. Numerical simulation of the top coal caving process using the discrete element method. International Journal of Rock Mechanics and Mining Sciences 46 (6), 983–991.

Xin, Y.J., Gou, P.F., Ge, F.D., 2014. Analysis of stability of support and surrounding rock in mining top coal of inclined coal seam. International Journal of Mining Science and Technology 24 (1), 63–68.

Xu, Y.Y., Yu, G.M., Li, C.J., Xu, J.F., Zhang, M.P., Meng, D., 2006. Study on the laminated board-unloading arch model of the mining overburden. Mining Safety & Environmental Protection 33 (3), 7–10.

Yang, J.H., Wang, S.R., Wang, Y.G., Li, C.L., 2015. Analysis of arching mechanism and evolution characteristics of tunnel pressure arch. Jordan Journal of Civil Engineering 9 (1), 125–132.

Yang, J.P., Chen, W.Z., Zheng, X.H., 2008. Stability study of deep soft rock roadways with weak intercalated layers. Rock and Soil Mechanics 29 (10), 2864–2870.

Yang, R., Scovira, D.S., 2007. Using blast vibration measurements to estimate rock triaxial strains/stresses and dynamic rock strength for blast damage evaluation. 1st Canada-US Rock Mechanics Symposium-Rock Mechanics Meeting Society's Challenges and Demands 2 (12), 1547–1552.

Zhao, B.Y., Liu, B.X., Wan, Y.P., 2008. Numerical simulation of blasting vibration on slope stability. Journal of Engineering Geology 16 (1), 59–62.

Zhan, D.L., Wang, H.Y., Qu, T.Z., 2000. Influence analysis of interface on stability stratified rock mass. Chinese Journal of Rock Mechanics and Engineering 19 (2), 140–144.

Zhan, Z.P., Li, N., Chen, F.F., Swoboda, G., 2007. Influence of different distance of weak interface on stability of underground openings. Rock and Soil Mechanics 28 (7), 1363–1368.

Zhang, B.S., Kang, L.X., Yang, S.S., 2006. Numerical simulation on roof separation and deformation of full seam roadway with stratified roof and large section. Journal of Mining & Safety Engineering 23 (3), 264–267.

Zhang, X.Y., 2009. Analysis of creep damage fracture of upper roof. Journal of Liaoning Technical University (Natural Science) 28 (5), 777–780.

Zhu, F.C., Cao, P., Wan, W., 2006. Determination of safe roof thickness of underground shallow openings based on axisymmetric thick plate mode. Ground Pressure and Strata Control 23 (1), 115–118.

Further Reading

Ding, X.M., Huang, R.Q., Liu, G.S., 2004. Development of pre- processing software for FLAC3D and its application to engineering. Journal of Geological Hazards and Environment Preservation 15 (2), 68–73.

Liao, Q.L., Zeng, Q.B., Liu, T., Lu, S.B., Hou, Z.S., 2005. Automatic model generation of complex geologic body with FLAC3D based on ANSYS platform. Chinese Journal of Rock Mechanics and Engineering 24 (6), 1010–1013.

Wang, S.R., Jia, H.H., 2012. Analysis of creep characteristics of shallow mined-out areas roof under low stress conditions. Applied Mechanics and Materials 105–107, 832–836.

Wang, S.R., Wang, Z.Q., 2013. Analytical solution to the roof bending deflection with mixed boundary in the mined-out areas under uniform load. Applied Mathematics & Information Sciences 7 (2), 579–585.

Wang, S.R., Li, N., Li, C.L., Cao, C., 2015a. Distribution characteristics analysis of the pressure-arch in horizontal stratified rocks under coal mining. Technical Gazette 22 (4), 997–1004.

Wang, S.R., Li, N., Li, C.L., Zou, Z.S., Chang, X., 2015b. Instability mechanism analysis of pressure-arch in coal mining field under different seam dip angles. DYNA 90 (3), 279–284.

Yan, C.B., Xu, G.Y., Yang, F., 2007. Measurement of sound waves to study cumulative damage effect on surrounding rock under blasting load. Chinese Journal of Geotechnical Engineering 29 (1), 88–91.

Index

'*Note*: Page numbers followed by "f" indicate figures and "t" indicate tables.'

A

Accumulated damage effects analysis
 computational model, 363–364, 364f
 deformation characteristics analysis, 367–369, 368f–372f
 determining dynamic load and boundary conditions, 364–365, 365f
 overview, 363
 parameter selection, 364, 364t
 plastic zones analysis, 366, 366f–367f
 simulation programs, 365, 365f
Anchorage performance test
 failure modes types, 191–192, 193f
 method, 191, 192f
 MW9 modified bulb cable, 193–195, 194f–195f
 overview, 187–191
 rock strength effect, 195
 superstrand cable, 192–193, 193f
Anchor tube installation, 175
Antirotation devices, 173, 175f
Apparent dilation angle, 108–109
Australian Coal Association Research Program (ACARP), 159
Axial performance
 anchor section, 199
 borehole size influence, 202, 202f, 202t, 203f
 embedment length influence, 200–201, 200f–201f
 embedment section, 198–199, 198f–199f, 199t
 new LSEPT apparatus, 197–199
 overview, 196–197
 results analysis, 199–202
 test sample strength influence, 201–202, 201f–202f, 201t

B

Bearing angle, 76–77
Bearing plate, 173, 174f
Bolt load transfer
 application example, 138–140, 139f–140f
 failure surface, stress tensor on, 137–138
 fully grouted bolt profiles, 134–138, 134f–135f
 methodology and governing equations, 132–134, 132f–134f
 Mohr–Coulomb failure, 138, 139f
 normal load, failure plane, 135–136, 135f
 overview, 128–132, 129f–132f
 shear load, failure plane, 136–137, 137f
40 Nm bolt torque confinement, 211, 211f
80 Nm bolt torque confinement, 211, 211f
Bond mechanism, 61–62
Bond strength model (BSM), 161, 163–164

C

Cable bolting system, 116–121, 116f–117f, 119f
Cable installation, 175
Cementitious grout
 cubic samples, 182, 188f
 cylindrical samples, 182, 186f–187f
 overview, 177–178
 processing procedures, 180
 sample preparation, 179, 179f, 180t
 sample size effect, 182–186
 stress-strain relationships, 181, 181f, 182t–186t, 183f–185f
 testing process, 179–180, 180f
 W/C ratio effect, 182, 189f
Circular tunnel
 computation model, 251, 252f
 energy balance equation, 251
 energy release quantity, 252–253, 252f
 energy release rate, 253, 253f
 energy release rate calculation, 251
 maximum/minimum principal stress distribution, 253–254, 254f
 overview, 250–251
 principal stress difference, 254–255, 255f
 simulation analysis schemes, 251–252
Coal mining field
 failure models analysis, 360–361, 361f
 instability characteristics, 361–363
 internal forces variation, 362–363, 362f
 mechanical model, 359, 359f–360f
 overview, 357–359, 357f–358f
 variation of characteristics, 361–362, 361f
Compressive failure
 fracture properties on experiment design, 53, 53f
 overview, 52–53
 procedure, 54
 sample collection, 53–54
 specimen preparation, 54, 54t
Cone bolts, 73
 mechanical model development, 76–80
 geometric face angles, 84, 84f–85f
 overview, 73–74, 74f
 problem description and approach, 74–76, 74f–76f
 theoretical and experimental results, 81–84, 81t, 82f–83f
Constant normal load (CNL), 87–88, 165–166
Constant normal stiffness (CNS), 87–88, 165–166
Continuous Frictionally Coupled (CFC) systems, 112

Continuous Mechanically Coupled (CMC) systems, 112
Cut- and-cover tunnels
  building computational model, 316–317, 317f, 317t
  computational model, 312–313, 313f, 313t
  engineering background, 312, 312f
  grouting effect analysis, 317–318, 317f–318f
  liquid-solid coupling, 315–316
  mechanics effect, 314–315, 314f–316f
  overview, 310–312
  simulation analysis scheme, 313–314, 317

D

D-bolts, 73
Deformed bolt, 144, 144f–145f, 144t
Dilational slip failure
  defined, 95–97
    governing equation, 95–96
Dilation limit, 163
Discontinuous deformation analysis (DDA), 165–166
Discreetly Mechanical or Frictionally Coupled (DMFC) system, 112
Double arch tunnel, pressure-arch in
  computational model, 268–269, 269f, 269t
  construction sequence effect, 273, 274f
  different conditions analysis, 269–270
  evolution process, 270–271, 270f–271f
  geometric size effect, 272–273, 273f
  overview, 266–267
  pressure-arch morphological characterization, 267–268, 268f
  simulation analysis schemes, 269
  skewed effect analysis, 271–272, 272f
  stress state effect, 274, 275f
Double-embedment pull test, 157, 157f
Double-layer rock plates
  boundary conditions change effect analysis, 31, 31f
  cohesive strength effect, 24–25, 24f, 31–32, 32f
  computational model, 22, 22f, 28, 28f
  ends design and connection method, 13, 14f

energy dissipation characteristics, 26, 26f
fixed condition ends, 27–32
fracture characteristics, 28–30, 29f
fracture instability models, 17–18, 17f–18f
friction coefficient effect, 24, 24f
geometry size effect analysis, 30, 30f
load–displacement curves, 14–15, 15f
loading devices and loading modes, 12–13, 13f
loading equipment, 19, 20f
load-time acoustic emission event rate curves, 16–17, 16f
microparameters, 27–28, 28f
numerical simulation results, 22–24, 23f–24f
overview, 11–12
parameters calibration, 7t, 21
rock particle radius changing analysis, 30, 30f
rupture and energy analysis, 18–27
sandstone plates, 19
sandstone samples and test programs, 12
size effect, 25, 25f
strain energy entropy, 25–26
test loading and data acquisition system, 13–14, 14f
test procedure, 14
test results and analysis, 19–21, 20f–21f
Double shear test setup, 142, 142f, 142t

E

Effective face angle, 76–77
Exponential decay model, 105–107, 106f

F

Fiberglass Cable (FCB), 151
Finite Element Method (FEM), 1
Fracture process analysis
  engineering mechanical model, 298, 298f
  example, 301–303
  formula derivation, 298–301
  key stratum settlement analysis, 302–303, 303f
  key stratum stress analysis, 301–302, 302f, 302t
  overview, 295–297

surface-underground combined mining, 297–298
Friction-dilation procedure, 161
Fully mechanized mining field
  dip angle variation, 398–399, 398f–399f
  mining advancing distance variation, 394–396, 395f–396f
  overview, 392
  pressure-arch geometry parameters, 393, 393f
  rock strength variation, 396–398, 396f–397f
  three-dimensional computational model, 394, 394f, 394t
  velocity variation, 397f, 398

G

Gloving, 89–90
Grouted cable
  anchorage performance test
    failure modes types, 191–192, 193f
    method, 191, 192f
    MW9 modified bulb cable, 193–195, 194f–195f
    overview, 187–191
    rock strength effect, 195
    superstrand cable, 192–193, 193f
  axial performance
    anchor section, 199
    borehole size influence, 202, 202f–203f, 202t
    embedment length influence, 200–201, 200f–201f
    embedment section, 198–199, 198f–199f, 199t
    new LSEPT apparatus, 197–199
    overview, 196–197
    results analysis, 199–202
    test sample strength influence, 201–202, 201f–202f, 201t
  cementitious grout
    cubic samples, 182, 188f
    cylindrical samples, 182, 186f–187f
    overview, 177–178
    processing procedures, 180
    sample preparation, 179, 179f, 180t
    sample size effect, 182–186
    stress-strain relationships, 181, 181f, 182t–186t, 183f–185f
    testing process, 179–180, 180f
    W/C ratio effect, 182, 189f

design factors
  anchor length section, 172–173, 173f
  antirotation devices, 173, 175f
  bearing plate, 173, 174f
  cable and anchor tube installation, 175
  concrete sample preparation, 174–175, 175f–176f
  hydraulic and monitoring system, 175–176, 176f–177f
  overview, 168–169
  test rig components, 169, 170f
  test sample dimensions, 169–172, 171f–172f
load transfer mechanism
  computational formula, 161–166, 161f–164f
  discussion, 158–159, 166–167
  double-embedment pull test, 157, 157f
  laboratory short encapsulation pull test, 157–158, 158f
  overview, 151–152
  single embedment pull test, 154–157, 154f–156f
  split-pull/push tests, 152–154, 152f
  test verification, 166, 167f
  theoretical analysis, 159–168, 160f
sample dimensions
  discussion, 211–212, 212f
  methodology, 205–209
  40 Nm bolt torque confinement, 211, 211f
  80 Nm bolt torque confinement, 211, 211f
  overview, 203–205, 204f–205f
  sample preparation, 205–208, 206f–208f
  test arrangement, 208–209, 208f–209f
  test variables, 209
  unconfined conditions, 209–210, 209f–210f
  zero torque confined, 210, 210f
Grouting material
  conceptualization, 88–89, 89f
  dilatancy behaviors accompanying shearing, 87–88, 88f, 88t
  experimental study, 89, 89f–90f
  failure mode, 87, 87f
  fully grouted bolting, 86–87
  gloving, 89–90
  overview, 86, 86f
Grout–rock interface failure, 90

H
H-bolts, 73
Horizontal stratified rocks, pressure-arch analysis in
  bedding planes spacing S variation, 391–392, 391f
  coal seam H variation, 387, 388f
  computational model, 384, 384f, 385t
  interlayer variation number N, 386–387, 386f–387f
  lateral pressure coefficient variation, 389–390, 390f
  mechanical characteristics, 384–386, 384f–385f
  mechanism, 383–384, 383f
  numerical simulation programs, 386, 386t
  overview, 381–383
  weak interlayer thickness variation, 387–388, 389f
Hydraulic/monitoring system, 175–176, 176f–177f
Hydraulic tunnel, lining reliability analysis
  engineering application, 238, 238t
  engineering situation, 234–235, 234t
  key parts, 235, 235f
  overview, 233–234
  reliability analysis, 236–238
  reliability analysis equation, 235–236

I
Interfacial shear stress (ISS) model, 91

L
Laboratory short encapsulation pull test (LSEPT) apparatus, 197–199
Landslide forecast method
  catastrophe mechanism analysis, 327–328, 328f
  engineering application, 328–331, 329f, 330t
  forecast method, 328
  longitudinal data analysis method, 326–327
  overview, 326
  regression analysis, 326–327
  regression function, 326
Linear decay model, 105, 106f
Linear Elastic Fracture Mechanics (LEFM), 1
Load transfer mechanism
  computational formula, 161–166, 161f–164f
  discussion, 158–159, 166–167
  double-embedment pull test, 157, 157f
  laboratory short encapsulation pull test, 157–158, 158f
  overview, 151–152
  single embedment pull test, 154–157, 154f–156f
  split-pull/push tests, 152–154, 152f
  test verification, 166, 167f
  theoretical analysis, 159–168, 160f

M
Metropolitan Colliery roadway, 98–101, 99f, 100t
Mined-out regions, tunnel and bridge
  bridge and tunnel engineering, 378–381, 378f–379f, 380t, 381f–382f
  computation model, 375–376, 376f, 377t
  engineering background, 374–375, 374f
  overview, 372–374
  simulation analysis schemes, 376–377
  surface subsidence, 377–378, 378f
Mining geomechanics
  accumulated damage effects analysis
    computational model, 363–364, 364f
    deformation characteristics analysis, 367–369, 368f–372f
    determining dynamic load and boundary conditions, 364–365, 365f
    overview, 363
    parameter selection, 364, 364t
    plastic zones analysis, 366, 366f–367f
    simulation programs, 365, 365f
  coal mining field
    failure models analysis, 360–361, 361f
    instability characteristics, 361–363
    internal forces variation, 362–363, 362f

Mining geomechanics (*Continued*)
  mechanical model, 359, 359f–360f
  overview, 357–359, 357f–358f
  variation of characteristics, 361–362, 361f
 fully mechanized mining field
  dip angle variation, 398–399, 398f–399f
  mining advancing distance variation, 394–396, 395f–396f
  overview, 392
  pressure-arch geometry parameters, 393, 393f
  rock strength variation, 396–398, 396f–397f
  three-dimensional computational model, 394, 394f, 394t
  velocity variation, 397f, 398
 horizontal stratified rocks, pressure-arch analysis in
  bedding planes spacing S variation, 391–392, 391f
  coal seam H variation, 387, 388f
  computational model, 384, 384f, 385t
  interlayer variation number N, 386–387, 386f–387f
  lateral pressure coefficient variation, 389–390, 390f
  mechanical characteristics, 384–386, 384f–385f
  mechanism, 383–384, 383f
  numerical simulation programs, 386, 386t
  overview, 381–383
  weak interlayer thickness variation, 387–388, 389f
 mined-out regions, tunnel and bridge
  bridge and tunnel engineering, 378–381, 378f–379f, 380t, 381f–382f
  computation model, 375–376, 376f, 377t
  engineering background, 374–375, 374f
  overview, 372–374
  simulation analysis schemes, 376–377
  surface subsidence, 377–378, 378f
 roof-bending deflection
  analytical solution verification, 343, 343f
  basic equation, 337
  boundary conditions, 343, 344f–345f
  computational model, 336–337, 336f–337f
  deflection equation, 338–343, 342t
  overview, 335–336
  thick plate basic system, 337–338, 337f
  thick plate real system, 338
 roof safe thickness
  basic equation, 347–348
  dynamic fundamental solutions, 348–349, 348f
  engineering application, 350, 350t
  generalized mechanical model, 346, 346f–347f
  overview, 345–346
  stress expression, 349–350
  stress function, 349
 stratified rock roof
  engineering mechanical model, 351
  overview, 350–351
  rock roof bending deformation evolution, 353
  rock roof deformation energy analysis, 351–353
  rock roof stochastic chaos characteristics, 355–356, 355f–356f
  rock roof stochastic resonance phenomenon, 353–355, 354f
Modified Hoek cell (MHC), 154
Mohr–Coulomb failure, 138, 139f
Mohr–Coulomb material, 74–75
MW9 modified bulb cable, 193–195, 194f–195f

P

Parameters optimization
 assessment criteria, 307
 calculation results and analysis, 308, 308f–311f
 computational model, 306–307, 306f, 307t
 engineering background, 304, 304f
 mining boundary, 308–309
 overview, 303–304
 physical model test and numerical simulation, 304–306, 305f–306f
 simulation schemes, 306
Particle flow code (PFC), 1–2

Poisson's ratio effect
 dilational slip failure, 108–109, 110f
 experimental data, analytical results with, 109, 110f–111f, 111t
 exponential decay model, 105–107, 106f
 linear decay model, 105, 106f
 overview, 102–103
 parallel shear failure, 103–108, 105t
 peak load, 107–108, 107f–108f
Pressure-arch evolution and control technique
 active control technology, 263–266, 264f
 arching mechanism, 255f, 257
 boundary parameters, 257–258, 257f
 combined bolts support, 264–265, 265t, 265f–266f
 computational model, 258, 258f, 258t
 different stress states, 263, 264f
 hydrostatic stress state, 262–263, 263f
 overview, 256–257
 research method, 258
 three-dimensional computational model, 262, 262f
 x–z plane, 259–261, 259f–261f

R

Rebar bolts
 axial loading failure modes, 92, 92f–93f
 case studies, 93–95, 94f, 95t
 dilational slip failure, defined, 95–97
 experimental study, 97, 97t, 98f
 governing equation, dilational slip failure, 95–96
 Metropolitan Colliery roadway, 98–101, 99f, 100t
 optimum rebar profile, 97–98
 overview, 90–92, 91f
 parallel shear failure formulation, 92–93, 93f
 vulnerable slipping surface, 96
Resin–bolt interface dilation, 110
Roadway excavation water inrush
 boundary conditions, 229, 229t
 calculation model, 228–229, 228f
 engineering situation, 226
 fault activation mechanical model, 226–227, 226f
 fault activation mechanics analysis, 227, 227f

fault dip angles, 232–233, 233f
fault displacements, 231, 232f
impermeable rock thickness, 232, 232f
overview, 225–226
process, 229–231, 230f–231f
roadway water inrush mode analysis, 227–228
simulation analysis programs, 229, 230t
water pressure conditions, 231, 232f
Rockbolting
  bolt load transfer
    application example, 138–140, 139f–140f
    failure surface, stress tensor on, 137–138
    fully grouted bolt profiles, 134–138, 134f–135f
    methodology and governing equations, 132–134, 132f–134f
    Mohr–Coulomb failure, 138, 139f
    normal load, failure plane, 135–136, 135f
    overview, 128–132, 129f–132f
    shear load, failure plane, 136–137, 137f
  cone bolts mechanical model
    development, 76–80
    geometric face angles, 84, 84f–85f
    overview, 73–74, 74f
    problem description and approach, 74–76, 74f–76f
    theoretical and experimental results, 81–84, 81t, 82f–83f
  failure modes studies, 121–122
    behavior analysis, 122–128, 123f–124f, 126f
    cable bolting system, 116–121, 116f–117f, 119f
    interfacial shear failure, 121–128
    overview, 112–113, 112f
    two-phase material system, 113–116, 114f–115f
  grouting material
    conceptualization, 88–89, 89f
    dilatancy behaviors accompanying shearing, 87–88, 88f, 88t
    experimental study, 89, 89f–90f
    failure mode, 87, 87f
    fully grouted bolting, 86–87
    gloving, 89–90
    overview, 86, 86f
  Poisson's ratio effect
    dilational slip failure, 108–109, 110f
    experimental data, analytical results with, 109, 110f–111f, 111t
    exponential decay model, 105–107, 106f
    linear decay model, 105, 106f
    overview, 102–103
    parallel shear failure, 103–108, 105t
    peak load, 107–108, 107f–108f
  rebar bolts
    axial loading failure modes, 92, 92f–93f
    case studies, 93–95, 94f, 95t
    dilational slip failure, defined, 95–97
    experimental study, 97, 97t, 98f
    governing equation, dilational slip failure, 95–96
    Metropolitan Colliery roadway, 98–101, 99f, 100t
    optimum rebar profile, 97–98
    overview, 90–92, 91f
    parallel shear failure formulation, 92–93, 93f
    vulnerable slipping surface, 96
  slip face angle
    failure modes IIA and IIB, 71–73, 72f
    governing equations, 67
    literature experimental investigations, 62–65, 64f–66f
    overview, 61–62, 62f
    problem description and assumptions, 65–67, 67f
    rib face angle, 67–68
    rib geometry, 68–69, 69f, 70t, 71f
    vulnerable slip face angle, 69–71, 71f
  tensile stress mobilization
    axial stress distribution, 144–145, 145t
    deformed bolt, 144, 144f–145f, 144t
    double shear test setup, 142, 142f, 142t
    failure type, 145–146, 145f
    overview, 141
    shear load and normal load, 143, 143f–144f
    strain gauge installation, 142–143, 142f–143f
Rock testing
  compressive failure, fracture properties on
    experiment design, 53, 53f
    overview, 52–53
    procedure, 54
    sample collection, 53–54
    specimen preparation, 54, 54t
  double-layer rock plates
    boundary conditions change effect analysis, 31, 31f
    cohesive strength effect, 24–25, 24f, 31–32, 32f
    computational model, 22, 22f, 28, 28f
    ends design and connection method, 13, 14f
    energy dissipation characteristics, 26, 26f
    fixed condition ends, 27–32
    fracture characteristics, 28–30, 29f
    fracture instability models, 17–18, 17f–18f
    friction coefficient effect, 24, 24f
    geometry size effect analysis, 30, 30f
    load–displacement curves, 14–15, 15f
    loading devices and loading modes, 12–13, 13f
    loading equipment, 19, 20f
    load-time acoustic emission event rate curves, 16–17, 16f
    microparameters, 27–28, 28f
    numerical simulation results, 22–24, 23f–24f
    overview, 11–12
    parameters calibration, 7t, 21
    rock particle radius changing analysis, 30, 30f
    rupture and energy analysis, 18–27
    sandstone plates, 19
    sandstone samples and test programs, 12
    size effect, 25, 25f
    strain energy entropy, 25–26
    test loading and data acquisition system, 13–14, 14f
    test procedure, 14

Rock testing (*Continued*)
    test results and analysis, 19–21, 20f–21f
  energy dissipation characteristics
    computational model, 47, 49f
    cutting debris specific energy analysis, 47, 48f, 48t
    cutting depth, acoustic emission variation, 51–52, 51f
    cutting depth, energy variation, 49–50, 50f
    cutting velocity, acoustic emission variation, 50–51, 51f
    cutting velocity, energy variation, 49, 50f
    modeling process, 47–49
    overview, 46–47
  results and analysis
    clean fracture surface, 55–56, 55f, 55t
    failure mechanism, 56–57, 57f
    intermediate layer, 56, 56f, 56t
  sandstone cutting fracture
    cutting debris fractal analysis, 43–45, 43t–44t, 45f
    cutting debris/screening, 42, 42f–43f
    overview, 39–46
    rock fragmentation fractal analysis, 40
    test equipment and samples, 40–41, 41f
    test methods and test procedures, 41–42, 41f
  single sandstone plate, 1–11
    acoustic equipment, 2, 3f
    computational model, 5, 8f
    data acquisition system, 2, 3f
    force–displacement curve, 3–4, 3f
    geometry size effect, 8, 10f
    loading equipment, 2
    loading rate/initial horizontal force effect, 10–11, 11f
    material parameter effect, 8, 9f
    numerical simulation results, 6–8, 8f–9f
    overview, 1–2
    parameters calibration, 4–5
    rock-arch instability, 8–11
    rock-plate failure, acoustic characteristics, 4, 5f–7f
    rock plate samples, 2
  viscoelastic attenuation properties
    experiment principle, 34
    loss factor analysis, 36, 37f
    loss modulus analysis, 36, 36f
    microstructure characteristics analysis, 36–39, 37f–38f
    overview, 32–33
    storage modulus analysis, 34–35, 34f–35f
    test equipment and materials, 33–34, 33f, 34t
Roof-bending deflection
  analytical solution verification, 343, 343f
  basic equation, 337
  boundary conditions, 343, 344f–345f
  computational model, 336–337, 336f–337f
  deflection equation, 338–343, 342t
  overview, 335–336
  thick plate basic system, 337–338, 337f
  thick plate real system, 338

S

Sample dimensions
  discussion, 211–212, 212f
  methodology, 205–209
  40 Nm bolt torque confinement, 211, 211f
  80 Nm bolt torque confinement, 211, 211f
  overview, 203–205, 204f–205f
  sample preparation, 205–208, 206f–208f
  test arrangement, 208–209, 208f–209f
  test variables, 209
  unconfined conditions, 209–210, 209f–210f
  zero torque confined, 210, 210f
Short encapsulation push testing, 102
Simulation analysis schemes, 269
Single embedment pull test, 154–157, 154f–156f
Single sandstone plate, 1–11
  acoustic equipment, 2, 3f
  computational model, 5, 8f
  data acquisition system, 2, 3f
  force–displacement curve, 3–4, 3f
  geometry size effect, 8, 10f
  loading equipment, 2
  loading rate/initial horizontal force effect, 10–11, 11f
  material parameter effect, 8, 9f
  numerical simulation results, 6–8, 8f–9f
  overview, 1–2
  parameters calibration, 4–5
  rock-arch instability, 8–11
  rock-plate failure, acoustic characteristics, 4, 5f–7f
  rock plate samples, 2
Skewed effect analysis, 271–272, 272f
Slip face angle
  failure modes IIA and IIB, 71–73, 72f
  governing equations, 67
  literature experimental investigations, 62–65, 64f–66f
  overview, 61–62, 62f
  problem description and assumptions, 65–67, 67f
  rib face angle, 67–68
  rib geometry, 68–69, 69f, 70t, 71f
  vulnerable slip face angle, 69–71, 71f
Slope engineering
  cut- and-cover tunnels
    building computational model, 316–317, 317f, 317t
    computational model, 312–313, 313f, 313t
    engineering background, 312, 312f
    grouting effect analysis, 317–318, 317f–318f
    liquid-solid coupling, 315–316
    mechanics effect, 314–315, 314f–316f
    overview, 310–312
    simulation analysis scheme, 313–314, 317
  fracture process analysis
    engineering mechanical model, 298, 298f
    example, 301–303
    formula derivation, 298–301
    key stratum settlement analysis, 302–303, 303f
    key stratum stress analysis, 301–302, 302f, 302t
    overview, 295–297
    surface-underground combined mining, 297–298
  landslide forecast method
    catastrophe mechanism analysis, 327–328, 328f
    engineering application, 328–331, 329f, 330t

forecast method, 328
longitudinal data analysis method, 326—327
overview, 326
regression analysis, 326—327
regression function, 326
parameters optimization
 assessment criteria, 307
 calculation results and analysis, 308, 308f—311f
 computational model, 306—307, 306f, 307t
 engineering background, 304, 304f
 mining boundary, 308—309
 overview, 303—304
 physical model test and numerical simulation, 304—306, 305f—306f
 simulation schemes, 306
tailings dam risk analysis
 computational model, 320—321, 321f, 322t
 deformation characteristics, 323—324, 323f—324f
 engineering overview, 319—320, 320f
 FLAC3D liquid-solid coupling, 321—322
 overview, 319
 pore pressure distribution, 323, 323f
 safety risk analysis, 324—325, 325f, 328f—329f
 simulation analysis schemes, 322
three-dimensional deformation effect, 281—289
 computation model and mechanical parameters, 282—284, 284f, 285t
 designs optimization analysis, 285—288, 287f—289f
 engineering background, 282, 283f
 excavated designs, 285, 286f
 overview, 281—282
 slope, 284—285
three-dimensional slope engineering stability analysis
 analysis schemes, 291—292, 292f
 computation model, 291
 displacements, 293, 293f
 engineering background, 291—292
 geological model, 291
 mining sequences optimization, 293—294, 294f

overview, 289—291, 290f
parameters model, 291, 292t
prediction analysis, 294—295, 295f—296f
step-by-step excavation and backfilling foot, 292—293, 293f
Soft rock tunnel, construction optimization
 climatic conditions, 218
 conversion technology, deformation mechanism, 220—221, 221f
 deformation failure characteristics, 219—220, 219f—220f
 deformation mechanism, 220
 excavation methods, field test, 223—224, 223f
 excavation methods, numerical analysis, 221—223, 221t, 222f—223f
 geological structure, 218, 219f
 geostress, 218
 overview, 217—218
 parameters optimization, 224—225, 224f—225f, 224t
 rock lithology, 218
 tunnel design, 218, 218f
Split-pull tests, 152—154, 152f
Split-push tests, 152—154, 152f
Strain gauge installation, 142—143, 142f—143f
Stratified rock roof
 engineering mechanical model, 351
 overview, 350—351
 rock roof bending deformation evolution, 353
 rock roof deformation energy analysis, 351—353
 rock roof stochastic chaos characteristics, 355—356, 355f—356f
 rock roof stochastic resonance phenomenon, 353—355, 354f
Stress analysis, 133
Superstrand cable, 192—193, 193f

T

Tailings dam risk analysis
 computational model, 320—321, 321f, 322t
 deformation characteristics, 323—324, 323f—324f
 engineering overview, 319—320, 320f

 FLAC3D liquid-solid coupling, 321—322
 overview, 319
 pore pressure distribution, 323, 323f
 safety risk analysis, 324—325, 325f, 328f—329f
 simulation analysis schemes, 322
Tensile stress mobilization
 axial stress distribution, 144—145, 145t
 deformed bolt, 144, 144f—145f, 144t
 double shear test setup, 142, 142f, 142t
 failure type, 145—146, 145f
 overview, 141
 shear load and normal load, 143, 143f—144f
 strain gauge installation, 142—143, 142f—143f
Trilinear bond-slip model, 107, 108f
Tunnel disturbance deformation
 defined, 242—248, 244t
 depths variations, 244, 247f—248f
 dip angles variations, 245—248, 249f—250f
 discussion, 248
 engineering background, 240, 240t
 lithologies variations, 244, 246f
 numerical simulation verification, 240t, 241—242, 243f—244f
 overview, 239—240
 similar model test, 240—241, 240f, 242f
 spacing variations, 244, 245f—246f
Tunnel engineering
 circular tunnel
  computation model, 251, 252f
  energy balance equation, 251
  energy release quantity, 252—253, 252f
  energy release rate, 253, 253f
  energy release rate calculation, 251
  maximum/minimum principal stress distribution, 253—254, 254f
  overview, 250—251
  principal stress difference, 254—255, 255f
  simulation analysis schemes, 251—252
 double arch tunnel, pressure-arch in
  computational model, 268—269, 269f, 269t

Tunnel engineering (*Continued*)
  construction sequence effect, 273, 274f
  different conditions analysis, 269–270
  evolution process, 270–271, 270f–271f
  geometric size effect, 272–273, 273f
  overview, 266–267
  pressure-arch morphological characterization, 267–268, 268f
  simulation analysis schemes, 269
  skewed effect analysis, 271–272, 272f
  stress state effect, 274, 275f
hydraulic tunnel, lining reliability analysis
  engineering application, 238, 238t
  engineering situation, 234–235, 234t
  key parts, 235, 235f
  overview, 233–234
  reliability analysis, 236–238
  reliability analysis equation, 235–236
pressure-arch evolution and control technique
  active control technology, 263–266, 264f
  arching mechanism, 255f, 257
  boundary parameters, 257–258, 257f
  combined bolts support, 264–265, 265f–266f, 265t
  computational model, 258, 258f, 258t
  different stress states, 263, 264f
  hydrostatic stress state, 262–263, 263f
  overview, 256–257
  research method, 258
  three-dimensional computational model, 262, 262f
  x–z plane, 259–261, 259f–261f
roadway excavation water inrush
  boundary conditions, 229, 229t
  calculation model, 228–229, 228f
  engineering situation, 226
  fault activation mechanical model, 226–227, 226f
  fault activation mechanics analysis, 227, 227f
  fault dip angles, 232–233, 233f
  fault displacements, 231, 232f
  impermeable rock thickness, 232, 232f
  overview, 225–226
  process, 229–231, 230f–231f
  roadway water inrush mode analysis, 227–228
  simulation analysis programs, 229, 230t
  water pressure conditions, 231, 232f
soft rock tunnel, construction optimization
  climatic conditions, 218
  conversion technology, deformation mechanism, 220–221, 221f
  deformation failure characteristics, 219–220, 219f–220f
  deformation mechanism, 220
  excavation methods, field test, 223–224, 223f
  excavation methods, numerical analysis, 221–223, 221t, 222f–223f
  geological structure, 218, 219f
  geostress, 218
  overview, 217–218
  parameters optimization, 224–225, 224f–225f, 224t
  rock lithology, 218
  tunnel design, 218, 218f
  tunnel disturbance deformation
    defined, 242–248, 244t
    depths variations, 244, 247f–248f
    dip angles variations, 245–248, 249f–250f
    discussion, 248
    engineering background, 240, 240t
    lithologies variations, 244, 246f
    numerical simulation verification, 240t, 241–242, 243f–244f
    overview, 239–240
    similar model test, 240–241, 240f, 242f
    spacing variations, 244, 245f–246f

U

Unconfined conditions, 209–210, 209f–210f
Uniaxial compressive strength (UCS), 77

V

Viscoelastic attenuation properties
  experiment principle, 34
  loss factor analysis, 36, 37f
  loss modulus analysis, 36, 36f
  microstructure characteristics analysis, 36–39, 37f–38f
  overview, 32–33
  storage modulus analysis, 34–35, 34f–35f
  test equipment and materials, 33–34, 33f, 34t

W

W/C ratio effect, 182, 189f

Edwards Brothers Malloy
Ann Arbor MI. USA
October 27, 2016